THE EVOLUTION OF
SCIENTIFIC THOUGHT

FROM NEWTON TO EINSTEIN

By A. d'Abro

Second Edition Revised and Enlarged

Dover Publications, Inc., New York

Published in Canada by General Publishing Company, Ltd., 30 Lesmill Road, Don Mills, Toronto, Ontario.

This Dover edition, first published in 1950, is a revised and enlarged version of the work originally published in 1927 by Liveright Publishing Corporation. This work is reprinted by special arrangement with Liveright Publishing Corporation.

International Standard Book Number: 0-486-20002-7
Library of Congress Catalog Card Number: 50-9480

Manufactured in the United States of America
Dover Publications, Inc.
180 Varick Street
New York, N. Y. 10014

PREFACE

TO THE FIRST EDITION

ALTHOUGH in the course of the last three centuries scientific theories have been subject to all manner of vicissitude and change, the governing motive that has inspired scientists has been ever the same—a search for unity in diversity, a desire to bring harmony and order into what might at first sight appear to be a hopeless chaos of experimental facts.

In this book the essential features of Newton's great discoveries, the apparent inevitableness of absolute space and time in classical science, are passed in review. Then we come to Riemann, that great mathematician who wrested the problem of space from the dogmatic slumber where it had rested so long. Finally we see how Einstein succeeded in transporting to the realm of physics the ideas that Riemann had propounded, giving us thereby that supreme achievement of modern thought, the theory of relativity.

Although I have used non-technical language, great care has been given to an accurate presentation of facts. In certain parts, however, notably in those devoted to non-Euclidean geometry and to the principle of Action, a looseness of presentation has appeared unavoidable owing to the extreme technicality of the subjects discussed. But as it was a question of presenting these subjects loosely or leaving them out of the picture entirely, it appeared preferable to sacrifice accuracy to general comprehensiveness.

Here, however, the reader may be reminded that even for those who are interested solely in trends of thought or in the evolution of ideas, no popular or semi-popular book can ever aspire to take the place of the highly technical mathematical works. The superiority of the latter lies not in the bare mathematical formulæ which they contain. Rather does it reside in the power the mathematical instrument has of giving us a deeper insight into the problems of nature, revealing unsuspected harmonies and extending our survey into regions of thought whence the human intelligence would otherwise be excluded. Thus the sole rôle a semi-popular book can hope to perform is to serve as a general introduction, to whet the appetite for further knowledge, if a craving for knowledge is within us. To presume, as the philosophers do, that a vague understanding of a highly technical subject, gleaned from semi-popular writings, or from the snatching at a sentence here and there in a technical book, should enable them to expound a theory, criticise it, and, worse still, ornament it with their own ideas, is an opinion which has done much to create a spirit of distrust towards their writings. The answer Euclid gave to King Ptolemy, "There is no royal road, no short cut to knowledge," remains true to-day, still truer than in the days of ancient Alexandria, when science had not yet grown to the proportions of a mighty tree.

I wish to take this opportunity to express my gratitude to Prof. Leigh Page of Yale University for his kindness in looking over the manuscript and offering many valuable suggestions.

A. D' ABRO

New York, 1927.

TO THE SECOND EDITION

Slips and errors that were present in the first edition have been corrected in the present edition, and some unnecessary repetitions have been eliminated. The chapter on the finiteness of the Universe has been re-written entirely, and has been supplemented by a brief discussion of the Expanding Universe of the Abbé Lemaitre.

A. D' ABRO

New York, 1949.

CONTENTS

PART III
THE GENERAL THEORY OF RELATIVITY

PART IV
THE METHODOLOGY OF SCIENCE

APPENDIX

FOREWORD

"And now, in our time, there has been unloosed a cataclysm which has swept away space, time and matter, hitherto regarded as the firmest pillars of natural science, but only to make place for a view of things of wider scope, and entailing a deeper vision."

H. WEYL ("Space, Time and Matter").

THE theory of relativity represents the greatest advance in our understanding of nature that philosophy has yet witnessed.

If our interest is purely philosophical, we may wish to be informed briefly of the nature of Einstein's conclusions so as to examine their bearing on the prevalent philosophical ideas of our time. Unfortunately for this simplified method of approach, it is scarcely feasible. The conclusions themselves involve highly technical notions and, if explained in a loose, unscientific way, are likely to convey a totally wrong impression. But even assuming that this first difficulty could be overcome, we should find that Einstein's conclusions were of so revolutionary a nature, entailing the abandonment of our ideas on space, time and matter, that their acceptance might constitute too great a strain on our credulity. Either we would reject the theory altogether as a gigantic hoax, or else we should have to accept it on authority, and in this case conceive it in a very vague and obscure way.

From a perusal of the numerous books that have been written on the subject by a number of contemporary philosophers, the writer is firmly convinced that the only way to approach the theory, even if our interest be purely philosophical, is to study the scientific problem from the beginning. And by the beginning we refer not to Einstein's initial paper published in 1905; we must go back much farther, to the days of Maxwell and even of Newton. Here we may mention that however revolutionary the theory of relativity may appear in its philosophical implications, it is a direct product of the scientific method, conducted in the same spirit as that which inspired Newton and Maxwell; no new metaphysics is involved. Indeed Einstein's theory constitutes but a refinement of classical science; it could never have arisen in the absence of that vast accumulation of mathematical and physical knowledge which had been gathered more especially since the days of Galileo. Under the circumstances it is quite impossible to gain a correct impression of the disclosures of relativity unless we first acquaint ourselves with the discoveries of classical science prior to Einstein's time. It will therefore be the aim of this book, before discussing Einstein's theory proper, to set forth as simply as possible this necessary preliminary information.

Modern science, exclusive of geometry, is a comparatively recent creation and can be said to have originated with Galileo and Newton. Galileo was the first scientist to recognise clearly that the only way to further our understanding of the physical world was to resort to experiment. However obvious Galileo's contention may appear in the light of our present knowledge, it remains a fact that the Greeks, in spite of their proficiency in geometry, never seem to have realised the importance of experiment (Democritus and Archimedes excepted).

To a certain extent this may be attributed to the crudeness of their instruments of measurement. Still, an excuse of this sort can scarcely be put forward when the elementary nature of Galileo's experiments and observations is recalled. Watching a lamp oscillate in the cathedral of Pisa, dropping bodies from the leaning tower of Pisa, rolling balls down inclined planes, noticing the magnifying effect of water in a spherical glass vase, such was the nature of Galileo's experiments and observations. As can be seen, they might just as well have been performed by the Greeks. At any rate, it was thanks to such experiments that Galileo discovered the fundamental law of dynamics, according to which the acceleration imparted to a body is proportional to the force acting upon it.

The next advance was due to Newton, the greatest scientist of all time if account be taken of his joint contributions to mathematics and physics. As a physicist, he was of course an ardent adherent of the empirical method, but his greatest title to fame lies in another direction. Prior to Newton, mathematics, chiefly in the form of geometry, had been studied as a fine art without any view to its physical applications other than in very trivial cases.* But with Newton all the resources of mathematics were turned to advantage in the solution of physical problems. Thenceforth mathematics appeared as an instrument of discovery, the most powerful one known to man, multiplying the power of thought just as in the mechanical domain the lever multiplied our physical action. It is this application of mathematics to the solution of physical problems, this combination of two separate fields of investigation, which constitutes the essential characteristic of the Newtonian method. Thus problems of physics were metamorphosed into problems of mathematics.

But in Newton's day the mathematical instrument was still in a very backward state of development. In this field again Newton showed the mark of genius, by inventing the integral calculus. As a result of this remarkable discovery, problems which would have baffled Archimedes were solved with ease. We know that in Newton's hands this new departure in scientific method led to the discovery of the law of gravitation. But here again the real significance of Newton's achievement lay not so much in the exact quantitative formulation of the law of attraction, as in his having established the presence of law and order at least in one important realm of nature, namely, in the motions of heavenly bodies. Nature

---

* As exemplified in the Pythagorean discovery of the relationship between the length of a vibrating string and the pitch of its note, a discovery utilised in musical instruments. Another example is represented by Archimedes' solution of the problem of Hieron's gold tiara.

thus exhibited rationality and was not mere blind chaos and uncertainty. To be sure, Newton's investigations had been concerned with but a small group of natural phenomena (planetary motions and falling bodies), but it appeared unlikely that this mathematical law and order should turn out to be restricted to certain special phenomena; and the feeling was general that all the physical processes of nature would prove to be unfolding themselves according to rigorous mathematical laws.

It would be impossible to exaggerate the importance of Newton's discoveries and the influence they exerted on the thinkers of the eighteenth century. The proud boast of Archimedes was heard again—"Give me a lever and a resting place, and I will lift the earth."—But the boast of Newton's successors was far greater—"Give us a knowledge of the laws of nature, and both future and past will reveal their secrets."

To-day these hopes appear somewhat childish, but this is because we have learnt more of nature than was ever dreamt of by Newton's contemporaries. Nevertheless, although we recognise that we can never be demigods, the mathematical instrument in conjunction with the experimental method, still constitutes our most fruitful means of progress.

Now Newton, in his application of mathematics to the problems of physics, had been concerned only with the very simplest of physical problems—planetary motions, mechanics, propagation of sound, etc. But when it came to applying the mathematical method to the more intricate physical problems, a considerable advance was necessary in our scientific knowledge, both mathematical and empirical. Thanks to the gradual accumulation of physical data, and thanks to the efforts of Newton's great successors in the field of pure mathematics (Euler, Lagrange, Laplace), conditions were ripe in the first half of the nineteenth century for a systematic mathematical attack on many of nature's secrets.

The mathematical theories constructed were known under the general name of *theories of mathematical physics*. In so far as they represented a mere application of mathematics to natural phenomena, they had their prototype in Newton's celestial mechanics. The only difference was that they dealt with a wide variety of physical phenomena (electric, hydrostatic, etc.), no longer with those of a purely mechanical nature. The most celebrated of these theories (such as those of Maxwell, Boltzmann, Lorentz and Planck) were concerned with very special classes of phenomena. But with Einstein's theory of relativity, itself a development of mathematical physics, the scope of our investigations is so widened that we are appreciably nearer than ever before to the ideal of a single mathematical theory embracing all physical knowledge. This fact in itself shows us the tremendous philosophical interest of the theory of relativity.

Now in all these theories of mathematical physics, the same type of procedure is invariably followed. Experimenters establish certain definite facts and detect precise numerical relationships between magnitudes, for example, between the intensity of an electric current flowing along a wire and the intensity and orientation of the magnetic field surrounding the wire. The mathematical physicist then enters upon the scene, assigns certain letters of the alphabet to the physical entities involved (in the present case electric current designated by i and magnetic intensity designated by H) and by this means translates the numerical relationships discovered by the experimenter into mathematical form. He thus obtains a mathematical relationship or equation a which is assumed to constitute the mathematical image of the concrete physical phenomenon A. His task will now be to extract from his mathematical equation or equations a all their necessary mathematical consequences. In this way, provided his technique does not fail him, he may be led to new equations β. These new equations β, when translated back from the mathematical to the physical, will express new physical relationships B.

The mathematician assumes that just as his equations β were the necessary mathematical consequences of his original equations a, so also must the physical translation of β constitute a physical phenomenon B, which follows as a necessary consequence of the existence of the physical phenomenon A. If A occurs, B must ensue.

We thus understand the significance of a theory of mathematical physics. Its utility is to allow us to foresee and to foretell physical phenomena. In this way it suggests definite experiments which might never have been thought of, and permits us to anticipate new relationships and new laws and to discover new facts. From a philosophical point of view, by establishing a rational connection between seemingly unconnected phenomena, it enables us to detect the harmony and unity of nature which lie concealed under an outward appearance of chaos.

Of course the experimenter in the first place must be very careful to give accurate information to the mathematician; for if by any chance his information should be only approximately correct, the mathematical translation a would likewise be lacking in accuracy, and the mathematical consequences of a might be still further at variance with the world of physical reality. It is as though, when firing at a distant target, we were to point the rifle a wee bit too far to one side; the greater the range, the wider would be the divergence. Dangers of this sort are of course inevitable, for human observations are necessarily imperfect. In any case, therefore, the mathematician's physical anticipations will always require careful checking up by subsequent experiment. Obviously, however, something much deeper is at stake than mere accuracy of observation.

Mathematical deductions are mind-born; they pertain to reason and are not dependent on experience. When, therefore, we assume that our mathematical deductions and operations will be successful in portraying the workings of nature, we are assuming that nature also is rational, and that therefore a definite parallelism or correspondence exists between the two worlds, the mathematical and the physical. *A priori,* there appears to be no logical necessity why any such parallelism should exist. Here, however, we are faced with a situation over which it is useless to philosophise. Success has attended the efforts of mathematical physicists in so large a number of cases that, however marvellous it may appear, we can scarcely escape the conclusion that nature must be rational and susceptible to mathematical law. In fact, were this not the case, prevision would be impossible and science non-existent.

It may be that nature is only approximately rational; it may be that her appearance of rationality is due to the very crudeness of our observations and that more refined experiments would yield a very different picture. Heisenberg and Bohr have suggested that the difficulties which confront us in the study of quantum phenomena may indeed be due to the fact that nature is found to be irrational when we seek to examine her processes in a microscopic way. This is a possibility which we cannot afford to reject. But at any rate, as long as our theories appear to be verified by experiment, we must proceed as though nature were rational, and hope for the best.

Now it must not be thought that the introduction of the mathematical instrument into our study of nature creates any essential departure from the commonplace method of ordinary deductive and inductive reasoning. There is no particular mystery about mathematical analysis; its only distinguishing feature is that it is more trustworthy, more precise, and permits us to proceed farther and along safer lines.

Consider, for example, the well-known change of colour from red to white displayed by the light radiated through an aperture made in a heated enclosure, as the temperature increases. From this elementary fact of observation Planck, thanks to mathematical analysis, was able to deduce the existence of light quanta and thence the possibility that all processes of change were discontinuous, and that a body could only rotate with definite speeds. Obviously, commonplace reasoning unaided by mathematics would never have led us even to suspect these extraordinary results.

Now when we say that a theory of mathematical physics is correct, all we mean is that the various mathematical consequences we can extract from its equations call for the existence of physical phenomena which experiment has succeeded in verifying. On the other hand, if our mathematical anticipations do not tally with experimental verification, we must recognise that our theory is incorrect. This does not mean that it is incorrect from a purely mathematical point of view, for in any case it exemplifies a possible rational world; but it is incorrect in that it does not exemplify our real world. We must then assume that our initial equations were in all probability bad translations of the physical phenomena they were supposed to represent.

In a number of cases, however, it has been found unnecessary to abandon a theory merely because one of its anticipations happened to be refuted by experiment. Instead, it is often possible to assume that the discrepancy between the mathematical anticipation and the physical result may be due to some contingent physical influence, which, owing to the incompleteness of the physical data furnished us by the experimenters, our equations have failed to take into consideration. A case in point is afforded by the discovery of Neptune.

The Newtonian mathematical treatment of planetary motions assigned a definite motion to the planet Uranus. Astronomical observation then proved that the actual motion of Uranus did not tally with these mathematical anticipations. Yet it was not deemed necessary to abandon Newton's law; Adams and Leverrier suggested the possibility that an unknown planet lying beyond the orbit of Uranus might be responsible for the deviations in Uranus' motion. Taking the existence of this unknown planet into consideration in his mathematical calculations, Leverrier succeeded in determining the exact position which it would have to occupy in the heavens at an assigned date. As is well known, at the precise spot calculated, the elusive planet (presently named Neptune) was discovered with a powerful telescope.

This procedure of ascribing discrepancies in our mathematical anticipations to the presence of contingent influences rather than to the falsity of our theory is only human. There is no inclination, merely because the hundredth case turns out to be an exception, to abandon a theory which has led to accurate anticipations in 99 cases out of 100. But we must realise that this procedure of appealing to foreign influences, while perfectly legitimate in a tentative way, must be applied with a certain amount of caution; in every particular case it must be justified by *a posteriori* determination of fact. Thus Leverrier was also the first to discover certain irregularities in the motion of the planet Mercury. As in the case of Uranus, he attempted to ascribe these discrepancies to the presence of an interior planet which he called Vulcan and which he assumed to be moving between the orbit of Mercury and the sun. Astronomers have, however, failed to find the slightest trace of Vulcan, and a belief in its existence has been abandoned. If contingent influences are to be invoked for Mercury's anomalies, we must search for them in some other direction.

In this particular case all other suggestions were equally unsatisfactory. Hence even before the advent of Einstein's theory, doubts had been raised as to the accuracy of Newton's law of gravitation. The procedure of patching up a mistaken theoretical anticipation with hypotheses *ad hoc* has not much to commend it. Yet when, as was the case with Vulcan and Neptune, the influence we appeal to is of a category susceptible of being observed directly, the method is legitimate. But when our hypothesis *ad hoc* transcends observation by its very nature, and when, added to this, its utility is merely local, accounting for one definite fact and for no other, it becomes worse than useless.

This abhorrence of science for the unverifiable type of hypothesis *ad hoc* so frequently encountered in the speculations of the metaphysicians is not due to a mere phenomenalistic desire to eliminate all that cannot be seen or sensed. It arises from a deeper motive entailing the entire *raison d'être* of a scientific theory. Suppose, for instance, that our theory had led us to anticipate a certain result, and that experiment or observation should prove that in reality a different result was realised. We could always adjust matters by arbitrarily postulating some local invisible and unverifiable influence, which we might ascribe to the presence of a mysterious medium—say, the ether A. We should thus have added a new influence to our scheme of nature.

If we should now take this new influence into consideration, the first numerical result would, of course, be explained automatically, since our ether A was devised with this express purpose in view. But we should now be led to anticipate a different numerical result for some other phenomenon. If this second anticipation were to be disproved by experiment we could invoke some second unverifiable disturbing influence to account for the discrepancy, while leaving the first result unchanged. Let us call this new influence the ether B. We might go on in this way indefinitely.

But it is obvious that our theory of mathematical physics whose object it was to allow us to foresee and to foretell would now be useless. No new phenomenon could be anticipated, since past experience would have shown us that unforeseen influences must constantly be called into play if theory were to be verified by experiment. Under these circumstances we might just as well abandon all attempts to construct a mathematical model of the universe.

Suppose now that by modifying once and for all our initial premises we are led to a theory which allows us to foresee and foretell numerical results that are invariably verified with the utmost precision by experiment, without our having to call to our assistance a number of foreign hypotheses. In this case we may assume that the new theory is correct, since it is fruitful; and that our former theory was incorrect, because it led us nowhere.

The considerations we have outlined have an important bearing on the understanding of the outside world as shared by the vast majority of scientists. If we hold that the simplest of all the mathematical theories which finds itself in accord with experiment constitutes the correct theory, giving us the correct representation of the real world, we shall recognise that it would be a dangerous procedure to saddle ourselves with a number of hypothetical presuppositions at too early a stage of our investigations.

To be sure, we may have to make a certain number of fundamental assumptions, but we must regard these as mere working hypotheses which may have to be abandoned at a later stage if peradventure they lead to too complicated a synthesis of the facts of experiment. We shall see, for instance, that relativity compels us to abandon our traditional understanding of space and time. It is this fact more than any other which has been responsible for the cool reception accorded to the theory by many thinkers. When, however, we become convinced that Einstein's synthesis is the simplest that can be constructed if due account be taken of the results of ultra-refined experiment, and when we realise that a synthesis based on the classical understanding of separate space and time would be possible only provided we were willing to introduce a host of entirely disconnected hypotheses *ad hoc* which would offer no means of direct verification, we cannot easily contest the soundness of Einstein's conclusions.

There are some, however, who argue that we have an *a priori* intuitional understanding of space and time which is fundamental, and that we should sacrifice simplicity of mathematical co-ordination if it conflicts with these fundamental intuitional notions. Needless to say, no scientist could subscribe to such views. Quite independently of Einstein's discoveries, mathematicians had exploded these Kantian opinions on space and time many years ago. As Einstein very aptly remarks in his Princeton lectures:

"The only justification for our concepts and system of concepts is that they serve to represent the complex of our experiences; beyond this they have no legitimacy. I am convinced that the philosophers have had a harmful effect upon the progress of scientific thinking in removing certain fundamental concepts from the domain of empiricism where they are under our control to the intangible heights of the *a priori*. For even if it should appear that the universe of ideas cannot be deduced from experience by logical means but is in a sense a creation of the human mind without which no science is possible, nevertheless this universe of ideas is just as little independent of the nature of our experiences as clothes are of the form of the human body. This is particularly true of our concepts of time and space which physicists have been obliged by the facts to bring down from the Olympus of the *a priori* in order to adjust them and put them in a serviceable condition."

In the passage just quoted, Einstein argues from the standpoint of the physicist, but the opinions he expresses will certainly be endorsed by pure mathematicians. They, more than all others, have been led to realise how cautious we must be of the dictates of intuition and so-called common sense. They know that the fact that we can conceive or imagine a certain thing only in a certain way is no criterion of the correctness of our judgment. Examples in mathematics abound. For example, before the discoveries of Weierstrass, Riemann and Darboux the idea that a continuous curve might fail to have a definite slant at every point was considered absurd; and yet we know to-day that the vast majority of curves are of this type. In the same way our intuition would tell us that a line, whether curved or straight, being without width, would be quite unable to cover an area completely; yet once again, as Peano and others have shown, our intuition would have misled us. Many other examples might be given, but they are of too technical a nature and need not detain us. At all events, mathematicians, as·a whole, refused to question the soundness of Einstein's theory on the sole plea that it conflicted with our traditional intuitional concepts of space and time, and we need not be surprised to find Poincaré, one of the greatest mathematicians of the nineteenth century, lending full support to Einstein when the theory was so bitterly assailed in its earlier days.

We have now to consider in a very brief way certain of the philosophical problems which antedate Einstein's discoveries, but with which his theory is intimately connected. Long before the advent of Einstein, problems pertaining to the relativity of motion through empty space had occupied the attention of students of nature. There were some who held·that empty space, and with it all motion, must be relative; that states of absolute motion or absolute rest through empty space were meaningless concepts. According to these thinkers, in order to give significance to motion and rest, it was necessary to refer the successive positions of the body to some other arbitrarily selected body taken as a system of reference. We should thus obtain relative rest or motion of matter, with respect to matter. All these views were in full accord with our visual perceptions and they were expressed by what is known as the **visual or kinematic principle of the relativity of motion.** Other thinkers preferred to uphold the opposing philosophy. They assumed that space was absolute; that all motion must be absolute; that there was meaning to the statement that a body was in motion or at rest in space, regardless of the presence of other bodies to be used as terms of comparison. The controversy might have continued indefinitely, had it not been for the appearance of the scientist with his empirical methods of investigation.

Galileo and Newton were the first to recognise in a clear way that, provided certain very plausible assumptions were made, the dynamical evidence adduced from mechanical experiments proved the relativistic philosophy to be untenable. Classical science was therefore compelled to recognise the absoluteness of space and motion. It is true that many philosophers still defended the relativity of all motion. But their failure to take into consideration the real obstacles that seemed to bar the way to a relativistic conception of motion, coupled with the looseness of their scientific arguments, precluded their opinions from exercising any influence on scientific thought. Now it is to be noted that notwithstanding the absolute nature which Newton attributed to all states of motion and of rest in empty space, a certain type of absolute motion called Galilean * or again uniform translationary motion (defined by an absolute velocity but no absolute acceleration), was recognised by him as being incapable of detection, so far as experiments of a mechanical nature were concerned. This complete irrelevancy of absolute velocity or absolute Galilean motion to mechanical experiments was expressed in what is known as the **Galilean or Newtonian or classical or dynamical principle of the relativity of Galilean motion** through empty space. The existence of such a principle of relativity created a duality in the physical significance of motion, hence of space, but the philosophical importance of the principle, as referring to the problem of space, was lessened by the fact that this relativity applied solely to experiments of a mechanical nature. It was confidently assumed that electromagnetic and optical experiments would be successful in revealing the absolute Galilean motions which had eluded mechanical tests.

Such was the state of affairs when Einstein, in 1905, published his celebrated paper on the electrodynamics of moving bodies. In this he remarked that the numerous difficulties which surrounded the equations

---

* The appellation *Galilean motion* does not appear to have been adopted generally. However, as it is shorter to designate "uniform translationary motion" under this name, we shall adhere to the appellation.

of electrodynamics, together with the negative experiments of Michelson and others, would be obviated if we extended the validity of the Newtonian principle of the relativity of Galilean motion (which applied solely to mechanical phenomena) so as to include all manner of phenomena: electrodynamic, optical, etc.

When extended in this way the Newtonian principle of relativity became **Einstein's special principle of relativity.** Its significance lay in its assertion that absolute Galilean motion or absolute velocity must ever escape all experimental detection. Henceforth absolute velocity should be conceived of as physically meaningless, not only in the particular realm of mechanics, as in Newton's day, but in the entire realm of physical phenomena. Einstein's special principle, by adding increased emphasis to this relativity of velocity, making absolute velocity metaphysically meaningless, created a still more profound distinction between velocity and accelerated or rotational motion. This latter type of motion remained absolute and real as before. From this we see that the special principle was far from justifying the ultra-relativistic belief in the complete relativity of all motion, as embodied in the kinematical or visual principle of relativity.

It is most important to understand this point and to realise that Einstein's special principle is merely an extension of the validity of the classical Newtonian principle to all classes of phenomena. Apart from this extension there is no essential difference between the two principles, since both deal exclusively with Galilean motion or velocity. Much of the criticism that has been directed against Einstein's special theory, and in particular against its so-called paradoxes, can be traced to a confusion on the part of the critic between the very wide visual or kinematical principle of the relativity of all motion, including acceleration as well as velocity, and the much more restricted special principle which deals solely with velocity.

Owing to this initial error it has been assumed by some that the theory of relativity professes to have established the complete relativity of all motion. So far as the special theory is concerned, this assumption is obviously untenable; and a very marked distinction is maintained, as in classical science, between the significance of acceleration and of velocity. This distinction is by no means new, since it forms part and parcel of classical science; but in view of its importance to any one who wishes to obtain any insight into the philosophical implications of Einstein's more difficult theory, we have deemed it advisable in the course of this book to devote a preliminary chapter to Newtonian mechanics. We may add that Einstein's paper created an uproar, for it compelled us to reorganise in a radical way those fundamental forms of human representation which we call space and time.

After discussing the special theory we shall proceed to examine Einstein's general theory, which he began to develop in 1912 and which he completed in 1916. The general theory deals more especially with gravitation, though it also sheds new light on the problem of the relativity of

motion. It is our opinion that the reader will save himself much unnecessary trouble if he realises from the start that Einstein, even in the general
theory, does not succeed in establishing the complete relativity of all
motion. While it is true that acceleration loses much of its absoluteness,
we are still far from being able to subscribe to the very wide visual or
kinematical principle which the complete relativity of all motion would
necessitate.

Following the general theory, Einstein entered into cosmological considerations on the form of the universe as a whole. This part of the
theory is still highly speculative; and until such time as astronomical
observations conducted on the remoter regions of the universe have given
their verdict, nothing definite can be said.

From the standpoint of the relativity of motion, this last part of Einstein's
theory is of fascinating interest; for had it been proved correct, the relativity of all motion would have been established and the bugaboo of absolute
rotation dispelled forever. We would then have been led to a modified
form of the visual or kinematical principle of the complete relativity of
motion: namely, to Mach's mechanics.

In addition to all this preliminary physical information with which it is
necessary to be acquainted (Newtonian mechanics, electrodynamics)
before proceeding to a study of Einstein's theory, a considerable amount
of purely mathematical knowledge must be mastered. Here we refer
not to the actual technique of calculation, but to the general significance
of the mathematical doctrines developed by the great mathematicians of
the past.

Einstein's theory, more especially the second part (the general theory),
is intimately connected with the discoveries of the non-Euclidean geometricians, Riemann in particular. Indeed, had it not been for Riemann's work, and for the considerable extension it has conferred upon
our understanding of the problem of space, Einstein's general theory
could never have arisen. As Weyl expresses it:

"Riemann left the real development of his ideas in the hands of some
subsequent scientist whose genius as a physicist could rise to equal
flights with his own as a mathematician. After a lapse of seventy years
this mission has been fulfilled by Einstein."

In a general way it may be said that prior to the discovery of non-
Euclidean spaces two conflicting philosophies held the field. Some thinkers inclined to Berkeley's views and maintained that the concept of space
arose from the complex of our experiences, and chiefly from a synthesis
of our visual and tactual impressions. Others, such as Kant, argued
that the concept of three-dimensional Euclidean space was antecedent
to all reason and experience and was essentially *a priori,* a form of
pure sensibility. As in the case of the relativity of motion, discussions
might have gone on indefinitely had it not been for the work of the psycho-

physicists and mathematicians.   The latter settled the question by proving that Berkeley had guessed correctly, at least in a general way.

The essence of Riemann's discoveries consists in having shown that there exist a vast number of possible types of spaces, all of them perfectly self-consistent.   When, therefore, it comes to deciding which one of these possible spaces real space will turn out to be, we cannot prejudge the question.   Experiment and observation alone can yield us a clue.   To a first approximation, experiment and observation prove space to be Euclidean, and this accounts for our natural belief in the truth of the Euclidean axioms, accepted as valid merely by force of habit.   But experiment is necessarily inaccurate, and we cannot foretell whether our opinions will not have to be modified when our experiments are conducted with greater accuracy.   Riemann's views thus place the problem of space on an empirical basis excluding all *a priori* assertions on the subject.

Of course, these discoveries on the part of mathematicians precede Einstein's theory by fully seventy years.   They are the direct outcome of non-Euclidean geometry and would in no wise be affected by the fate of Einstein's theory.   But, on the other hand, the relativity theory is very intimately connected with this empirical philosophy; for, as will be explained later, Einstein is compelled to appeal to a varying non-Euclideanism of four-dimensional space-time in order to account with extreme simplicity for gravitation.   Obviously, had the extension of the universe been restricted on *a priori* grounds, by some ukase, as it were, to three-dimensional Euclidean space, Einstein's theory would have been rejected on first principles.   On the other hand, as soon as we recognise that the fundamental continuum of the universe and its geometry cannot be posited *a priori* and can only be disclosed to us from place to place by experiment and measurement, a vast number of possibilities are thrown open.   Among these the four-dimensional space-time of relativity, with its varying degrees of non-Euclideanism, finds a ready place.

PART I
PRE-RELATIVITY PHYSICS

CHAPTER I

In this chapter we shall be concerned solely with manifolds and their dimensionality; and not with their metrics, which pertains to a different problem entirely.

We all possess a certain instinctive understanding of what is meant by continuity. We notice, for example, that sounds, colours or tactual sensations merge by insensible gradations into other sounds, colours or tactual sensations, without any abrupt transitions. An aggregate of such continuous sensations constitutes what is called a *sensory continuum or continuous manifold*. That continuity is a concept which springs from experience can scarcely be doubted, and it can be accounted for by the inability of our crude senses to differentiate between impressions which are almost alike.

Consider, for example, the succession of musical notes exhibited in the chromatic scale on the piano. Here we are not in the presence of a sensory continuum, for the successive sounds do not merge into one another by insensible degrees. Even an untrained ear can differentiate between a C and the C-sharp immediately following it. But we can conceive of a piano in which a sufficient number of semitones and intermediary notes have been interposed so that every note would be indistinguishable from its immediate successor and immediate predecessor, although we should still be able to differentiate between non-contiguous notes. It would thus be possible for us to pass through a continuous chain of sounds from any one musical sound to any other without our ear's ever being able to detect a sudden jump; and this is what we mean by calling our aggregate of sounds *a sensory continuum*.

Suppose now that we were to remove any one of these notes from our piano (excluding the two extreme ones). The continuity of our chain of sounds would be broken, for when we reached the missing note we should detect a sudden variation in pitch as we passed from the sound immediately preceding the removed note to the one immediately following it. In short, the removal of one of the notes would cut our continuous chain of sounds in two.

The dimensionality of our continuum of sounds is obviously unity, for we can assign successive numbers to the successive notes, starting from some standard note, and by this means determine them without ambiguity.

Let us complicate matters somewhat by assuming that every individual note may be sounded with various intensities, but always in such a way that a note of given intensity can never be distinguished from the one sounded just a little louder or just a little softer. Once again it will be possible for us to pass in a continuous way from a note of feeble intensity to the same note sounded with louder intensity, without our ear's ever being able to detect a variation in intensity between two successive sounds. More generally we shall be able to pass in a continuous way from a note of given pitch and given intensity to one of some other pitch and some other intensity.

In the case assumed we should be dealing with a two-dimensional continuum or continuous manifold of sounds; for in order to locate a definite sound it would be necessary for us to designate it by two numbers, one specifying its pitch and the other its intensity. We may also notice that whereas in the first case, by removing one of the notes, we were able to cut the manifold of successive pitches into two parts, the removal of one particular note of definite pitch and intensity will now be incapable of effecting this separation.

For instance, if we were to remove the note D sounded with a definite intensity it would still be possible to pass in a continuous way from any one note of given intensity to any other note of our continuum by circumscribing the missing note; namely, by choosing some route of transfer which would pass through a D differing in intensity from that of the D we had removed.

In the present case, if we wished to divide our two-dimensional manifold into two parts, it would be necessary to remove some one-dimensional continuum of sounds, for example the one-dimensional continuum formed by all the notes of a given pitch D, but varying in intensity, or again of given intensity but varying in pitch. If this were done, it would be impossible for us to pass in a continuous way from a note of definite pitch and intensity of our two-dimensional continuum to a note of any other definite pitch and intensity; for we could never get past the removed line of sounds without our ear's detecting a sudden change.

We might complicate matters still further by taking into consideration variations in tonality, as for instance the variation which our ear can detect between two given notes of the same pitch and intensity sounded by two different instruments, such as a violin and an organ. Assuming that every one of our notes of given pitch and intensity in our two-dimensional manifold could also vary in tonality by imperceptible degrees, we should be dealing with a three-dimensional sensory continuum in which every note of given pitch, intensity and tonality could be defined unambiguously by the choice of three numbers.

As before, we should find that the removal of a single note of given pitch, intensity and tonality, or even the removal of a one-dimensional continuum of notes such as all those of given intensity and pitch, but varying in tonality, was quite insufficient to effect a separation in our three-dimensional manifold. In the present case we should have to remove some two-dimensional continuum—say, all notes of given intensity but varying in pitch and tonality. Only then should we have effected a separation between any given element and any other one, rendering it impossible for a continuity of sound impressions to extend between the two elements. By proceeding in this way indefinitely it is obvious that we can conceive of sensory continua of any number of dimensions; there is no need to limit ourselves to three.

In a general way we may say, therefore, that a sensory continuum is n-dimensional when, in order to render a path of continuous passage impossible between any two of its elements, it is necessary to remove a sub-continuum of $n - 1$ dimensions. This sub-continuum itself is known to be $n - 1$ dimensional because we can separate it into two parts only by removing from it a sub-continuum of $n - 2$ dimensions, etc., till we finally get a sub-sub-sub . . . continuum, which can be separated by removal of a single element. But such an element no longer constitutes a continuum, since passage in it is excluded; its dimensionality is then zero; so that the continuum it separates in two is obviously one-dimensional.

A sensory continuum would also be given by a succession of weights placed on one's hand, each only slightly heavier than the preceding one. Again our tactual impressions might yield a sensory continuum. In view of the fact that it is possible to account for the rise of the concept of space, even in the consciousness of a blind man, through the sole means of his tactual impressions, it may be of interest to discuss briefly an illustration of a tactual continuum—that obtained by exploring the surface of our skin by means of pinpricks. If these pinpricks are sufficiently close to one another, it will be impossible for us to differentiate between them and we shall always experience the sensation of one solitary pinprick. We can thus consider the sensory continuum obtained by some definite chain of pinpricks extending, let us say, from our elbow to our hand. This particular chain of sensations exhibits all the characteristics of a one-dimensional sensory continuum, since every one of its elements is indistinguishable from its immediate neighbours and since the removal of one of these elements (pinpricks) would create a hiatus rendering it impossible for us to pass in a continuous way from elbow to hand, along the chain.

But if we should now consider all the possible chains of pinpricks extending from a point on our elbow to a point on our hand, the mere removal of one particular pinprick from one particular chain would be insufficient to interfere with the continuous passage. We might always follow one of the other chains, or even, following the same chain up to the missing element, skip round the latter without sensory continuity being interfered with. In the present case the only way to render this continuous passage impossible would be to remove some continuous chain of pinpricks—say, those circling round our wrist. The sensory skin-continuum would now be divided into two parts, and as the continuum removed was one-dimensional, we should conclude that the skin continuum as manifesting its sensitivity to tactual stimuli was two-dimensional.

In a similar way, in crude geometry we recognise a wire as one-dimensional, since by removing a point of the wire our finger cannot pass in a continuous way from one extremity to the other. Likewise, a surface is regarded as two-dimensional because only by cutting it along a line is it possible to interrupt the smooth passage of our finger from any one point to any other. The mere removal of a point on the surface would not interfere with the continuous passage as it did in the case of the wire. It is the same for a volume. Only a surface can divide it in two; hence volume is three-dimensional.*

When we seek to determine the dimensionality of perceptual space, itself a sensory continuum produced by the superposition of the visual, the tactual and the motive continua, the problem is more difficult. It would be found, however, that perceptual space has three dimensions; but as the necessary explanations would require several chapters we must refer the reader to Poincaré's profound writings for more ample information.

Summarising, we may say that our belief in the tri-dimensionality of space can be accounted for on the grounds of sensory experience.

Now the subject of our investigations up to the present point has been the dimensionality of sensory continua and the general characteristics of sensory continuity; considerations relating to measurement, or to the extensional equality of two continuous stretches in our continua, have not been entered upon. Neither has any definition of what is meant by a straight line been introduced at this stage. As a result, metrical geometry, which deals with measurements, and projective geometry, which deals with the projections of points, cannot be discussed. The only type of geometry we can consider at this stage is that purely qualitative non-metrical type called *Analysis Situs,* which deals solely with problems of connectivity.

Connectivity relates to the types of paths of continuous passage from one part of a continuum to another. Manifolds may possess the same dimensionality and yet differ in connectivity. Thus, the connectivity of a sphere differs from that of a torus or doughnut; since the doughnut, in contrast to the sphere, presents a hole or discontinuity through its centre. Yet both sphere and doughnut are two-dimensional surfaces.

In Analysis Situs, metrical considerations obviously play no part. From a metrical point of view, although a sphere differs in shape from an ellipsoid, yet the connectivity or Analysis Situs of the two surfaces is exactly the same. We may add that there exists an Analysis Situs for every continuous manifold, so that we may conceive of an Analysis Situs of n dimensions corresponding to an n-dimensional manifold.

---

* We need not discuss here the difficult problems that relate to the connectivity of the continua. For instance, a point on a closed line does not divide the line into two parts, and yet the line remains one-dimensional. Problems of this sort pertain to one of the most difficult branches of geometry, namely, Analysis Situs, with which the names of Riemann, Betti and Poincaré are associated. (See note on page 70.)

Our next task is to determine how a metrics can be established in a sensory continuum. Consider, for example, a continuous stretch of shades of grey passing from white to black. What do we mean exactly when we say that some definite shade is twice as dark as another? Obviously no definite meaning can be assigned to this statement until we have posited some convention permitting us to establish comparisons.

As another instance, take the case of a continuous stream of sounds varying in pitch. What do we mean by saying that some particular musical note is twice as high in pitch as some other, or that the interval between two notes is equal to the interval between two others? If we were to be guided solely by our ear we might assert that as certain musical notes, though differing in pitch, yet appear to present a certain undefinable similarity (the successive octaves), the intervals between these successive similar notes should be considered equal or congruent. We should thus define as equal the extension of notes subtending the successive octaves of a given musical note.

But if now we had learnt to measure the frequencies of vibration of the various musical sounds, a new type of measurement would immediately suggest itself. Starting from any musical note—say, the middle A of the piano, we should find that the octave of A was vibrating twice as fast, the following E three times as fast, and the second octave of our original A four times as fast. It would then appear plausible to define equal intervals between musical notes by the differences in their rates of vibration, and we should infer that the distance between A and its first octave was equal to the distance between this last note and the following E, and equal again to the distance between this E and the following superoctave A.

We should thus have obtained a definition of equal stretches of sounds which was at variance with our original definition, in which all octave intervals were regarded as equal or congruent. In view of these conflicting results, we could not well escape the conclusion that a sensory continuum of itself offers us no precise means of defining equal stretches, and that whatever definition we might finally select would be a mere matter of choice, an arbitrarily posited convention.

And yet in the case of space, itself a sensory continuum, men have found no difficulty in agreeing on a common system of measurements. As we shall see in the following chapters, the definition of the equality of different stretches of space to which men were unavoidably led was imposed upon them by the behaviour of certain bodies located in space, bodies which were deemed to remain rigid, hence to occupy equal volumes and equal lengths of space wherever they were displaced. For the present, however, we may leave these metrical considerations aside and confine our attention to the general concept of mathematical or geometrical space, which is the subject of study of the pure mathematician.

The concept of a sensory continuum, hence of perceptual space, as presented to us by crude experience, contains certain contradictions and peculiarities which it was necessary to eliminate before it could be subjected to rigorous mathematical treatment. In the first place, this perceptual space is not homogeneous, and the principle of sufficient reason demands that pure empty conceptual space be homogeneous and isotropic, the same everywhere and the same in all directions.

This homogeneity of space permits us to foresee that it must be unbounded, since a boundary would suggest a discontinuity of structure, defining an inside and an outside, hence a lack of homogeneity. Prior to Riemann's discoveries it was thought that the absence of a boundary would necessitate the infiniteness of space. To-day we know that this belief is unjustified, for a space can be finite and yet unbounded; and two major varieties of such spaces have been discovered by mathematicians.

But the inherent inconsistencies which endure in all sensory continua constituted a still more important reason for compelling mathematicians to idealise perceptual space. In a sensory continuum, as we have seen, a sensation A cannot be distinguished from its immediate successor, the sensation B; neither can B be differentiated from C. Yet no difficulty is experienced in differentiating A from C. Expressed mathematically, these facts yield the inconsistent series of relations $A = B$; $B = C$; $C \neq A$. Now, an inconsistency of this sort precludes all mathematical treatment. In mathematics magnitudes cannot be both equal and unequal; they must either be one or the other. The mathematician is therefore compelled to idealise the sensory continuum of experience by assuming that were it not for the crudeness of our senses, the points or sensations A, B and C would all be distinguishable, and that in place of $A = B$; $B = C$ and $A \neq C$, we should have $A \neq B$; $B \neq C$; $C \neq A$.

But it is obvious that a continuum idealised in this way becomes atomic or discrete, since between A and B, as between B and C, no intermediary points have been mentioned. In order to re-establish continuity the mathematician is forced to postulate that between any two points A and B there exist an indefinite number of intermediary points, such that no one of these points has an immediate neighbour. In other words, the continuum is infinitely divisible.

Thus the magnitudes 1 and 2 are not neighbours, since a number of rational fractions separate them. And no two of these fractional numbers are immediate neighbours, since whichever two such numbers we choose to select, we can always discover an indefinite number of other fractional numbers existing between them. Between any two points on a line in our continuum, however close together they may be, we have thus interposed an indefinite number of rational fractions defining points; yet, despite this fact, we have by no means eliminated gaps between the various points along our line.

The Greek mathematician Pythagoras was the first to draw attention to this deficiency after studying certain geometrical constructions. He remarked, for instance, that if we considered a square whose sides were of unit length, the diagonal of the square (as a result of his famous geometrical theorem of the square of the hypothenuse) would be equal to $\sqrt{2}$. Now, $\sqrt{2}$ is an irrational number and differs from all ordinary fractional or rational numbers. Hence, since all the points of a line would correspond to rational or ordinary fractional numbers, it was obvious that the opposite corner of the square would define a point which did not belong to the diagonal. In other words, the sides of the square meeting at the opposite corner to that whence the diagonal had been drawn, would not intersect the diagonal; and we should be faced with the conclusion that two continuous lines could cross one another in a plane and yet have no point in common.

The only way to remedy this difficulty was to assume that the point corresponding to $\sqrt{2}$ and in a general way points corresponding to all irrational numbers (such as π, e and radicals) *were after all present* on a continuous mathematical line. Accordingly, mathematical continuity along a line was defined by the inclusion of all numbers whether rational or irrational, and a similar procedure was followed for a mathematical continuum of any number of dimensions. In this way mathematicians obtained what is known as the **Grand Continuum, or Mathematical Continuum.***

Now, it is obvious that although the mathematical continuum is still called a continuum, it differs considerably from the popular conception of a continuum, where every element merges into its neighbour. However this may be, the mathematical continuum, and with it mathematical continuity, are as near an approach to the sensory continuum and to sensory continuity as it is possible for mathematicians to obtain. The sensory continuum itself is barred from mathematical treatment owing to its inherent inconsistencies.

And here an important point must be noted. In a sensory continuum considered as a chain of elements, an understanding of nextness or *contiguity*, hence an understanding of order, was imposed upon us by judgments of identity in our sensory perceptions. But the same no longer holds in the case of a mathematical aggregate of points, owing to the absence of that merging condition which guided us in the sensory manifolds. Theoretically, we may with equal justification order these points in whatever way we choose, and by varying the order in which we pass from point to point, we should find that the dimensionality of the aggregate varied in consequence.†

Dimensionality is thus a property of order, and order must be imposed before dimensionality can be established.

---

* The discoveries of du Bois Reymond have shown that we could proceed still further and interpose an indefinite number of additional points, but in the present state of mathematics this extension is viewed only as a mathematical curiosity. We may mention, however, that the rejection by Hilbert of the axiom of Archimedes (to be discussed in the following chapter), leading as it does to the strange non-Archimedean geometry, would be equivalent to considering the mathematical continuum as of this more general variety.

† That the choice of an ordering relation is all-important for the determination of dimensionality may be gathered from the following examples. Consider a number

In practice, the geometrician will retain the type of ordering relation imposed by our sensory experience and will conceive, by abstraction, of a mathematical space of points which will manifest itself as three-dimensional when this ordering relation is adhered to.   There is nothing to prevent him, however, from conceiving of mathematical spaces of any whole number of dimensions, either by modifying the ordering relation or again by modelling his mathematical manifold on some n-dimensional sensory continuum.

Let us suppose, then, that we have conceived of a three-dimensional mathematical space, obtained as an abstraction from the three-dimensional space of common experience.   Just as was the case with our sensory continuum, the mathematical continuum will be amorphous; no intrinsic metrics will be inherent in it, hence it will present us with no definite geometry.   The definition of congruence, that is, of the equality of two spatial stretches and more generally of two volumes, and the identity of

of diapasons emitting notes of various pitch. We should have no difficulty in ranging these various diapasons in order of increasing pitch. This ordering relation would be instinctive and not further analysable, since it would issue from that mysterious human capacity which enables all men to assert that one sound is shriller than another. Human appreciation is thus responsible for this definition of order in music; and we may well conceive of men whose reactions to sonorous impressions might differ from ours and who in consequence would range the diapasons in some totally different linear order. As a result, a note which we should consider as lying "between" two given notes might, in the opinion of these other beings, lie outside them. In either case, however, if we should assume that the various notes differed in pitch from one another by insensible gradations, a sensory continuum of sound pitches would have been constructed, though what we should call a continuum of pitches would appear to the other beings as a discontinuity of notes, and vice versa. In either case the respective sensory continua would be one-dimensional, since by suppressing any given note, continuity of passage would be broken. And now let us suppose that these strange beings were still more unlike us humans. Let us assume that not only would every note of their continuum appear to them as indiscernible from the one immediately preceding it and the one immediately following it, but that it would also be impossible for them to differentiate each given note from two additional notes. This assumption is by no means as arbitrary as it might appear, for we know full well that even normal human beings experience considerable difficulty in differentiating extreme heat from extreme cold and we know, too, that people afflicted with colour blindness are unable to distinguish red from green. But then, to return to our illustration, we should realise that the sound continuum, when ordered with this novel understanding of nextness or contiguity, would no longer remain one-dimensional.

In other words, in virtue of this new ordering relation, imposed by the idiosyncrasies of their perceptions, the same aggregate of notes would range itself automatically into a two-dimensional sensory continuum. This is what is meant when we claim that an aggregate of elements has of itself no particular dimensionality and that some ordering relation must be imposed from without. We must realise, therefore, that when we speak of the space of our experience as being three-dimensional in points, no intrinsic property of space can be implied by this statement. It is only when account is taken of the complex of our experiences interpreted in the light of that sensory order which appears to be imposed upon our co-ordinative faculties that the statement can acquire meaning. These conclusions are by no means vitiated by such facts as our inability to get out of a closed room or to tie a knot in a space of an even number of dimensions. Unfortunately, owing to lack of time, we cannot dwell further on these difficult questions.

shapes and sizes, will remain as conventional as before; and it will be only after we have introduced measuring conventions into an otherwise indifferent mathematical space that Metrical Geometry as opposed to Analysis Situs will be possible. However, the discussion of these points will be reserved for the next chapters.

CHAPTER II

THE BIRTH OF METRICAL GEOMETRY

IN the preceding chapter, we mentioned that the concept of continuity was suggested by our sense experiences; and that our understanding of space as a three-dimensional continuum had arisen from a synthesis of our various sense impressions. In the present chapter, we shall be concerned more especially with the geometry of the space continuum; and we shall show how measurement in turn would appear to have arisen from experience, more particularly visual experience.

For this purpose, let us consider the case of a motionless observer, rooted to the earth ever since his birth—a species of man-plant. Viewing the world whence he stood, he would notice that whereas certain visual impressions manifested a property of what he would recognise as "permanence," others would appear as squirming forms moving across his field of vision. In order to simplify this discussion, we shall omit to take into consideration any awareness of focussing efforts, on the part of our observer, as also any appreciation on his part of the convergence of his eyes. Under the circumstances, his visual perceptions would reveal a world of two dimensions, "up and down, right and left"; the third dimension with which we are all familiar, *i.e.*, "away from and towards," would be lacking. As a result, the squirming forms passing through his field of vision would be interpreted as betraying a two-dimensional world of changing forms, which would in no wise be connected with the existence of a third dimension. In particular, there would be no reason for him to attribute these changes to the variations in the distances of rigid material objects from his post of observation.

But suppose, now, that concomitant with the activity of his will, our observer were to become aware of certain muscular exertions that accompanied a variation in the shapes of those forms which had hitherto remained fixed and undeformed in his field of vision. In ordinary parlance, our observer would be displacing his post of observation, that is to say, "walking." He would no longer remain fixed like a tree. As a result of these displacements, which he would end by recognising as such, not only would forms erstwhile fixed in shape appear to vary, but, vice versa, certain forms hitherto squirming could be made to maintain an unchanging appearance. Eventually, he would recognise that in those cases where variations of shape and size could be counteracted by suitable displacements of his post of observation, he had been observing rigid bodies varying their distances along a third dimension with respect to him. In this way, there would arise an understanding both of rigid bodies and of a third dimension. Furthermore, owing to his being able to repeat his experiments *here* as *there*, a realisation of the homogeneity of space would ensue.

We see, then, that the three-dimensional space of experience appears to have arisen as the result of a synthesis of private views, each one of which would be that of an observer unable to move from a certain fixed spot. This synthesis would be extremely complex; unfortunately we have no time to mention the various conditions that would have to be taken into consideration. Suffice it to say that our senses of sight, of touch, of muscular effort, of sound and of smell, to which should also be added the action of the semi-circular canals, would all play a part, dovetailing one into the other. Further considerations would also show that there is nothing mysterious in the fact that these various data should yield concordant results, rather than an incompatible set of conflicting spaces.

All we wish to point out is that by the physical space of experience, we do not merely wish to imply space with the objects located therein, such as it would appear from some definite point of observation. We do not mean the private vision in which rails converge and distant objects appear smaller; we mean a synthesis of these private perspectives, yielding us a common public space.

One private perspective with its converging rails taken by itself and considered without reference to other perspectives could not contain sufficient data to enable us to conceive of three-dimensional space, homogeneous and isotropic.* That this synthesis has been arrived at without the conscious effort of reason is granted. Nevertheless, though instinctive, the co-ordination of private experiences and perspectives is of great complexity; and it would not be impossible to conceive of this co-ordination as having followed other lines, just as an aggregate of books may be arranged in alphabetical order, or in order of size or of content, and so forth. With a change in our ordering relation we might have obtained a space of a greater number of dimensions. Undoubtedly, however, when account is taken of the facts of experience, the three-dimensional co-ordination is by far the simplest; hence there is no reason to be surprised at its having imposed itself with such force. These too brief indications must suffice for the present.

And now let us return to the problem of measurement. We have mentioned that in certain cases it would be possible, by changing our post of observation, to counteract the apparent modifications in the visual shapes of bodies that had suffered a displacement in our field of vision.

For instance, if we received a visual impression corresponding to a circle, and if this impression were followed by one corresponding to a triangle, and if it were impossible for us to re-establish the circular impres-

---

* Here again we are neglecting, from motives of simplicity, our awareness of the focussing effort and of that of convergence. If we take these into account, we are in all truth considering not merely one private perspective, but a considerable number. Even so, our data would be incomplete.

sion, we should have to assume that the body had changed in shape; whereas, if it were possible to re-establish the circular impression by exerting certain efforts (which would finally be interpreted as a displacement of our point of observation), we should end by assuming that we had witnessed a partial rotation of a rigid cone-shaped object.

This discovery of rigid objects in nature is of fundamental importance. Without it, the concept of measurement would probably never have arisen and metrical geometry would have been impossible. But with the discovery of objects which were recognised as rigid, hence as maintaining the same size and shape wherever displaced, it was only natural to appeal to them as standards of spatial measurement.

Measurements conducted in this way would soon have proved that between any two points a certain species of line called the straight line would yield the shortest distance; and this in turn would have suggested the use of straight measuring rods. Henceforth, two straight rods would be considered equal or congruent if, when brought together, their extremities coincided. As for a physical definition of straightness, it could have been arrived at in a number of ways, either by stretching a rope between two points or by appealing to the properties of these rigid bodies themselves. For instance, two rods would be recognised as straight if, after coinciding when placed lengthwise, they continued to coincide when one rod was turned over on itself. Finally, parallelograms would be constructed by forming a quadrilateral with four equal rods, and parallelism would thus have been defined.

Equipped in this way, the first geometricians (those who built the Pyramids, for instance) were able to execute measurements on the earth's surface and later to study the geometry of solids, or space-geometry. Thanks to their crude measurements, they were in all probability led to establish in an approximate empirical way a number of propositions whose correctness it was reserved for the Greek geometers to demonstrate with mathematical accuracy. Thus there is not the slightest doubt that geometry in its origin was essentially an empirical and physical science, since it reduced to a study of the possible dispositions of objects (recognised as rigid) with respect to one another and to parts of the earth. In fact, the very word geometry proves this point conclusively.

Now, an empirical science is necessarily approximate, and geometry as we know it to-day is an exact science. It professes to teach us that the sum of the three angles of a Euclidean triangle is equal to 180°, and not a fraction more or a fraction less. Obviously no empirical determination could ever lay claim to such absolute certitude. Accordingly, geometry had to be subjected to a profound transformation, and this was accomplished by the Greek mathematicians Thales, Democritus, Pythagoras, and finally Euclid.

The difficulty that Euclid had to face was to succeed in defining exactly what he meant by a straight line and by the equality of two distances in space. So long as geometry was in its empirical stage these definitions were easy enough. All that men had to say was, "Two solid rods will be recognised as straight if after turning one of them over they still remain in perfect contact," or again, "The distance between the two extremities of a material rod remains the same by definition wherever we may transport the rod."

Euclid, however, could not appeal to such approximate empirical definitions; for perfect rigour was his goal. Accordingly he was compelled to resort to indirect methods. By positing a system of axioms and postulates, he endeavoured to state in an accurate way properties which were presented only in an approximate way by the solids of nature. Euclid's geometry was thus the geometry of perfectly rigid bodies, which, though idealised copies of the bodies commonly regarded as rigid in the world of experience, were yet defined in such a manner as to be untainted by the inaccuracies attendant on all physical measurements.

But this empirical origin of Euclid's geometrical axioms and postulates was lost sight of, indeed was never even realised. As a result Euclidean geometry was thought to derive its validity from certain self-evident universal truths; it appeared as the only type of consistent geometry of which the mind could conceive. Gauss had certain misgivings on the matter, but did not have the courage to publish his results owing to his fear of the "outcry of the Bœotians." At any rate, the honour of discovering non-Euclidean geometry fell to Lobatchewski and Bolyai.

To make a long story short, it was found that by varying one of Euclid's fundamental assumptions, known as the **Parallel Postulate**, it was possible to construct two other geometrical doctrines, perfectly consistent in every respect, though differing widely from Euclidean geometry. These are known as the non-Euclidean geometries of Lobatchewski and of Riemann.

Euclid's parallel postulate can be expressed by stating that through a point in a plane it is always possible to trace one and only one straight line parallel to a given straight line lying in the plane. Lobatchewski denied this postulate and assumed that an indefinite number of non-intersecting straight lines could be drawn, and Riemann assumed that none could be drawn.

From this difference in the geometrical premises important variations followed. Thus, whereas in Euclidean geometry the sum of the angles of any triangle is always equal to two right angles, in non-Euclidean geometry the value of this sum varies with the size of the triangles. It is always less than two right angles in Lobatchewski's, and always greater in Riemann's. Again, in Euclidean geometry, similar figures of various sizes can exist; in non-Euclidean geometry, this is impossible.

It appeared, then, that the universal absoluteness of truth formerly credited to Euclidean geometry would have to be shared by these two other geometrical doctrines. But truth, when divested of its absoluteness, loses much of its significance, so this co-presence of conflicting universal truths brought the realisation that a geometry was true only in relation to our more or less arbitrary choice of a system of geometrical postulates.

From a purely rational point of view, there was no means of deciding which of the several consistent sets was true. The character of self-evidence which had been formerly credited to the Euclidean axioms was seen to be illusory.

However, there are a number of rather delicate points to be considered, and these we shall now proceed to investigate. Euclid's parallel postulate and the alternative non-Euclidean postulates reduce to indirect definitions of what we intend to call a *straight line* in the respective geometries mentioned. If there existed such a universal as *absolute straightness*, represented, let us say, by a Euclidean straight line, we might claim that Euclidean geometry constituted the true geometry, since its straight line conformed to the ideal of absolute straightness. But this existence of a universal representing absolute straightness is precisely one of the metaphysical cobwebs of which the discovery of non-Euclidean geometry has purged science. To illustrate this point more fully, let us assume that we think we know what is implied by a straight line. Whether we merely imagine a straight line or endeavour to realise one concretely, we are always faced with the same difficulty. For instance, we consider that a rod is straight when it can be turned over and superposed with itself, or else we place our eye at one of its extremities and note that no bumps are apparent. Again, we may realise straightness by stretching a string, viewing a plumb line or the course of a billiard ball. We may also execute measurements with our rigid rods; the straight line between any two points will then be defined by the shortest distance. But whatever method we adopt, it is apparent that our intuitive recognition of straightness in any given case will always be based on physical criteria dealing with the behaviour of light rays and material bodies. We may close our eyes and think of straightness in the abstract as much as we please, but ultimately we should always be imagining physical illustrations.

Suppose, then, that material bodies, including our own human body, were to behave differently when displaced. If corresponding adjustments were to affect the paths of light rays, we should be led to credit rigidity to bodies which from the Euclidean point of view would be squirming when set in motion. As a result, our straight line, that is, the line defined by a stretched rope, our line of sight, the shortest path between two points, would no longer coincide with a Euclidean straight line. From the Euclidean standpoint our straight line would be curved, but from our own point of view it would be the reverse; the Euclidean straight line would now manifest curvature both visually and as a result of measurement. A super-observer called in as umpire would tell us that we were arguing about nothing at all. He would say: "You are both of you justified in regarding as straight that which appears to you visually as such and that which measures out accordingly. It will be to your advantage, therefore, to reserve your definitions of straightness for lines which satisfy these conditions. But you are both of you wrong when you attribute any absolute significance to the concept, for you must realise that your opinions will always be contingent on the nature of the physical conditions which surround you."

Incidentally, we are now in a position to understand why the Euclidean axioms appeared self-evident or at least imposed by reason. They represented mathematical abstractions derived from experience, from our experience with the light rays and material bodies among which we live. We shall return to these delicate questions in a subsequent chapter. For the present, let us note that since our judgment of straightness is contingent on the disclosures of experience, even the geometry of the space in which we actually live cannot be decided upon *a priori*. To a first approximation, to be sure, this geometry appears to be' Euclidean; but we cannot prophesy what it may turn out to be when nature is studied with ever-increasing refinement. It was with this idea in view that Gauss, who had mastered in secret the implications of non-Euclidean geometry, undertook triangulations with light rays over a century ago. Furthermore, even were the geometry to be established for one definite region of space, we could not assert that our understanding of straightness, hence of geometry, might not vary from place to place and from time to time; hence we cannot assert with Kant that the propositions of Euclidean geometry possess any universal truth even when restricting ourselves to this particular world in which we live.

Such discussions might have appeared to be merely academic a few years ago; and non-Euclidean geometry, though of vast philosophical interest, might have seemed devoid of any practical importance. But to-day, thanks to Einstein, we have definite reasons for believing that ultra-precise observation of nature has revealed our natural geometry arrived at with solids and light rays to be slightly non-Euclidean and to vary from place to place. So although the non-Euclidean geometers never suspected it (with the exception of Gauss, Riemann and Clifford), our real world happens to be one of the dream-worlds whose possible existence their mathematical genius foresaw.

Now, all these investigations initiated by attempts to prove the correctness of the parallel postulate led mathematicians to further discoveries.

A more thorough study of Euclid's axioms and postulates proved them to be inadequate for the deduction of Euclid's geometry. Euclid himself had never been embarrassed by the incompleteness of his basic premises, for the simple reason that although he failed to express the missing postulates explicitly, he appealed to them implicitly in the course of his demonstrations. The great German mathematician Hilbert and others succeeded in filling the gap by stating explicitly a complete system of postulates for Euclidean and non-Euclidean geometries alike. Among the postulates missing in Euclid's list was the celebrated postulate of Archimedes, according to which, by placing an indefinite number of equal lengths end to end along a line, we should eventually pass any point arbitrarily selected on the line. Hilbert, by denying this postulate, just as Lobatchewski and Riemann had denied Euclid's parallel postulate, succeeded in constructing a new geometry known as non-Archimedean. It was perfectly consistent but much stranger than the classical non-Euclidean varieties. Likewise, it was proved possible to posit a system of postulates which would yield Euclidean or non-Euclidean geometries of any number of dimensions; hence, so far as the rational requirements of the mind were concerned, there was no reason to limit geometry to three dimensions.

Incidentally, we see to what rigour of analysis and to what profound introspection the mathematical mind must submit; for the implicit postulates appealed to unconsciously by Euclid are so inconspicuous that it is only owing to the dialectics of modern mathematicians that their presence was finally disclosed and the deficiency remedied by their explicit statement.

From all this rather long discussion on the subject of postulates and axioms we see that the axioms or postulates of geometry are most certainly not imposed upon us *a priori* in any unique manner. We may vary them in many ways and, as regards real space, our only reason for selecting one system of postulates rather than another (hence one type of geometry in preference to another) is that it happens to be in better agreement with the facts of observation when solid bodies and light rays are taken into consideration. Our choice is thus dictated by motives of a pragmatic nature; and the Kantians were most decidedly in the wrong when they assumed that the axioms of geometry constituted *a priori* synthetic judgments transcending reason and experience.

CHAPTER III

THE procedure of presentation of non-Euclidean geometry which we have followed to this point hinges on the parallel postulate, hence on the definition of the straight line. In many respects, a much deeper method of investigation was that pursued by Riemann, founded on the concept of *congruence*. By congruence we mean the equality of two distances and more generally of two volumes in space. As we have explained elsewhere, the two methods lead to the same results. Indeed, once a metrical geometry has been defined, whether by the method of postulates or by any other means, a corresponding definition of a straight line and of equal or congruent distances is entailed thereby.

Thus, with Euclidean geometry, congruent lengths at different places are exemplified by the lengths spanned by a material rod transported from one place to another. Congruent or rigid objects having thus been defined, a straight line is given by the axis of rotation of a material body, two of whose points are fixed, or again by the shortest distance between two points measured with our rigid rod.

Nevertheless, although the various methods of presentation are equivalent, it may be of advantage to make the definition of congruence fundamental rather than that of the straight line. Such was the procedure followed by Riemann.

When we revert to experience for an understanding of congruence, we find it exemplified in the rigid bodies of nature, whose geometrical dispositions yield, more or less precisely, the results of pure Euclidean geometry. If we idealise congruence, as thus defined, and express it mathematically, we may say that perfectly rigid bodies are those whose measurements would yield Euclidean results with absolute precision. But though the mathematician has thereby eliminated from his definitions the inaccuracies attendant on physical measurements, his understanding of congruence reduces to a mere idealised copy of the behaviour of special bodies found in nature. While he has thus obtained a possible mathematical definition of congruent bodies (that given in nature), it remains to be seen whether other types of congruence would not also be rationally possible. His aim must therefore be to define congruence mathematically, without appealing to experience.

When, however, we discard the empirical criterion which prompted us to define material bodies as rigid, we find that a unique mathematical definition of rigidity eludes us. For to say that a body remains rigid or congruent with itself during displacement means that the spatial distance between its extremities remains ever the same. But our only means of disclosing this fact is by measuring the body with an admittedly rigid rod at successive intervals of time and noting the continued identity in our numerical results. Hence it follows that the value of these results would be nullified were we to cast any doubt on the maintenance of the rigidity of our measuring rod. And how could we ever justify its rigidity unless we were to compare it with some other rod regarded as rigid, and so on *ad infinitum?* From all this it appears that a body can be regarded as rigid only with respect to our measuring rod; and in order to ascribe any significance to rigidity we must first admit that our measuring rod is rigid by definition or by convention. We have no other means of establishing this rigidity.

We should reach the same conclusions were we to compare two lengths *AB* and *CD* situated in different parts of space. We could not say that the two lengths were equal or congruent in any absolute sense merely because our measuring rod could be made to coincide now with *AB*, now with *CD*. A definition of this sort would obviously presuppose that our measuring rod had remained undeformed or congruent with itself when displaced from *AB* to *CD*. It would reduce to testing the congruence of *AB* and *CD* by presupposing that we knew how to recognise the maintenance of congruence in our rod during displacement. In thus defining congruence in terms of congruence our argument would be circular.

The whole trouble arises from the continuity of space which precludes us from attributing any absolute meaning to the statement that two lengths situated in different parts of space are equal or unequal. There is no absolute significance in stating that there is as much space between *A* and *B* as between *C* and *D*. We cannot compare lengths by counting the number of points that they contain, since there are just as many points between the extremities of an inch as between the extremities of a mile—an infinite number in either case.*

In short, we see that in mathematical continua, just as in sensory ones, measurement and comparison of lengths can be considered only as a result of some convention; there is nothing in the continua themselves to suggest any definite metrics or geometry. This is what is meant by saying that space is amorphous and presents us with no means of determining absolute shape and size.

* This difficulty of counting points might be obviated to a certain degree were space to be considered discrete or atomic; for in that case we might count the atoms of space separating points and thereby establish absolute comparisons between distances. But here again the procedure would be artificial, for it would be nullified unless we were to assume that the spatial voids separating the successive atoms were always the same in magnitude; furthermore (as Dr. Silberstein points out in his book, "The Theory of Relativity"), we should be in a quandary to know how a succession of atoms would have to be defined, since this definition would depend on a definition of order. At any rate, we need not concern ourselves with atomic or discrete manifolds, for Riemann assumed that mathematical space was a continuous manifold. In view of the quantum phenomena we may eventually be led to modify these views and to attribute a discrete nature to space, but this is a vague possibility which there is no advantage in discussing in the present state of our knowledge.

Now, it might appear from the preceding discussions that so far as the mathematician is concerned, since a definition of equal distances is purely conventional, any conceptual rod might be chosen for this purpose. We might take a rod which, when moved about, would squirm like a worm, and elevate it to the position of a standard rod to which all other lengths would have to be compared. But this would be too extreme; for although mathematical space possesses no inherent metrics, yet certain requirements are demanded of rods susceptible of being considered as remaining congruent when displaced. It is these requirements, as postulated by Riemann, which we shall now proceed to discuss. Riemann assumed that the necessary requirements would be as follows:

In the first place, he assumed that if we restricted ourselves to infinitesimal volumes of space, congruence would be established by means of Euclidean solids and measuring rods. This postulate is called the *postulate of Euclideanism in the infinitesimal*.

Furthermore, if our rods remained congruent and we made certain constructions with them (pyramids, etc.), we should be able to transfer these rods and make exactly the same constructions, presenting exactly the same numerical relationships, in any other part of space. This postulate is a mere reflection of our belief in the homogeneity and isotropy of measurements in space, the same everywhere and the same in any direction. When presented in a slightly different form, it is often referred to as *the postulate of free mobility*.

We may also say that if two rods coincide in a certain region of space, they should continue to coincide when separated, then brought together again in any other region.*

* Were it not for this restriction, comparisons of distance in space situated in different places could never be obtained by transporting rods from one place to another; for since the measurements yielded by our standard rods when they had reached their point of destination would depend essentially on the route they had followed, they could scarcely be called rigid. When the restriction is adhered to we obtain the most general type of Riemannian spaces or geometries exemplified by the three major types, the Euclidean type, the Riemannian type and the Lobatchewskian type. Weyl, however, dispenses with Riemann's postulate and thereby obtains a more generalised type of space or geometry.

The non-mathematical reader is likely to become impatient at this rejection on the part of mathematicians of apparently self-evident postulates. But it must be remembered that a postulate which can be rejected and whose contrary leads to a perfectly consistent doctrine can certainly not be regarded as rationally self-evident; so that in a number of cases the legitimacy of so-called self-evident propositions can only be discussed *a posteriori* and not *a priori*. As a matter of fact our belief in self-evident propositions is derived in the majority of cases from crude experience and we cannot exclude the possibility that more refined observations may compel us to modify our opinions in a radical way. Einstein's discovery that Euclid's parallel postulate would have to be rejected in the world of reality is a case in point.

So far as Weyl's exceedingly strange geometry is concerned, it is conceivable that it also may turn out to represent reality after all, for Weyl found it possible to account for the existence of electromagnetic phenomena in nature by assuming that the space-time of relativity was of the more general Weylian variety and not, as Einstein had assumed, of the more restricted Riemannian type.

With these restrictions imposed, Riemann discovered that the Euclidean type of measurement was only one among others. Two other types, namely, the Lobatchewskian and the Riemannian, were also possible.

Sophus Lie considered the same problem, though from a different standpoint. He argued that a rigid body would be such that when one of its points was fixed, any other of its points would describe a surface. When one of its lines was fixed, any other of its points would describe a curve, and finally, when three points were fixed, the rigid body would be unable to move.

As a result of these investigations it was proved that there existed bodies which, as contrasted with Euclidean solids, would squirm when displaced. Yet, in spite of this fact, these non-Euclidean bodies would present all the mathematical requirements of congruent bodies. These alternative types of bodies may be called Riemannian bodies and Lobatchewskian bodies, respectively.

If we select Euclidean congruence for the purpose of measuring space, non-Euclidean bodies will appear to squirm as they move; if we select non-Euclidean congruence, it will be the Euclidean bodies which appear to vary in form when displaced. It is usual, however, to reserve the term **rigid** for the Euclidean bodies; but this is only in order to conform to the ordinary understanding of rigidity as derived from experience. If we omit to take into consideration the physical behaviour of material bodies which has forced a certain conception of rigidity upon us, there appears to be no mathematical reason for assigning greater importance to Euclidean measurements than to non-Euclidean ones. It follows that the non-Euclidean bodies are just as much entitled to the appellation "rigid" as are the better-known Euclidean ones. When it comes to deciding which of the types of rigidity represents conservation of absolute shape and size, the problem appears to be entirely meaningless. Accordingly we shall refer to the various types of bodies as Euclideanly or as non-Euclideanly rigid.

Now it is perfectly obvious that when we measure lengths and ratios of lengths with one type of standard rods or another, we shall obtain conflicting numerical ratios and values. Thus, if we fix one extremity of our standard rod and allow it to rotate in all directions, its free extremity will describe points on a spherical surface regardless of the type of rod or geometry with which we are dealing. In a similar way we might obtain a circumference. If, then, we abide by the Euclidean system of measurement, the value of the ratio of the lengths of circumference and diameter would remain ever the same however great the diameter; we should always obtain the same constant number, known as π—equal to $3.141592 \ldots$, first calculated by Archimedes. On the other hand, with Riemannian or Lobatchewskian measurements, we should obtain a variable value for this ratio, always smaller than π in the case of Riemannian geometry, decreasing as the diameter increased; and always greater than π in the event of our having selected Lobatchewskian measurements. Similar discrepancies would attend the measurement of all other geometrical figures and angles.

And here there is a further point to be considered. Even when we have defined a particular type of measurement, be it Euclidean or non-Euclidean, we have by no means fixed the behaviour of our rods and bodies in any absolute sense; and an understanding of absolute rigidity still escapes us. We can only define Euclidean bodies, for example, as those whose laws of disposition yield Euclidean numerical results. The following illustration will show us that these bodies might all vary in absolute shape when displaced, yet still yield the same Euclidean results.

All we have to do is to consider a transparent plane and a man executing geometrical constructions with Euclidean rods on its surface. If we consider a point-source of light casting the shadows of these rods on some other surface, whether plane or curved, these shadows, considered as measuring rods in turn, will still yield exactly the same Euclidean results, their laws of disposition will remain unchanged, and hence they will be rigid Euclidean rods exactly to the same extent as the original ones. Yet as contrasted with these original rods the shadows would squirm when displaced varying in shape and size, and the Euclidean straight lines would appear curved.

In short we see that absolute shape, straightness, size and rigidity in conceptual mathematical space escape us completely, and the significance of rigidity as portraying the maintenance of an unchanging volume of space, even when considering Euclidean bodies, is indeterminate. In its most extended sense, this is what mathematicians mean by the relativity of space. The sole justification for the general acceptance of the concept of rigidity in the popular sense is due to the presence of material bodies in our universe which we agree to accept as standards. These, incidentally, yield Euclidean results.

Summarising, we see that mathematical space is amorphous; it has no particular metrics, no particular geometry. According to our methods of measurement, we may obtain one geometry or another in the same space. It is often convenient to express all these results by saying that space is Euclidean, Riemannian or Lobatchewskian; but we must be careful to note that space itself has very little to do with the matter.

Let us now examine another aspect of the problem. Until such time as we have fixed our choice on a system of measurements, the amorphous nature of space forbids us to attach any determinate significance to a distance between two points. This, in turn, prevents us from attaching any significance to what may be considered the *shortest* distance between two points. However, when we have adopted one of the three measuring conventions, the significance of a shortest distance becomes determinate and a definite line joining the two points will be found to embody this shortest distance. Lines of this type are known under the name of *geodesics;* and in any of the geometries with which we shall be dealing, they play the part taken by the straight line in Euclidean geometry.*

Corresponding to every one of the three congruence definitions we have mentioned, there exists a definite type of straight line or geodesic between any two points. With Euclidean congruence we obtain the Euclidean straight line which satisfies Euclid's parallel postulate, and vice versa. Again, if we adopt one of the two non-Euclidean types of congruence we are led to the non-Euclidean straight lines or geodesics, those which satisfy the non-Euclidean postulates, and vice versa. In short, we see that the two methods of presenting non-Euclidean geometry, either through the metrical or congruence method or through the parallel-postulate method, are in the main equivalent.

In a more thorough treatment of non-Euclidean geometry other non-metrical methods of presentation are sometimes adhered to. One of these is based on the theory of groups and projective geometry; another is due to the discoveries of Levi-Civita and Weyl and depends on the fundamental concept of an infinitesimal parallel displacement; it opens the way to Weyl's still more general geometry. Still another mathematical method of exploring space is that of continuous tracks which was investigated by Eisenhart and Veblen. In this book we shall refrain from discussing these more difficult methods, for they present too technical an aspect; † but it would be a great mistake to assume that these imaginative flights of the pure mathematicians were of no utility to physical science. Apart from the deep philosophical light which they throw on the entire problem of space, we know from past experience that what was at its inception a pure mathematical dream has more than once become at some later date the image of physical reality. Non-Euclidean geometry, and possibly Weyl's still stranger geometry, are cases in point.

We have now to consider a number of questions pertaining to location and motion in space. Mathematical space, as we have seen, is n-dimensional and amorphous. But in order to bring it into closer contact with the physical space of experience we may assume the necessary postulates of order and contiguity to have been specified. We thus obtain three-dimensional mathematical space. Now mathematical space, in view of its amorphous nature, is essentially relative. The very homogeneity of space, its sameness "here" as "there," precludes our being able to differentiate one absolute position from another. It will follow that absolute motion, *i.e.*, a variation of absolute position in empty amorphous mathematical space, can have no significance.

Our sole means of giving significance to position and motion will be to select some three-dimensional frame of reference as standard. For

---

* The association of a straight line with the shortest distance between two points only holds, however, provided the several dimensions of the space are of an identical nature. When we consider continua in which the several dimensions differ in nature, as in the space-time of relativity, the straight line may turn out to be the longest distance between two points.

† In the chapter on Weyl's theory we shall mention Weyl's method briefly.

instance, we might consider three mutually perpendicular axes meeting at a point. A more concrete illustration would be afforded by considering the two adjoining walls and the floor of our room. Then, after having selected the standard rods which we intend to employ, and also a clock, we should proceed to measure the position and change of position of an object with respect to our room. This would yield us the relative position and motion of the object (relative to our room).

Now, if our choice of a frame of reference, and our methods of measurement, were not arbitrary, but were imposed in some inevitable way by the texture or nature of space, we might still claim that position and motion, though relative to our frame, were yet absolute, since they could always be determined without ambiguity. But this inevitability of a particular frame of reference is precisely what the amorphous nature of mathematical space renders impossible, for there is nothing in mathematical space which permits us to single out any particular frame. Hence no absolute unique determination of position and motion can be contemplated. In fine, we must conclude that in mathematical space position and motion, like distance and shape, are entirely relative.

Thus far we have been considering mathematical space; and mathematical space is but an abstraction from the space of experience. It remains to be seen, therefore, whether our conclusions will apply without modification when we consider the physical space of experience. Here we may state that so far as absolute position is concerned there is nothing to change in our conclusions. Thus, in everyday life, when we speak of returning to the same point in space, we do not really mean the same point in *empty space*. We mean rather the same point on the earth's surface; and we do not stop to consider whether the earth has moved through space in the interval. As a matter of fact it would be totally impossible for us to discover exactly what distance the earth had travelled. Granting that we might ascertain the displacement of the earth along its orbit round the sun, all further progress would be arrested by our ignorance of the velocity of the sun through space. We might even measure the velocity of the sun with respect to the stars, but this again would not yield us its velocity through empty space; for we should still remain in complete ignorance of the velocity of the stars as a whole through space. So we see that when we speak of the same point in physical space we are always referring to the same relative position with respect to some observable frame of reference. It is not absolutely necessary that the frame be material, and we might perfectly well conceive of it as being defined by three perpendicular light rays. The essential requirement is that it should be observable, so that all men should agree on the frame they were talking about.

Owing to the important part the earth's crust occupies in our daily lives, the impossibility of defining the same point at different times without the help of a frame of reference is somewhat obscured. But it is imperative to master this aspect of the elusiveness of space before proceeding to the study of the more complex questions raised by the theory of relativity.

When, however, we consider the problem of motion, we shall find that in contradistinction to what holds for mathematical space, motion in physical space appears to be absolute.* While it is true that from a purely visual standpoint the motion of a body will depend essentially on the frame of reference we select, and will vary when we change our frame from the earth to, say, the moon or the sun, yet owing to the generation of dynamical effects it is possible to decide in an absolute way whether the body is rotating or not. This is due to the fact that these telltale dynamical effects are present just the same regardless of our choice of a frame of reference. It was owing to these empirically discovered dynamical facts that Newton felt compelled to accept the absolute significance of motion, hence of space. We may gather from what has been said that we may not extend lightly to the physical space of experience the conclusions derived from conceptual mathematical space.

We now come to a last point. Suppose that in mathematical space we wish to measure the distance between two objects moving in various ways through our frame. For our measurements to have any significance, we must of course specify that the positions of the two objects correspond to the same instant of time. If, therefore, a change in the relative motion of our frame of reference were assumed to cause a modification in our understanding of the sameness of a time at two different places, the distance between the two objects would vary with the frame in which we were reckoning it. But here we are discussing pure mathematical space, which we may always conceive of as divorced from time. Hence we may assume that the sameness of a time at two different places remains unaffected by any alteration in the choice of our frame.

Classical science believed that the space of experience would also satisfy these conditions. It thought that time was absolute, that its rate of flow was ever the same, and that the concept of the sameness of a time at two different places was an absolute and unambiguous one. This assumption was tantamount to divorcing space from time and regarding them as two essentially different categories. It has been one of the triumphs of the relativity theory to disprove this fundamental thesis, and to show that in the real world of experience a continuum of space by itself is but a shadow. Henceforth the continuum of reality is space-time, a four-dimensional continuum of events, and no longer, as in classical science, a three-dimensional one of points, coupled with a one-dimensional continuum of instants. But inasmuch as the necessity for these new conceptions will be justified at length in future chapters, we shall refrain from discussing them at this stage.

---

* Accelerated motions appear to be absolute.

CHAPTER IV

THUS far we have discussed more especially that abstraction from the space of experience which we called mathematical space; we have seen it to be entirely amorphous. In the present chapter we shall have to consider the space of experience and determine to what extent it differs from conceptual mathematical space. An important difference as regards the problem of motion has already been noted, for we remember that in physical space, motion often manifests itself as absolute. For the present, however, we shall confine our attention to the problem of congruence.

Mathematical space is amorphous; it possesses no intrinsic metrics, and our choice of standards of measurement is largely arbitrary. As a result, absolute shape, size and straightness are meaningless concepts. But in physical space, it is a matter of common knowledge that men have no difficulty in agreeing at least approximately on the sameness of two shapes or of two sizes. They agree that a stone remains undeformed when displaced, hence declare the stone to be rigid, whereas they recognise that an object behaving like a worm is of the squirming variety.

So here there appears to exist an important difference between mathematical space, where no particular definition of congruence is suggested, and physical space, where a definite type seems to impose itself naturally and is accepted unanimously. After all, there is nothing very mysterious about this unanimous agreement; for had men refused to be guided in their definition of congruence by their sense of sight, they would have been led into all manner of difficulties. They would have had to assume that a stone carried in their hands, though appearing unchanged visually, was yet squirming, and, vice versa, that bodies which appeared to squirm were yet rigid. The difficulties in reaching some common understanding of measurement would thus have been hopelessly great. In fact, the definition of congruence, which we have mentioned, imposes itself so irresistibly that it is only since the discovery of non-Euclidean geometry that its absolute validity has been refuted.

Viewing the situation as it now stands, we may say that material rods, visually recognised as rigid, will be taken as norms of measurement in physical space. In other words, it will be assumed that rods which coincide when brought together will continue to remain congruent when transferred, independently of one another, to other regions of space. Of course, the physicist will guard himself as much as possible against local contingent influences, such as variations of pressure and temperature, which might influence the behaviour of his rods. But when all these elementary precautions have been observed, the geometry determined by our rods will automatically become the geometry of space. This physical definition of congruence may be termed *practical congruence*, as distinguished from *theoretical congruence*, which is embodied by the mathematical types we have discussed.

With his standard rigid bodies defined in this way by physical objects, the physicist can perform measurements in the space of his frame and in consequence obtains the numerical results of three-dimensional Euclidean geometry. Inasmuch as the more careful he is to guard against such contingent influences as variations of temperature, the more accurately will his numerical results approximate to those of pure Euclidean geometry, he feels justified in stating that rigid objects behave like rigid Euclidean bodies and that the space of our experience is rigorously Euclidean.

We may also recall that his rigid bodies having been defined, the definition of a straight line as the axis of rotation of a revolving solid, two of whose points are fixed, or as the shortest distance between two points in space, follows immediately; and of course the straight line thus defined satisfies Euclid's parallel postulate, since it is derived from the behaviour of Euclidean solids.

At this stage, a number of popular exponents of non-Euclidean geometry have fallen into a rather unfortunate error. They have argued that material bodies under perfect conditions must necessarily behave like Euclidean solids, for if they behaved like non-Euclidean bodies when displaced they would squirm and change in shape. As it would be inadmissible to credit any such distorting influence to that void which we call empty space, a non-Euclideanism of material bodies would be debarred on first principles. But these men overlook the fact that Riemann's and Lobatchewski's geometries do not in any way refer to bodies which squirm and are distorted in any absolute sense as they move about. The non-Euclidean bodies are merely distorted when contrasted with Euclidean bodies taken as standards; but it would be equally true to state that Euclidean bodies likewise would squirm when displaced if we were to contrast them with non-Euclidean bodies taken as standards. In any case, both Euclidean and non-Euclidean bodies behave in a homogeneous way throughout space.* By this we mean that wherever they might be situated in empty space, measurements computed with them would yield the same numerical results. As for Euclidean congruence and Euclidean rigidity, it is by no means more representative of real rigidity than are the non-Euclidean varieties. There is therefore no reason to appeal to a distorting effect of empty space in order to account for a possible non-Euclidean behaviour of our material solids when displaced from point to point. Non-Euclideanism may or may not exist in real space, but this is a point for physical measurement and not for philosophy or mathematics to decide.

* We are referring solely to those bodies which would yield the numerical results of Riemann's or Lobatchewski's geometry—both these types of geometry being compatible with the homogeneity of space.

All we can say is that the principle of sufficient reason compels us to credit empty space with a sameness throughout, and that our measuring rods and material bodies must also behave homogeneously and isotropically, as indeed they do in the three geometries discussed. Only if measurements undertaken with our rods in different parts of space yielded variable non-homogeneous numerical results should we have to assume that space was not really empty and that our rods were subject to local influences.

At any rate, the early non-Euclidean geometers, realising that space as a result of measurement might turn out to be non-Euclidean, busied themselves with devising means of settling the question once and for all. Now, as a result of their measurements with material rods there was no doubt that space was very approximately Euclidean. But here it must be realised that the two non-Euclidean geometries as opposed to the Euclidean variety are not unique. We may conceive of various intensities of non-Euclideanism of both types, merging by insensible gradations into Euclidean geometry. It was therefore still an open question whether space, in spite of its apparent Euclidean characteristics, might not betray a slight trace of non-Euclideanism. A simple illustration will make this point clearer.

We saw that in Euclidean geometry the ratio of the length of a circumference to its diameter was always the same number, π. In Riemann's geometry this number was always smaller than π, and decreased progressively from the value π to the value zero as the diameter of the circle increased. But there was nothing definite about this rate of decrease; it might be very rapid, just as it might be exceedingly slight. We must conceive, therefore, of varying intensities of non-Euclideanism, or of departure from Euclideanism. Hence, if the non-Euclideanism of real space were exceedingly slight, it might require measurements extending over a circumference of gigantic proportions, reaching as far as the stars in order to detect it; and measurements conducted in restricted areas could not be considered conclusive. The only means of disclosing slight traces of non-Euclideanism would therefore be obtained by having recourse to measurements conducted over very large distances.

Of course, in an attempt of this sort, measurements with material rods were out of the question and it was necessary to appeal to other methods of exploration. These were obtained by taking advantage of the propagation of light rays in empty space. It was known that over the limited extensions of ordinary experience the paths of light rays coincided with the straight lines determined by rods or stretched strings, so that rays of light could be used in place of stretched strings or rods for the purpose of defining geodesics. Triangulations effected with light rays should therefore yield the same results as measurements performed with rods, and so should reveal the geometry of space with as much accuracy as would the more conventional kinds of measurement.

Gauss appears to have been the first to undertake space explorations of this sort, when he performed triangulations with light rays from one mountain top to another. But his observations were too crude and executed over too small an area to detect any trace of non-Euclideanism. Lobatchewski suggested astronomical observations conducted on the course of rays of starlight through interstellar space. For instance, if two light rays emitted from a very distant star and striking the earth at two different points of its orbit appeared to manifest converging directions, we should know that space was Riemannian. If· the two rays appeared to diverge, however distant the star, space would be Lobatchewskian; and, finally, if for very distant stars the two directions appeared identical, space would be truly Euclidean. Yet the most refined astronomical measurements of stellar parallaxes failed to reveal the slightest trace of non-Euclideanism. Hence it was assumed that if any trace of non-Euclideanism was present in real space it was without doubt exceedingly slight, so that for all practical purposes the geometry of space might be regarded as Euclidean. Such were the results obtained by a physical exploration of space.

From all this we see that the physicist, basing his exploration of space on empirical methods, is perfectly justified in stating that its geometry can be determined, that a true definition of congruence can be arrived at, and that the equality of two lengths and of two spatial configurations has a definite significance in nature.

And yet, when we submit all these various examples to a critical analysis, we cannot help but see that this determination of the geometry of space is essentially physical and is, therefore, contingent on the behaviour of material objects and of rays of light. Had the behaviour of material bodies when displaced been regulated by other physical laws, had rays of light followed different courses, the geometry we should have attributed to space might have been entirely different. And we may well wonder what the behaviour of physical objects should have to do with the geometry of space. We shall return to this aspect of the problem later.

Also, it has sometimes been argued that our recognition of shape and size must possess a much deeper significance and cannot be attributed merely to the laws of behaviour of material objects and of rays of light. For instance, it is pointed out that even a child who knows nothing of measurement judges, on simple visual inspection, that a coin (when viewed from a perpendicular direction) is round and an egg oval. He does not feel it necessary to verify this fact by applying a ruler. However, regardless of what opinions we may eventually defend on the subject of a geometry intrinsic to physical space, it can scarcely be held that this last argument of the critic proves his point in the slightest degree.

It may be instructive to consider this illustration in greater detail. In the first place, we must realise that all we are in a position to appreciate when merely viewing an object, is its image on our retina. Indeed, were we to interpose a microscope or a deforming lens between object and eye, the object would suffer no change; but its image cast on our retina would, of course, be modified. As a result, our judgment of its shape and size would vary. Hence, when we decide unhesitatingly that the coin is round, the only inference to be drawn is that the coin's image on the retina is judged by the brain (or the mind, or whatever we decide to call it) to be circular. We do not wish to intrude on the ground of the eye-specialist or physiologist, who is better equipped than we are to proceed farther in this analysis, but we may point out that the problem which we are now considering is of an entirely different nature from the one whence we started. There we were considering whether shape in empty space was absolute; here we are considering whether an image cast on our retina should impress itself upon our recognition with any definite shape.

The difference amounts to just this: In mathematical space, and even in physical space, absolute measurement seemed to elude us, since in view of the continuity of space it appeared impossible to proceed with an enumeration of points. But in the case of the retina, its surface is no longer homogeneous; it possesses a heterogeneous structure like all tissues, probably a discrete one forming a pattern. In all such cases a definite metrics suggests itself naturally, just as on a net, in the absence of a ruler, we would compare lengths instinctively by counting the holes separating our points.

Whether or not, in the case of the retina, our appreciation of the coin's roundness comes from some unconscious counting process is a problem for the specialist to decide; but at all events, when we consider that the retina and the eye are Euclidean bodies just like our rulers, the concordance between our computations of shape and size, as determined by rods, and our direct visual appreciation of shape does not appear to present much of a mystery.

But there is still a further point to be considered. Were we to have a direct intuition of congruence and of absolute shape or length, our measurements with rods would have to be adjusted so as to conform to this intuition. The introduction of rods would thus be contemplated merely as an adjunct, in order to obtain greater definiteness. But it so happens that such is not the case.

Our intuitive visual appreciations yield results which differ from results obtained with rods, not merely *accidentally* as a result of the imperfection of human observation, but *systematically*.

A coin that measures out as round will appear flattened to the eye.* This phenomenon is illustrated by the well-known optical illusion wherein two rigid rulers which coincide when placed side by side, appear of unequal magnitude when placed horizontally and vertically, respectively. It is a well-known fact that the vertical appears to be longer than the horizontal. For this reason, vertical stripes on a cloth cause the wearer to appear thinner and taller, whereas horizontal stripes produce the reverse effect. Now the mere fact that we have agreed to accept such

---

* In this discussion we are always assuming that the observer is standing right above the coin; hence we are not considering the variations in apparent shape due to a slanting line of sight. This latter problem is of a totally different nature.

discrepancies as due to optical illusions rather than to the untrustworthiness of our rods proves that we deliberately reject our intuitive judgment of shape and size in favour of more sophisticated rules of measurement. In other words, we have abandoned direct intuition for physical determinations, hence for convenient but conventional standards.

In short, we may state that we possess no direct intuition of shape and length, and that what little we appear to have is traceable in the final analysis to the properties of material bodies and light rays.

Let us consider a last example relating to distance. We realise, without resorting to measurement, that the distance across the street is less than the entire length of the street. A number of different reasons conspire to account for this conviction. First, the muscular sensations which accompany the convergence of the eyes and the focussing of images on the retina would in themselves suggest the origin of our appreciation of distance. In addition, we know from experience that it requires a greater effort and a longer time to travel along the entire length of the street than to step across from one sidewalk to the other. And if we enquired why it took a shorter time to cross the street, we should find that it was due to the fact that we advanced with definite steps.

Although in mathematical space the two distances would be equal or unequal according to our measuring conventions, in practical life we have unwittingly posited our measuring convention by walking. Our successive steps henceforth define in our estimation congruent distances; and under the circumstances distances become measurable in terms of these steps. But these steps are themselves controlled by our human frame; hence all we have done has been to measure space in terms of our legs. That measurement originated in this way is indicated clearly by such words as *foot* and *cubit,* or again by the definition of a yard as expressed by the distance between the tip of a certain king's nose and the extremities of his fingers. And it is because our human limbs manifest the same type of congruence as the material bodies around us that this type of measurement once again imposes itself so strongly upon us. Indeed, measurements as computed with human limbs appear to be instinctive and are exemplified in children who, having grown, are surprised to find the rooms and buildings they have not seen for some years, appear considerably smaller. Instinctively, the child is measuring size in terms of his own height.

Thus, in whatever way we examine the matter, there appears to be nothing mysterious in our natural belief in the absoluteness of shape and size or in a definite geometry pertaining to space. None of the examples mentioned thus far entitle us to maintain that physical space manifests a definite metrics and that congruence is other than conventional.

If we consider the problem in its present state, we see that it is the physical behaviour of material bodies and light rays which is in the final analysis responsible for our natural belief in absolute shape. But this realisation brings with it the assurance that space itself has eluded us entirely in our discussions. Such was indeed Poincaré's stand. He maintained that though for purposes of convenience it was only natural for us to measure space as we do, yet if needs be we could disregard the behaviour of material bodies entirely, adopt non-Euclidean standards of measurement, and proceed as before.

In spite of this change we could construct exactly the same engineering works, in fact rewrite the whole of physics. Needless to say, everything would be extremely complex, all our known laws would be disfigured, and hypotheses *ad hoc* would have to be introduced. But when all is said and done, the task would be theoretically possible; so that if we disregard the criteria of convenience and simplicity, there is nothing to choose between the various types of measurements, any more than between the metric system and the British units.

As a further illustration of the elusiveness of absolute shape and size, Poincaré asks us to conceive of a hollow spherical volume placed anywhere in space, and to assume that the temperature in the sphere decreases progressively from the centre, becoming absolute zero at the surface. He assumes that this hollow sphere is peopled by imaginary beings whose bodies expand and contract with the temperature and that all material bodies in the sphere behave in a similar manner. If we should supplement these suppositions by assuming that the refractive index of the medium in the sphere's interior varies in a certain definite way, the rays of light in this hypothetical world would describe circles.

This closed universe would of course appear infinite to its inhabitants, since as they proceeded from the centre to the surface their bodies would grow smaller, their steps shorter, so that it would be impossible for them to reach its boundary however long they walked. The geometricians of this imaginary world would feel justified in proceeding exactly as we have done ourselves. They would define as remaining congruent when displaced, hence as rigid, those bodies which appeared to them to remain the same wherever they carried them. Owing to the paths devised for the light rays and to the sameness in the reduction of the sizes of all objects as the centre was left behind, the expanding and contracting bodies of this universe would present all the characteristics of rigidity. On conducting measurements with their rigid rods the hypothetical beings would obtain Lobatchewskian results, their entire world would appear to them as non-Euclidean, and non-Euclidean geometry would be as inevitable to them as Euclidean geometry is to the average layman. Some Kant among the hypothetical beings would surely arise and explain that non-Euclidean space was the *a priori* form of pure sensibility, transcending reason and experience. Then eventually some great mathematician would come along, sweep all those cobwebs aside, and prove that there existed other perfectly consistent types of geometries and that the ingrained preference of his fellow citizens for non-Euclidean geometry was due to the dictates of common experience and constituted by no means the *a priori* form of pure sensibility.

It is not merely in its philosophical aspect that Poincaré's illustration is interesting. The major point is the following: The hypothetical beings would be just as much entitled to assert that space was non-Euclidean as we are to assert that it is Euclidean. It is true that if we could look into their world we should say that they had a wrong understanding of measurement, and were totally in error when they assumed that their bodies were rigid, since we could see them getting smaller and smaller as they neared the surface. But we must not forget that the imaginary beings, in turn, could they but view our Euclidean bodies, would return the compliment and accuse us of having a wrong understanding of rigidity and measurement.

From all this it follows that by a mere variation in physical conditions the same space would be considered non-Euclidean or Euclidean. Obviously, by reason of this contradiction, space itself can have nothing to do with the problem; the type of space which physicists are discussing reduces therefore to a relational synthesis of physical results. Space itself remains amorphous.

Poincaré develops analogous arguments when he discusses the parallax observations conducted on the rays of starlight. Euclidean geometry, for instance, regarded purely as a system of measurement, is from a mathematical point of view the simplest type of geometry for the same reason that a monosyllable is simpler than a polysyllable. It is therefore obviously to our interest to retain it if possible. Of course if, as in the hypothetical world discussed previously, material bodies behaved like non-Euclidean solids and if light rays followed appropriate courses, we should have to abandon Euclidean geometry for reasons of practical convenience. But since, in the world we live in, our habitual solids behave to a high order of approximation as do Euclidean solids, our preference for Euclidean geometry seems perfectly legitimate even from the standpoint of physics.

Suppose now that the parallaxes of the very distant stars turned out to be negative: would Euclidean geometry and Euclidean space have to be abandoned? As Poincaré points out, this would by no means be necessary or even advisable. It is true that if we assumed, as is the custom, that rays of starlight follow geodesics through space, negative parallaxes would imply a trace of Riemannianism in space; but the primary point to decide is, "How do we know that rays of light follow geodesics?" Obviously this is capable neither of proof nor of disproof. An empirical proof that such a contention was correct or incorrect would be possible only were we to know beforehand how the geodesics of space were situated, for then we could determine by observation whether rays of light followed them or not. But how could we establish the way the geodesics lie unless we were already apprised of this geometry which we now proposed to determine? Obviously, our procedure would be circular. Can we at least assume that rays of light must inevitably follow geodesics? Would any other assumption be impossible? Certainly not. A denial of the assumption would modify our understanding of optical phenomena; but what if it did? We could always get out of the difficulty by varying the laws of optical transmission, and still retain Euclidean geometry.   In other words, the geometry the physicist credits to space is contingent on his acceptance of a number of physical laws; and by varying these laws in an appropriate way he could still account for observed facts and credit corresponding types of geometry to space. Since all these various systems of physical laws would account for the facts of experience, how can we ever hope to decide which one of these systems corresponds to reality?   And under the circumstances, what use is there in discussing the *real* geometry of space?   All we can discuss is expediency.

In other words, Poincaré, by divorcing space from its material content, geometry from physics, places space and its geometry beyond the control of experiment; so that there is really nothing left for the physicist to argue about.

Poincaré's attitude towards space is that of the pure mathematician, and not that of the physicist.   If, however, we consider space and its physical content jointly, and no longer view space as a mathematical abstraction, the geometry of space acquires a definite significance.   It is that geometry which permits the simplest co-ordination of the facts of experience.   If, therefore, a co-ordination of the facts of experience presents greater simplicity when we assume space to be Euclidean or non-Euclidean, then space *is* Euclidean or non-Euclidean in spite of the fact that phenomena might just as well have been co-ordinated (though in a less simple way) had some other hypothesis been selected.

Riemann expresses the selective rôle of the criterion of simplicity when he writes:

"Nevertheless it remains conceivable that the measure relations of space in the infinitely small are not in accordance with the assumptions of our geometry [Euclidean geometry], and, in fact, we should have to assume that they are not if, by doing so, we should ever be enabled to explain phenomena in a more simple way." *

The two conflicting attitudes towards the problem of space are both of them defensible.   Einstein, himself, whose entire theory centers around a definite geometry being ascribed to the spatio-temporal background writes: "*Sub specie aeterni,* Poincaré in my opinion is right."   We must, however, be careful to distinguish between the two attitudes, or equivalently between the two conceptions of space: amorphous mathematical space, and physical space endowed with a metrics.

At all events, in what follows, we shall concern ourselves solely with the real space of the physicist, that is, with the space to which he is led when he seeks to co-ordinate the phenomena of the physical world with the maximum of simplicity.   With this understanding of space in our minds, a first reason for rejecting the concept of an amorphous space arises when we find that a large number of different methods of investigation all point to the same definite metrics for space.   Thus, the various material bodies we

* Riemann's views on space and its geometry were expressed in his famous memoir of 1854, subsequently translated into English by Clifford.

encounter are by no means identical in nature; some are light, others are heavy, and their chemical and molecular constitutions are certainly not the same. And yet in every case, whether our rods be of wood, of stone, or of steel, we obtain the same Euclidean results provided we operate as far as possible under the same conditions of temperature and pressure. In other words, there appears to be a sameness in our determinations of congruence regardless of the material bodies to which we appeal.

This uniqueness of the geometry of space is still further exemplified in the following example: Here are two totally different methods of exploring space, one with material rods giving us a physical definition of congruence, and one with light-ray triangulations giving us a physical definition of geodesics. In either case we are led to the same Euclidean geometry, and this concordance appears rather strange, for we might have expected that if the geometry we credited to space were irrelevant to space, the type of geometry obtained would have varied according to the physical exploration method considered. Besides, if space were amorphous, hence possessed no geodesics, it would be inconceivable that a free body or a light pulse should know how and where to move. The very definiteness and Euclidean straightness of the paths of free bodies and light rays, when referred to a certain frame of reference, would seem to indicate that space had a structure and was not amorphous.

To be sure, in view of modern discoveries there is nothing very strange in the fact that the courses of free bodies should coincide with the paths of light waves, since light has been proved to possess momentum just as matter does. But even so, it appears strange that the courses defined by moving bodies should yield the same geometry as measurements conducted with bodies at rest.

Then again, there are the dynamical properties of space, which we cannot afford to neglect. If physical space were amorphous, all paths through space should be equivalent, and yet centrifugal force and forces of inertia manifest themselves for certain paths and motions and not for others. Whence could these forces arise if not from the structure of space itself? Such was indeed Newton's contention.

In view of all these occurrences, difficult to account for if we believe in the amorphous nature of space, unless we appeal to some miraculous pre-established harmony, it appears as though space must be credited with a definite structure or metrics which, in the light of experiment, turns out to be Euclidean, at least to a first approximation. Expressed in a different way, real space appears to be permeated by an invisible field, **the Metrical Field**, endowing it with a metrics or structure.

Now, in view of the fact that the mathematical requirements of that void we call empty space preclude it from having a metrical field (*i.e.,* a structure or a metrics *per se*), the simplest way out of the difficulty is to assume that real space is not truly empty, but is filled with some mysterious physical medium, which we may call the "ether," * and that

---

* This metrical ether must not be confused with the classical ether of optics and electromagnetics.

it is this physical medium and not space itself which possesses a Euclidean structure. Henceforth it will be this ether structure which will be responsible for the apparent metrics or metrical field of space, which will cause material bodies to settle into definite shapes, and which will regulate the courses of light rays and free bodies far from matter.

For the physicist, however, this ether will be inseparable from real space; so that, to all intents and purposes, when the physicist discusses real space he will be referring to space together with its ether content. Were this ether structure to vanish, space would become amorphous, bodies would not know what shape to take, and light rays would not know where and how to move.

At this stage we must mention the premonitions of Riemann on the subject of the metrical field of real space. Riemann did not attribute this structure of space to the presence of some invisible medium, the ether, possessing a structure of its own. According to him the origin of the metrical field should be sought elsewhere. He felt that the metrical field of space should be compared to a magnetic or an electric field pervading space. And just as a magnetic field exists in the space surrounding a magnet, Riemann searched for the physical cause of the metrical field. With characteristic boldness, he found it in the matter of the universe; the metrical field thus became a species of material field. If Riemann's ideas are accepted we can understand how a redistribution of the star matter in the universe, altering as it would the lay of the metrical field, would produce deformations in the shape of a given body and variations in the paths of light rays. As Weyl tells us, a spherical ball of clay compressed into any other form might again be made to appear spherical were all the matter in the universe to be redistributed in a suitable way.

It would also follow that were all the matter in the universe to be annihilated, and as a result the metrical field to vanish, space (assuming that any physical meaning were left the term) would become completely amorphous, just like mathematical space; light rays would not know where to move, all geodesics having disappeared; and were one lone material body to be introduced into otherwise empty space, it would not know what shape to take. Without the metrical field, physical space would be unthinkable.

Still further important consequences follow from this matter-moulding hypothesis of Riemann. Prior to these views, the principle of sufficient reason appeared to imply that physical space would always turn out to be homogeneous—the same in all places. This does not necessarily mean Euclidean, for, as we know, Riemann's and Lobatchewski's geometries also correspond to homogeneous spaces; but all varying degrees of non-Euclideanism from place to place were thought to be excluded *a priori*. Of the vast realm of possible types of geometries or spaces discovered by Riemann, only the homogeneous types survived; a situation which Weyl finds appropriately expressed by the classical line: *"Parturiunt montes, nascetur ridiculus mus."*

But with the new views advocated by Riemann the situation changes entirely; for now the texture, structure or geometry of space is defined by the metrical field, itself produced by the distribution of matter. Any non-homogeneous distribution of matter would then entail a variable structure or geometry for space from place to place.

Now the question arises, What is the nature of this metrical field? Is it merely a name we are giving to the structure of space caused by matter? This view appears impossible, for space of itself, being a mere void, is not amenable to structure. Is the metrical field a direct emanation from matter, a rarefied form of matter? Or again is it a reality of a category differing from matter (call it the ether), which in the absence of matter would be amorphous, and which could only be forced into a structure by the influences due to matter and transmitted from place to place through the ethereal substance? In this case we should again be led to the view that what we commonly call the structure or geometry of real space reduces to the structure of the matter-moulded ether-filling space.

Riemann's exceedingly speculative ideas on the subject of the metrical field were practically ignored in his day, save by the English mathematician Clifford, who translated Riemann's works, prefacing them to his own discovery of the non-Euclidean Clifford space. Clifford realised the potential importance of the new ideas and suggested that matter itself might be accounted for in terms of these local variations of the non-Euclideanism of space, thus inverting in a certain sense Riemann's ideas. But in Clifford's day this belief was mathematically untenable. Furthermore, the physical exploration of space seemed to yield unvarying Euclideanism. And here the remarkable irony of the whole situation must be noted. Although experimenters had utilised the most refined apparatus for detecting a possible non-Euclideanism of space and had failed in their efforts, it was reserved for the theoretical investigator Einstein, by a stupendous effort of rational thought, based on a few flimsy empirical clues, to unravel the mystery and to lead Riemann's ideas to victory.* Nor were Clifford's hopes disappointed, for the varying non-Euclideanism of the continuum was to reveal the mysterious secret of gravitation, and perhaps also of matter, motion and electricity.

Before solving the problem, however, Einstein had been led to recognise that space of itself was not fundamental. The fundamental continuum whose non-Euclideanism was to be investigated was therefore not one of space but one of **Space-Time,** a four-dimensional amalgamation of space and time possessing a four-dimensional metrical field governed by the matter distribution. Einstein accordingly applied Riemann's ideas to space-time instead of to space, and attempted to explore the geometry of space-time by a purely rational co-ordination of known empirical facts. He discovered that the moment we substitute space-time for space (*and not otherwise*), and assume that free bodies and rays of light follow geodesics *no longer in space but in space-time,* the

---

* In all fairness to Einstein, however, it should be noted that he does not appear to have been influenced directly by Riemann.

long-sought-for local variations in geometry become apparent. They are all around us, in our immediate vicinity; and yet we had never realised it. We had called their effects gravitational effects, ascribing them to forces foreign to the geometry of the extension, and never suspecting that they were the result of those very local variations in the geometry for which our search had been vain. Indeed, it may be said that the theory of relativity is the theory of the space-time metrical field.

CHAPTER V

THE type of geometry we obtain is dependent, as we know, on our definition of congruence. If we define as congruent displacements the displacements of those bodies which in ordinary life we consider rigid and undeformable, we obtain Euclidean geometry. If, on the other hand, we define as congruent the displacements of those bodies which in ordinary life we should regard as of changeable shape, we generally obtain some non-Euclidean type of geometry.

In order to simplify what is to follow, we shall first consider the particular case of two-dimensional geometry. Consider, then, a plane surface. If on this plane surface we effect measurements with Euclidean rods, we obtain Euclidean results; if we employ rods which according to the Euclidean point of view squirm in an appropriate way when displaced, we obtain the geometry of Riemann or of Lobatchewski, as the case may be. It is to be noted that the plane is the same in all three cases, and yet the geometry we obtain on its surface may be Euclidean or non-Euclidean. It is not the plane itself but our methods of measurement conducted thereon which have changed.

Now it may be mentioned that there exist a number of alternative ways of representing both Euclidean and non-Euclidean geometries. Which way is to be preferred depends largely on the scope of our investigations. One of these alternative procedures has often been appealed to in popular writings (by Helmholtz in particular) because it enables us to visualise the sequence of theorems involved, without abandoning thereby our habitual Euclidean representations.

Thus three-dimensional Euclidean space is taken as a starting point. It is then proposed to show how, in this three-dimensional Euclidean space, a two-dimensional non-Euclidean geometry can arise. The fundamental space which we are here postulating being Euclidean, distances in this space must be computed with rigid Euclidean rods. Then it is shown that if we apply tiny Euclidean rods to the surface of a sphere and conduct measurements on this surface, we shall obtain a series of numerical results which are representative of Riemann's geometry. The chief advantage of this method of presentation is that it allows us to foresee at a rapid glance that Riemann's geometry must be consistent,* since in the present case it reduces to Euclidean geometry on a sphere, which is a particular case of Euclidean geometry in three-dimensional

---

* A fact by no means evident *a priori.*

space; and Euclidean geometry is known to be consistent. But aside from this advantage the procedure is to be avoided, for it tends to obscure the philosophical importance of non-Euclidean geometry.

In the first place it is not always possible to follow this method. Consider, for example, the case of Lobatchewski's geometry. Just as Riemann's geometry was that of the spherical surface, so Lobatchewski's geometry turns out to be that of a peculiar saddle-shaped surface called the pseudosphere, as was proved by Beltrami. Hilbert, however, has shown that there cannot exist a surface free from singularities which would represent the total spread of Lobatchewski's plane geometry; hence here is a first reason for being on our guard against a number of unsuspected difficulties.

But this is not all. If we adopt the method of representing two-dimensional non-Euclidean geometry as the geometry obtained by taking Euclidean measurements on a curved surface, how are we to conceive of three-dimensional Riemannian geometry? Obviously we must start with a four-dimensional Euclidean space in which a three-dimensional spherical surface is embedded. But as a four-dimensional Euclidean space transcends our immediate experience and is in the nature of a mathematical fiction, we should be led to the erroneous conclusion that three-dimensional non-Euclidean geometry must likewise be a purely conceptual construction having nothing in common with measurements which might be conducted with material rods.

We should be led into still greater difficulties if we wished to represent a three-dimensional non-Euclidean space possessing various degrees of non-Euclideanism from place to place. Reasoning by analogy, we should be tempted to say that just as an irregularly curved surface embedded in three-dimensional Euclidean space yielded a two-dimensional geometry of varying degrees of non-Euclideanism, so now all we should have to do would be to conceive of an irregularly curved three-dimensional surface embedded in a four-dimensional Euclidean space. But the analogy would be deceptive; for calculation shows that it would be impossible to represent an arbitrarily curved three-dimensional surface in a four-dimensional Euclidean space. In the general case we should have to situate our variously bumped three-dimensional surface in a six-dimensional Euclidean space. In the same way, a four-dimensional non-Euclidean space of variable curvature could be represented only in a ten-dimensional Euclidean space; and so on.*

Now in view of these difficulties, in view of the fact that to represent a non-Euclidean space of three dimensions in terms of Euclidean space, we may be compelled to appeal to a Euclidean space of six dimensions, it is certainly exalting Euclidean space unduly to regard it as *The Space*. This is especially obvious when we remember that we could have con-

* Calling n the dimensionality of the non-Euclidean space the number of dimensions of the generating Euclidean space would be $\dfrac{n(n+1)}{2}$ and not $n+1$, as at first sight we might be inclined to believe.

ceived of our three-dimensional non-Euclidean space directly, as a result of measurements with squirming rods, without any reference to Euclidean geometry, and without ever having had to introduce any greater number of dimensions.

At this juncture the beginner is often inclined to argue as follows: "You say that non-Euclidean geometry of two dimensions is the geometry obtained by executing Euclidean measurements on a suitably curved surface. But your concept of curvature is meaningless unless we conceive it as contrasted with some pre-existing standard of straightness. Does not your presentation prove, therefore, that Euclidean space as representative of flatness or straightness is logically antecedent to non-Euclidean space, which connotes curvature?"

This argument is radically incorrect, and arises from too loose an understanding of non-Euclidean geometry. In the first place, to assume that the concept of curvature presupposes the concept of straightness is no more inevitable than to assume that the concept of straightness presupposes that of curvature. For if it is true to say that that which is curved is that which is not straight, it is equally true to say that that which is straight is that which is not curved; so that in the absence of curvature, straightness would in turn be meaningless. The question of deciding, for example, whether a circle or a straight line is the more fundamental is strictly a matter of opinion. The Greeks held circular motion to be the noblest of all motions; and between the designations "noblest" and "most fundamental," the distinction is exceedingly slight. And even this is not all, for thus far we have been arguing as though Euclidean space were truly flat, and non-Euclidean space truly curved. But we must remember that this curvature of which we speak is not to be construed as representing anything absolute; it arises solely from the particular type of representation we have agreed to select.

Thus, in the presentation of non-Euclidean geometry which we have just been discussing, we started from three-dimensional Euclidean space as a basis and considered curved surfaces embedded in this space. This form of presentation was equivalent, as we know, to stating that Euclidean measurements must be adhered to on the curved surface. But we might have proceeded otherwise. We might have started with three-dimensional non-Euclidean space and considered the geometry of curved surfaces (of a sphere, for example) embedded in this space. This would have been equivalent to applying non-Euclidean rods on our curved surface.

Under suitable conditions the geometry of our curved surface (or sphere) would then become Euclidean. Euclideanism itself would thus be linked with curvature, while non-Euclideanism, which would here be the type of geometry obtained on the plane embedded in our three-dimensional non-Euclidean space, would accordingly be linked with flatness. In short, we must not allow the word "curvature" to mislead us into forming a false impression about non-Euclidean geometry. In many respects Riemann's choice of the word "curvature" has proved unfortunate and the more scientific appellation non-Euclideanism should be adhered to.*

There is still another point which we must mention. It might be argued that in Riemann's geometry of two dimensions, that is to say, in the geometry of the spherical surface when Euclidean measurements are adhered to, a straight line on the sphere, being a great circle, constitutes a closed curve, whereas in Euclidean geometry a straight line can never constitute a closed curve. Here, however, the difficulty which arises does not pertain so much to Euclideanism and non-Euclideanism as to another branch of geometry, Analysis Situs, on which we shall make a few brief remarks in a note at the end of the chapter. The fact is that the geometry of the cylinder embedded in Euclidean space is Euclidean to the same extent as the geometry of the plane, and yet on a cylinder certain straight lines, namely, the rings which surround the cylinder, constitute closed curves.

It might be objected that these rings are not Euclidean straight lines. But here the critic would be confusing geometry and dimensionality. From the standpoint of two-dimensional Euclidean geometry, the rings on the cylinder are perfectly straight lines, and it would never enter the mind of the flat being moving over the cylindrical surface to view them in any other light. Only when we represent the cylinder's surface in a three-dimensional Euclidean space must we regard the rings as curves. In short, lines that would be straight in two dimensions might be curved when viewed from the standpoint of a higher dimensionality. By analogy, lines which would be regarded as straight in three-dimensional space

* An equivalent form of the criticisms discussed in the text consists in assuming that as Euclidean geometry is associated with the number zero, it must be logically antecedent to the non-Euclidean varieties since these are associated with non-vanishing numbers. But here, again, it is only because we take "curvature" as fundamental that Euclidean geometry is connected with zero. Should we choose to take "radius of curvature," Euclidean geometry would be associated with infinity, and the non-Euclidean varieties with finite numbers. Hence, the problem would now resolve itself into determining whether "curvature" or "radius of curvature" was the more fundamental; and, in point of fact, "curvature" is a more complex conception than "radius of curvature."

Then again, we might characterise the various geometries by means of the parallel postulate. In this case, Riemann's geometry would be associated with the number zero, Euclidean geometry with the number one, and Lobatchewski's geometry with infinity. Still another method of presentation would be to approach the problem through projective geometry. Then we should find that Lobatchewski's geometry was associated with a circle, and Euclid's with the intersection of the line at infinity with two imaginary lines, yielding the "circular points at infinity," whereas Riemann's geometry would be connected with a circle of imaginary radius. With this method of representation, zero would never enter into our discussions. And no one would maintain that the concept of imaginary points at infinity (the circular points) was logically antecedent to a real circle of finite radius. In short, we see that there are a large number of different methods of representing the various geometries; and, according to the method selected, the number zero may be associated with Euclideanism or with non-Euclideanism, or, again, with neither. All that we are justified in saying is that Euclidean geometry is the easiest of the geometries, but not necessarily the most fundamental.

might be recognised as curved were we to appeal to a fourth dimension. Thus, whichever way we choose to investigate the concept of straightness, we find that it eludes us more and more.

It appears unnecessary to dwell further on these rather special problems, for they would lead us too far afield.

Now, finally, there is still another reason, on which we have lightly touched, which must make us very wary of being led astray by faulty analogies. When, for instance, we consider the two-dimensional geometry of a spherical surface embedded in a three-dimensional Euclidean space, we realise that the surface divides space into an outside and an inside. We should then be inclined to argue that a three-dimensional Riemannian or spherical space must also divide four-dimensional space into an inside and an outside.

This line of reasoning would be correct in the present case, but it is important to note that this conception of an inside and an outside is in no wise essential to non-Euclidean geometry. Had we adopted the more rational method of presenting three-dimensional non-Euclidean space as due to the peculiar behaviour of our measuring rods, the conception of an inside and an outside would have been meaningless. We must understand that this notion arose merely as a result of our particular choice of a mode of presentation, and in no wise constitutes an intrinsically necessary condition.

If these rather delicate points have been understood, no harm can be done by discussing non-Euclidean geometries as the geometries obtained by applying Euclidean rods on curved surfaces. In many respects, indeed, this method of presentation is helpful; for accustomed as we are in everyday life to effect measurements with Euclidean rods, we are able to visualise more easily a series of abstract investigations when this familiar procedure is followed.

Let us now proceed to a more thorough study of the preceding method. We have said that non-Euclidean results are obtained when we confine ourselves to Euclidean rods while conducting our measurements on curved surfaces. In the case of Riemannian geometry of two dimensions, the required surface is that of a sphere embedded in three-dimensional Euclidean space. In the case of Lobatchewski's geometry, it is a saddle-shaped surface known as a pseudosphere (subject to the limitations mentioned previously).

We will confine ourselves to the study of Riemann's geometry for the time being. Such a study in two dimensions permits easy visualisation, owing to the fact that since the earth is spherical in shape it is precisely this type of geometry which we obtain when we conduct measurements on its surface with rigid Euclidean measuring rods (assuming the surface to be perfectly smooth, like that of the ocean on a calm day). In fact, we shall see that the postulates of Riemann's geometry are immediately verified on the sphere.

To begin with, the shortest distance, or, better still, the most direct distance, between two points on the earth's surface, when Euclidean

rigid rods or inextensible tape measures are employed as a means of measurement, is an arc of a great circle. Great circles, such as meridians
on the surface of our planet, constitute therefore the straight lines or
geodesics of Riemann's geometry; and when discussing Riemann's geometry we may call them straight lines.

Consider, then, one such straight line or geodesic, and let it be that
particular great circle which we call the equator. According to the axioms
of Euclid's geometry two perpendiculars drawn at two different points
of the same straight line can never meet; they lie parallel to one another.
On the sphere, on the other hand, two perpendiculars to the equator are
two meridians and these meridians always meet at the pole. In fact it is
impossible to draw any two great circles on the sphere which do not
intersect. Hence we see that no parallel geodesics can exist in Riemann's
geometry. Likewise, whereas in Euclid's geometry the sum of the three
angles of a triangle is equal to two right angles, in Riemann's geometry
this sum is always greater than two right angles. This we could easily
verify by stretching ropes between three distant points on the earth's
surface so as to form a triangle, and then summing the values of the
angles at the three corners of our triangle.

A similar verification would ensue if we were to effect measurements
along a circumference and its diameter. Thus, if at a central point on
the earth's surface we fix one end of a rope and cause the other extremity
to rotate around the central point, it will describe a circle on the earth's
surface. We shall then find, upon measuring the length of the circumference marked out by the extremity of the rope, that the ratio of the
length of this circumference to the length of the diameter is a *variable*
number, always smaller than π and depending on the area covered by our
circle; whereas in Euclidean geometry this ratio would be the constant
number π.

There is still another result of importance which must be mentioned.
On a plane, the farther we wander along a straight line, the farther we
move from our starting point, whereas on our sphere we see that if we
follow a straight line or geodesic we shall finally return to our starting
point after having circled the earth. We express this fact by saying that
in Riemann's geometry space is finite, yet unbounded. It is finite since
we cannot wander away indefinitely from our starting point; it is unbounded because however far we go, we never come to a stone wall or
a gap beyond which we can proceed no farther. On the other hand,
Euclidean space, such as that of the plane surface, is *infinite* and
unbounded.

And now suppose the earth on which we were conducting our measurements were to swell indefinitely. All the characteristics of Riemann's
geometry which we have discussed would gradually fade away, provided we limited our measurements to the same restricted area of the
surface. Our measurements would still yield Riemannian results, of
course, since however large our earth had grown, it would still remain
spherical. But the results of our measurements over a definite area

would approximate more and more to those of Euclidean geometry. The reason is obvious, since the greater the volume of the sphere, the more nearly would a given area of its surface approximate to an ordinary plane.

In fact a plane surface may be assimilated to the surface of a sphere of infinite radius; so that if our sphere were to grow indefinitely we should find that our measurements lost little by little their Riemannian characteristics, until it would be impossible to distinguish them from Euclidean measurements.

Suppose now that our sphere continued to change in shape after having attained an infinite radius; in particular, suppose that our surface gradually began to assume the shape of a saddle, curved downwards if we intersected it from east to west and curved upwards if we intersected it from north to south. As this double curvature became gradually more pronounced we should find that our measurements lost their Euclidean character and became more and more conspicuously Lobatchewskian.

Thus we see that Euclidean geometry stands at the dividing line of Riemannian and Lobatchewskian geometry. When our surface changes in shape the geometry remains Riemannian so long as there subsists the least trace of sphericity, that is, of positive curvature. Likewise it remains Lobatchewskian so long as there subsists the least trace of saddle-shapedness, or negative curvature. Finally, it is strictly Euclidean only when the surface has become a plane, that is, exhibits zero curvature.

For this reason Euclidean geometry has a uniqueness about it which is denied to the non-Euclidean varieties. These latter constitute a class of geometries; and the precise type of the geometry we may happen to be discussing is determined by the intensity of curvature of the sphere or pseudosphere to which it pertains. We also see why it is so difficult to determine by empirical methods whether the space of the universe is truly Euclidean or not. It is because both Riemannian and Lobatchewskian geometry merge by insensible gradations into Euclidean geometry.

The analysis we have given refers to non-Euclidean geometries as derived from the geometry of surfaces; that is, to the numerical results obtained when we use rigid Euclidean rods to effect measurements on curved surfaces. It must be noted, however, that the surfaces we have discussed are of a very special kind. Both spheres and pseudospheres are known as surfaces of constant curvature. There is no need to go into the mathematical definition of what is meant by this; we will limit ourselves to stating the principal characteristic of such surfaces.

Consider, for example, a net made of inextensible threads (inextensible or undeformable being used in the Euclidean sense). If this net can be applied with perfect contact to any portion of a plane, it can be slid over the entire plane without ever losing its perfect contact. The same is true of the sphere and pseudosphere. Thus, a net which could be fitted with perfect contact to any part of a sphere could be slid over any portion of the surface without ever losing its perfect contact.

The only surfaces which possess this property are precisely the sphere, the plane and the pseudosphere (and those derived therefrom without stretching); and these surfaces are the surfaces of constant positive curvature, of constant zero curvature, and of constant negative curvature, respectively. So far, then, as the shape of nets applicable to the surface is concerned, there is no means of distinguishing one part of the surface from any other part, since the net can be slid over the surface, preserving its dimensions and never losing its perfect contact with the surface. In other words, the surface appears to be isotropic and homogeneous when measurements with rigid Euclidean rods are conducted over it, or when the nets we slide over it are Euclideanly inextensible. This is the property called *free mobility*.

Such would no longer be the case were we to consider surfaces of variable curvature, *e.g.*, the surface of an ellipsoid or of a trumpet. A net which could be applied with perfect contact to one part of the surface would lose its perfect contact when slid over the surface, unless we deformed it by stretching it or causing it to shrink.

It is because all three types of geometry we have mentioned (Euclidean, Riemannian and Lobatchewskian) hold in exactly the same way for all parts of the space in which we are operating, that the surfaces which portray them in two dimensions are necessarily of the constant-curvature type.

We have yet to show the connection between non-Euclidean geometry of two dimensions defined as the geometry obtained with Euclidean rods on a surface of constant curvature, and non-Euclidean geometry obtained with squirming Euclidean rods, *i.e.*, rigid non-Euclidean ones. For this purpose consider a sphere resting on a plane. We may call the south pole that point of the sphere which stands in contact with the plane; and we shall assume that a point-source of light is located at the north pole. If the sphere is transparent to rays of light all figures traced on the surface of the sphere will cast shadows on the plane. But the same will be true of our little Euclidean rods which we place alongside these figures on the sphere for purposes of measurement.

It is easy to see that a circle, for instance, traced on the spherical surface will, generally speaking, have for its shadow a circle on the plane. As this circle is displaced on the sphere towards the north pole, its shadow will grow larger and larger, tending finally, when any point of the circle coincides with the north pole, to become a straight line extending to infinity. But inasmuch as the shadows of the little rods with which we measure lengths on the sphere vary in exactly the same way as do the shadows of the figures measured, we shall obtain on the plane exactly the same Riemannian results that endure on the sphere.

In short, if we assume that on the plane our standard measuring rods are given by the shadows of the standard measuring rods used on the sphere, we shall obtain Riemannian geometry on the plane. It is scarcely necessary to state that with our Euclidean ideas of congruence it would be impossible for us to accept the statement that these successive squirming shadows of a Euclidean length displaced on the surface of the sphere were all congruent to one another; but this is not the point at issue. If we were perfectly flat, two-dimensional beings, living on the plane, and if not only all the flat bodies sliding over the plane behaved as did the shadows in our previous example, but also our measuring rods and our own living bodies behaved likewise, we should have no other alternative than to assume that these shadows represented the displacements of bodies that moved while remaining rigid or without change. We should accordingly regard the geometry of our two-dimensional space as Riemannian.

It is impossible to show more clearly that the geometry we attribute to space is nothing but an expression of the properties of our bodies and measuring rods when displaced, since here we have a plane which is Euclidean or Riemannian in its geometry according to the behaviour of the bodies that glide over its surface. And so we see that two alternative methods of interpreting non-Euclidean geometry are open to us. Either we may regard our rods as Euclidean and applied to a curved surface, or else we may attribute non-Euclideanism to the behaviour of our rods on a flat surface.*

Now in this chapter we have discussed only non-Euclidean geometries of two dimensions. Riemann's geometry and Lobatchewski's geometry are more particularly these same geometrics, extended to the case of three dimensions. We can easily see how Riemann's geometry of three dimensions would arise.

We have only to suppose that, as contrasted with Euclidean solids, all material bodies expand as did the shadows when displaced from a fixed centre. But the Riemannian observer of course would have no realisation of this expansion, owing to the modified laws of light propagation and owing to his own body's expanding in company with all other bodies. So he would be of the opinion that all these bodies maintained their same shape and size, that is, remained congruent when displaced. In fact, if he applied the test of practical congruence he would find that these bodies which the Euclidean observer would claim were expanding were, so far as he himself was concerned, always the same both in visual appearance and as regards exploration by the sense of touch. If our Riemannian being were to view bodies which the Euclidean man considered rigid and undeformable, he would assert without hesitation that those bodies were decreasing in size as they receded from a certain point. Once again maintenance of shape and size is essentially relative.

All the conclusions which we reached when discussing a Riemannian

---

* It is to be noted that had the source of light been placed at the centre of the sphere instead of at the north pole, all the great circles on the sphere would have cast straight lines for shadows on the tangent plane. Still, the shadows on the plane would, as before, have yielded Riemann's geometry (of the elliptical variety). We see then, once again, that absolute shape, size and straightness escape us in every case, and we cannot even say that Reimann's straight lines are curved with respect to Euclidean ones. All that is relevant is the mutual behaviour of bodies, their laws of disposition.

world of two dimensions can be extended to the case of three dimensions. Thus, a Riemannian universe or spherical space of three dimensions is finite but unbounded. Travelling for a sufficient length of time along a geodesic or straight line, we should return finally to our starting point, and rays of light, which follow geodesics, would circle round our space.

Just as in the case of two-dimensional geometry, we may also conceive of a three-dimensional non-Euclidean space as of constant or variable curvature. Prior to Einstein's discoveries mathematicians concerned themselves more especially with the homogeneous types of geometry, because it was assumed (notwithstanding Riemann's premonitions) that whatever the practical geometry of real space might turn out to be, it would always remain the same throughout, there being no palpable reason for it to vary *here* from *there*. But with the advent of the material or metrical field, conditioned by a more or less capricious distribution of matter, the possibility of a variation in the geometry of space from place to place had to be taken into consideration, and it then became impossible to limit our analysis to spaces of constant curvature. It has been proved that the variable non-Euclideanism of the space surrounding matter in Einstein's theory—say, over the equatorial plane of the sun—would be represented by the geometry of a surface of variable curvature, which turns out to be that of a paraboloid of revolution.

There is one additional aspect of Riemann's geometry which it may be of interest to mention on account of its possible bearing on the shape of the universe in Einstein's theory. We refer to elliptical space. The type of Riemann's geometry which we have discussed so far is known as Riemann's spherical geometry, because in the case of two dimensions it turns out to be the Euclidean geometry of a spherical surface. But there exists another type of Riemannian geometry discovered by Klein. It is called elliptical space, though it has nothing to do with the surface of an ellipsoid. It corresponds to exactly the same geometry as that of Riemann proper, the only difference consisting in the connectivity of the space. That is to say, the paths of continuous passage from a point A to a point B are different in the two spaces.

In a similar way the geometry of the cylinder is Euclidean, as is that of the plane (since a plane sheet of paper can be rolled round a cylinder); but its connectivity is different since we can go round the cylinder by following a geodesic and yet return to our starting point.

In order to illustrate elliptical space in two dimensions, as we have illustrated spherical space, we must assume that we limit the surface of our sphere to one of its hemispheres. It might appear that in this case the surface would no longer be unbounded since it would stop abruptly at the equator, but the connectivity of elliptical space is such that every point on the equator is identical with its antipodal point on the other side of the equator. It would be as though, in our concrete representation, there were cross-connections of zero length uniting into one sole point, each pair of points situated diametrically opposite one another on the equator.

THE EVOLUTION OF SCIENTIFIC THOUGHT

It is exceedingly difficult to visualise how it can be that a diameter joining two antipodal points of the equator should be vanishing in length, but the difficulty arises from our endeavouring to represent the distance between two points by a line joining these two points. Such a procedure may often be successful, but in the more complex cases of connectivity it breaks down completely. As Eddington remarks, for obscure psychological reasons we find it natural to represent a distance between two points by a line joining them, believing that this line represents a something fundamental in nature. But it is probable that the line we trace does not have the fundamental importance with which we credit it, and that it cannot be regarded as having any intrinsic significance outside of the ease with which it often introduces itself into our graphical description of natural phenomena. Distance would be more akin to a difference of temperature between two points, and it would never enter our minds to represent this difference of temperature by a line joining the two points. Why in the case of distance such a graphical representation appears obligatory is more or less of a mystery. At any rate the graphical representation is not always possible, and our description of elliptical space gives us an illustration to this effect.

Incidentally we see that the problem of space, even when restricted to the particular cases studied by Riemann, is not exhausted by the sole concept of congruence. Connectivity or routes of continuous passage between points must also be taken into consideration. The study of connectivity opens up a new branch of geometry also created by Riemann and known as *Analysis Situs*.*

* For the sake of completeness we may mention that connectivity can be studied by the same procedure as that by which dimensionality was investigated. Thus, a space was considered to be two-dimensional when it was possible to intercept the path of continuous transfer between any two points by tracing a continuous line throughout the space. If we apply this test to the surface of a doughnut, it will be seen that a continuous line, such as a circumference enclosing part of the doughnut, like a ring, would be insufficient to divide the doughnut into two separate parts which could not be connected by a route of continuous transfer over the surface. This property would differentiate the surface of a doughnut from that of a sphere or an ellipsoid, and yet the surface of the doughnut would still be two-dimensional, but its connectivity would be different owing to the hole passing through its substance.

CHAPTER VI

TIME

OUR awareness of the passage of time constitutes one of the most fundamental facts of consciousness, and our sensations range themselves automatically in this one-dimensional irreversible temporal series. In this respect we must concede a certain difference between space and time, the former being, as we have seen, a concept due in the main to a less inevitable synthetic co-ordination of sense impressions. Yet, on the other hand, there exists a marked similarity between space and duration, in that they both manifest the characteristics of sensory continuity.

Thus, in the case of time, two sensations A and B may be so close together that it will be impossible for us to determine which of the two was sensed first; the same may happen in the case of two sensations B and C and yet we might have no difficulty in ascertaining that A was prior to C. These facts are expressed by the inconsistent relations $A = B$, $B = C$, $A \neq C$, which, as we have seen, are characteristic of sensory continuity, be it in space, or in time, or in whatever domain we please. In the case of time, as in that of space, only by a process of abstraction can we remedy this inconsistency. We thereby conceive of mathematical duration, a duration in which mathematical continuity is assumed to have replaced sensory continuity.

Now, owing to the ordering relation or path of continuous transfer which appears to be imposed upon us by nature, perceptual time must be regarded as uni-dimensional, and by retaining the natural ordering relation, we will attribute the same uni-dimensionality to mathematical duration. It is, of course, to this mathematical duration that we appeal in all theoretical investigations in mechanics and physics.

Primarily the only time of which we are conscious is an "egoistic" time, the time of *our* stream of consciousness. But as a result of our perception of change and motion in the exterior universe of space, it becomes necessary to extend the concept of duration throughout space, so that all external events are thought of as having occurred at a definite instant in time as well as at a definite point in space.*

Now, when discussing space, we saw that the concept of the same point in space, considered at different times, was ambiguous; it was necessary to specify the frame of reference by which space was defined, and according to the frame selected, an object would continue to occupy the same point or else manifest motion. The analogous problem in the case of

---

* An instant of time is of course a mere abstraction, but so is a mathematical point in space.

71

time would be to conceive of the same instant in time at two different places, *i.e.*, of the simultaneity of two spatially separated events.

If we allowed ourselves to be guided by our analysis of position in space, we should be inclined to say that a determination of the simultaneity of spatially separated events must necessitate the choice of a frame of reference. The observer who was to pass an opinion on the simultaneity of the two events would then select that particular frame in which he stood at rest. But from this it would follow that a change in the motion of the observer, hence a change of frame, might bring about a modification in his understanding of simultaneity. Simultaneity would thus manifest itself as relative, just like position.

Although these possibilities cannot be discarded *a priori* on any rational grounds, classical science and common-sense philosophy refused to entertain them. The fact is that this thesis would have tended to surround with ambiguity the concept of time in physics, since the duration of the same event would vary with the motion of our frame of reference. It had always been felt that although, through an act of our will, we might change our motion through space, yet on the other hand the flowing of time transcended our action. These views were confirmed by the behaviour of physical phenomena known to classical science; hence the uniqueness of time and the absolute character of simultaneity had been accepted. In other words, the simultaneity and the order of succession of two spatially separated events were assumed to constitute facts transcending our choice of a system of reference.

This stand was fully justified in Newton's day; for inasmuch as *a priori* arguments could not decide the question, so long as experiment did not compel us to recognise the relativity of simultaneity and the ambiguity of duration there appeared to be no good reason to hamper science with an unnecessary complication. But the point we wish to stress is that *if* perchance experiment were ever to suggest that simultaneity was not absolute, no rational argument could be advanced to prove the absurdity of this opinion. As we shall see, Einstein's interpretation of certain refined electromagnetic experiments consists precisely in recognising the relativity of simultaneity and the ambiguity of duration.

Let us now pass to the problem of congruence for time. As is the case with space, mathematical time is an amorphous continuum presenting no definite metrics. The concept of congruence or of the equality of successive durations has no precise meaning; and a definition of time-congruence can be obtained only after we have imposed some conventional measuring standard on an indifferent duration. We might define as equal the durations separating the rising and the setting of the sun, or, again, the durations separating successive rainstorms, etc. Theoretically, all such definitions would be legitimate, since in the final analysis duration, just like space, presents no intrinsic metrics.

In common practice, however, we find that just as in the case of space, a definite species of congruence appears to be imposed on human beings and animals alike. For instance, so long as we are in a conscious state, we all agree that the successive tickings of the clock correspond to what our psychological sense of duration would qualify as congruent intervals. In other words, we appear to be gifted with a species of psychological time-keeping device which allows us to differentiate between durations which we have come to call "seconds" and those which we have come to call "hours." What is commonly called the sense of rhythm, which manifests itself in the primitive music of drum beats and in the dances of savages, is nothing but a consequence of this intuitive understanding of time-congruence.

A superficial analysis of the situation might lead us to assume that, contrary to our previous assertions, a definite species of congruence was intrinsic to duration. This opinion, however, is quite untenable unless supplemented by additional explanations. The problem with which we are confronted is in all respects similar to the one we mentioned when discussing space. Of itself, space, or at least mathematical space, is amorphous. In spite of this fact, we have no difficulty in agreeing that a coin is round while an egg is oval. But we saw elsewhere that our unreasoned predilection for a definite type of congruence in space was imposed by the behaviour of material bodies when displaced. Visually these bodies could be made to present us with the same appearances, whether situated here or there, and for that reason they imposed themselves as defining congruent space intervals. Furthermore, our own human limbs, the steps we take, the ground on which we walk, all suggest the same definition of congruence.

Likewise in the case of time-congruence. The human organism is replete with rhythmical motions such as the beating of the heart, the periodic contractions of the arteries, or the respiratory motions of the lungs. Again, an infant feels the pangs of hunger at periodic intervals. Quite independently of these internal processes and changes, certain external mechanical motions affecting our bodies must be considered. Thus, when we walk with our normal gait, our successive steps follow one another in time in a certain definite way, and the same applies to the swinging of our arms. By an effort of our will we might vary the rhythm of our steps, alternating short steps with long ones, now tarrying, now hastening. But we should soon be aware that something was wrong, for we should find that our irregular motions demanded additional effort, so that on relaxing our will we should relapse automatically into our normal gait. It may be that in addition to these periodic motions we have mentioned, cyclic processes occur in our brains, conferring on our consciousness direct norms of time-congruence. This, however, is a point for physiologists to decide. At any rate, when it is realised that these various methods, whether they appeal to our respiratory motions or to our successive steps when walking normally, all issue in the same definition for time-congruence, there is no cause to be surprised at a definite sense of time-congruence imposing itself naturally upon us, and this in spite of the fact that pure duration of itself possesses no metrics.

Thus far we have been considering time-congruence from the standpoint of the savage and the animal. For the most rudimentary purposes, a definition of equal durations obtained in this crude way would suffice; yet it is obvious that no great precision could thus be arrived at. Our crude psychological clock is subject to variations according to our mental and physical condition; furthermore, though it can differentiate between what we have come to call "one second" and "one minute," it would be incapable of differentiating between fifty-nine minutes and one hour. In the case of animals this imprecision in their sense of duration is manifested when shepherd dogs lead the flocks back during an eclipse or hens lay eggs under the influence of artificial light, believing that day has come. However, as these last illustrations permit of other interpretations, we need not dwell on them and may confine ourselves to the well-known fact that our human sense of duration varies between one individual and another, and between one psychological state and another.

In addition to these difficulties, our understanding of equal durations is entirely subjective and can give rise to no common knowledge. We must needs objectivise it, hence appeal to some objective phenomenon which all men can observe. Let us assume, for the sake of argument, that the monotonous dripping of water drops falling on a pan may have served as a primitive clock. Still, something would be lacking. We should not be completely satisfied, for we might sense now and then an irregularity in the successive thuds of the drops falling on the pan. Furthermore, if we were to establish two such water clocks we should observe that the drops sometimes fell in unison and sometimes in succession. We should wonder then which of the clocks was most reliable and be in a quandary to know to which to appeal. Just as in the case of space, where we substituted wooden rods for the lengths of our steps, then metallic rods maintained at constant pressure and temperature, so now, in the case of time-keeping devices, we must find some way of selecting clocks which will satisfy all our requirements.

We thus arrive at more scientific methods of time-determination. The psychological clock, water clock, sand clock and burning of candles, which had served men in the earlier days of civilisation, have now been discarded in favour of more precise mechanisms. These, as we shall see, will not differ perceptibly in their indications from our psychological clock, but their measurements will be given with increased refinement. We are thus led to examine what are the conditions that must be satisfied for a time-keeping device to be considered perfect.

In a general way we obtain a definition of time-congruence by appealing to the principle of causality. The following passage, quoted from Weyl, expresses this fact:

"If an absolutely isolated physical system (*i.e.*, one not subject to external influences) reverts once again to exactly the same state as that in which it was at some earlier instant, then the same succession of states will be repeated in time and the whole series of events will constitute a cycle. In general, such a system is called a **clock**. Each period of the cycle lasts **equally** long."

We shall obtain a more profound understanding of the nature of the problem by considering the various methods of time-determination that have been contemplated by physicists.

We remember that Newton refers to absolute time as "flowing uniformly." But of course this allusion to time does not lead us very far, for a rate of flow can be recognised as uniform only when measured against some other rate of flow taken as standard. Obviously some further definition will have to be forthcoming. Now both Galileo and Newton recognised as a result of clock measurements that approximately free bodies moving in an approximately Galilean frame described approximately straight lines with approximately constant speeds. Newton then elevates this approximate empirical discovery to the position of a rigorous principle, the *principle of inertia*, and states that absolutely free bodies will move with absolutely constant speeds along perfectly straight lines, hence will cover equal distances in equal times. When expressed in this way as a rigorous principle, the space and time referred to are the absolute space and time of Newtonian science. In other words, it is the principle of inertia coupled with an understanding of spatial congruence that yields us a definition of congruent stretches of absolute time.

In practice, a definition of this sort entails measurements with rods; that is to say, we measure off equal distances along the straight path of a body and then define as congruent, or equal, the times required by the body to describe these congruent spatial distances. It is obvious that, however perfect our measurements, errors of observation will always creep in. Furthermore, a body moving under ideal conditions of observation can never be contemplated; hence all we can hope to obtain is the greatest possible approximation. But although physical measurements become inevitable as soon as we wish to obtain a concrete definition, an objective criterion of equal durations, we are able to proceed in our mechanical deductions in a purely mathematical way without further appeal to experiment. The principle of inertia, together with the other fundamental principles of mechanics, enables us, therefore, to place mechanics on a rigorous mathematical basis, and rational mechanics is the result.

It will be observed that science, in the case of mechanics, has followed the same course as in geometry. Initially our information is empirical and suffers from all the inaccuracies of human observation and all the various contingencies that characterise physical phenomena. But this empirical information is idealised, then crystallised into axioms, postulates or principles susceptible of direct mathematical treatment. To be sure, as we proceed in our purely theoretical deductions, unless we are to lose all contact with reality, we must check our results by physical experiment; and this necessity is still more apparent in rational mechanics than in geometry. If peradventure further experiment were to prove that our mathematical deductions in mechanics were not borne out in the world of reality, we should have to modify our initial principles and postulates or else agree that nature was irrational. With mechanics, the necessity of modifying the fundamental principles became imperative when it was recognised that the mass of a body was not the constant magnitude we had thought it to be; hence it was experiment that brought about the revolution.   On the other hand, in the case of geometry, it was the mathematicians themselves who foresaw the possibility of various non-Euclidean doctrines, prior to any suggestion of this sort being demanded by experiment.

And now let us revert once again to the problem of time.   Theoretically, the law of inertia should permit us to obtain an accurate determination of congruent intervals, but as it was quite impossible to observe freely-moving bodies owing to frictional resistances as also to the gravitational attraction of earth and sun, it was advantageous to discover some other physical method of determination.   This was soon obtained.   The principles of mechanics enable us to anticipate that if a perfectly rigid sphere, submitted to no external forces, is rotating without friction on an axis fixed in a Galilean frame, its angular rate of rotation will be uniform or constant, as measured in the frame.   By "constant," we mean *constant as measured in terms of the standards of time-congruence defined by a free body moving under ideal conditions according to the principle of inertia.*   Hence we are in possession of a method of measuring time, more convenient than that afforded by the motions of free bodies along straight courses.   It may be noted that the definition of congruent durations as given by the rotating sphere is in perfect accord with the definition in terms of causality as formulated in the passage quoted from Weyl.

Unfortunately, here again, ideal conditions can never be established, for we can never obtain perfect isolation, perfect rigidity and complete absence of friction.   But a near approach to ideal conditions would appear to be given by the phenomenon of the earth's rotation.   And so the rotation of the earth, defining astronomical time, was regarded as furnishing science with the most reliable objective criterion of congruent time-intervals that it was possible to obtain.   Nevertheless, it was well known that a definition of this sort was still far from perfect, since the rate of rotation of the earth could not be absolutely uniform, when "uniform" is defined by the standards of the principle of inertia.   This realisation was brought home when the frictional resistance generated by the tides was taken into consideration.*   The tidal friction would have for effect the slowing down of the earth's rate of rotation (when measured against the accurate mathematical definition given by a perfectly rigid and isolated body rotating without friction).   In addition to this first perturbating influence, our planet is suspected of varying in shape, expanding and contracting periodically, a phenomenon known as *breathing*. According to the laws of mechanics, a change in shape of this sort would entail periodic fluctuations, successive retardations and accelerations in the rate of the earth's rotation.

It is instructive to understand how this discovery of the earth's breathing was brought about.   It was noticed that the moon's motion exhibited

* We are not referring solely. to the tides of the sea, but also to the solid tides generated in the earth's substance.

variations which it seemed impossible to account for under Newton's law, even when due consideration was given to the perturbating effects of the sun and other planets. Not only did the moon's motion appear to be gradually increasing, but in addition, superposed on this first uni-directional effect,* periodic accelerations and retardations had been observed. The gradual slowing down of the earth's rate of rotation under the influence of the tides accounted for the apparent acceleration of the moon's motion, but the other periodic changes in the velocity of our satellite could not be explained away in so simple a manner. Yet everything would be in order were we to assume that the earth's rate of rotation was subjected to appropriate periodic changes; and the principles of mechanics then showed that a phenomenon of this kind could only be brought about by periodic variations in the earth's shape.

In all these corrections we are dealing with highly complex phenomena involving Newton's law of gravitation and the laws of mechanics. Our astronomical time-computations have thus been adjusted so as to satisfy the requirements of these laws, regarded as of absolute validity. We may recall that it was this belief in the accuracy of Newton's law that led Römer in 1675 to the discovery of the finite velocity of light. He noticed that Jupiter's satellites appeared to emerge from behind the planet's disk several minutes later than was required by the law of gravitation. Rather than cast doubts on the accuracy of this law, he ascribed a finite velocity to the propagation of light. Needless to say, his previsions have since been justified by direct terrestrial measurements. The calculations leading to the discovery of the planet Neptune were also based on the assumption that Newton's law was correct.

And now let us suppose that we had elected to define time by entirely different methods. We might, for instance, appeal to the vibrations of the atoms of some element, such as cadmium. Then the intervals of time marked out by the successive beats of the atom at rest in our Galilean frame would be equal or congruent by definition. These successive congruent intervals of time would be measured by the wave length of the radiation, since a wave length is the distance covered by the luminous perturbation during the interval separating two successive vibrations. We might again proceed in still another way in terms of radioactive processes. It is well known that, when measured by ordinary clocks, the rate of disintegration of radium appears to follow an exponential law. And it is a remarkable fact that this rate of disintegration does not appear to be modified in the slightest degree by the surrounding conditions of temperature or of pressure. Thus the phenomenon may be regarded as being in an ideal state of isolation. Curie had therefore suggested defining time in terms of this rate of disintegration.

We see, then, that a great variety of methods for determining time have

---

* The problem is of course much more complex than would appear from the present analysis. For the slowing down of the earth's rate of rotation, owing to tidal action, would also result in causing a retreat of the moon, producing thereby a variation in the period of its revolution round the earth.

been considered. Some were mechanical, others astronomical, others optical, while still others appealed to radioactive phenomena. In each case our definitions were such as to confer the maximum of simplicity on our co-ordination of a certain type of phenomena. For instance, if in mechanics and astronomy we had selected at random some arbitrary definition of time, if we had defined as congruent the intervals separating the rising and setting of the sun at all seasons of the year, say for the latitude of New York, our understanding of mechanical phenomena would have been beset with grave difficulties. As measured by these new temporal standards, free bodies would no longer move with constant speeds, but would be subjected to periodic accelerations for which it would appear impossible to ascribe any definite cause, and so on. As a result, the law of inertia would have had to be abandoned, and with it the entire doctrine of classical mechanics, together with Newton's law. Thus a change in our understanding of congruence would entail far-reaching consequences.

Again, in the case of the vibrating atom, had some arbitrary definition of time been accepted, we should have had to assume that the same atom presented the most capricious frequencies. Once more it would have been difficult to ascribe satisfactory causes to these seemingly haphazard fluctuations in frequency; and a simple understanding of the most fundamental optical phenomena would have been well-nigh impossible.

Since we realise that the definitions of time-congruence we ultimately adopt are governed by our desire to obtain the maximum of simplicity in the co-ordination of some particular group of phenomena, considerable divergences in our definitions might be expected when we considered in turn phenomena of a different nature. Why, indeed, should a determination arrived at in terms of the vibrations of atoms agree in any way with the astronomical definition issuing from the earth's rotation? Even when restricting our attention to the vibrations of the atoms of the same element, it might have been feared that different atoms of cadmium would have beaten out time at different rates, since, after all, even though identical in many respects, yet their life-histories must have differed greatly. Instead of this we find that all the methods of time-congruence mentioned lead to exactly the same determinations so far as experiment can detect. It would appear as though a common understanding existed between the rates of evolution of the various phenomena mentioned. The situation is similar to that presented in the problem of space. There we saw that rods, whether of platinum or of wood, whether situated here or there, appeared to yield the same geometrical results (provided they were maintained at constant temperature and pressure). Again, it seemed as though some common understanding existed among the various rods.

Just as this same mysterious situation in the case of space suggested that the uniformity of our spatial measurements was due to the presence of an all-embracing physical reality, the metrical field of space, regulating the behaviour of material bodies and light rays, so now, as we shall see in later chapters, the general theory of relativity will compel us to extend the same ideas to time. We shall find that there exists a space-time metrical field which splits up into a temporal and a spatial field. The temporal metrical field will then be responsible for that apparent concordance which seems to exist as regards time-congruence among the various phenomena (whether spatially separated, or not).

These new ideas modify the position of the problem of time-congruence. While it is still correct to state that pure duration as a mathematical continuum is amorphous and possesses no intrinsic metrics, yet from the standpoint of the physicist we must concede a certain physical reality to this temporal metrical field which controls the processes of all isolated phenomena.

We may summarise the preceding statements by saying that we define time-congruence so as to introduce the maximum of simplicity in our coordinations of physical phenomena, regardless of the particular type of phenomenon we may be considering. That a unique definition of time-congruence should be possible, one holding for mechanics, optics, electromagnetics and astronomy, is a fact that could scarcely have been anticipated *a priori;* hence we must recognise its undeniable existence as establishing the presence of unity in nature.

And here we may mention a type of objection which is often encountered in people who do not trouble to differentiate between physical time and psychological duration. It has been claimed, for instance, that it is absurd to maintain that our sense of time originated from a desire to retain harmony between the planetary motions and the requirements of Newton's law, since men possessed an intuitive understanding of equal durations long before Newton's law or any other natural law was discovered. We trust that those who have taken the trouble to follow the course of this analysis will realise that the criticism is quite beside the mark. No scientist has ever contended that we have derived our sense of time from astronomical observations, for we know that even savages are not lacking in a sense of rhythm. All that is claimed is that the accurate determination of time-congruence must be based on physical processes or laws, since our crude time-sense is too vague to be of any use in the precise problems with which science is concerned. The fact that our crude appreciation of time agrees in a more or less approximate way with the more refined methods of scientific determination offers no great mystery, for, after all, the laws which govern material processes in the outside world would in all probability also govern that which is material in our organisms. Regardless of our ultimate views as to matter and mind, we cannot banish matter from our anatomy.

At any rate, once the existence of a universal type of time-congruence is conceded, the problem is to select some particular phenomenon which can be studied with a minimum of contingency and a maximum of certainty. Now, when we consider the astronomical definition in terms of the earth's rotation, we realise that were it not for the fact that we had started by assuming the absolute accuracy of Newton's law, the corrections we made in order to account for the peculiarities of the lunar motion would have to be discarded or replaced by others. If, therefore, any doubt were to be cast on the accuracy of Newton's law or on those of mechanics, corresponding modifications would ensue for our definition of time-congruence. But it is obvious that we have no right to assume that Newton's law or those of mechanics are above all suspicion. The laws of physical science present no *a priori* inevitableness; and our sole reason for accepting them is that they appear to be borne out to a high degree of approximation when we compare their rational consequences with the results of observation. Approximation is thus all we can aspire to in physical science; and even classical scientists recognised the possibility that more precise observation might prove Newton's law and those of mechanics to stand in need of correction. All that could be asserted was that the corrections would certainly be very slight, far too slight to entail any perceptible change in our understanding of time-congruence when applied to the processes of commonplace observation.

We now come to Einstein's definition. As we shall see, it does not differ in spirit from the definitions of classical science; its sole advantage is that it entails a minimum of assumptions, and is susceptible of being realised in a concrete way permitting a high degree of accuracy in our measurements. Einstein's definition is, then, as follows: If we consider a ray of light passing through a Galilean frame, its velocity in the frame will be the same *regardless of the relative motion of the luminous source and frame, and regardless of the direction of the ray*. It follows that a definition of equal durations in the frame will be given by measuring equal spatial stretches along the path of the ray, and asserting that the wave front will describe these equal stretches in equal times.

As can be seen, the definition is simple enough, but the major question is to decide on its legitimacy. Classical science would have rejected the definition. Why? Because it would have maintained that Einstein's definition was equivalent to defining time in terms of a rigid body rotating round an axis, but *submitted to frictional forces*. It would have claimed that the velocity of the frame through the stagnant ether would have introduced complications entailing anisotropy, that is to say, a variation of the speed of light according to direction. It would then be necessary to take this perturbating effect into account just as it was necessary to take into consideration the frictional effect of the tides on the earth's angular rate of rotation.

But when it was found that contrary to the anticipations of classical science not the slightest trace of anisotropy could be detected even by ultra-precise experiment, the objections which classical science might have presented against Einstein's definition lost all force. Henceforth it was not necessary to take into consideration the velocity of the frame through the stagnant ether, since this ether drift appeared to exert no influence one way or the other.

Now, it may not be easy to understand how isotropic conditions could be demonstrated by experiment, for isotropy signifies that the velocity of light is the same in all directions. And how can we ascertain the equality of a velocity in all directions when we do not yet know how to measure time? Experimenters solved the difficulty by appealing to the observation of coincidences.

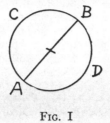

Fig. I

The following elementary example will make this point clear. Consider a circle along which two bodies are moving in opposite directions. We shall assume that the two bodies leave a point A at the same instant of time, hence in coincidence, and then meet again at a point B halfway round the circle, hence diametrically opposite A. There is no difficulty in ascertaining that the point B is diametrically opposite A, since we know how to measure space. All we have to do is to verify the fact that the length of the curve $A\,C\,B$ is equal to the length $A\,D\,B$. Neither does the observation of a coincidence, that is, of the simultaneous presence of two objects at the same point of space, entail any knowledge of a measure of duration. Hence, although we know nothing of the measurement of time, we cannot escape the conclusion that the two bodies have taken the same time to pass from A to B by different routes. But these spatial routes are of equal length; hence we must assume that the speeds of the bodies along their respective paths have been the same. In all rigour, we may only claim that the mean speed has been the same, since one body may have slowed down, then spurted on again, making up for lost time. But if we repeat the experiment with circles of different sizes, and if in every case the bodies meet one another at the opposite point, we are justified in asserting that not only the mean speed, but also the instantaneous speed at every instant, is the same for either body. In short, thanks to spatial measurements coupled with the observation of coincidences, we have been able to establish the equality of two velocities, even though we knew nothing of time-measurement.

A more elaborate presentation of the same problem would be given as follows: Waves of light leaving the centre of a sphere simultaneously are found to return to the centre *also in coincidence*, after having suffered a reflection against the highly polished inner surface of the sphere. As in the previous example, the light waves have thus covered equal distances in the same time; whence we conclude that their speed is the same in all directions.

Inasmuch as this experiment has been performed, yielding the results we have just described, even though the ether drift caused by the earth's motion should have varied in direction and intensity, the isotropy of space to luminous propagations was thus established. (The experiment described constitutes but a schematic form of Michelson's.) It is to be noted that in this experiment the observation of coincidences is alone appealed to (even spatial measurements can be eliminated).* When it is realised that coincidences constitute the most exact form of observation, we understand why it is that Einstein's definition is justified.

The optical definition presents a marked superiority over those of classical science. Whereas, with the rotation of the earth, we had to assume the correctness of Newton's law and of those of mechanics in order to determine the compensations necessitated by the earth's breathing or by the friction of the tides, in the present case we assume practically nothing, and what little we do assume issues from the most highly refined experiments known to science.

Now, the importance of Einstein's definition lies not so much in its allowing us to obtain an accurate definition of time in our Galilean frame as in its enabling us to co-ordinate time reckonings in various Galilean frames in relative motion. So long as we restrict our attention to space and time computations in our frame, we may, as before, appeal to vibrating atoms for the measurement of congruent time-intervals and to rigid rods for the purpose of measuring space. It is when we seek to correlate space and time measurements as between various Galilean frames in relative motion that astonishing consequences follow. We discover that the concepts of spatial and temporal congruence of classical science must be modified to a very marked degree. They lose those attributes of universality with which we were wont to credit them. It is then found that congruence can only be defined in a universal way when we consider the extension of four-dimensional space-time.

As we have mentioned, any change in our concepts of space and time congruence is not merely a local affair. Owing to the unity of nature, we shall be led to modify our entire understanding of the relatedness of phenomena, and must not be surprised to learn that a totally new science is the necessary outcome of Einstein's new definitions. Old laws are cast aside, new ones take their place; and Einstèin, by following his mathematical deductions with rigorous logic, *without introducing any* hypotheses *ad hoc*, proved that if his new definition of space-time congruence was physically correct, a whole series of hitherto unsuspected natural phenomena should ensue and should be revealed by precise experiment. Wonderful as it may appear, Einstein's previsions have thus far been verified in every minute detail.

---

* This is because in Michelson's experiment it is not necessary to consider a sphere. The two arms of the apparatus may be of different lengths; and all that is observed is the continued coincidence of the interference-bands with markings on the instrument.

CHAPTER VII

WE have mentioned in a general way the significance of congruence and of spatial distance. It now remains for us to find a means of defining these concepts in a rigorous mathematical form. We remember that the equality or congruence of two spatial distances between two point pairs *AB* and *CD* was an indeterminate concept, depending essentially on the behaviour of our measuring rods. An alternative presentation of non-Euclideanism (in the case of two-dimensional geometry) was then found to be afforded by assuming that the distance between points could in all cases be determined by measurements with rigid Euclidean rods; but that, whereas in the case of Euclidean geometry all the points should be considered as existing in the same plane, in the case of non-Euclidean geometry it would be as though the points were situated on a suitably curved surface. Thus, in the case of the earth, the non-Euclidean distance between two points, say New York and Paris, would be given by the Euclidean length of a great circle extending between these points, hence by a curved line following the contour of the earth's surface. On the other hand, the Euclidean distance between these two same points would be given by the Euclidean length of the straight line joining them and passing, of course, through the earth's interior.

We now propose to investigate the mathematical expression of distance in two dimensions; we will assume that we are discussing it from the standpoint of Euclidean measurements conducted on surfaces. Let us first consider the case where the surface is an unlimited plane. If we wish to define the position of a point of the plane, we must refer it to some system of reference. Three centuries ago Descartes devised a method whereby this result could be accomplished. He considered two families of Euclidean straight lines which we may call horizontals and verticals, respectively. The lines of these two intersecting families are equally spaced (Euclideanly speaking), so that they form a mesh-system or network of equal Euclidean squares. The scientific name for a mesh-system is *co-ordinate system,* but the appellation "mesh-system" introduced by Eddington has the advantage of giving a more graphic picture of what is involved. The type of mesh-system constituted by horizontals and verticals introduced by Descartes is called a *Cartesian co-ordinate system.*

If to each vertical and to each horizontal of the mesh-system we assign consecutive whole numbers from zero on indefinitely, we see that the point of the plane which happens to coincide with the intersection of some particular horizontal and some particular vertical is defined by the numbers which represent the two lines, respectively. In this way *every point of intersection* is defined by two numbers, and these numbers are called the *Cartesian co-ordinates* of the point. Of course, by this method we are unable to define the positions of points which do not happen to coincide with the corners of our squares. But there is nothing to prevent us from assuming that between the verticals and horizontals we have mentioned there lie an indefinite number of other similar lines, to which intermediary fractional numbers will be assigned. Henceforth every point of the plane can be regarded as defined by the intersection of some particular horizontal and some particular vertical.

To what extent is it permissible to say that points on the plane have been defined by this method? If we disregard the existence of the co-ordinate system, nothing has been defined, but if we consider the co-ordinate system *as given,* then every point of the plane can be considered as defined unambiguously. In short, the points are defined *not* in the abstract, but in relation to the co-ordinate system. There is nothing mysterious about this method of defining the positions of points. Thus, in everyday life, when we agree to meet a friend at the corner of Fifth Avenue and Forty-second Street, we are inadvertently locating our point of meeting in terms of the Cartesian co-ordinate system defined by the avenues and streets. In the present case the co-ordinate system is not strictly Cartesian, since the streets and avenues may not enclose perfectly equal Euclideanly square blocks, but the general principle involved is the same. Needless to say, the definition of our point of meeting would convey no significance were the avenues and streets non-existent. Hence once again we see that it is only relative to the co-ordinate system that points can be defined.

Now, the essential characteristic of the Cartesian procedure is its use of a system of reference represented by separate families of intersecting lines. The fact that the lines we have considered are mutually perpendicular straight lines forming a network of Euclidean squares is of no particular importance. It would be just as feasible, in place of our horizontals and verticals, to select two families of intersecting curves, which we might call the u and v curves. Of course, our mesh system would now be curvilinear and the spaces enclosed by the meshes would no longer be Euclidean squares, nor even necessarily equal in area. This generalization of Descartes' method was introduced by Gauss, and for this reason curvilinear mesh-systems are also called *Gaussian mesh-systems.* As before, every point will be defined by the numbers designating the two curves of either family, that is, by the numbers designating the u curve and the v curve which intersect at this point; and these two numbers will be called the *Gaussian numbers or co-ordinates of the point.*

The necessity of generalising Cartesian co-ordinates by introducing Gaussian ones arises from the fact that Cartesian mesh-systems of equal squares can be traced only on a plane and could never be drawn on a curved surface, like that of a sphere, for example. Hence, were we to ignore the use of Gaussian mesh-systems, it would be impossible for us to localise points on a curved surface, by means of a mesh-system applied on the surface. The nearest approach to a network of squares on the surface of a sphere would be a network of meridians and parallels, and such a network is not one of equal Euclidean squares; it is a curvilinear or Gaussian mesh-system tapering to points at the North and South Poles.

As a matter of fact, we also make use of Gaussian co-ordinates in everyday life. Such is the case when we state that the position of a ship is so many degrees of latitude and so many degrees of longitude; our mesh-system, being one of meridians and parallels, is a Gaussian one, and the latitude and longitude of the ship constitute its Gaussian co-ordinates.

Having determined how the positions of points may be defined on any surface in terms of some mesh-system, let us now see how it is proposed to express the distance between two points, as measured over the surface with rigid Euclidean rods. Here we must proceed with the utmost caution. Let us recall exactly what is involved. Any definite u line is one along which u remains constant while v varies continuously, and inversely any definite v line is one along which v remains constant while u varies continuously. In much the same way, on the earth's surface a latitude line or parallel is one along which the latitude remains constant and the longitude varies, whereas a longitude line or meridian is one along which the reverse is true.

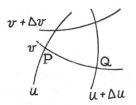

FIG. II

Suppose, then, we wish to express the distance between two points P and Q, defined by the intersections of a certain v line with two successive u lines (Fig. II). The co-ordinates of the two points are (u, v) and $(u + \Delta u, v)$, respectively, where Δu represents the increase in the value of u as we pass from P to Q on the v line.

On no account may we say that the distance PQ (which we shall call Δs) is given by the value Δu, for Δu is nothing but a difference between numbers serving to localise points; it has nothing in common with a distance. In order to dispel any doubts on this score, we may notice that if we compare the v lines to parallels and the u lines to meridians, the points P and Q will be given by the intersections of two consecutive meridians with the same parallel. In this case Δu would correspond to the difference in longitude between the two points; and obviously a difference in longitude is no criterion of a distance, since for the same difference in longitude the distance decreases from equator to pole. In short, the distance Δs between P and Q cannot be fully determined by Δu.

Yet, on the other hand, just as the distance between two points lying on the same parallel is affected by their difference in longitude, so now the distance Δs between P and Q must be a function of the co-ordinate difference Δu. A sufficiently general expression of a quantity Δs when defined in terms of another quantity Δu on which it depends is given by

$$\Delta s = A \Delta u + B \Delta u^2 + C \Delta u^3 + \ldots$$

where A, B, C are appropriate magnitudes. If, however, we consider points exceedingly close together, Δu becomes exceedingly small in value, and as a result Δu^2 and Δu^3, etc., are many times smaller still. At the limit, therefore, when the two points are at an infinitesimal distance apart, the higher powers of Δu become so insignificant that they can be neglected in comparison with Δu. We thus obtain $\Delta s = A \Delta u$, and in order to specify that we are considering infinitesimal distances we replace the symbol Δ by d and obtain

$$ds = A\,du.$$

It is customary to designate A by $\sqrt{g_{11}}$, so that, using squares, our formula becomes

$$ds^2 = g_{11} du^2.$$

And now we have to consider an important question. What is g_{11}? What does it represent? We do not propose to enter into its full mathematical significance, but we may mention certain of its important characteristics. In our illustration of meridians and parallels, let us consider the various points of intersection defined by the intersections of the successive parallels with two fixed meridians. The difference in longitude between these various point pairs will, of course, be constant and will be given by the same invariable quantity du. Inasmuch as the distance ds between these point pairs varies from pole to equator, we see from $ds^2 = g_{11} du^2$ that g_{11} must vary in an appropriate way as we consider various portions of the sphere's surface. In fact, owing to the constancy of du, we must have g_{11} proportional to ds^2. If, then, the meridians were parallel lines instead of tapering together towards the poles, g_{11} would remain constant. Whence it follows that a knowledge of the way g_{11} varies from point to point yields us information on the shape of the mesh-system, and inversely the lay of the mesh-system yields us information as to the value of g_{11} from place to place.

Then again, if, leaving our two points P and Q fixed, we select some new mesh-system, du may change in value, and in fact a difference dv may also appear, since with a change of mesh-system there is no reason why P and Q should still lie on the same v curve. Hence we must conclude that even at a fixed point of our surface the value of g_{11} will be subject to change when we vary our mesh-system.

Precisely the same arguments would apply were we to consider the expression of the distance between two infinitely close points on the same u curve. We should then obtain $ds^2 = g_{22}dv^2$, where g_{22} is a magnitude generally differing from g_{11}.

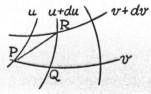

Fɪɢ. III

We must now consider the more general case where our two points lie on different u and v curves (Fig. III). From what precedes we know that the squared distance PQ is given by $g_{11}du^2$ and that the squared distance QR is expressed by $g_{22}dv^2$. But this information is obviously insufficient to tell us what the squared distance PR will be, since this distance will also be affected by the slant of the lines PQ and QR. Some new magnitude will obviously have to be introduced in order to specify the value of this slant; and the new magnitude considered will be called g_{12} and g_{21}. Under the circumstances, the squared distance of the infinitesimally near points P and R can be written

$$ds^2 = g_{11}du^2 + g_{12}du\,dv + g_{21}du\,dv + g_{22}dv^2$$

but as g_{12} and g_{21} are found to be always identical we have

$$ds^2 = g_{11}du^2 + 2g_{12}du\,dv + g_{22}dv^2.$$

We may write this formula more concisely by referring to u as u_1, and to v as u_2. In this case our formula becomes

$$ds^2 = \sum g_{1k}du_1\,du_k$$

where $\sum$ indicates summation and where for i and k we substitute the values 1 and 2 in all possible ways. This most important mathematical expression was discovered by Gauss; it is destined to play a part of paramount significance in the physical theory of relativity.

In order to understand the geometrical meaning of g_{12}, let us consider the special case of a diamond-shaped mesh-system, where the co-ordinate lines make an angle θ (Fig. IV). Elementary geometry teaches us that

$$PR^2 = PQ^2 + 2\cos\theta.PQ.QR + QR^2.$$

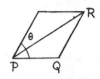

FIG. IV

The lines being always equally spaced, g_{11} and g_{22} remain constant throughout the mesh-system; hence we may put them equal to unity and write

$$ds^2 = du^2 + 2\cos\theta\, du\, dv + dv^2.$$

In this formula we have

$$g_{11} = 1; \quad g_{12} = \cos\theta; \quad g_{22} = 1$$

from which we see that g_{12} represents the cosine of the angle formed by the u and v lines, at the point where g_{12} is calculated.* Inasmuch as $\cos\dfrac{\pi}{2} = 0$, we see that where the u and v lines are perpendicular, g_{12} vanishes. Hence, whenever we have an expression of ds^2, such as

$$ds^2 = g_{11}du^2 + g_{22}dv^2,$$

from which g_{12} is absent, we may be certain that the u and v lines are orthogonal at the point for which g_{12} has been calculated.

If the u and v lines remain perpendicular throughout the entire mesh-system, g_{12} will vanish not only for one particular point, but for all points of the surface.

Let us now consider the particular case of a Cartesian mesh-system of equal squares. In this case, the u and v lines being orthogonal and equally

* More precisely $\cos\theta = \dfrac{g_{12}}{\sqrt{g_{11}\, g_{22}}}$

spaced, we have g_{11} and g_{22} constant, with g_{12} vanishing. Hence we may replace g_{11} and g_{22} by unity, and we obtain

$$ds^2 = du^2 + dv^2,$$

a result in complete agreement with the Pythagorean theorem of the square of the hypothenuse.

Suppose, now, that on this same plane we were to trace another type of co-ordinate system, one known as a polar system (Fig. V). It is constituted by symmetrically disposed v lines radiating from a central point

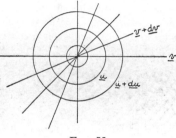

FIG. V

O. The u lines are then given by equally distanced concentric circles having O as centre. Under the circumstances the expression for ds^2 becomes

$$ds^2 = du^2 + u^2 dv^2.$$

In other words, we have

$$g_{11} = 1; \quad g_{12} = 0; \quad g_{22} = u^2,$$

proving once again that the values of the g's vary with our choice of a co-ordinate system. As before, g_{12} vanishes for the same reasons as previously stated, namely, because the lines of our mesh-system are orthogonal at their intersections.

Finally, we may consider the mesh-system defined by the parallels and meridians on a sphere, where we may assume that $u =$ constant gives a parallel and $v =$ constant gives a meridian. Here we have

$$ds^2 = du^2 + \cos^2 u \, dv^2,$$

so that

$$g_{11} = 1; \quad g_{12} = 0; \quad g_{22} = \cos^2 u;$$

From these various examples the following conclusions may be drawn:

$1°$. When a definite mesh-system has been traced on a surface and its lines numbered consecutively, every point of the surface is defined unambiguously by its two numbers (its two Gaussian numbers or co-ordinates). These numbers represent, of course, those of the respective *u* and *v* curves at whose intersection the point stands. A variation in the shape or in the numbering of the *u* and *v* lines entails a change in the co-ordinate numbers of any given point on the surface.

2°. Alongside of these two Gaussian numbers at every point defining the positions of points in the mesh-system, there exist four *g* numbers, at every point, namely g_{11}, g_{12}, g_{21}, g_{22}. But as g_{12} and g_{21} are always identical, we have only three of these *g* magnitudes to consider. The values of these *g*'s at a point vary when we change our mesh-system. In the majority of cases they also vary from place to place throughout the same mesh-system. If, however, the meshes are always orthogonal at their points of intersection, g_{12} vanishes at every point of the surface. If the *u* and *v* lines always have the same slant over the surface and are always equally spaced, the three *g*'s will remain constant throughout; their values being given by 1, $\cos\theta$, 1. If the mesh-system is Cartesian, hence forms equal squares g_{12}, of course, vanishes while, as before, g_{11} and g_{22} remain constantly equal to unity.

3°. The value of the square of the distance between two infinitesimally distant points is given by

$$ds^2 = g_{11}\,du^2 + 2g_{12}\,du\,dv + g_{22}\,dv^2$$

where the values of the *g*'s and of *du* and *dv* will vary when the mesh-system is changed or when we consider different regions of the same mesh-system. The only case in which the *g*'s will remain constant is when the mesh-system is of the uniform type, that is, a network of two families of parallel lines intersecting one another.

It is a remarkable fact that although everything entering into the expression of ds^2 varies when we change our mesh-system, yet the value of ds^2 as defining the value of the square of the distance between two infinitesimally distant points remains unchanged. In other words, ds^2 is a *scalar*, an *invariant*. This allows us to place a new interpretation on the *g* numbers; they appear to act as correctives counterbalancing the variations of *du* and *dv*. If we compare the variations in the values of *du* and *dv* when the mesh-system is changed, to the advance of a squirrel in a drum, we see that the action of the *g*'s is similar to a backward revolution of the drum, offsetting the squirrel's advance, so that the squirrel remains motionless in space.*

* When we consider the *g*'s in this light, we realise that the expression for ds^2 might have been anticipated directly without any regard to measurement. We might, for instance, have attempted to construct in a purely mathematical way the possible expressions containing variables such as *du* and *dv*, together with corrective factors whose rôle it would be to ensure invariance. Riemann remarked that in addition to the classical expression, a number of such expressions could be constructed. But if we wish our value of ds^2 to be compatible with the existence of the Pythagorean theorem, namely $(ds^2 = du^2 + dv^2$ for a right triangle), the classical expression of ds^2 must be adhered to. For this reason the type of space which is obtained under these conditions .is called Pythagorean space; and in what is to follow, we shall have no occasion to consider any other variety.

Now, up to this point we have been considering the expression of a distance between infinitely close points, but in practice we also wish to establish the length of an extended curve traced on the surface. In this case we proceed from point to point along the curve, computing the successive infinitesimal distances ds, then summating them, or, as it would be more proper to say, integrating them. We thus obtain $s = \int ds$. Also we may state that the area $d\mathcal{A}$ of an infinitesimal parallelogram formed by two-line elements du and dv is given by

$$d\mathcal{A} = \sqrt{g_{11}g_{22} - g_{12}{}^2}\,du\,dv.$$

Here again we may calculate any finite area $\mathcal{A}$ by a process of integration, so we see that the finite geometry of the surface can be studied by concentrating our attention on infinitesimal portions and then extending our results from place to place. In short, the method reduces to an application of the differential calculus to geometrical problems, and for that reason is named **differential geometry.**

Powerful as this method of differential geometry has proved to be, there are cases in which it cannot be applied. However, as in the problems of physics with which we shall be concerned, difficulties do not arise, we need not dwell on a number of special cases which in the present state of our knowledge are of interest only to the mathematician.*

And now we come to the main body of Gauss' discoveries. We have seen that on a given surface the values of the three g's at any point, or, more correctly, their variations in value from point to point, are defined by our choice of a mesh-system. But we know that a mesh-system, though in large measure arbitrary, is yet not completely independent of the nature of the underlying surface. For instance, a Cartesian mesh-system of equal squares, or again a diamond-shaped one, both of which hold on a plane, cannot be traced on a sphere. Neither can a network

---

*In a very brief way the difficulties are as follows: Differential geometry involves continuity; hence in a discrete continuum it would lose its force. But even this is not all, for there are various kinds of continuity, and continuity must be of a special type for differential geometry to remain applicable. For instance, in the foregoing exposition of the method we assumed that the expression of Δs would tend to a definite limit ds when the points were taken closer and closer together. In particular we assumed that for infinitesimal areas of the surface the Pythagorean theorem $ds^2 = du^2 + dv^2$ would hold. This was equivalent to stating that infinitesimal areas of our curved surface could be regarded as flat or Euclidean, hence as identified with the plane lying tangent to them. Inasmuch as the restriction of Euclideanism in the infinitesimal is precisely one that Riemann has imposed on space, we need have no fear of applying the differential method to the types of non-Euclidean space thus far discussed.

Nevertheless, we may conceive of spaces where Δs would not tend to a limit, and where, however tiny the area of our surface, we should still be faced with waves within waves *ad infinitum.* The situation would be similar to that presented by curves with no tangents at any point. However, such cases are only of theoretical interest.

of meridians and parallels which holds on a sphere be traced on a plane. For this reason the representation of the disposition of oceans and continents is necessarily distorted in some way or other when given on a flat map.

In short, every species of surface possesses an infinite aggregate of possible mesh-systems, but those systems which are applicable to one type of surface are never applicable to surfaces of any other type. Inasmuch as the Cartesian mesh-system and the diamond-shaped variety are the only ones that entail the constancy of the three g's throughout the surface, and inasmuch as such co-ordinate systems can be traced only on a plane or on surfaces derived therefrom without stretching (cylinder, cone), we see that the constancy of the three g's is characteristic of Euclideanism. This does not mean, of course, that all co-ordinate systems traced on a plane yield constant values for the three g's; it simply means that on a plane it is always possible to trace a Cartesian mesh-system, whereas on all other types of surfaces the task is impossible.

All this goes to prove that the curvature of a surface must exert a modifying influence on the g-distribution, since when the surface is curved no constant distribution is possible. We must infer, therefore, that the g-distribution is governed by two separate influences; first, by the intrinsic curvature of the surface from place to place; and secondly by our choice of the mesh-system over the surface. Gauss realised the importance of separating these two influences and of determining in what measure respectively they affected the g-distribution.

Obviously, if it were possible to discover some mathematical expression connecting the g's at one point with the g's at neighbouring points, and if this mathematical expression remained invariant in value to a change of mesh-system in spite of the variations of the individual g's which must accompany the change of mesh-system, we should be in the presence of a magnitude which, transcending our choice of a mesh-system, would refer solely to the shape of the surface itself, *i.e.*, to its curvature at the point considered. But before we investigate the nature of Gauss' discoveries, certain elementary notions must be recalled.

We know that the curvature of a circle at every point of its circumference is a constant given by $\dfrac{1}{R}$, where R is the radius of the circle. But if in place of a circle we trace any arbitrary curve on our plane, the curvature will vary from point to point along the curve. The curvature of the curve at a given point A is then defined by the curvature of a certain circle which is tangent to the curve at the point A. As a matter of fact there exist an indefinite number of circles of varying radii lying tangent to the curve at A; but among these circles one stands out prominently in that it is, so to speak, more perfectly tangent than all the others. Whereas a tangent circle intersects the curve in two coincident points at A, the privileged circle intersects it in three, or more, points. It is called the *osculating* circle (*osculare* meaning to kiss in Latin). The curvature of the curve at A is defined by the curvature of its osculating circle at A. Calling R the radius of this osculating circle, the curvature of the curve at A is thus given by $\dfrac{1}{R}$ (Fig. VI).

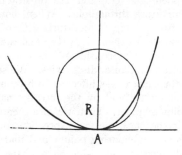

FIG. VI

We must now pass to the curvature of a surface. At a point A on the surface where the curvature is to be computed we trace a normal to the surface. Then through this normal we trace a plane, which of course intersects the surface along a plane curve. We assume this normal plane to revolve round the normal as axis and we thus obtain a series of plane curves of intersection defined by the normal plane and surface. Each one of these curves passing through the point A possesses a definite curvature at A, and this curvature can be computed through the medium of the corresponding osculating circle.

Here a geometrical fact is evidenced. It is found that in the general case there exist two remarkable positions of the intersecting normal plane, perpendicular to one another and therefore sectioning the surface along two curves (1) and (2) orthogonal (i.e. perpendicular) to each other at A. These two curves are characterized by the fact that their curvatures are relative maxima or minima, as compared with the curvatures of all other curves of intersection obtained by rotating the normal plane about A. The two privileged curvatures are called the two *principal curvatures* of the surface at the point A and are designated by $\dfrac{1}{R_1}$ and $\dfrac{1}{R_2}$ respectively.

We then define the *total curvature* or the *Gaussian curvature* of the surface at the point A by the product $\dfrac{1}{R_1 R_2}$. If our surface is a sphere we have $R_1 = R_2$, hence the Gaussian curvature becomes $\dfrac{1}{R_1^2}$ or more simply $\dfrac{1}{R^2}$, where R is the radius of our sphere; but in the general case, R_1 and R_2 are unequal. According to the nature of the surface the two principal curvatures may be of the same or of opposite signs. When of the same sign, the total curvature $\dfrac{1}{R_1 R_2}$ is obviously positive; hence the surface is said to manifest positive curvature at the point considered. The sphere and ellipsoid are illustrations of surfaces presenting a positive curvature throughout. When, however, the surface is saddle-shaped, the two principal curvatures are of opposite sign; the total Gaussian curvature is then negative, and we have a surface of negative curvature at the point considered.

And now let us return to Gauss' discoveries. We saw that the distribution of the g's over the surface was affected both by our choice of a mesh-system traced on the surface and by the intrinsic nature or curvature of the surface from point to point. Then we also mentioned that if it were possible to discover some mathematical expression connecting the g's at a point A with the g's at neighbouring points, and that if this mathematical expression remained invariant in value to a change of mesh-system in spite of the variations of the individual g's which accompany the change of mesh-system, we should be in the presence of a magnitude which, transcending our choice of mesh-system, would refer solely to the shape of the surface itself, *i.e.*, to its curvature at the point considered. This important mathematical invariant, built up with the g's, was discovered by Gauss; it is generally designated by the letter G and is referred to as the *scalar* or *invariant of curvature* at the point A. Gauss then proved that this scalar of curvature G was none other than minus twice the total curvature defined previously. Hence we may write:

$$\text{Total curvature at } A = -\frac{G}{2} \text{ (computed at } A)$$

Aside from a constant factor, these two curvatures are thus the same, so that we shall often refer to G as the Gaussian curvature, even though the appellation is not strictly accurate.*

* It should be mentioned that there exists another type of curvature, different from the Gaussian curvature. This other type of curvature is called the *mean curvature* at a point and is given by $\cdot \dfrac{1}{2}\left(\dfrac{1}{R_1} + \dfrac{1}{R_2}\right)$

In this chapter we have mentioned only the Gaussian curvature, for, as Gauss discovered, it is this type of curvature alone which characterises the geometry of the surface when explored with Euclidean measuring rods. This discovery is a direct consequence of the following considerations:

Two surfaces which have the same Gaussian curvature can be superposed on each other without being torn or stretched, whereas two surfaces which have different Gaussian curvatures can never be applied on each other unless we tear them or stretch them. Obviously the metric relations of a surface are not disturbed so long as we do not stretch the surface; and this is why surfaces which have the same Gaussian curvature and which can therefore be superposed without being stretched, must necessarily possess the same geometry.

A few illustrations may be helpful: A plane sheet of paper has a zero Gaussian

Before proceeding farther, we must recall that the method we have followed of investigating the geometry of surfaces and using Euclidean rigid rods for the purpose of conducting measurements over the surface, leads to the same geometrical results as would be obtained by an exploration of a two-dimensional space (of a plane, for example) by means of appropriate non-Euclidean rods. If, therefore, we conducted measurements with non-Euclidean rods over a plane, we would of course obtain a non-Euclidean geometry; and the Gaussian curvature of the plane would no longer vanish, as it would were we to make use of the Euclidean measuring rods.*

In terms of our non-Euclidean measurements the same plane would be curved. We see, therefore, that the non-vanishing of the Gaussian curvature does not necessarily represent curvature in the usual visualising sense. It represents more truly a relationship between the surface and the behaviour of our measuring rods; in other words, it represents non-Euclideanism, and the word "curvature" is apt to be misleading. In a general way, therefore, we may state that the type of geometry of our two-dimensional space from place to place is defined by the value of the Gaussian curvature from point to point, hence by the law of g-distribution throughout the space; and that when the space is Euclidean, the g-distribution is always such that the Gaussian curvature vanishes at all points, regardless of the particular mesh-system selected.

Now all these discoveries of Gauss relating to a two-dimensional space were extended by Riemann to spaces of any number of dimensions. Riemann found that for spaces of more than two dimensions the results became very much more complex. In the case of a space of n dimensions we of course require n Gaussian co-ordinates to define the position of a point in our mesh-system, which now becomes n-dimensional. As before,

curvature and a zero mean curvature at every point. A cylinder, a cone or a roll has a zero Gaussian curvature but a positive mean curvature. It is the differences in the mean curvatures that cause these surfaces to appear to us visually as differing from the plane. But owing to the fact that the plane, the cylinder, the cone and the roll have the same zero Gaussian curvature, we can wrap the plane on to the cylinder or cone without stretching it.

For this reason all these surfaces have the same Euclidean geometry (Euclidean measuring rods being used). On the other hand, a *minimal surface* (such as the surface defined by a soap film stretched from wires situated in different planes) has a zero mean curvature just like the plane, but in contra-distinction to the plane it has a negative Gaussian curvature; hence, cannot be applied to a plane, and its geometry is therefore non-Euclidean.

For the same reason a sphere, whose Gaussian curvature is a constant positive number, or that saddle-shaped surface called the pseudosphere whose Gaussian curvature is a constant negative number, can neither of them be flattened out on to a plane without being stretched. As a result their geometries differ from the Euclidean geometry of the plane. We have seen that these geometries are Riemannian and Lobatchewskian, respectively.

* If we were to represent non-Euclidean geometry as arising from the behaviour of our measuring rods, squirming when displaced as compared with Euclidean rods, we should see that Euclideanism in the infinitesimal implies that when our rods are of infinitesimal length and are displaced over infinitesimal distances, they behave in the same way as rigid Euclidean rods.

we can conceive of Cartesian and Gaussian mesh-systems, the former being a generalisation to n dimensions of our network of Euclidean squares. As for the invariant mathematical expression of the square of a distance between two points, it now contains a greater number of terms. In the case of three-dimensional space we have no longer three separate g quantities at every point; this number is increased to six. In the case of a four-dimensional space it is increased to ten, and in the case of an

n-dimensional space to $\dfrac{n(n+1)}{2}$.

Owing to the importance of a four-dimensional extension in the theory of relativity, we shall write out the expression of the square of the distance for a four-dimensional space. If we call du_1, du_2, du_3, du_4 the four differences between the four Gaussian co-ordinates of our two infinitely close points A and B and designate the ten g numbers at every point by g_{11}, g_{12}, etc., we have

$$ds^2 = g_{11} du_1{}^2 + 2g_{12} du_1 du_2 + 2g_{13} du_1 du_3 + 2g_{14} du_1 du_4$$
$$+ g_{22} du_2{}^2 + 2g_{23} du_2 du_3 + 2g_{24} du_2 du_4$$
$$+ g_{33} du_3{}^2 + 2g_{34} du_3 du_4$$
$$+ g_{44} du_4{}^2$$

or, more concisely,

$$ds^2 = \Sigma g_{ik} du_i du_k$$

where we give to i and k all whole values from 1 to 4, permuting them in all possible ways. Accordingly, in what follows, we shall refer to the g's as the g_{ik}'s.

Proceeding by the same method as we did in the case of two dimensions we may construct, as before, an invariant relation between the g_{ik}'s at any point of the continuum and their values at neighbouring points; the invariance of this relation implies that its magnitude remains unchanged when the mesh-system is altered, and so this magnitude defines some intrinsic property of the space itself. The invariant magnitude here considered is the generalization of the Gaussian curvature G of surfaces, extended to four dimensions. We shall represent it by the same letter G as before. Except for the more generalized form of the Gaussian curvature, the situation seems to be much the same as for a two-dimensional surface. But a novel feature arises when we wish to determine the geometry of the space.

Thus we recall that in the case of a two-dimensional surface, the invariant of curvature G (or Gaussian curvature) fully defines the geometry of the space. For instance, if the Gaussian curvature vanishes throughout, the surface is Euclidean or at least flat.* In the same way, if the Gaussian

---

* It is necessary to make this distinction for a space may be flat and yet only semi-Euclidean, as will be understood in later chapters when discussing space-time.

curvature is an invariable positive or negative number throughout, the surface is one of constant curvature, either positive, as with a sphere, or negative, as with a pseudo-sphere. But when we consider spaces of more than two dimensions, a knowledge of the generalised Gaussian curvature throughout the space is no longer sufficient to fix its geometry. While this curvature may vanish or have the same non-vanishing value throughout, we cannot infer therefrom that the space is necessarily flat or of constant curvature. All that we can say, in the case of a space of more than two dimensions, is that the vanishing (or the constancy) of the generalised Gaussian curvature G is a necessary condition, but not a sufficient one, for the space to be flat (or to be of constant curvature). There still remains a large measure of indeterminateness in the actual geometry of the space.

Fortunately, for spaces of more than two dimensions, another magnitude serving to define the geometry of the space is at our disposal. This other magnitude, discovered by Riemann, is called a *tensor,* and is usually referred to as the Riemann-Christoffel tensor. It is no longer an invariant expressible by a single number at each point of the space and indifferent to our choice of mesh-system. Instead it has a number of components, which vary in value when the mesh-system is changed. Each component defines a relation between the g_{ik}'s at a point and their values at neighbouring points. The Riemann-Christoffel tensor is denoted by B_{ikst} or R_{ikst}. We choose the first symbol.

The various components of this tensor are represented by giving to the four subscripts i, k, s, t, all integral values from 1 to n, where n is the dimensionality of the space. In the case of four-dimensional space, there are 256 of these components, but as some of the components always vanish and others are repetitions, we are left with only 20 independent components. As we have said, the values of these components at the point of our space vary when the mesh-system is changed, and so it might appear that our tensor, depending as it does on the mesh-system, cannot express an intrinsic property of the space. But this is not the case, for if all the components are known in a given mesh-system, we may derive the values which these components would have in another mesh-system. In other words, the components, though variable individually, constitute in their aggregate a definite entity; and this entity, *i.e.* the tensor itself, represents an intrinsic quality of the space at the point considered.

If in any given mesh-system the values of the 20 components of the Riemann tensor are known at each point of the space, the geometry of the space is completely determined. In particular, if all the 20 components vanish in one mesh-system, they will continue to vanish in all other mesh-systems, and the space will be flat. In short, we see that in dealing with spaces of more than two dimensions, the Riemann tensor plays the role that was played by the Gaussian invariant G in two-dimensional space.

The Riemann tensor, which has made its appearance in connection with spaces of more than two dimensions is also present for a two-dimensional surface. In the case of a surface, however, the tensor has only one independent component, which when suitably combined, with the g_{ik}'s of the surface at the point considered, turns out to be the ordinary Gaussian curvature of the surface at the point considered. For this reason, in the case of a surface, the role of the Riemann tensor is adequately taken over by the Gaussian curvature, and so in the theory of surfaces, the Riemann tensor becomes inconspicuous.

In addition to the Riemann tensor B_{ikst}, Einstein has made use of another structural tensor G_{ik}, derived from the Riemann tensor. The new tensor G_{ik} has 10 independent components in a four-dimensional space. Like the components of the Riemann tensor, the components of G_{ik} express relationships between the g_{ik}'s at a point and their values at neighbouring points. Unlike the Riemann tensor, however, a knowledge of the values of the 10 components of the new tensor at each point of space in any given mesh-system does not suffice to determine the exact geometry of the space. In particular, the vanishing of the 10 components, though a necessary condition for the space to be flat, is insufficient in itself to insure perfect flatness, since it is also consistent with a wide variety of non-Euclidean spaces. This wide variety of possibilities is, however, more restricted than would be required by the vanishing of the generalised Gaussian invariant, G, alone.

Turning to the physical significance of the g_{ik} numbers, we may note that it is these g_{ik} magnitudes which enter into the expression of a distance, hence which serve to define congruence. Consequently the g_{ik}-distributions arise only when a definite metrics is imposed upon the continuum. These g_{ik}-distributions will then characterize the continuum in the manner explained previously, and in their aggregate they will define the geometry, or metrical properties, of the continuum.

The same arguments would apply to the structural tensors of curvature of the continuum, namely, to G, to G_{ik} and to B_{ikst}. If the continuum is amorphous, these structural tensors, together with the g_{ik}'s that compose them, are meaningless until we have decided upon some theoretical type of congruence, or geometry; and even then they represent characteristics of structure that are as conventional as the geometry we have selected. But if, on the other hand, a definite metrics exists in the continuum, then these structural tensors represent curvatures of the continuum, which we may regard as physically existent and no longer as purely conventional.

CHAPTER VIII

THE MEANING OF THE WORD RELATIVITY

ALL our previous discussions and explanations have been made in order to prepare the ground for Einstein's theory of relativity. However, before we proceed to discuss Einstein's theory proper, it may be useful to reach a more precise understanding of the scientific meaning of the words "relativity" and "absoluteness."

In traditional philosophy an absolute reality is the primary, self-existing reality, the substance, the true being or the metaphysical reality, whereas a relative reality is a secondary, dependent reality owing its form of being to a relation of the genuinely real.*

Now, so far as scientific philosophy is concerned, absolute reality is a myth. All we can ever become cognisant of in nature reduces in the final analysis to structure, that is, to relationships. Hence, if we adopt this very general understanding of the word "relativity," we may anticipate even at this stage that Einstein's theory can add nothing to our belief in the fundamental relativity of all knowledge. Obviously, then, when physicists speak of absolutes and relatives, they are referring to categories of a nature differing from those of philosophers.

Consider, for instance, the length of an object, or again the comparison of two lengths situated in different parts of space. We have seen, when discussing mathematical space, that according to our measuring standards, which were conventional, these lengths might be equal or unequal; and it was this aspect of relativity which expressed the fundamental mathematical relativity of length. But we also saw that although in theory the definition of congruence was conventional, in practice, if we wished to define congruence so as not to enter into conflict with our sense perceptions, a definite type of congruence (practical congruence) was imposed upon us by nature. We saw further that bodies which were thereby defined as congruent constituted what are known as Euclidean solids; and that these were assumed to remain absolute in length, regardless of the observer's motion.

Under these circumstances, when a physicist such as Einstein, in contradistinction to a mathematician, asserts that length is relative, he is referring to the actual length *measured out by a so-called rigid body;* and he wishes to imply that this length will turn out to be indeterminate and to depend essentially on our conditions of observation. We see, therefore, that the relativity of the physicist is not inevitable, as is that of the mathematician. It is not an *a priori* type of relativity expressing a

* Quoted from one of the standard textbooks of philosophy.

trivial relativity to standard. It is essentially an empirical type of relativity, always subject to proof or disproof by precise empirical tests. The two following illustrations may be helpful:

Consider a telegraph pole rising vertically from the ground. The visual angle under which this pole will appear to us is essentially relative, since it varies in value with the distance of the observer from the pole. Of course, this visual angle is fundamentally relative, since its value depends on the system of geometry we may adopt, but we are no longer discussing this mathematical aspect of relativity. We are assuming that our system of measurements has been fixed according to the requirements of practical congruence, *i.e.*, that our measurements are performed with ordinary rigid rods and hence are Euclidean. In this event the visual angle under which we perceive a given pole is perfectly determinate when our distance from the pole is specified. But, on the other hand, it is relative, inasmuch as it is not immanent in the pole; it varies with our relative distance and thereby reduces to a mere relationship between the pole and our distance therefrom.

In contradistinction to the relativity of the visual angle, classical science considered that the length of the pole itself was absolute, being irrelevant to our relative position as observers. Once again, if we argue as mathematicians, we may claim that this length in turn is relative, since it depends on our measuring conventions. But if we assume that our measuring conventions have been fixed by the requirements of practical congruence, this type of relativity is excluded. Such being the case, classical science believed that the length of the pole was an absolute, inasmuch as its numerical value was so many feet or so many metres regardless of our position and regardless of our motion. It is this last assumption that is contested by Einstein. The absolute magnitude is not the length of the pole; for this, analogously to visual angle, is relative. The absolute magnitude is rather a something which transcends space and time, and relates to the absolute world of space-time.

A second illustration may be given by the concepts of **mass and weight.** As a result of Newton's discoveries, the weight of a body was realised to be relative, since its value varied with the proximity and distribution of matter. On the earth's surface the value was so many tons, on the surface of the moon it was considerably reduced, and in interstellar space it vanished completely. A body of itself had no weight. Weight was not immanent in the body; it reduced to the mere expression of a physical relationship between the position of the body and that of surrounding matter.

In contradistinction to the relativity of weight, classical science assumed that mass was absolute. The mass of a body, wherever situated and in whatever circumstances observed, was supposed to be an invariant remaining always the same. For this reason, mass, in contrast to weight, was assumed to be immanent in matter. It was not the result of a physical relationship, but was a reflection of the very existence of matter. True, when Bucherer's experiments had shown that the mass of an electron appeared to increase with its velocity, mass unquestionably appeared to be relative. But it was assumed that this apparent increase in the mass of the electron was to be ascribed *not* to any variation in the mass of the electron itself, but to a deformation of its electromagnetic field brought about by motion through the ether. It was as though the ether was dragged along by the electron in its motion, just as a ship moving through the ocean carries along a more or less considerable volume of surrounding water according to the speed with which it is moving. In other words, there had been no real change in the mass of the electron itself, and this belief of classical science was expressed by the principle of the conservation of mass.

The preceding examples teach us two things: First of all, it is not sufficient to state that a concept is a relative; we must complete our statement by stipulating with respect to what surrounding conditions this concept happens to be relative: whether it be, *e.g.*, to the distance of the observer, to his motion, or to the distribution of matter. Secondly, we see that the only experimental means we have of distinguishing a relative quantity from an absolute is by varying surrounding conditions and by modifying the nature of the motion and the position of the observer. Were the value of the quantity to vary under these conditions, we should be assured that it was a relative. But obviously it may be impossible for us to vary all surrounding conditions; we cannot vary, for instance, the star distribution. Inasmuch as in this universe the whole influences the part and the part influences the whole, it would appear that our differentiations between absolutes and relatives might reduce to a mere expression of our limitations.

This view is undoubtedly correct in theory, but it must be remembered that physics can proceed only by successive approximations. For example, experiment proves that the action of a certain phenomenon on another decreases with the distance. The physicist is therefore perfectly entitled to assume that if the disturbing cause is far enough removed its influence can be neglected. In this way, by a process of progressive elimination, he is enabled to conceive phenomena to proceed in complete isolation from all foreign disturbing influences. Henceforth it will be these ideally isolated phenomena which he will subject to the tests of relativity by submitting them in succession to those foreign influences over which he can exercise some control. Under these conditions the question of the absoluteness or the relativity of a given concept or magnitude acquires a definite physical meaning.

In classical science it was always assumed that a distance in space, a duration in time and a simultaneity between distant events were absolute concepts. Regardless of the relative motion or position of the observer, regardless of the distribution of matter, these concepts remained unchanged. Einstein succeeded in proving that these opinions must be erroneous, since they were incompatible with certain refined optical experiments. He showed that the relative motion of the observer could not help but exert a modifying influence. Henceforth, distance, duration and simultaneity become relatives expressing relationships between the magnitudes measured and the relative motion of the observer. Just as the visual angle under which an object appeared to an observer was in no wise immanent in the object itself, since it varied with the observer's position, so now length, duration and simultaneity are in no wise immanent in the real world, since they vary with the observer's motion.

Once more we must draw attention to the fact that the relativity we are discussing is essentially physical; it is posterior to our definitions of practical congruence. It follows that this relativity of distance and duration must correspond to variations which would be seen and felt. These variations are not due to mere arbitrary changes which we may introduce into our mathematical systems of measurement. They are not merely conceptual; they are perceptual.

The fact is that the systems of measurement are imposed upon us by nature, and by that we mean that they translate what we actually see and feel and not that which we might see and feel if the world happened to be such as some nebulous mathematician might posit it to be. In a similar way, Einstein in the special theory proved that mass, which classical science had considered absolute, was in reality relative, since it depended on the relative motion of the observer. The general theory of relativity went still farther, proving that mass was relative in yet another sense. It established that not only the relative motion of the observer, but also the distribution of surrounding matter, would influence the mass of a given body. To this extent Mach's premonitions were vindicated. Following the relativity of mass, a number of other concepts, such as temperature, pressure, electric and magnetic forces, and gravitational force, were in turn proved to be relative, depending on the motion of the observer.

Now, although the epithet *relative* in its widest sense implies relative to surrounding conditions, we shall find that in Einstein's theory it is employed most generally with reference to the observer's motion. *A magnitude which is relative will, therefore, mean one whose value depends on the relative motion of the observer; a magnitude which is absolute will imply one whose value remains unaffected by this motion.* The major aim of the theory will be to separate those magnitudes which are relative from those which are absolute. The absolute magnitudes will then be representative of the absolute, common world of which the various observers will obtain but private perspectives. This absolute world will be the world of space-time standing in contrast to the relative world of space and time.

Among the absolutes which the theory has discovered, we may mention electric charge, entropy, the velocity of light *in vacuo* (in the special theory, at least) and the length of what is known as the *Einsteinian interval* between any two points in four-dimensional space-time; this latter type of absolute holding throughout the entire theory, both special and general.

CHAPTER IX

THE PRINCIPLES OF RELATIVITY

MUCH of the difficulty which philosophers and laymen experience in understanding Einstein's theory arises from a confusion between the different meanings that may be attributed to the concepts "relativity of space and motion." For the student who already possesses some knowledge of classical science, confusions of this type, of course, are not to be feared. But in view of the general misconceptions on the subject we will mention briefly the various principles, elucidating them further as we come across them in the course of this book. We must name here:

1. *The primordial mathematical relativity of space and time.*
2. *The kinematical or visual principle of relativity.*
3. *The dynamical or classical Galilean and Newtonian principle of relativity.*
4. *Einstein's special principle of relativity.*
5. *Einstein's general principle of relativity.*
6. *The radical Mach-Einstein principle of relativity.*

Let us consider these various principles in their order. **The mathematical type of relativity** is the one we have already had occasion to mention when discussing mathematical space and time. It implies that a distance in space has a purely relative magnitude; so that two distances may be congruent or unequal according to our measuring conventions. Even after we have decided upon our measuring conventions, the magnitude of a spatial distance can only be expressed by the magnitude of our measuring rod; and if during the night all lengths were to contract in the same way, no difference could be detected when we awoke on the following day. Similar conclusions apply to a duration in time.

The kinematical or visual principle of relativity, which is at least as old as the Greeks, states that a body can be considered in motion only when referred to some other body. For instance, if we consider the particular case of the earth and sun, we would conclude, according to whether we referred motion to the earth or sun, either that the sun was rotating round the earth every twenty-four hours, or else that the earth was rotating on its axis.

The two methods of presentation would be equivalent. Again, it would be impossible for us to state whether a body was approaching us faster and faster with some accelerated motion or whether it was *we* who were travelling with accelerated motion towards the body, visual appearances being the same in either case. Obviously this kinematical or visual principle would connote the complete relativity of all motion and rest, hence the complete relativity of space.

But a closer study of the dynamics of material systems proves the visual principle to be untenable. It was found that the behaviour of material systems, and thus the results of mechanical experiments, were influenced by absolute acceleration and rotation through space, although they remained totally unaffected by absolute velocity.

The Newtonian or Galilean or classical or dynamical principle of relativity expresses this elusiveness of absolute velocity or Galilean motion through space so far as mechanical experiments are concerned, while it stresses by contrast the physical significance of absolute acceleration and rotation. As is well known, it was this absoluteness or physical significance of acceleration and rotation which compelled Newton to recognise that space could not be relative. That the Newtonian principle of relativity will considerably restrict the scope of the visual principle can be gathered easily from the following example:

If a body were moving with respect to our frame of reference with a relative velocity, *but with no relative acceleration*, we should be justified according to the Newtonian principle in maintaining that it would be impossible to decide *by mechanical experiments* whether it was *we* who were moving towards the body or *the body* that was moving towards us. To this extent the visual principle is satisfied. But, on the other hand, if a relative acceleration or rotation existed as between ourselves and the body, mechanical experiments *would* enable us to ascertain what fraction of the relative acceleration or rotation was due to our own absolute acceleration or rotation in empty space, hence also what fraction was due to the body's motion.

We now come to **Einstein's special principle of relativity.** It is an exact replica of the Newtonian principle, upholding the relativity of velocity and the absoluteness of acceleration.* The sole difference between the two principles is that in Einstein's, the elusiveness of absolute velocity through space is no longer restricted to mechanical experiments and to material systems; electromagnetic and optical experiments are now placed on exactly the same footing.

While we are on the subject of the various principles of relativity, we may discuss two further types which play an important part in that extension of Einstein's special theory known as the general theory of relativity. There, we are introduced to the **General Principle of Relativity.** This principle states that the mathematical expressions of the laws of

* We are endeavouring to explain things as simply as possible, but as a matter of fact the statement we are making that acceleration remains absolute in Einstein's special theory is not quite correct. The acceleration of a body which in Newtonian science remained the same regardless of our selection of one Galilean frame or another, varies in value in the special theory under similar circumstances. Nevertheless, inasmuch as a sharp distinction still persists between velocity and acceleration, we have felt justified for reasons of simplicity in presenting the problem as we have done.

nature must maintain the same form regardless of our choice of a frame of reference, be it Galilean (*i.e.*, unaccelerated through empty space), accelerated, or even squirming like an octopus, while our clocks situated throughout the frame may beat at the most capricious rates. The classical principle of relativity conceded this invariance of natural laws in the case of Galilean frames, and synchronised clocks alone, and then only in the case of the laws of mechanics. The special principle recognised that this invariance held true even for the electromagnetic laws, but, as before, only for Galilean frames. The general principle of relativity, by extending the invariance of the laws of nature to all types of motions of the frame of reference, marks the starting point for the possible relativisation of acceleration, which had heretofore stood out as aloof and as distinctly absolute.

The radical **Mach-Einstein principle of relativity** is the result of a natural desire to bring about the complete relativisation of all manner of motion, rotationary and accelerated, as well as uniform. This is achieved by ascribing all the dynamical effects which accompany the acceleration and rotation of material and electrodynamic systems, to motion with respect to the material universe as a whole. According to this principle, which is still highly speculative, there can exist no observable difference between the rotation of a body with respect to the universe of stars and the rotation of the stars round the body; exactly the same dynamical effects of centrifugal force would be set up in either case, so that no trace of absolute motion through empty space would be left. We see that Mach's principle constitutes an attempt to vindicate the kinematical principle in spite of the difficulties of a dynamical nature which had been the cause of its rejection. It was in part with a view to satisfying Mach's principle that Einstein elaborated the hypothesis of the cylindrical universe.

CHAPTER X

CLASSICAL MECHANICS AND THE NEWTONIAN PRINCIPLE OF RELATIVITY

WE mentioned, when discussing amorphous mathematical space, that position and hence motion could have no meaning other than that of expressing conditions with respect to some particular frame of reference. In short, the visual principle of relativity would have to be adhered to.

We will accordingly select some arbitrary system of reference, which we will assume to be defined by three mutually perpendicular planes meeting at a common point O. The instantaneous position of a body will henceforth be defined when we have measured with rigid rods its distances from the three planes of our frame; and the only type of position and of motion of the body which we can define in an unambiguous way will be its position and motion as referred to our frame.

The measurement of the velocity of the body with respect to our frame necessitates the addition of a clock. We will assume that this clock beats out congruent intervals of time, where congruence is, of course, defined by the requirements of practical congruence for time, that is, by the beatings of some isolated periodic mechanism.

Now, it is perfectly apparent that unless referred to some particular frame, the path followed by a body and its motion along this path can have no determinate meaning. Thus, if in a uniformly moving train we throw a ball into the air and catch it again in our hands, then, as referred to the train, the ball will have followed an up-and-down motion along the same vertical; on the other hand, as referred to the embankment, the ball will have described a parabola. Again, if a stone is allowed to fall from a great height, its motion with respect to the earth will be accelerated, whereas, with respect to a frame of reference falling together with the stone, it will have remained at rest.

It would thus appear that we might credit to the path followed by a body any arbitrary shape we wished; all we should have to do would be to select some appropriate frame of reference. The same indeterminateness would, of course, apply to the motion of the body along its path. Thus the shape of a path in space (rectilinear or curved) or the species of a motion (uniform or accelerated) could have no absolute significance. From the standpoint of mathematical space these would all be essentially relative to the reference frame we had selected.

But it is obvious that this attitude into which our understanding of mathematical space has forced us presupposes that all the frames of reference we might select were on exactly the same footing, that is, were indistinguishable from one another. Galileo and Newton noticed that such was not the case. In certain frames we were pulled hither and thither by strange forces unsymmetrically distributed, and objects placed on the floor did not remain where they had been put. Space around us appeared to be in a chaotic condition, as, for example, on a rotating disk or in a train rounding a curve. In other frames no such forces were apparent (excluding the force of gravitation), and space appeared stagnant and quiet, the same everywhere. These latter frames are called **Galilean or Inertial.**

Now, a Galilean frame did not manifest itself as unique. So far as the most delicate mechanical experiments could detect, there appeared to exist an indefinite number of such frames, all moving in respect to one another with various constant speeds along straight lines, without suffering any relative rotation during their motion. Mechanical experiments conducted in any one of these Galilean frames showed that free bodies followed straight lines with constant speeds (Newton's law of inertia), and when referred to these frames, mechanical phenomena, including the planetary motions, were susceptible of being formulated by very simple laws. As viewed from Galilean frames, the other frames, the non-Galilean ones (those filled with strange forces), were found to be moving with accelerated or rotationary motions.

Mechanical experiments conducted in the non-Galilean frames proved that the disposition of the strange unsymmetrical so-called **inertial** forces bore a close relationship with the apparent motions of these frames as viewed from the Galilean ones. It was possible, therefore, for an experimenter situated in one of these strange frames to anticipate, without looking outside, exactly how his frame would appear to be moving when viewed by a Galilean observer (*i.e.*, an observer in a Galilean frame), at least so far as acceleration and rotation were concerned. Thus, in a non-Galilean frame moving with respect to a Galilean frame along a straight line with some definite acceleration, the forces of inertia would be disposed in a parallel way, pulling in the direction opposite to that of the relative acceleration. In a non-Galilean frame rotating with respect to a Galilean observer, the disposition of the inertial forces would be much more complicated. There would be the **centrifugal force** pulling outwards and the so-called **Coriolis forces** pulling sideways.

As viewed from all these non-Galilean frames, the laws of mechanics would appear hopelessly complex. Free bodies would no longer appear to follow straight lines with constant speeds, but would describe capricious curves with varying velocities. Newton's Law of Gravitation would no longer account for the planetary motions as observed from these frames, and a study of mechanics would present tremendous difficulties. In short, all the simple laws observed from Galilean frames would have to be compounded with the effects of these strange inertial forces when we wished to formulate laws controlling phenomena as viewed from the non-Galilean frames.

No surprise need be felt, therefore, when it is said that classical science selected the Galilean frames as standard frames. All the laws of mechanics and of physics were referred to them, and this, indeed, was the essence of Copernicus' discovery. Incidentally, we see that we are in possession of another definition of a Galilean frame; for we may say that a Galilean frame is one with respect to which the laws of classical mechanics (law of inertia, Newton's law, etc.) hold true.

We may also mention that the very possibility of there being definite laws of mechanics, such as the law of inertia, is contingent upon the existence of certain fundamental differences between the various frames. If all frames were identical, as the complete relativity of motion would demand, the law of inertia would express nothing at all. In every case we might make the motion of a body appear anything we wished by selecting our frame suitably; and if we had no means of distinguishing one frame from another the straight path of a body through empty space would convey no physical meaning.

We may now give still another definition of a Galilean frame. It follows from the law of inertia, according to which free bodies, when viewed from Galilean frames, will describe straight lines with constant speeds that the stars, being presumably free bodies, will obey this law and, owing to their remoteness, appear to suffer very slight displacements even in the course of a century This fact led to a more easily obtainable definition of a Galilean or inertial frame, namely, a frame with respect to which the stars would appear fixed. This definition is not meant to supplant the one given previously of a frame in which no forces of inertia are experienced, for theoretically both definitions are equivalent. Its sole advantage is that it leads to a more accurate empirical determination of a Galilean frame and shows us immediately that our earth does not constitute such a frame, since the stars appear to circle round the Pole star. As viewed from a true Galilean frame, our earth would therefore be rotating round the polar axis every twenty-four hours.

Had we sought to discern the non-Galilean nature of the earth frame by endeavouring to detect traces of inertial, centrifugal and Coriolis forces, the task would have been more difficult owing to the minuteness of these forces. Nevertheless, the forces have been detected, and they account for the protuberance of the equator, the clockwise and anti-clockwise motions of cyclones, the rotation of the plane of Foucault's pendulum, the directions of trade winds and of ocean currents, etc.

Now, thus far we have not attempted to draw any general conclusions from the existence of these various frames. We have merely stated that they undeniably exist, since they have been revealed by experiment, and that it is simpler to refer the laws of physics, and particularly those of mechanics, to Galilean or inertial frames. When it comes to interpreting the deeper significance of all these facts, various alternatives present themselves, and in this chapter we will discuss Newton's solution at some length.

The problem to determine is, "Why do different types of frames exist when our understanding of amorphous mathematical space would seem to preclude any such differences?" Here we may make one of two assumptions. Either we may assume that the differences in conditions which exist in the various frames (differences which are exemplified by varied distributions in the lay of the inertial forces, or again in their total absence) are intrinsic, or else we may assume that they are due to foreign influences. Newton adopted the first alternative, which we may. call the postulate of isolation; Mach adopted the second.

If we follow Newton we must assume that the peculiarities existing in our frames are not to be attributed to causes existing beyond them. We may therefore assume that the inertial forces would be exactly the same were the balance of the universe to be annihilated, so that to all intents and purposes we might consider our frames isolated from all external influences. We assume, then, that the various distributions of the inertial forces manifesting themselves in our various frames must be attributed to the conditions of rest and of motion of these frames, *not* with respect to the remainder of the universe, which would be as good as annihilated, but with respect to those limited portions of the ether or of space in which the frames are situated.

Newton thus objectivises space and, owing to the symmetrical conditions which endure in Galilean frames, assumes these latter to be at rest in space (or in uniform rectilinear motion), whereas the non-Galilean frames are assumed to be accelerated or rotating. Since a correspondence exists between the centrifugal and Coriolis distribution of forces in a non-Galilean frame, and its rate of rotation as viewed from a Galilean one, we may assume that the space in which a Galilean frame is at rest is also that in which a non-Galilean frame is accelerated or rotating. There is, then, but one space, an absolute space with respect to which motion and rest have a physical significance. Galilean frames are those which are non-accelerated; non-Galilean ones are those which are accelerated or rotating in this absolute space. This was Newton's solution, and provided we accept his postulate of isolation there appears to be no way of escaping it.

Under the circumstances Newton's law of inertia, according to which a free body follows a straight course with constant speed in relation to a Galilean frame, acquires a more determinate significance. The law now implies that this motion is being described in absolute space—in a space, therefore, that possesses a definite structure and which is thereby able to guide the body along its straight course. Had space been entirely amorphous and relative, as we assumed originally, there would have been nothing to guide the body; and even had the latter possessed enough intelligence to guide itself, there would have been nothing in amorphous space for it to direct itself by.

A number of philosophers attacked the Newtonian hypothesis of absolute motion and absolute space. For instance, some of them remarked that this so-called absolute space might well be assimilated to a bubble of ether floating and rotating in a more embracing space, this other in yet another, *ad infinitum;* so that Newon's absolute rotation would be meaningless. Needless to say, arguments of this sort are totally irrelevant to Newton's stand and to science generally.

Classical science, by absolute space, meant one with respect to which motion, or at any rate accelerated motion and rotation, appeared to manifest themselves dynamically; and experiment proved conclusively that such a space existed. There was no use, and in fact no sense, in wondering whether this absolute space should be conceived of as rotating in some other space; for even if a hypothesis of this sort were ever to be vindicated, the fact remained that rotation appeared to be absolute in a certain space regardless of whether or not this space was the super-space *par excellence.* This was all that was meant by the absoluteness of rotation so far as science was concerned. In short, as scientists clearly recognised, there was only one way to refute Newton's absolute space and motion, and that was to deny his postulate of isolation. Such, indeed, was the course adopted by Mach.

If we follow Mach and Neumann and assume that the various internal dynamical conditions which distinguish the various frames are due to influences of an extrinsic nature, we shall no longer be able to subscribe to Newton's conclusions; for we shall no longer be able to assume that were the entire outside universe to be annihilated, the internal conditions in our frames would endure as before. Henceforth, when ascribing the appearance of centrifugal and Coriolis forces to the rotation of the body, we should have to imply that this rotation was relative to a something in the universe that *was not* the empty space in which the frame was situated. The natural conclusion was that this rotation was relative to that other form of existence, namely, to matter. Neumann suggested a mysterious body called the "Alpha Body," to whose dynamical influence he ascribed the rise of inertial forces; but, as Poincaré remarked, "Of this mysterious Alpha Body all we can ever know is its name." A hypothesis such as Neumann's, which is incapable of verification, reduces to a mere metaphysical speculation of no scientific value whatsoever.

Mach adopted a much more plausible alternative. He assumed that the rotation of the frame which appeared to accompany centrifugal and Coriolis forces was a rotation with respect to the remainder of the physical universe, hence to the totality of the matter of the universe. As the vastest agglomeration of matter in the universe is given by the totality of the star masses, we are led to Mach's conclusions, namely, that the forces of inertia arise from a relative rotation or acceleration of the frame with respect to the universe of stars. Were the stars, then, to be annihilated, centrifugal force and all forces of inertia would vanish, a rotating mass of water would not splash over the sides of a bucket, and so forth. It is, of course, scarcely necessary to mention that the actual visibility of the stars or our mere ability to imagine their existence has nothing to do with the problem; it is solely the causal influences produced by their masses and relative acceleration which are relevant.

As a matter of fact, there is no very essential difference between the solutions of Mach and Newton. Even if we follow Mach, unless we are to believe in action at a distance, we must assume that the stellar influences can reach us only as a result of a propagation from place to place through space. We may perfectly well assume that as a result of this influence space receives a definite structure, so that in fine the physical reality of rotation may still be ascribed to a rotation in space. There is a difference, however; for with Newton this structure of space is intrinsic to space itself and has nothing to do with the contingent presence of matter generally, whereas if we follow Mach this structure of space must be conceived of as extrinsic and caused by the matter of the universe. In the absence of matter, the structure of space would automatically disappear and the various frames become indistinguishable.

As we can see, Mach's views lead to a partial vindication of the visual or kinematical principle of relativity. But the vindication is only partial, since in all cases the relativity of acceleration and rotation exists only as between the totality of the stars and a body, and not between any two bodies taken arbitrarily, as in the visual principle.

Against Mach's solution there is the argument that the stars are too remote to produce any such palpable effects as the bursting of a rotating flywheel or the splashing of a revolving mass of water over the sides of the bucket. But, still more important, it appeared impossible for classical science to account for these dynamical actions attributed to the stars.

Let us now return to Newton's solution of the problem, since it constitutes the embodiment of classical mechanics. As we have seen, Newton, by positing his postulate of isolation, rendered inevitable a belief in some real absolute medium, space or the ether, from which rotation and acceleration would derive a real meaning. But here we must recall that mechanical experiments, though clearly differentiating a Galilean frame from a non-Galilean frame, were totally incapable of differentiating one Galilean frame from another. It appeared impossible to discover, therefore, which particular Galilean frame was to be considered at absolute rest in stagnant absolute space and which ones were to be regarded as moving through space with uniform translationary motions. In other words, when we confined ourselves to mechanical experiments, velocity through space was relative, the only velocity susceptible of mechanical detection being relative velocity, that is, velocity of one frame with respect to another. Acceleration and rotation, on the other hand, were absolute, and could be detected and measured mechanically without our having to appeal to a Galilean frame taken as term of comparison.

Were we to appeal to a Galilean frame in order to check up our computations, it would be a matter of complete indifference which particular Galilean frame we might select. As referred to any one of these Galilean frames, a body would manifest exactly the same acceleration and rate of rotation, whereas its velocity would, of course, vary with the Galilean frame to which it was referred. This is what is meant by the *absolute character of acceleration* in contradistinction to the *relative character of velocity.*

Of course, even an accelerated body could be made to appear at rest or to be moving with uniform motion if we selected a suitable frame. But the frame would have to be non-Galilean or accelerated in order to ensure this result, and we have seen that non-Galilean frames can never (at this stage, at least) be regarded as being at rest in space. It follows that the appearance of a body's motion relatively to any of these non-Galilean frames could never be considered representative of the real motion of the body in space.

The Principle of Relativity of Classical Science merely summarises these discoveries by stating that **it is impossible for an observer situated in a Galilean frame to ascertain by any mechanical experiment whether he is at rest in space or in a state of uniform translationary motion.** A simple illustration will clarify the meaning of these statements.

Consider a train moving with constant speed along a straight course; obviously there is no trace of acceleration in the train's motion, since no forces of inertia are felt in its interior. In common practice, we should speak of the train as having a definite speed of so many miles per second. But if we reflect we shall see that this constant speed is computed with respect to the earth's surface, considered as defining a Galilean frame of reference. Were the earth suddenly to become invisible, we should have no physical means of determining the exact speed of the train through space. But suppose, now, that the train were slowing down, speeding up or rounding a curve. In this case, even if the earth were invisible, the train would still possess at each instant a definite acceleration and this acceleration would still be perceptible, and would be measured by the forces of inertia pulling and pushing the passengers about in the train. We see, then, that acceleration has an existence *per se* and that, in contrast to velocity, it is determinate even in the absence of a frame of reference. Now, it follows from the classical principle of relativity that, absolute velocity being meaningless (at least so far as mechanics is concerned), the only type of uniform translationary motion which can be detected by mechanical means reduces to relative uniform translationary motion of matter with respect to matter; that is, motion with respect to something observable. The magnitude of this relative motion depends on the Galilean frame we may select as frame of reference.

Relative uniform translationary motion can, of course, be detected visually, but it can be detected also in a number of other ways. Thus the relative motion of a razor blade on a strop causes heat due to friction. The relative displacement of a knife between the two poles of an electromagnet produces induced currents which manifest themselves by rendering it difficult to cut through the lines of force of the field. The relative motion of an incandescent atom produces a shift in the spectral lines (Doppler effect), and it is thanks to this shift that astronomers are able to determine the radial velocities of the stars with respect to the earth. In other words, though absolute velocity can never be detected by mechanical experiments and is therefore meaningless in mechanics, yet relative velocity of matter with respect to matter *can* be detected in a number of ways; so that this relative type of velocity is the only one that has any physical significance, or at least any mechanical significance.

We have now to consider a number of difficulties that arise from Newton's conception of absolute space. In the first place, although Newton had postulated absolute space and motion, the fact remained that of all the absolute motions, absolute acceleration and rotation were the only ones that had ever been detected, at least by mechanical means. It seemed strange, if absolute velocity were to have a meaning, as indeed it must if space were absolute, that it should be so obstinate in refusing to reveal itself. We could not hold that our inability to discover which one of the Galilean frames was truly at rest was due to the crudeness of our mechanical experiments; for the entire structure of Newtonian mechanics would have come crashing down had any such suggestion been proved correct.

The reason for this statement is easily understood. The fundamental law of Newtonian mechanics states that the force acting on a body is equal to the mass of the body (an invariant) multiplied by the acceleration. Now, the acceleration of a body as computed in any Galilean frame is always the same; for the various Galilean frames differ only in their velocity through absolute space, and a velocity added to or subtracted from an acceleration can never modify an acceleration. The mass of the body being an invariant according to classical mechanics, the force must also measure out the same in all Galilean frames. In other words, it was essential to classical science that the fundamental law of mechanics should remain the same in all Galilean frames, and similar conclusions would have to apply to all the mechanical laws. But then no mechanical experiment, however precise, executed in one Galilean frame or another, could ever yield varying results; so that absolute velocity could never be known. In short, if Newtonian science were to stand, the failure of mechanical experiments to detect the absolute velocity of our Galilean frame through space could never be attributed to their lack of refinement, since any positive results would have overthrown the entire structure of Newtonian dynamics.

Here, then, was an absolute velocity which was a reality since space was absolute, but which on no account could ever manifest itself!

It was, of course, possible to suggest that space had a dual structure, partly absolute and partly relative; absolute for rotation and acceleration, but relative for velocity. But this intrinsic duality in the nature of space and motion was hard to accept; it seemed as though space and motion should be one thing or the other: either entirely absolute or else entirely relative. Moreover, it was extremely difficult to conceive of a duality in the structure of space, so the entire situation was most mysterious.

There was, however, a possible way out of the difficulty, and that was to assert that absolute velocity was real, truly enough, and that it might perfectly well be detected provided we appealed to other types of physical experiment—to electrodynamic and optical experiments, for instance, and not solely to mechanical ones. But we shall see that precise experiment has since proved that these expectations were also doomed to disappointment; hence it appeared perfectly clear that no manner of experiment, mechanical, optical or electrodynamic, would ever reveal this elusive absolute velocity.

Even had optical and electromagnetic experiments succeeded in detecting absolute velocity, everything would not have been plain sailing. It would then always have been possible to assert that the absolute velocity detected in this way was not absolute velocity through space, but velocity through the ether floating in space. But we may point out that this last objection carries very little weight. For reasons which we shall explain in the next chapter, it does not appear legitimate to dissociate the ether from space. Had optical experiments detected velocity through the ether, we should have had to assume that they had also disclosed the long-sought-for absolute velocity through space; and the Newtonian belief in absolute space would have been considerably strengthened.

However, it is unnecessary to dwell on what might have happened, seeing that optical and electromagnetic experiments have failed to reveal the slightest trace of absolute velocity. Accordingly we are led to **Einstein's special principle of relativity,** which states that not only mechanically, but in every way, absolute velocity through space must escape empirical detection. Under the circumstances, unless we wish to retain in science an absolute velocity which can never be detected and which is therefore any one's guess, there is no alternative but to state that space together with its ether has a dual structure: relative for velocity, absolute for acceleration. Einstein appears to have thrown us back on the extremely difficult task of conceiving of this dual nature for space. But the solution of this particular problem was forthcoming as soon as Minkowski discovered space-time. With four-dimensional space-time the reason for the duality becomes apparent; for the absoluteness of space-time (as will be explained in a later chapter) accounts for the relativity of velocity and the absoluteness of acceleration.

Thus in Einstein's theory, in its original form at least, velocity remains relative and acceleration remains absolute, as in classical science,* the only difference being that the relativity of velocity is now more thorough, applying as it does to all manner of experiments and not merely to mechanical ones. It is only when we consider the more speculative part of the theory, that pertaining to the form of the universe, that the complete relativisation of all motion in the Machian sense occurs.

* Subject to the restrictions mentioned in a note in the previous chapter.

Space-time is then found to be no longer an absolute existing *per se* but to be conditioned entirely by the matter of the universe. But, as we have already had occasion to mention, this part of the theory is still highly speculative, and in order to avoid unnecessary confusion, we would strongly advise the reader to eliminate it entirely from his mind until such time as he has acquired a proper understanding of the special and of the first part of the general theory. Accordingly we will ignore it for the present; and we may state that in Einstein's theory motion possesses a dual nature, just as was the case in classical science.

CHAPTER XI

THE ETHER

CLASSICAL science assumed that those manifestations we called electricity, magnetism, and light were nothing but strains, compressions and wavelike motions in an imponderable medium, the stagnant ether, floating in space. This belief was not altogether blind guesswork. In view of the stresses and strains that clearly appeared to surround electrified bodies, electric currents and magnets, and in view of the well-established wave nature of light propagation, some medium had to be postulated in order to give these stresses, strains and waves physical significance. Of this mysterious ether itself, practically nothing was known; but from what scant information could be gathered, it appeared to possess strange and contradictory properties.

Its resistance to motion was practically nil—far less than that of a gas like the atmosphere; since the frictional effects due to the atmosphere are sufficient to cause shooting stars to become incandescent, while no such effect was observed in those regions of ether-filled space extending beyond the limits of the atmosphere. Besides, if motion through the ether generated friction, mechanical phenomena should be influenced by the velocity of our Galilean frame through space and hence through the ether; but so far as experiment was able to decide, no such influences had ever been detected. (This was indeed the essence of the Newtonian principle of relativity.) It appeared, therefore, that the ether must be assimilable to some ultra-rarefied gas. On the other hand, the transverse vibrations of light waves were compatible only with the existence of a rigid medium; but between the properties of rigidity and of extreme rarefaction there was an obvious contradiction. Yet, however baffling and artificial the ether appeared to be, it was extremely hard to banish it from science owing to the great mass of experimental evidence that appeared to lend support to its existence.

At any rate, once the existence of a semi-material ether was accepted, it was natural to suppose that the motion of our optical and electrical apparatus through its substance should exert a perceptible influence on the results of experiment. Just as the velocity of the wind adds itself to the velocity of sound waves propagated through the atmosphere, so was it logical to anticipate, in view of our understanding of light waves and of electrical phenomena, that an ether wind should carry the light waves and ether strains along with it. Optical and electrical experiments conducted with sufficient refinement should therefore be susceptible of detecting the velocity of our planet through the ether; so that in con-

adistinction to the relativity of Galilean motion for mechanical
1enomena with which the ether was not concerned, Galilean motion
ould be anything but relative were electromagnetic or optical experi-
ients to be attempted.

Now this duality which was thus assumed to exist in the nature of
ialilean motion or velocity, making it relative for mechanical phenomena
nd absolute for electrodynamic ones, was perfectly acceptable, provided
: was permissible to draw a sharp distinction between mechanics and
lectrodynamics, between space and the ether. But the entire trend of
ecent research had been to show that this distinction was unjustified.
n the first place, matter seemed to be constituted by electrons and
rotons, hence to be reducible to phenomena of an electromagnetic nature.
'hen again, electromagnetic phenomena were known to develop mechan-
:al forces. Where, then, was the essential difference between mechanical
nd electromagnetic phenomena? Under the circumstances how could
7e account for this anticipated duality in the palpable effects caused by
. velocity through the ether, relative or physically meaningless for
nechanical experiments, and capable of producing physical effects in
lectromagnetic ones?

It appeared as though mechanical and electromagnetic phenomena
hould be susceptible of being placed on the same footing. We should
hen have to assume that mechanical phenomena together with electro-
nagnetic ones would be affected by velocity through the ether, hence
:hrough space, or else to recognise that both electrodynamic and mechanical phenomena could never be influenced by this type of motion. If we
should accept the first hypothesis, the Newtonian or mechanical principle
of relativity would be overthrown, and with it the entire structure of
Newtonian mechanics; and if we should accept the second hypothesis,
our understanding of electromagnetic processes and of the ether would
have to be modified profoundly. In either case, whether or not electromagnetic and optical experiments succeeded in detecting absolute Galilean
motion, science seemed to be faced with a difficult situation.

Classical physicists held to the view that optical and electromagnetic
experiments would surely reveal absolute Galilean motion, and they were
not perturbed by the fact that as a result Newtonian mechanics might
have to be abandoned. Indeed, doubts as to the validity of Newtonian
mechanics already presented themselves when Kaufmann and Bucherer
showed that electrons in very rapid motion appeared to increase in mass
—a phenomenon which was of course incompatible with the Newtonian
belief in the invariance of mass, unless some hypothesis *ad hoc* was to be
invoked. It is true that Michelson's very refined test attempted in 1887
failed to detect any influence of absolute Galilean motion in a certain
optical experiment; but it was always assumed that an explanation for
this failure would be forthcoming sooner or later, and that other experiments would be successful where Michelson's had failed. When, therefore, Einstein suggested as a principle of nature that, regardless of
whether or not the ether was a necessary hypothesis, Galilean motion

through the ether or through space would never be detected by any type of experiment, whether optical or electromagnetic or mechanical, the majority of physicists paid little heed to the new doctrine.

It must not be thought that this hostility of classical physicists to the Einstein theory in its initial stages was due solely to their belief in the absolute and objective nature which it appeared inevitable to credit to the ether. It was due also to other reasons which we shall now proceed to investigate.

From a mathematical point of view, a physical phenomenon reduces to the systematic variation in intensity and in orientation from place to place and from time to time of a number of magnitudes. A physical phenomenon as referred to some arbitrary frame of reference can therefore be expressed by mathematical equations in which the space and time co-ordinates with respect to the frame figure as variables. Now a fundamental problem of theoretical physics is to determine how the equation of a phenomenon referred to a frame B, which is in motion with respect to a frame A, can be obtained from its equation referred to the frame A, by purely mathematical means without any further appeal to experiment. The solution of this problem would enable us to anticipate the change in appearance of a phenomenon when observed, first from the frame A, then from the frame B. Obviously some kind of mathematical operation would have to be performed on the equation as referred to A. This operation is called a **mathematical transformation;** and the equation operated upon is said to have been transformed. In the type of problem we are considering, where one frame of reference is replaced by another, the transformation will bear on the space and time variables or co-ordinates present in the equation; and transformations of this character are termed **space and time transformations.** The entire problem reduces, therefore, to the discovery of those space and time transformations which hold in this world of ours. These transformations cannot be guessed at *a priori;* and a study of the behaviour of physical phenomena is a prerequisite condition for their determination. Inasmuch as all the experiments known to classical science appeared to corroborate the impression that a distance in space and a duration in time were absolutes, remaining unaltered when we changed our frame of reference, a certain definite species of space and time transformations followed as a matter of course.

Consider, for example, a train moving with constant speed along an embankment. If an event occurs in the moving train at any instant of time, say at two o'clock, this event will obviously have occurred at precisely the same instant (two o'clock) when referred to the embankment, since time is assumed to be absolute, the same for all. In other words, time and duration for the observer in the moving train are identical in all respects with time and duration for the observer on the embankment. This fact is expressed by $t' = t$, where t' is the time for the train, and t the time for the embankment.

Now suppose that a ball is rolling along the moving train with a speed u, and suppose that owing to the train's motion the train observer is carried past the embankment observer at the instant of time when the clocks mark noon. If the ball leaves the train observer at this precise instant of time, its distance from the train observer at a time t' (*i.e.*, t' seconds after the noon instant) will be $x' = ut' = ut$ (since t' and t are identical). But then its distance from the embankment observer will be the sum of the distance between the two observers, and of this distance ut we have just expressed (distances being absolute). It will be, therefore, $x = x' + vt'$ (where v is the velocity of the train along the embankment).* This last deduction is obvious provided we assume that a distance is an absolute concept which would turn out the same whether measured by one observer or another, and such was precisely the assumption that classical science always made. In short, knowing the point x' where an event takes place at a given instant as referred to the train we can deduce immediately the point where the same event takes place when referred to the embankment. All we have to do is to replace x' by $x' + vt'$, and the trick is done.

These space and time transformations permitting us to pass from one Galilean frame to another were the celebrated **Galilean transformations** of classical science, and they were, as we have seen, the necessary consequences of our belief in absolute duration and distance. They were given by:

$$x = x' + vt'; \quad t = t';$$

or again by

$$x' = x - vt; \quad t' = t.$$

Applying them to particular problems, as, *e.g.*, to the change in the shape of the trajectory of a falling body when viewed successively from a moving train and from the embankment, or to the change in the colour of a monochromatic source of light, or in the pitch of a musical note, the anticipations of the transformations were always found to be verified, to the order of precision of our experiments. Corroborations of this sort, among others, were considered to prove the correctness of the transformations, hence also of our traditional belief in the absoluteness of duration and distance.

Now if, instead of restricting ourselves to any particular phenomenon, we consider all phenomena of the same type, such as all mechanical or all electrodynamic phenomena, we know that these phenomena in their aggregate satisfy the requirements of certain general laws or equations: the laws of mechanics or the laws of electrodynamics, as the case may be. To say, therefore, that Galilean motion or velocity is relative so far as a certain type of phenomenon is concerned means that the general

---

* This formula $x = x' + vt'$ or $x' + vt$ can also be written: $x = (u + v)t$, since $x' = ut$. We may conclude that the velocity of the ball with respect to the embankment is equal to its velocity in the train plus the velocity of the train with respect to the embankment. This expresses the classical belief that co-directional velocities add up like numbers.

laws governing such types of phenomena remain unchanged in form when we pass from one Galilean frame to another. Expressed mathematically, this means that when the space and time transformations are applied to the general laws, these laws or equations are transformed into equations possessing exactly the same mathematical appearance; they are then said to have been transformed into themselves.

When the transformations were applied to the general laws of mechanics, these laws suffered no change in form; and this fact constituted the mathematical expression of the Newtonian or mechanical principle of the relativity of Galilean motion. But when the transformations were applied to the equations or laws of electrodynamics, a distinct change in form was noted; the laws or equations lost their simple appearance. This suggested that the electrodynamical laws as accepted by classical science held only in their simple form for a privileged Galilean frame, which for reasons of symmetry was naturally assumed to be one at rest in the stagnant ether. But then, the laws changing in form according to the Galilean frame to which they were referred, it followed as a necessary consequence that electromagnetic experiments should pursue a different course and yield different results according to the frame in which the observer and his instruments were placed. It should then be possible to discover the magnitude of the velocity of the frame through the ether by means of electromagnetic or optical experiments, so that once again, independently of any particular ether hypothesis, mathematical reasoning as well as commonplace reasoning confirmed the physical reality of velocity through the ether.

Of course these mathematical anticipations were dependent on a twofold hypothesis: first, that the classical transformation-formulæ based on absolute distance and duration were accurate; and, secondly, that the equations or laws of electrodynamics were correct. To question the first assumption appeared unjustified, and to question the validity of the mathematical expression of the laws led to conflict with experiment. True, we could modify the laws of electrodynamics (by suppressing those mathematical terms which were the cause of their variability), rendering them thereby invariant under the Galilean transformations, but these terms turned out to be those responsible for the phenomenon of electromagnetic induction. To suppress them would have been equivalent to denying the existence of induction, hence of wireless, radio, dynamos and light propagation. Obviously, a solution of this kind could not be accepted, since dynamos, etc., were known to exist.

In view of all these facts, not the slightest doubt was entertained by scientists that experiments sufficiently precise in nature would yield us the velocity of our instruments, hence of our planet, through the stagnant ether.

We must now pass on to a rapid survey of those so-called negative experiments which were primarily responsible for all the trouble. Let us recall once more the point at issue. We wish to ascertain to what extent the velocity of our Galilean frame through the ether is capable of modifying the results of our electromagnetic and optical experiments.

First in order among these tests we must mention the well-known phenomenon of astronomical aberration, according to which the direction of incidence of a ray of starlight varies annually as the earth circumnavigates the sun. The obvious analogy of this phenomenon is the slanting direction followed on the window pane of a moving train by raindrops which to an observer stationed on the embankment would appear to be falling vertically. This analogy might tempt us to assume that inasmuch as the slant of the raindrops on the pane permits us to deduce the speed of our train through the stagnant atmosphere, so now the slant of the ray of starlight resulting from the earth's motion should yield us our velocity through the stagnant ether. But the conclusion would be faulty; for if we reflect we shall see that we were able to deduce our motion through the atmosphere, solely because we had reason to assume that had our train stood motionless in the atmosphere the raindrops would have appeared to fall vertically. But in the case of astronomical aberration we have no means of basing our deductions on any such preliminary information. We cannot tell from what directions the rays of starlight would appear to issue were we at rest in the ether; for we ignore the true positions of the stars. And so all that astronomical aberration can yield us is the annual *variation* in our velocity through the ether, hence our velocity relative to the sun,* In short, astronomical aberration does not enable us to detect our absolute speed through the ether, and we must rely on other means.

In all the experiments we shall have occasion to mention, the velocity through the ether which we shall seek to determine will be that of our earth; for, attached as we are to the earth's surface, this will be the speed we shall best be in a position to determine. Unfortunately, the earth does not constitute a truly Galilean frame on account of its acceleration or rotation round its axis. However, this acceleration of the points on the earth's surface is very small and the centrifugal force to which it gives rise is scarcely perceptible. Calculation shows that, so far as optical and electromagnetic experiments are concerned, we may regard a frame attached to the earth as constituting a Galilean frame, possessing velocity but no acceleration, at least as a first approximation. Nevertheless this statement is not altogether correct. Experiments can be devised both in mechanics and in optics which would reveal the earth's acceleration or rotation. Of course this is only natural, since *acceleration,* in contradistinction to *velocity* through space, is known to be absolute.

A proper understanding of Einstein's theory requires that this point be fully understood, so we must be excused for repeating once more that all we are concerned with at present is to reach a decision on the

---

* In the classical theory, however, aberrational observations of very high refinement should reveal our speed through the ether, but observations of this sort are beyond our present powers; and Einstein's theory has since proved that however precise our experiments, this velocity could never be revealed.

relativity or the absoluteness of velocity or Galilean motion through the ether. In the case of the earth there is no danger of our confusing any possible effects of acceleration with those due to velocity because, owing to the comparative slowness of the earth's rotationary motion, any effects that might arise from acceleration would in any case be far feebler than those arising from velocity through the ether.

Summarising, we may say that a veritable ether hurricane must be blowing over the earth, varying both in intensity and in direction at the different points of the earth's surface and also at the various times of the day and year. It is this ether hurricane or ether drift that we must proceed to detect. Now all experiments, whether optical or electrical, obstinately refused to detect the slightest trace of this ether drift; so that one of two things was obvious. Either there was no drift; or else, if drift there was, it was completely concealed by compensating phenomena of which as yet nothing was known. Stokes adopted the first alternative and suggested that the earth in its motion round the sun might possibly drag the ether along with it, in its immediate vicinity at least. It would follow that there could never be any drift on the earth's surface; and all the negative experiments which had failed to detect the drift would be explained. The obvious objection to Stokes' hypothesis was the existence of astronomical aberration, which seemed to imply that the earth was moving through a stagnant ether. Stokes, however, proved mathematically that aberration was no argument against his hypothesis; that this phenomenon would still be in order, provided the earth's motion did not set up whirlpools in the convected ether.

In addition to a number of other difficulties to be explained shortly, we may mention that for the motion of the ether to be such as to satisfy Stokes' requirements, hence the observed phenomenon of astronomical aberration, calculation proved that the ether could not be incompressible. In fact, Planck showed that on the earth's surface the ether would have to be 60,000 times denser, and thus correspondingly more compressed than in distant regions; and it would seem strange indeed that this considerable increase in the density of the ether should fail to manifest itself when we compared the results of optical experiments performed on the summit of a high mountain and at sea level. Moreover, light vibrations being transversal, the hypothesis of a rigid ether appeared to be inevitable, so long as we wished to conceive of light as due to vibrations in the ether. Here again the Stokes-Planck theory was difficult to accept.

Lodge endeavoured to subject Stokes' hypothesis to a direct experimental test. He argued that if the earth were capable of dragging the ether along with it in its motion, the same should be true of other moving bodies. But when he investigated this anticipated drag effect in the case of rapidly revolving disks of steel, he found it entirely absent.

Even if we ignore these contradictions and assume that the terrestrial atmosphere is able to drag the ether along bodily, a new series of difficulties awaits us. A total drag of this sort would be incompatible with a general phenomenon predicted by Fresnel and verified by Fizeau according to which dielectrics in motion (electrical non-conductors), such as water, glass and air, drag the ether along only in a partial way; the percentage of drag increasing with the refractivity of the dielectric. Inasmuch as the refractivity of the atmosphere is comparatively feeble it would be impossible to credit the total drag of the ether to the atmosphere.

A quite recent experiment of Michelson (not to be confused with the celebrated experiment of Michelson and Morley) also renders Stokes' hypothesis extremely unlikely. The essence of this particular experiment is to send two light rays round the earth in opposite directions. Its result was to prove that the ray travelling eastward required a longer time to complete its circuit than the one travelling westward. Such would not be the case were the earth to drag the ether around with it in its rotation on its axis. True, this experiment did not prove that the ether was not dragged along by the earth's motion along its orbit, but, on the other hand, it did prove that the earth's rotationary motion produced no such drag. This, of course, rendered extremely improbable the hypothesis that the earth's motion along its orbit should be any more successful in producing an ether drag. The aforementioned experiment of Michelson is an apt illustration of the type of experiment which enables us to detect by optical means the earth's rotation or acceleration through the ether, and is therefore irrelevant to the problem we are here investigating, namely, that of its velocity through the ether. The only reason we mentioned the experiment was because of its bearing on Stokes' hypothesis.

Stokes' hypothesis was thus discarded, and the most satisfactory explanation of all negative experiments that had been suggested was given by Fresnel. It must be realised that up to this stage all the experiments attempted were of an optical nature, and related to the transmission of light rays through dielectrics (glass, water, etc.). Such experiments had been performed by Foucault, Airy, and Arago. Fresnel succeeded in proving by mathematical analysis that if dielectrics carried the ether along in their interior in a partial manner, increasing in a definite degree with their refractive power, all the negative results attained up to his day would be explained, provided they were not pursued to too high an order of accuracy. This required percentage of drag of the ether by dielectrics was called **Fresnel's convection coefficient.**

Fresnel's hypothesis was subjected to a direct experimental test. Fizeau discovered, exactly as had been anticipated by Fresnel, that when a ray of light was propagated through running water, the total speed of the light ray with respect to the tube was equal to its speed through the water, when the water was stagnant, plus *only a fraction* of the water's velocity in the tube; and that this fraction was precisely equal to the convection coefficient predicted by Fresnel.

In view of this remarkable confirmation, Fresnel's hypothesis was considered fairly established, so that the situation confronting science was the following: The ether was stagnant in that the earth moved through it without disturbing it in any way. Velocity through the ether had therefore a definite physical significance, and a real ether hurricane must be blowing over the earth's surface in various directions according to the time of day. Experiments had failed to detect it, not because it did not exist, but on account of this compensating effect of the partial drag of the ether by dielectrics in motion. Such a failure, however, could only be regarded as temporary; for Fresnel's hypothesis specifically showed that this compensating effect was but partial, and that experiments of a still more refined order ought certainly to detect the ether drift on the earth's surface, revealing thereby the earth's absolute velocity.

Obviously Fresnel's hypothesis would fall if more refined experiments should also fail to detect the ether hurricane. Furthermore, even to the order of precision contemplated by Fresnel, if in our experiments we should appeal to phenomena in which dielectrics played no part, his hypothesis of an ether drag would obviously cease to apply. Now the experiment of Michelson and Morley was of this type. Not only was it far more precise than the ones hitherto attempted, but in addition refraction played no part in it. Yet, in spite of all, the Michelson and Morely experiment again gave a negative result.

The next great advance in our understanding of the problem was due to Lorentz. He, however, like all his predecessors, dared not take the great step, but still endeavoured to explain the negative results of experiment while holding to the view of a real stagnant ether. As might have been expected, Lorentz was compelled to appeal to new compensating influences. But before going farther it appears indispensable to devote a few pages to the general subject of electrodynamics; for as we proceed, we shall see that theoretical considerations are destined to play a part of ever-increasing importance.

CHAPTER XII

THE EQUATIONS OF ELECTROMAGNETICS AND LORENTZ'S THEORY

WHEN amber is rubbed, it develops the peculiar property of attracting small bodies, such as bits of paper, particles of dust and the like. This phenomenon, which was known to the Greeks, was described by saying that the amber had become electrified (*electron* meaning amber in Greek). To-day, however, we should say that it had received an electric charge. Soon it was found that electrified bodies did not always attract one another, but that in many instances they appeared to exert a repulsive action. For this reason it was assumed that there existed two different species of electricity, the positive and the negative; and the laws governing the phenomena involved were compressed into the statement that like charges repelled whereas opposite charges attracted. Similar conditions were found to endure between magnetic poles, so that the existence of two different types of magnetism was also assumed. But magnetism and electricity remained entirely distinct; no reciprocal action appeared to exist between an electric charge and a magnetic pole.

We may illustrate these phenomena of attraction and repulsion in a more concrete way by assuming that invisible fields of electric and magnetic forces surround electrified bodies and magnetic poles. To these fields we may ascribe the mechanical actions of attraction and repulsion detected by experiment; the electric fields act on electrified bodies, and the magnetic ones on magnetic poles. It was not until the early years of the nineteenth century that Coulomb submitted the mutual attractions and repulsions of charged bodies to a quantitative test. He found them to be expressed by a law (Coulomb's law) which was of the inverse-square variety; the same as Newton's law of gravitation, except that the electric charges took the place of masses, and that the actions could be either attractive or repulsive.

But with the invention of the electric cell by Volta a new manifestation of electricity presented itself for study and experiment, in the form of the electric current. Oersted discovered that an electric current was able to deflect a magnetized needle placed in its vicinity. This was of great significance, as showing that electricity and magnetism were closely allied phenomena. The nature of their relationship became better understood when, by placing iron filings round a wire, it was shown that an electric current was surrounded by a magnetic field. The deviation of the magnet in Oersted's experiment was accordingly ascribed to the existence of the magnetic field surrounding the current. In the following year Ampère gave an exact quantitative formulation of the laws involved.

Now the fact that an electric current generated a magnetic field rendered it legitimate to suspect that conversely a varying magnetic field should generate an electric current. If this were the case, a magnet displaced near a closed wire should generate or induce an electric current in the wire. This most important phenomenon, known as *electromagnetic induction,* was discovered by Henry and Faraday, and as a result the connection between magnetism and electricity became still more pronounced. Our present-day dynamos and generators are nothing but machines constructed with a view to utilising this phenomenon of induction to generate electric currents on a commercial basis.

Faraday was the first scientist to realise the enormous importance of the electromagnetic field. He saw in it a reality of a new category differing from matter. It was capable of transmitting effects from place to place, and was not to be likened to a mere mathematical fiction such as the gravitational field was then assumed to be. In his opinion, the phenomena of electricity and magnetism should be approached via the field rather than via the charged bodies and currents. In other words, according to Faraday, when a current was flowing along a wire, the most important aspect of the phenomenon lay not in the current itself but in the fields of electric and magnetic force distributed throughout space in the current's vicinity. It is this elevation of the field to a position of pre-eminence that is often called *the pure physics of the field.* Faraday was not a mathematician and was unable to co-ordinate the phenomena he foresaw in a mathematical way, and derive the full benefit from his ideas. Before dying, however, he entrusted this task to his colleague Maxwell; and one of the most astounding theories of science, eclipsed only in recent years by Einstein's theory of relativity, was the outcome.

In order to appreciate the nature of Maxwell's contributions, let us recall how matters stood in his day. If we confine our attention to regions of space where no electric currents, no charged bodies or magnets are present, so that solely the electric and magnetic fields need be considered, the fundamental law of electromagnetism is Faraday's law of induction. This law states that a variable magnetic field generates an electric field.

Maxwell, however, considered that this law, standing alone, lacked symmetry; so he formulated the hypothesis that conversely a variable electric field should generate a magnetic one, and proceeded to construct his theory with this idea in view.* In the particular case of free space in which only fields but no charges or currents are present, Maxwell's equations of electromagnetics, termed field-equations (since they describe the state of the electromagnetic field), can be written:

$$(1) \ \text{div } \mathbf{E} = 0 \qquad\qquad (3) \ \text{div } \mathbf{H} = 0$$

$$(2) \ \text{curl } \mathbf{E} = -\frac{1}{c}\frac{d\mathbf{H}}{dt} \qquad (4) \ \text{curl } \mathbf{H} = \frac{1}{c}\frac{d\mathbf{E}}{dt}$$

---

* Although no experimental results could be claimed to have justified any such assumption, Maxwell introduced it unhesitatingly into his theory. His celebrated equations of electromagnetics represented, therefore, the results of experiment, supplemented by this additional hypothetical assumption. The advisability of making this hypothesis was accentuated when it was found to ensure the law of the conservation of electricity.

where **E** represents electric field intensity, **H** magnetic field intensity, t denotes time, and c a most important constant about which we shall have much to say later. As for "div" and "curl," they represent mathematical operations of a peculiar sort. At any rate, the reader need not be alarmed at these equations, for it is not necessary that we understand them. Suffice it to say that (2) represents Faraday's law of induction, while (4) constitutes the hypothetical law postulated by Maxwell in order to ensure symmetry. Prior to Maxwell's investigations the fourth equation would have been written "curl **H** = 0"; hence it is the adjunction of the term $\frac{1}{c}\frac{dE}{dt}$ which constitutes the originality of the theory.

From his field equations, Maxwell by purely mathematical means was able to deduce the two additional ones:

$$(5) \quad \triangle\mathbf{E} = \frac{1}{c^2}\frac{d^2\mathbf{E}}{dt^2} \qquad\qquad (6) \quad \triangle\mathbf{H} = \frac{1}{c^2}\frac{d^2\mathbf{H}}{dt^2}$$

where $\triangle$ represents a particular species of mathematical operation. In ordinary language, these last two equations connote that varying electric and magnetic intensities will be propagated through the ether in wave form with a velocity c, where c is a magnitude yet to be determined. This discovery removed all possibility of action at a distance, since the field perturbations now appeared to be propagated from place to place with a finite velocity. It was, of course, of interest to determine the precise value of c. The most obvious means would have been to create electromagnetic disturbances and then measure their speed of propagation directly. Physicists, however, were unable, in Maxwell's day, to devise a means of performing such delicate experiments. But there existed another way of determining the value of c. Maxwell remarked that it would be given by the ratio of the magnitude of any electric charge, measured successively in terms of electrostatic units (based on electricity) and then of electromagnetic units (based on magnetism).* Precise measurements conducted on electrified bodies and magnets then proved that the value of this ratio was about 186,000 miles per second; whence it became necessary to assume that periodic perturbations in the strains and stresses of the field would be propagated in the form of waves moving through the ether with this particular speed. But this velocity was precisely that of light waves propagated through the luminiferous ether.

---

* If two magnetic poles of equal strength, situated in empty space at a distance of one centimetre apart, attract or repel each other with a force of one dyne, either pole is said to represent one unit of magnetic pole strength in the *electromagnetic system of units*. Owing to the interconnections between magnetism and electricity, we can deduce therefrom the unit of electric charge also in the electromagnetic system. Likewise, if two electric charges of equal strength, also situated in empty space at a distance of one centimetre apart, attract or repel each other with a force of one dyne, either charge is said to represent one unit of electric charge in the *electrostatic system of units*. From this we may derive the unit of magnetic pole strength in the electrostatic system.

The conclusion was obvious. Unless we were to assume that this extraordinary coincidence in the values of these two characteristic velocities, that of electromagnetic induction and that of light waves, was due to blind chance, there was no other alternative but to recognise that what we commonly called a ray of light was nothing else than a series of oscillations in the electromagnetic field, propagated from point to point. Electromagnetic waves and luminous waves were thus all one. Henceforth the two ethers, the electrical ether and the luminiferous ether, were seen to be the same continuum; electricity, magnetism and optics were merged into one single science.

Now, although experimenters had attempted by various means to submit Maxwell's views to a test, the technical difficulties were so great that no success had been achieved. It appeared clearly from Maxwell's equations that no appreciable effects could be anticipated unless $\frac{d\mathbf{E}}{dt}$ was very great; and this meant that the electric intensity $\mathbf{E}$ would have to vary with extreme rapidity. The simplest means of obtaining a result of this kind would be to produce an oscillating field of electric intensity in which the oscillations were extremely rapid, say, several millions a second. But no mechanical contrivance could yield such rapid vibrations, and apart from mechanical contrivances no other methods suggested themselves. For these reasons a number of years passed before Maxwell's views could be put to the test.

In 1885 Helmholtz directed the attention of his pupil, Hertz, to the problem. Hertz was one of the most remarkable experimenters of the nineteenth century; he succeeded at last in vanquishing the technical difficulties and in generating by purely electrical means an oscillating electric field of extremely high frequency. Electromagnetic waves of sufficient intensity were thus produced; and after having been sidetracked for a time by a secondary phenomenon whose nature was elucidated by Poincaré, Hertz verified the fact that the waves advanced with the speed of light and indeed possessed all the essential properties of light waves other than those of visibility to the human eye. Thus, as a result of Hertz's experiments, the foundations were laid for the commercial use of wireless and radio; but, more important still, Maxwell's electromagnetic theory of light establishing the intimate connection between electricity and optics had at last been vindicated.

Just as Hertz had proved that electromagnetic waves were of the same nature as light waves, so now, conversely, Lebedew succeeded in showing that light waves were of the same nature as electromagnetic ones. It is, indeed, a necessary consequence of Maxwell's theory that electromagnetic waves should exert a definite pressure on bodies upon which they impinge, and this pressure was verified and measured by Lebedew and again by Nichols and Hull in the case of light waves falling on matter.*

---

* It is to this pressure of light that the repulsion of comets' tails away from the sun is due.

Yet it must be realised that marvellous as were Maxwell's equations, so far as the field phenomena in empty space were concerned, they led to conflict with experiment in the presence of matter. The conflict was notably apparent when the phenomenon of dispersion was considered; and by dispersion we mean the splitting up of white light by a prism into a rainbow of colors. Maxwell's equations proved that on entering a transparent dielectric such as glass, the electromagnetic vibrations which constitute light would be slowed down; and this slowing down, as had long been known, would account for refraction. But dispersion demands that the angle of refraction, hence the slowing down, should depend on the frequency, hence on the colour of the light vibrations; and this is what Maxwell's theory failed to account for. His equations showed that the slowing down would be the same for all frequencies. A similar difficulty was encountered when the phenomenon of the magnetic rotation of the plane of polarisation was considered. Faraday had discovered the existence of this phenomenon, and any theory of electromagnetics would have to account for it. Nevertheless, it will be seen that all these difficulties which beset Maxwell's theory arose solely when experiments involving matter were taken into consideration. It appeared, therefore, that the fault resided not in the field equations themselves, but in the too crude assumptions that had been made with respect to the constitution of matter.

Furthermore, Maxwell had considered more especially the case of matter at rest in the ether, and it was necessary to investigate the general case of matter in motion. This extension of Maxwell's theory, known as the electrodynamics of moving bodies, was attempted by Hertz, who assumed that matter in motion through the ether would drag the ether along with it in its interior—a hypothesis not so radical, but somewhat in line with that of Stokes. So long as conductors were contemplated, Hertz's hypothesis seemed to be in accord with experiment; but when we came to consider dielectrics in motion, Hertz's anticipations were contradicted. Fizeau's experiment in particular, proving that the velocity of running water did not compound with the velocity of the light in stagnant water, was incompatible with Hertz's hypothesis of a total drag. Also, in an experiment of a totally different nature, dealing with the motion of a dielectric through a magnetic field, Wilson proved the existence of the same partial drag, given once again by Fresnel's convection coefficient. It appeared, therefore, that Fresnel's convection coefficient was a fact which would have to be fitted into any theoretical scheme of the electrodynamics of moving dielectrics.

Now, Fresnel's partial drag hypothesis, though seemingly confirmed by experiment, had yet to be accounted for theoretically, and this appeared impossible so long as we remained in such utter ignorance of the ultimate constitution of dielectrics and of matter in general. At this stage Lorentz started his theoretical investigations which were to lead him to a refinement of Maxwell's equations, extending their application to cases where matter was present, either at rest or in motion through the ether.

As we have mentioned, Maxwell's equations had proved themselves incapable of accounting for dispersion. It appeared necessary to conceive of some structure for dielectrics which would act selectively, imposing different degrees of retardation on light waves of different frequencies. Lorentz achieved this result by assuming that electricity was atomic and that matter was constituted by more or less complicated groupings of these electric atoms or *electrons*.

In the case of dielectrics or non-conductors such as glass, the electrons were assumed to be tied down to their atoms by elastic forces. When disturbed they would vibrate with some characteristic frequency depending on the magnitude of the elastic forces which held them in place. If, then, a wave of light were transmitted through the transparent dielectric, the electrons would be set into forced vibration and the retarding effect these vibrations would impose on the incident light would depend on the relative frequencies of the incident light and of the natural frequencies of the electrons. In this way it was possible to obtain a mathematical interpretation of dispersion.

Further phenomena were accounted for by taking into consideration the frictional resistances that would interfere with rapid vibrations of the electrons. When these frictional resistances were weak, oscillatory disturbances, such as rays of light, could be propagated through the dielectric, which was then termed transparent (glass). When these frictional forces were considerable, the light ray was unable to set the electrons into vibration; its energy was consumed in the attempt, and as a result it could not proceed; the dielectric was then opaque (ebonite, sulphur).

In the case of conductors such as metals, the electrons were assumed to be very loosely held to their atoms so that the slightest difference of potential would tear them away and cause them to rush in the same direction, thereby producing an electric current. It was precisely because electrons in conductors were not tied down to fixed positions by elastic forces that they were incapable of vibrating; and so conductors were necessarily opaque to electromagnetic vibrations or to light. Conversely, it was because the electrons were all tied down to fixed positions in dielectrics, that they could not rush along in one direction. As a result dielectrics were opaque to currents, and hence were non-conductors. According to these views of Lorentz, an electric current passing through matter was nothing but a rush of electrons. If this were the case, we should expect the motion of a charged body to generate the magnetic field of a current. Rowland had established the existence of this effect many years previously, so that Lorentz's theory conformed to facts in this respect. This same effect is illustrated in a more vivid way by the magnetic field which surrounds cathode rays. These are constituted by streams of electrons travelling with enormous velocities through a partial vacuum; and for this reason the cathode rays have much in common with a current passing through a metal wire.

In perfectly empty space no electrons were assumed to be present; and the propagation of electromagnetic disturbances through space was credited entirely to the oscillations of the field which stood out as a manifestation of energy differing radically from the substantial electrons.

In contradistinction to Hertz, Lorentz assumed that the ether was never disturbed by the motion of matter; and just as in Maxwell's theory, a luminous source in motion through the ether would never communicate its velocity to the light waves it emitted. With the constitution of matter postulated in this way, Lorentz set out to extend Maxwell's theory of electromagnetics in the free ether to cases where matter was present, either at rest or in motion. As already mentioned, he succeeded in giving a theory of dispersion, one of the phenomena which Maxwell had been unable to explain. Furthermore, such highly complicated phenomena as abnormal dispersion, absorption, metallic reflection, selective reflection, body colour and many others were accounted for by the theory with great precision.

The apparent partial drag of the ether verified in Fizeau's experiment and also in Wilson's (Fresnel's convection coefficient) was proved to be a necessary consequence of the theory. It had nothing to do with a partial dragging of the ether, since the ether was in no wise disturbed by the motion of the matter; the apparent drag was accounted for solely in terms of the electronic constitution of the moving matter. In this way Fresnel's convection coefficient received the rational explanation which it had thus far lacked. The most refined experiments confirmed Lorentz's theory in a number of minute details; and when the electron was finally isolated and studied, all doubts appeared to be removed, at least as to the general correctness of the doctrine.

Theories, however, are especially convincing, not so much when they account for what has been observed but has not been explained, as when they allow us to foretell phenomena which no one has anticipated. Triumphs of this sort were soon forthcoming in the case of Lorentz's theory. For instance, the theory required that light waves be produced by electrons vibrating in the interior of the atom. Inasmuch as the presence of a magnetic or an electric field affects the motion of electrified bodies, the vibrations of the electrons in the luminous atom should be modified by the presence of a strong magnetic field. Calculation then showed Lorentz that a magnetic field would cause an atom normally emitting a monochromatic light to emit two or three separate lights, according to the relative inclination of the magnetic field to the line of sight of the observer. This totally unexpected effect was soon verified by Zeeman, one of Lorentz's colleagues, and is now known as the Zeeman effect. It has since been observed in the light emitted by sunspots, proving that very intense magnetic fields must there be present.

It is quite unnecessary to dwell on a number of further verifications of Lorentz's theory. All that we have wished to show is that the empirical discoveries that lend weight to Lorentz's electronic theory are so numerous that it can certainly not be cast aside lightly. Although no achievement in theoretical physics can ever claim to be permanent, yet in view of the wonderful accuracy of Lorentz's previsions it would require some reason of a truly imperative nature for us to feel justified in tampering with it in any essential detail.

Lorentz's electronic theory is also perfectly consistent with the negative experiments we have heretofore mentioned, since it explains the existence of the Fresnel convection coefficient; and we have seen that this coefficient in turn accounts for the negative results of experiments conducted to the first order of approximation in dielectric media moving through the ether. But Lorentz's theory as it now stands suggests that experiments conducted to a higher order of precision, to the second order, for example, would cease to yield negative results; so that with experiments of this sort our motion through the ether would be detectible.*

Now, the Michelson and Morley experiment is precisely of this type. It is a second-order experiment. It is assumed that the reader is sufficiently familiar with it to render its detailed description unnecessary. Stated briefly, the experiment proves that if waves of light leave the centre of a sphere simultaneously, they will return in perfect unison 'o the centre of the sphere after having been reflected against the sphere i inner surface, no matter in what direction through the ether, or with wha constant velocity, the sphere attached to the earth may be moving. A explanation was suggested by both FitzGerald and Lorentz, and was to the effect that the sphere had contracted in a definite degree in the direction of its motion through the ether. The sphere accordingly became an ellipsoid, and Michelson's negative result was explained.

More generally, any body, a yardstick, for example, when lying in the direction of the earth's motion through the ether, would be contracted and become shorter than when placed perpendicularly to the ether hurricane. However, we should never have any means of detecting this contraction, for all things, the human body included, would participate with it, whence observationally, at least, nothing would be changed.

This Lorentz-FitzGerald contraction hypothesis was a hypothesis formulated *ad hoc* for the sole purpose of explaining the null result of Michelson's experiment; and like all such hypotheses, it could be accepted only provided it was justified by a wide variety of experiments. At the time this hypothesis was suggested, scientists were busying themselves with the study of the electron when in motion through the ether. Abraham assumed that an electron moving at high speed through the ether

---

* By a first-order experiment, we mean one that is refined enough to detect magnitudes of the order of $\dfrac{v}{c}$ where v is the velocity of the earth through the stagnant ether and c is the velocity of light. Likewise, by a second-order experiment, we mean one capable of detecting magnitudes of the order of $\left(\dfrac{v}{c}\right)^2$. Inasmuch as v is certainly very much smaller than c, $\dfrac{v}{c}$ is an extremely small quantity, and $\left(\dfrac{v}{c}\right)^2$ is very much smaller still. So we see that a second-order experiment is necessarily very much more precise than a first-order one. We may also mention that no experiments have yet been successful in exceeding those of the second order in precision.

would remain spherical and rigid; and he worked out mathematically the path a rigid electron would pursue when acted upon by electric and magnetic forces. Lorentz, holding to his view of the FitzGerald contraction, attacked the same problem, assuming the electron to contract more and more as its speed through the ether increased. Experimenters were called upon to decide between the hypotheses of the *rigid* and the *contractile* electron. Accordingly, highly refined experiments were conducted on the electrons emitted by radium (β rays). Kaufmann, by photographing the paths followed by the electrons, came to the conclusion that Abraham's rigid electron corresponded to reality. But, still more precise experiments conducted by Bucherer reversed the verdict, and little doubt remained that the electron suffered the FitzGerald contraction in the precise degree demanded by Lorentz.

Inasmuch as matter was assumed to be electronically constituted, these experiments of Bucherer lent weight to the FitzGerald contraction of matter, hence to the Lorentz-FitzGerald explanation of Michelson's null result. This contraction being seemingly established, attempts were made by experimenters to detect it in bodies which were attached to the earth. Could this result be achieved, we should have a means of ascertaining the earth's velocity through the ether; and it was confidently expected that this time success would attend us.

Now, as we have mentioned, if in our frame of reference, on the earth's surface, for instance, all things, including our measuring rods and our own bodies, contracted in exactly the same way owing to the motion of our planet through the ether, it would, of course, be idle for us to expect this contraction to be observable by direct measurement. But although not observable in a direct way, it was argued that this contraction of a body should modify its optical and electrical properties so that a study of the variations in these properties when the body was placed successively in different directions should yield us the earth's motion through the ether. In particular the contraction of a solid transparent body, owing to its motion through the ether, should give rise to a phenomenon of double refraction, much as would follow from an ordinary physical compression of the body. The experiment of Rayleigh and of Brace was based on this idea, but the result was a complete disappointment. The FitzGerald contraction gave rise to no such effect, and a negative result was registered. The experiment of Trouton and Rankine aimed to detect a variation in the electrical conductivity of a strip of metal as a result of the FitzGerald contraction. Once more a negative result was the outcome. These experiments having failed, still another was suggested and carried out by Trouton and Noble. The essence of this experiment was based on the well-known phenomenon verified by Rowland, that an electrified body in motion produces all the effects of an electric current. If, therefore, two electrified bodies were placed side by side, the motion of the earth through the ether should produce two parallel electric currents; and since parallel electric currents attract one another, a very sensitive torsion balance should be able to detect this attraction. Again, the result of this experiment was negative; the velocity of the earth through the ether evaded us completely.

Now, it should be noticed that the last two experiments pertain purely to electricity and electrodynamics; there is nothing of an optical nature about them. Hence, it would appear that all electromagnetic phenomena behave in exactly the same way, regardless of the velocity of the ether hurricane. In other words, they, too, yield negative results. But there is yet another aspect to these experiments.

If we regard the FitzGerald contraction as physically real—and we must do so in order to explain Michelson's negative result—we must recognise that it is certainly compensated in a miraculous way; since no experiments, whether optical or electrical, appear to be capable of revealing it. Furthermore, in the last experiment, that of Trouton and Noble, we are not dealing primarily with the FitzGerald contraction; we are here in the presence of an experiment of electrodynamics, and we are endeavouring to detect a purely mechanical force arising as a result of the motion of our electrified bodies through the ether. The laws of electrodynamics require that such a force be generated, and yet no force can be observed; so we must assume that the mysterious compensations that conceal our motion through the ether are of so perfectly general a nature that they affect mechanics as well as electromagnetics and optics.

True, none of the experiments we had been able to execute exceeded the second order in precision; and there was nothing in Lorentz's theory, even when the FitzGerald contraction was taken into consideration, to suggest that all experiments, whatever their nature and however precise they might be, would fail to detect our velocity through the ether. A very simple hypothetical experiment will make this point clear. Suppose, for instance, that the earth were rushing through the ether with a speed approximating that of light; and suppose that we suddenly stepped in front of a mirror. All that the FitzGerald contraction hypothesis would ensure would be that the time required by the light rays to leave our body and be reflected back to our eyes should be independent of the orientation of the mirror with respect to the earth's motion. But the actual duration which the light rays would require, hence the time it would take for our image to appear to us, would depend essentially on the earth's speed through the ether; so that theoretically, at least, a measure of duration should reveal this mysterious speed.

However, in view of the fact that the highly precise experiments that had been performed were incapable of detecting this velocity through the ether, it appeared legitimate to extrapolate, to assume that something deeper was at stake, and that in all probability this velocity would remain undetected, however far we might extend the precision of our electrodynamic or optical experiments, and however much we might vary their nature. Lorentz accepted what appeared to him as the inevitable, and asserted that the time had come to recognise that nature seemed to have entered into some giant conspiracy to defraud us of a knowledge of our velocity through the ether. Accordingly, he laid down his celebrated

principle of correlation, according to which adjustments were so regulated in nature that the velocity of our planet through the ether could never be detected, *however precise our experiments.*

Lorentz applied his mathematical talents to the discovery of the necessary adjustments which would have to exist in nature for this correlation to be satisfied completely so far as electrodynamics and optics were concerned. In order to understand the significance of his investigations, we must recall what we have said of the laws of nature and of the space and time transformations.

We remember that Newtonian relativity expressed the fact that mechanical phenomena, hence the laws of mechanics regulating these mechanical phenomena, remained unchanged when we passed from one Galilean frame to another. The significance of this statement is as follows:

To begin with, when we wish to fix the position of a body in our frame, it is also necessary to specify the instant of time at which the body occupies this position. We thus have four co-ordinates to take into consideration: three for space, x, y, z; and one for time, t. If, now, we change our frame of reference, taking our stand in some other Galilean system, which is in motion with respect to the first, these co-ordinates will change in value or will be transformed. For instance, if a body remains at rest at the meeting point of the three axes of our first frame, its co-ordinates will always be $x = 0$, $y = 0$, $z = 0$ in this frame for all values of the time t. But computed with respect to our second frame, the new co-ordinates represented by x', y', z' will vary with time, since with respect to this second frame the body is moving—*i.e.*, is adopting successive positions in time.

The classical Galilean transformations had for their object to tell us exactly how these co-ordinates would vary when we passed from one Galilean frame to another. As it was always assumed by classical science that time was absolute, the same for all systems, the only co-ordinates susceptible of variation as we changed frames were the space co-ordinates x, y, z. As for t, it had the same value in all frames, and this was expressed by writing $t' = t$. Now the laws or equations which govern phenomena contain in their expression, among other quantities, these co-ordinates, x, y, z, t. Hence, in the general case it is only to be expected that the laws will change in form when we refer them first to one Galilean frame, then to another; or again, when we replace x, y, z, t by x', y', z', t'. When we say that the laws suffer no change in form or are transformed into themselves we mean that this substitution of the new co-ordinates for the old does not modify the form of the equations or laws.

We can now understand what mathematical condition will have to be satisfied for Lorentz's principle of correlation to hold. The principle asserts that all things are so adjusted in nature that absolute velocity through the ether must ever elude us. In other words, all electrodynamic experiments will yield negative results. But this obviously makes it necessary that the laws of electrodynamics which govern all electrodynamic phenomena, hence all electrodynamic and all optical experiments, should remain unchanged in form when we pass from one Galilean system to another. In mathematical language, the laws of electrodynamics will remain invariant when we submit them to the space and time transformations.

Now we have said that when submitted to the classical space and time transformations, the invariance of the equations of electrodynamics does not hold. It appeared then that, contrary to the negative results of experiment and contrary to Lorentz's principle of correlation, velocity through the ether should be detectible by electromagnetic experiments. Science was thus placed in an exceedingly difficult position; the equations of electrodynamics had been so strongly confirmed by extremely accurate experiment that we could scarcely question their accuracy. There remained but one alternative, namely, to suspect the accuracy of the classical transformations.

This was indeed the course followed by Lorentz. He succeeded in establishing the transformations which would be in harmony with the invariance of the electrodynamic equations, and he found these to differ perceptibly from the classical ones; although reducing to the latter in the case of low velocities. The new transformations constitute the celebrated **Lorentz transformations.***

Translated into ordinary language, these transformations expressed the existence of two separate phenomena: first, the FitzGerald contraction of bodies moving through the ether, a phenomenon with which we are already acquainted; and, secondly, a new phenomenon consisting in the slowing down (with increase of velocity through the ether) of the rate of time-flow as applying to electromagnetic processes. Thus, contrary to the classical transformations, where time was absolute $(t' = t)$, we see that in the Lorentz transformations this absoluteness of time begins to be questioned. For psychological reasons, however, imbued as he was with the spirit of classical science, Lorentz was unable to realise the importance of his discovery; and he never succeeded in ridding himself of his belief in the absoluteness of time. For him, as also for Larmor, who contributed to these discoveries, this new species of variable duration, depending as it did on the motion through the ether of the Galilean frame, was not real time. It was a species of "local time"—a distortion of real time—and it was not assumed that this local time corresponded in any way to the time which the observer in the moving frame would live and sense.

---

* The classical or Galilean transformations were $x' = x - vt$, $y' = y$, $z' = z$, $t' = t$, where v was the velocity of the frame with respect to a frame at rest in the ether, hence with respect to the ether itself; this velocity being directed along the x axis and the two frames being assumed to have coincided at the initial instant $t = 0$. Under the same conditions the Lorentz transformations were $x' = \beta(x - vt)$,

$$y' = y, \ z' = z; \ t' = \beta \left(t - \frac{vx}{c^2} \right) \text{ where } \beta = \frac{1}{\sqrt{1 - \frac{v^2}{c^2}}}; \ c \text{ being the velocity of light. It}$$

is easy to see that when v is small compared to c, the difference between the two types of transformations becomes negligible.

And so Lorentz assumed that these new transformations applied only to purely electromagnetic quantities, and no reference was made to their being applicable to mechanical phenomena as well. Though, as a result of these transformations, the velocity of light proved to have always the same invariant value through all Galilean frames when measured by the observer in the frame, no suspicion was cast on the classical formula for the composition of the velocities of material bodies; and this in spite of the fact that these two circumstances were mutually incompatible.

It appeared then as if Lorentz still considered the classical transformation-formulæ to be the real ones, the new transformations being merely of limited application and referring solely to the electromagnetic quantities. When, therefore, in such experiments as that of Trouton and Noble, mechanical and purely electromagnetic effects were indissolubly connected, a large measure of obscurity was involved. Lorentz, however, succeeded in extricating himself from his difficulties and in accounting for all the negative experiments; but he was compelled to appeal to additional hypotheses, such as a modification in those elastic forces which cause the electrons to vibrate in transparent bodies (Rayleigh and Brace experiments). In a similar way it was possible to explain the negative result of the Trouton and Noble experiment. Nevertheless, we can understand the numerous difficulties into which Lorentz's theory was leading us. We knew very little about the constitution of matter, and here, in Lorentz's theory, we were compelled to account for negative experiments by taking this unknown constitution of matter into consideration.

Taking Lorentz's theory as it stands, one cannot help but recognise that this accumulation of hypotheses postulated *ad hoc* makes it painfully artificial. According to the theory, the ether must be regarded as stagnant and the earth as moving through it with some definite velocity. This velocity must, however, remain eternally unknowable to us. Its elusiveness arises from the slowing down of time and the Fitz-Gerald contraction, in the case of the simplest electromagnetic experiments; in the case of experiments dealing with dielectrics and conductors, we have to appeal to still more complicated adjustments pertaining to the constitution of matter. But although forever indiscernible, our velocity through the ether was a reality; it was there, and had it not been for all these compensating effects, it would have been observed. All the forces of the cosmos appeared to have conspired to deprive us of a knowledge we were so anxious to obtain, namely, the velocity of our planet through the ether.

The existence of the FitzGerald contraction, holding in exactly the same degree for all bodies, was an instance of this marvellous conspiracy of nature. Even if we admitted that all matter were electronically constituted, and that the constituent electrons became flattened as a result of their motion through the ether, it was still impossible to conceive of an identical flattening for all bodies, whether rigid or soft, unless we assumed that an appropriate adjustment had also taken place in the other factors entering into the constitution of matter (elastic forces, etc.).

The whole theory was unsatisfactory in the extreme. If Nature was blind, by what marvellous coincidence had all things been so adjusted as to conceal a velocity through the ether? And if Nature was wise, she had surely other subjects to attend to, more worthy of her consideration, and would scarcely be interested in hampering our feeble attempts to philosophise. In Lorentz's theory, Nature, when we read into her system all these extraordinary adjustments *ad hoc,* is made to appear mischievous; it was exceedingly difficult to reconcile one's self to finding such human traits in the universal plan.

Possibly more space than was needed has been devoted to a discussion of these difficulties of pre-Einsteinian physics, but the purpose of these lengthy explanations has been to show that science was confronted with an exceedingly difficult situation. Before proceeding to a study of Einstein's solution, we must mention a tentative hypothesis put forward by the Swiss physicist Ritz.

Ritz remarked that Michelson's experiment could be explained by assuming that, contrary to the requirements of the electromagnetic theory of light, waves of light were shot out from the luminous source just as bullets are shot out from a gun. With this species of ballistic theory, the speed of the light ray with respect to the observer would be given by the constant speed c of the ray with respect to its source, compounded with the relative speed of the observer and source. Now, there is not the slightest doubt that Ritz's hypothesis affords an explanation of Michelson's negative result, but, on the other hand, it leads us into serious difficulties when we wish to interpret a number of other experiments, Bucherer's, for example. This can be understood when we realize that the Ritz hypothesis compels us to deny the existence of the FitzGerald contraction, since were the contraction to subsist in addition to Ritz's assumption, Michelson's experiment would no longer yield a negative result. But, as we have said, the FitzGerald contraction permits a very simple explanation of the moving electron's increase in mass in Bucherer's experiment. This difficulty would not be insuperable, but still it would constitute an argument adverse to the acceptance of Ritz's idea. Other experiments which it would be extremely difficult to account for are those connected with the Fresnel convection coefficient, *e.g.,* Fizeau's experiment.

Furthermore, with Ritz's conception of the velocity of a light wave through the ether being compounded with the velocity of the source, the frequency of light issuing from a source, moving towards us through the ether, would be increased, whereas the wave length would remain unchanged. Now, it so happens that the angle of refraction of a monochromatic light ray passing through a prism depends on the frequency of the incident light, whereas the dispersion caused by a grating depends on the wave length. It should, therefore, be possible to determine whether frequency and wave length truly behaved like independent variables as demanded by Ritz's hypothesis. Experiments conducted by Stark in 1910 on canal rays disproved Ritz's anticipations. Subsequently other experiments led to the same results.

Besides these objections to Ritz's views, it must be realised that Michelson's experiment was only one of many. Here it cannot be emphasised too strongly that the gravity of the situation arose from the fact that as a result of a large number of different experiments (many of which we have not mentioned), the entire structure of classical electrodynamics seemed to be crumbling on every side. Ritz's hypothesis consists in shoving one individual brick back into place. Lorentz, thanks to his mathematical knowledge, realised that something much deeper was at the root of all the trouble and that more radical means would have to be adopted. Accordingly, he proceeded to tackle the very foundations of science, namely, the space and time transformations which had endured for centuries. But Lorentz only proceeded in a half-hearted way and did not have the courage to push his discoveries to their logical conclusion, contenting himself with patching up rather than reconstructing. Einstein, as we shall see, took the bull by the horns, abandoned the classical structure, and proceeded to build up an entirely new edifice, superbly coherent and free from all artificial support and scaffoldings. Thus it was that even before Ritz's hypothesis was disproved by numerous direct investigations (experiments of Majorana and of Sagnac, and astronomical observations on double stars by de Sitter), scientists had ceased to give it serious consideration.

PART II

THE SPECIAL THEORY OF RELATIVITY

CHAPTER XIII

EINSTEIN'S SPECIAL THEORY OF RELATIVITY

HOWEVER unsatisfactory, for the numerous reasons already discussed, Lorentz's theory might appear, it constituted for some years the only possible explanation of the negative results of all electromagnetic experiments performed on the earth's surface. But in 1905 Einstein published a paper on the "Electrodynamics of Moving Bodies," a document destined to prove one of the most epochal events in the history of human thought. In this paper, all the cobwebs of electrodynamics were swept aside.

The essence of Einstein's position was that the difficulties which had beset the study of electrodynamics (we have mentioned only a small number of them) had arisen from our retention of the stagnant ether as a fundamental frame of reference; but that in view of the anomalies this attitude had brought about, the time had come to submit it to a critical analysis. After all, we knew little or nothing about the ether; why then start by stating it to be an absolute medium floating in space and hampering thereby all future progress? Why state that velocity through the ether must have a physical significance, then under the evidence of the negative experiments proceed to postulate complicated hypotheses in order to explain away the absoluteness we had ourselves introduced? Would it not be simpler to adopt a more cautious attitude, deriving knowledge from experiment rather than trying to reconcile experiment with a series of *a priori* beliefs which, for all we knew, might be totally erroneous?

If with Einstein we follow this line of argument, we must assume that the large number of negative experiments prove conclusively that velocity through the stagnant ether or through space is physically meaningless, not only from a purely mechanical point of view, as was the belief of classical science, but from *every* point of view. If we accept this relativity of the ether and space for Galilean motion as a fundamental fact of nature, holding for all manner of experiments, our past difficulties are seen to be of our own making. We had endeavoured to discover that which was meaningless, and then blamed nature for tricking us and hiding it from us. Instead of saying that velocity through the ether appeared meaningless because it *was* meaningless, we had complicated matters by saying that velocity appeared meaningless but of course was *not* meaningless.

Quite apart from the negative experiments, Einstein lays special stress on another type of phenomenon: Thus, whether we displace a magnet before a closed circuit or the closed circuit before the magnet, the current induced in the wire is exactly the same in either case, so far as experiment can detect. That which appears relevant is the relative motion between magnet and circuit; the respective absolute velocities of magnet and circuit through the ether, which are of course different in both cases seem to be totally irrelevant.

In view of the importance of the question, let us mention another example. We remember that when discussing electrical phenomena in the first pages of the preceding chapter, we mentioned that an electrified body was surrounded by an electric field of force, whereas a current was surrounded by both an electric and a magnetic field. Subsequent experiments performed by Rowland confirmed the view that a charged body in motion generated the same electric and magnetic fields as a current flowing along a wire, so that we had to agree that a charged body in motion developed a magnetic field by the sole virtue of its motion.

Now here again we are in a quandary to understand what is to be meant by motion. Classical science assumed that motion in this case meant motion with respect to the stagnant ether, and that an electric current was constituted by the rushing of electrons through the ether. Numerous difficulties beset this view. Owing to the earth's motion through the ether, every charged body on the earth's surface would constitute an electric current; so that around every electrically charged body a magnetic field should be present. Yet experiment failed to detect this magnetic field. Was it due to the crudeness of our experiments? This solution was scarcely possible; for if we displaced our charged body before a magnetised needle (Rowland's experiment), the needle was deflected, proving that experiments were perfectly able to detect the magnetic field when it was truly produced. It appeared as though the type of velocity that constituted a current was not velocity through the ether at all, but velocity relative to the recording instrument or, more generally, to the observer. Once again, the only type of velocity which appeared to have any significance in nature was relative velocity, and never velocity through the stagnant ether or absolute space.

Hence, Einstein postulated his **special principle of relativity,** according to which Galilean motion through the ether or space is meaningless. This principle is, as we see, merely an extension of the Newtonian mechanical principle to the case of the ether, or in other words to electromagnetic and optical experiments.

What is called the **special theory of relativity** concerns the rational consequences that must follow from the special principle. These consequences may be anticipated as follows: If Galilean motion through the ether is meaningless, the laws of electrodynamics must remain invariant in form when we change our Galilean system of reference. But this implies that the space and time transformations which the co-ordinates of all points undergo, when we change systems, must be given by the Lorentz transformations.

In Einstein's theory, however, the significance of the transformations is entirely different. It will be remembered that in ordinary language these transformations implied the FitzGerald contraction and the slowing down of time. In Lorentz's theory these modifications arose as a result of a real velocity through the ether; and though it was impossible for the observer in the moving frame to detect their presence, yet in spite of all they were physically real. In Einstein's theory, on the other hand, velocity through the ether being meaningless, the modifications arise only as a result of the relative motion of one frame with respect to the other, and are thus due to conditions of observation and not to any intrinsic change. For instance, if in any two Galilean frames originally at relative rest, two identical clocks and two identical cubes are placed, and if then the two frames are set in relative motion, either observer would discover as a result of his measurements that the other man's clock had slowed down and that his cube had become flattened in the direction of the relative motion. Nothing would have happened to the cubes and clocks; only the conditions of observation would have been changed.

Expressed mathematically, this means that the variable v which enters into the Lorentz transformations now stands for the relative velocity of one frame with respect to the other, and no longer for velocity through the ether. In this measure the transformations of Einstein's theory, though identical in appearance with those of Lorentz's, are far different in meaning. To avoid confusion, the appellation **Einstein-Lorentz** or **Lorentz-Einstein transformations** is accordingly often made use of.*

There is still a further difference, a most important one, distinguishing the implications of the transformations in the two theories. In Lorentz's theory there was a privileged observer situated in the frame at rest in the ether. In Einstein's theory there is no such privileged observer; for since a privileged state of rest in the ether is meaningless, all observers in Galilean frames are on the same footing. Hence, Lorentz's distinction between the true time of the frame at rest and the distorted, local or electromagnetic times of the frames in motion through the ether loses its significance. Henceforth, all these so-called local times given by the Lorentz transformations constitute the true times lived, sensed and experienced by the respective observers in their respective Galilean frames.

There being no longer any reason to subordinate one frame to another, the descriptions of *all phenomena* in terms of space and time are on the same footing in *all Galilean frames;* and there is no longer any reason to follow Lorentz in limiting the validity of the transformations to the sole case of electromagnetic phenomena. Henceforth, these transformations appear of universal application, holding for mechanics, for the com-

---

* As a matter of fact the appellation "FitzGerald contraction" should also be abandoned in Einstein's theory, since FitzGerald and Lorentz had always regarded the contraction in the light of a real physical contraction in the stagnant ether. Nevertheless, if the reader realises the difference in the two conceptions of the contraction, it will simplify matters to retain the original name.

position of velocities, and indeed for the entire universe of physical phenomena. This belief is not dependent on the fact that, matter being constituted electronically, all mechanical experiments must be regarded as electrodynamic experiments viewed microscopically. Rather is it due to the fact that these changes in space and time are absolutely general. They arise from the relative motion of the observer and not from the peculiar microscopic constitution of the phenomena observed.

The method of presentation of Einstein's theory we have followed up to this stage is not the one usually adhered to, but it has appeared preferable to proceed as we have done in order to bring into prominence certain important features which are obscured in the customary presentation (these features will be better understood later). We will now fall in with the usual procedure.

As Maxwell proved many years ago, a necessary consequence of the laws of electromagnetics was that light waves *in vacuo* in the stagnant ether should travel with a velocity of 186,000 miles per second. This was the celebrated law of light propagation of classical science. Now we have seen that the laws of electromagnetics (and with them their immediate consequence, the law of light propagation) held only for privileged observers, those situated in any frame at rest in the stagnant ether. As referred to all other Galilean observers situated in frames moving through the ether, an application of the classical transformations showed that the equations of electromagnetics, and with them the velocity of light propagation, would be modified. But since the essence of the relativity of Galilean motion is to deprive any particular Galilean observer of his privileged position, the laws of electromagnetics must now maintain the same form for all Galilean observers. Hence the law of light propagation, which is a mathematical consequence of these electromagnetic laws, must likewise hold in exactly the same way for all Galilean observers. Whence the new result: *"Light waves must travel with the same invariant speed of 186,000 miles through any Galilean frame when this speed is measured by the observer located in the frame."* It is this statement which Einstein has called the **principle** or the **postulate of the invariant velocity of light.**

If we combine this principle with the relativity of the ether for Galilean motion and endeavour to construct the transformations which will be compatible with these two principles, we are again led to the Lorentz-Einstein transformations.* This should not surprise us since, as we have seen, the principle of the invariant velocity of light is but a direct consequence of the invariance of the electrodynamical equations; so that the transformations ensuring invariance will be the same in either case.

---

* It is necessary to take both of these conditions into account. If, for instance, we limited ourselves to the invariance of light's velocity without taking into consideration the relativity of velocity, transformations such as $x' = x - vt;$ $t' = t - \frac{vx}{c^2}$ would also satisfy our requirements, and these would not be the Lorentz-Einstein transformations and would not be in accord with the relativity of Galilean motion.

In short, we see that without appealing directly to the equations of electrodynamics, we can deduce the Lorentz-Einstein transformations merely by taking one of their consequences into consideration, namely, the invariant velocity of light, and combining it with the relativity of Galilean motion. From a mathematical point of view, this procedure is the simpler. Hence, Einstein posits as his fundamental assumptions:

1. The relativity of space or of the ether for Galilean motion, or, more simply, the relativity of Galilean motion without particular reference to the ether or to space.

2. The postulate or principle of the invariant velocity of light.

This procedure followed by Einstein offers a number of advantages. In the first place, as we have said, the mathematical discovery of the Lorentz-Einstein transformations is considerably simplified. Secondly, as the classical law of light propagation, though a consequence of the laws of electromagnetics, is susceptible of being tested by direct experiment without any reference to the laws of electromagnetics, there is no necessity for dragging in these highly complex laws. In fact, even had the laws of electromagnetics been unknown, the relativity of velocity and the classical law of light propagation are all that would have been required to construct the theory. Hence, from the standpoint of mathematical elegance, the procedure is certainly one of extreme simplicity. Furthermore, inasmuch as in Einstein's theory the Lorentz-Einstein transformations are of general application and do not concern only electromagnetic phenomena, it is preferable to avoid conferring on them an exclusively electromagnetic aspect. This aim is achieved by deducing them from a general law of propagation, such as that of light.

On the other hand, this method of presentation, by appealing to the propagation of light, is likely to confuse the beginner, who is apt to assume that Einstein postulated the invariant velocity of light as a hypothesis *ad hoc* for the sole purpose of accounting for Michelson's negative experiment. In this way the entire theory is supposed to hinge on Michelson's experiment, and the critic assumes that could Michelson's experiment be explained in some other way, Einstein's theory would be obviated. This assumption appears all the more natural to the critic as Michelson's experiment is, nine times out of ten, the only negative experiment he is acquainted with. The result is that he assumes Einstein's theory to be nothing but a wild guess grafted on one of those highly delicate experiments where the chances of error are always great. As a matter of fact, by reasoning in this way, the critic loses sight of the entire *raison d'être* of the theory. It is safe to say that even had Michelson's experiment never been performed, Einstein's theory would have been forthcoming just the same (though, of course, had Michelson's experiment given a positive result, enabling us to measure our velocity through the ether, the theory of relativity would have been untenable).

This explains why, in presenting the theory, we started by showing how it arose as a necessary consequence of the irrelevance of absolute velocity in all electromagnetic experiments, hence sprang from the invariance of the equations of electrodynamics, which expresses mathematically the aggregate of all the negative results. If the reader has grasped the significance of these Lorentz-Einstein transformations, we may proceed to examine certain of their particular consequences, and to show how, quite apart from the negative experiments, the theory of relativity has cleared up a number of obscure points in our understanding of electromagnetics.

For instance, we mentioned that a charged body at rest was surrounded by an electric field of force, whereas the same body in motion (or an electric current) was surrounded both by an electric and a magnetic field. Ampère had given a formula describing the distribution and intensity of the electromagnetic field surrounding an electric current of any given intensity. But this formula was purely empirical, and it was felt that we should have been able to anticipate the existence of this electromagnetic field and derive an exact expression of its disposition and magnitude by purely rational methods. These hopes were disappointed; for there appeared to be no rational connection between the field developed by an electrified body when at rest and its field when set in motion.

Contrast our ignorance on this score with our ability to understand the nature of other problems attendant on relative motion. For example, a monochromatic source of light at rest appears yellow, whereas, when moving towards us with a sufficiently high speed, it appears greenish. But there is nothing mysterious about this change in colour; we could have foreseen it, knowing what we do of the wave nature of light. The situation was much more obscure in the case of electromagnetics.

Einstein's theory solves the entire problem. Consider the simple case of a charge in uniform motion. We have said that owing to its motion it develops a magnetic field (just like a current), in addition to its electric field. But in Einstein's theory velocity through the ether is meaningless; hence there is no such thing as a charge in uniform motion in any absolute sense. If we were to rush after the electrified body and accompany it in its motion, it would cease, according to Einstein, to be a moving charge and would become a charge at rest in our frame; but then its erstwhile magnetic field would necessarily have vanished, the electric field alone remaining. We must assume, therefore, that the appearance of the magnetic field arises only because we have changed our Galilean frame of reference by passing from one fixed with respect to the charge to one *in relative motion*.

Now, the Lorentz-Einstein transformations tell us how our space and time measurements change under these circumstances; and by applying the transformations to the equations of electrodynamics we are able to ascertain how an electric field which is purely electric when viewed from one frame should appear to us when viewed from another Galilean frame. Proceeding in this way, we do in fact find that the electric field of the electrified body at rest will appear as an electromagnetic field when the body is set in relative motion. Likewise, the precise mathematical formula for the new field is obtained, proving incidentally that our classical empirical formula was only approximate. Thus a rational justification for the appearance of a magnetic field round an electrified charge in motion is finally obtained, and this field is seen to be a direct consequence of the variations in those fundamental space and time forms of perception—variations which are expressed by the Lorentz-Einstein transformations.

In exactly the same way a magnetic pole in motion develops an electric field around it in addition to its magnetic field when at rest, and it is this phenomenon that gives rise to the electromagnetic induction on which we base our generators and dynamos. Here again the Lorentz-Einstein transformations throw complete light on the subject.*

In a general way the transformations show us that a field which appears to be exclusively electric or exclusively magnetic in one Galilean frame will appear to be electromagnetic in another frame, thereby compelling us to recognise the relativity of electric and magnetic forces. We shall see in due course that this relativity of force is general.

Let us consider another type of phenomenon which is also explained by the transformations. We refer to the Fresnel convection coefficient illustrated by Fizeau's experiment. Experiment proves that the ray of light progressing through the running water is not carried along bodily by the water; it lags behind a little. Lorentz, who, though he had discovered the transformations, did not recognise their general validity, sought to explain this phenomenon by postulating a suitable electronic constitution for the water. But Einstein by merely applying the transformations, without having to appeal to our highly problematic knowledge of the precise electronic constitution of matter, found that they anticipated Fizeau's result with marvellous precision; for according to the transformations, velocities *do not* add up like numbers, as they did in classical science. It follows that the velocity of the light with respect to the tube must necessarily be less than the velocity of the light through the water plus the velocity of the water in the tube.†

Consider again the FitzGerald contraction. Here Lorentz thought it permissible to apply the transformations; but owing to the slight difference in their significance in his theory, he concluded that a body in motion was really contracted owing to its real motion through the ether. Although the observer carried along with the body could not detect the contraction, yet it was physically real and would be observed by the

---

* These results had already been anticipated by Lorentz, since as we have seen, he considered that his transformations were applicable to purely electromagnetic magnitudes. But Lorentz was always dealing with a fixed ether and with motion through the fixed ether; hence the interpretation of these discoveries was much less satisfactory.

† We must remember that it is only light *in vacuo* which is propagated with the speed *c*, hence which possesses an invariant velocity. Light passing through a transparent medium moves with a velocity which is less than *c;* hence in this case its velocity ceases to be invariant and is affected by the velocity of the medium, as also by that of the observer.

observer at rest in the ether. A similar interpretation would have to be placed on the slowing down of phenomena. In Lorentz's theory the difficulty consisted in accounting for an identical contraction manifesting itself in exactly the same way for all bodies, soft or hard. Lorentz again appeals to the electronic and atomic constitution of matter and has to take into consideration elastic forces.

With Einstein the explanation is simple. The contraction is due solely to a modification in our space and time measurements due to relative motion, and is completely irrelevant to the hardness or softness of the body, whose atomic or electronic structure need not be taken into consideration at all. In much the same way an object appears magnified under the microscope, and this magnification is independent of the body's nature.

In short, the modifications are due to variations (as a result of relative motion) in our space and time measurements and perceptions, and in every case they are irrelevant to the microscopic constitution of matter. We see, then, that so far as all these curious modifications are concerned, Einstein's theory does not require any particular knowledge of the microscopic constitution and hidden mechanisms that are assumed to underlie matter. Herein resides one of the principal advantages of Einstein's theory over that of Lorentz; for we know very little about the mysterious nature of electricity and matter, and were all progress to be arrested until such knowledge was forthcoming, we might have to wait many a day without result.

One of the chief difficulties attendant on an atomistic theory such as Lorentz's was due precisely to this interpretation of observable effects in terms of invisible ones about which practically nothing could be known and where hypotheses *ad hoc* had to be invoked at each stage. As for the mathematical difficulties, they, of course, grew in proportion to the complexity of the hidden mechanisms which we had to postulate. Einstein's theory is thus a return to universal principles induced from experiment, and in this respect is analogous to the physics of the general principles which appeared in the course of the nineteenth century. There again such comprehensive universal principles as those of entropy, of least action, of the conservation of energy, mass and momentum, took the place of the hidden mechanisms whereby the atomists had endeavoured to account for observed phenomena.

These periodic swings in the scientific viewpoint have always been necessitated by the continuous advance of knowledge. Both attitudes are fruitful and have yielded important results. In the present case Lorentz had progressed about as far as it was possible to go in his speculations on hidden mechanisms, and a theory such as Einstein's was the necessary antidote to the increasing difficulties which were hampering further developments.

We have now to consider certain important consequences of Einstein's theory which affect our traditional concepts of space and time. The Fitz-

Gerald contraction is no longer a real physical contraction, as it was assumed to be in Lorentz's theory. It no longer arises as a result of a perfectly concrete motion of a body through a stagnant ether; the stagnant ether with respect to which velocity acquired its significance having been banished by Einstein. The FitzGerald contraction is now solely due to the relative motion existing between the observer and the body observed.

If the observer remains attached to the body, there is no contraction; if the observer is moving with respect to the body, or the body moving with respect to him, the FitzGerald contraction appears. If the observer once more changes his velocity relative to the body, the FitzGerald contraction of the body will likewise change in magnitude. Nothing has happened to the body and yet its length has altered. Obviously, physical length is not what we once thought it to be; it can in no wise be immanent in the body, since a body has no determinate length until the relative motion of the observer has been specified. A length is therefore but the expression of a relationship between the observer and the observed, and the two partners of the relationship must be specified before the length can have any meaning.

In the same way the colour of an opal has no meaning. It is red from here, green from there, blue from elsewhere, and yet the opal has not changed. It is our position with respect to the opal which has changed, and the colour of the opal is indeterminate until such time as we have specified our relative position. In other words, length and the colour of the opal both express relations and not immanent characteristics.

Similar considerations apply to the slowing down of time. Duration and time are mere relatives, mere expressions of relationships, and have no absolute significance *per se*. This does not mean that the duration we sense is a myth; for as our consciousness always accompanies our human body wherever we go, we always experience the same flow of time. All that is implied is that this rate of time-flow cannot be credited with any unique significance in nature, and a comparison of time-flows characteristic of various Galilean frames will reveal differences in the rate of flow.

We have mentioned elsewhere a further illustration of relativity, when discussing electric and magnetic forces. We showed how it was that an electric field, as such, was indeterminate; how according to the observer it would present itself as a purely electric or as an electromagnetic field. As all these conditions of observation are on the same footing, there is no sense in distinguishing between apparent fields and real fields, or apparent lengths and real lengths, or apparent durations and real durations. All these concepts, of themselves, are mere phantoms, to which substance can be given only when the conditions of observation are specified.

A certain number of lay philosophers have been confused by this continual reference to the observer in Einstein's theory, and have assumed that all things occurred in the observer's mind, the outside common objective world of science playing no part.*   But this extreme idealistic interpretation cannot be defended.

The word "observer" is a very loose term and does not necessarily mean a living human being. We might replace the observer's eyes by a photographic camera, his computation of time-flow by a clock—in fact, all his senses and measurements by recording instruments of a suitable nature, whose readings any man situated in any frame could check later.   The results would still be the same.

Now it might be feared that, with this vanishing of the absoluteness of such fundamental forms of perception as duration and distance, the entire objective world of science would sink to a shadow; and without a common objective world, science would be impossible.   However, we need have no fear of any such catastrophe; for, as will be explained in later pages, a new common objective world of space-time will take the place of the classical one of space and time.   But even at the present stage, without appealing to space-time, we can see that objectivity is not denied us for the following reasons:

When two different Galilean observers measure the same object, or time the duration of the same phenomenon, their computations will differ according to the rules specified in the Lorentz-Einstein transformations. It follows that the observers may infer that they are discussing the same objects, *not* when their respective measurements agree according to the classical standards of absolute distance and duration, but when they agree with the new standards set by the Lorentz-Einstein transformations. Thus a definite criterion of objectivity is still possible, though its form differs from that of classical science.   So in spite of the indefiniteness of the concepts on which classical science was founded (space, time, force, mass), Einstein's theory does not deprive us of the possibility of conceiving of the existence of a common outside world.

Enough has been said even at this stage to show that Einstein's theory cannot be considered a mere mathematical dream.   The extraordinary difficulties with which classical science was confronted owing to the negative results of the experiments we have mentioned (and to numerous other problems) have disappeared.   With them have vanished the miraculous compensations which Lorentz was compelled to invoke in order to explain these negative results.   The complete relativity of Galilean motion explains all our troubles.   Nature is no longer mischievous, but the ether or space is relative for Galilean motion.   We had failed to recognise this relativity which was staring us in the face; in so doing we ourselves were the creators of our difficulties.   Yet rather than abandon the classical concepts of space and time, physicists in general refused to follow Einstein.   To-day criticisms have subsided (on the part of the great majority of scientists) as experiment after experiment has confirmed Einstein's previsions.   The majority of experiments, however, concerned

---

* The objective world of science has nothing in common with the world of things-in-themselves of the metaphysician.   This metaphysical world, assuming that it has any meaning at all, is irrelevant to science.

the consequences of the theory and not its foundations. The experiments of de Sitter and Majorana have filled this gap by proving that a ray of light always passes the observer with the same invariant speed regardless of the relative velocity of observer and source.

Now one of the most important criticisms that has been directed against Einstein's theory is that it deprives us of the possibility of conceiving of an objective ether. There is not the slightest doubt that Einstein's theory compels us to abandon our conception of the stagnant ether of classical science, with reference to which motion should be measurable. But this is not quite the same thing as holding that the theory banishes the ether entirely. Nevertheless, upon first inspection, the fact that whatever our Galilean motion may be, experiments conducted in our frame will always yield exactly the same results, would seem to relegate the ether to the realm of ghosts, making it a useless hypothesis. If this were the case we could no longer conceive of electromagnetic fields and light rays as expressing states of the ether, but should have to regard these fields as constituting independent realities of some new category, differing from matter but susceptible of existence in space without the support of a carrier, or without being the mere manifestations of its states. However, in the general theory which we shall discuss in the second part of this book we shall find that the ether is reinstated in the guise of the metrical field of space-time. But as this new ether has only its name in common with the stagnant Lorentzian ether, there does not appear to be much advantage in retaining the older appellation.

If, then, we wish to emphasise the great distinction between the classical view and the relativistic view, we must say: "According to classical science, the speed of light waves, and of electromagnetic waves generally, is constant in all directions with respect to the stagnant ether, hence also with respect to that particular observer who happens to be at rest in the stagnant ether. According to the relativistic point of view, the speed of light waves is constant throughout empty space when measured by any Galilean observer; that is, by any observer in any non-accelerated frame" (such frames being recognised by the absence of all centrifugal or inertial pushes and pulls).

For the present we are in no position to predict what would happen if the observer, instead of being posted in a Galilean frame, were situated in a non-Galilean, *i.e.*, accelerated or rotating, frame. All the statements we have made up to this point concern only Galilean observers. Such are the restrictions which the special principle and the special theory of relativity impose upon us. It is most important to understand this fact, as many of the criticisms levelled at Einstein's theory are due to a failure to grasp the point. As the theory now stands, acceleration and rotation remain absolute and are therefore excluded from the special principle of relativity, which refers solely to motions in space that are relative. These are Galilean motions.

Perhaps a definite illustration will make these points clearer. Consider, for example, Michelson's latest experiment (not the celebrated one), or, again, consider Sagnac's experiment. The essence of both these experiments is to show that a ray of light travelling round the earth in the direction of the earth's rotation requires a longer time to return to its starting point than would be the case for a ray travelling in the opposite direction. Obviously the velocity of the light waves with reference to the earth is not the same in all directions, so that we are able to detect the rotation of the earth on which we stand. The critic then infers that Einstein's principle of relativity is upset by experiment. But the critic fails to realise that the motion that has been detected is a rotation, hence an acceleration, and that Einstein's special principle confines itself to denying any significance to absolute velocities, that is, to motions which are not accelerated. Had this not been the case, Einstein's principle would have been untenable since it is a fact of common knowledge that a large number of experiments (Foucault's pendulum, the gyroscope, etc.) are capable of revealing the earth's rotation. *It is absolute velocity and not acceleration that experiment* has ever obstinately refused to reveal.

A very similar criticism consists in stating that inasmuch as the speed of light with reference to the earth's surface is greater from east to west than from west to east, the postulate of the invariant velocity of light is refuted. But once again the same error is involved. The postulate states that the velocity of light is invariant through any *Galilean frame* when measured by the observer in the frame. The postulate ceases to be true in an accelerated frame, and the rotating earth does constitute an accelerated frame. True, in the negative experiments, the earth was treated as a Galilean frame; but it was always stated that this attitude was only approximately correct, and that experiments could be devised to detect the earth's acceleration. However, the effects due to the earth's absolute velocity (which was then assumed to exist) would have been so much more pronounced than those due to acceleration that as a first approximation we could afford to neglect these weaker effects of acceleration.

There is still another point on which we must insist, as it has given rise to a number of criticisms. The Newtonian principle of relativity stated that the velocity of matter through empty space was meaningless or relative. Einstein's special theory of relativity, on the other hand, compels us to consider the velocity of light as an absolute. How, then, can Einstein's special principle be a mere extension to electrodynamics or to the ether of the Newtonian principle?

To answer this question we must realise that although velocity was a relative in Newtonian science, yet there did exist one definite velocity which was assumed to be absolute. This was the infinite velocity. It was assumed that a velocity that was infinite or instantaneous for one observer would remain infinite or instantaneous for all other observers. So far, therefore, as velocity is concerned, the sole difference between Einstein and Newton is that with Einstein the absolute or invariant velocity is no longer infinite. Though very great (186,000 miles a second), it is now finite. It is this difference between the invariant velocities of Newton and Einstein which is responsible for all the major differences between classical and Einsteinian science, as will be explained more fully in a following chapter. In particular, it is this finiteness of the invariant velocity which precludes us from attaching any absolute value to shape and distance.

Viewing the question in another way, we should notice that as a matter of fact the Newtonian principle of relativity merely states that velocity *of matter with respect to matter* alone has physical significance. Velocity of matter through empty space is physically meaningless. It is the same in Einstein's special principle. Einstein's theory proves (see next chapter) that molar matter can never move with the absolute speed of light. We are therefore perfectly justified in saying that the velocity of matter remains essentially relative, since it can never attain that critical velocity (*i.e.*, that of light) which is absolute.

CHAPTER XIV

WE saw in the preceding chapter that the time and space variables which entered into the Lorentz-Einstein transformations should be considered of absolutely general validity. They were assumed to represent the duration actually sensed and lived by the observer and the spatial measurements actually determined by his rigid rods in his Galilean frame. It was no longer a case, as in Lorentz's theory, of supposing that these space and time variables applied only in certain electromagnetic cases and not in others. They were now of absolutely general application, and in particular should hold for all mechanical phenomena.

But this realisation brought us face to face with a peculiar difficulty. Einstein's fundamental postulate was that Galilean motion was meaningless, or at least could never be detected by experiment—whether of a mechanical or electromagnetic nature. Now the complete relativity of Galilean motion is expressed mathematically by the invariance of form of the laws of nature for transformations corresponding to that type of motion. The classical laws of mechanics satisfied this requirement in classical science, since they remained invariant when submitted to the classical transformations. But if we accept Einstein's views, these laws must remain invariant no longer to the classical transformations but to the Lorentz-Einstein ones.

The classical laws of mechanics did not fulfil this new condition of invariance. From this it would appear that the relativity of Galilean motion which Einstein had succeeded in establishing for electrodynamic and optical experiments could be maintained only at the price of abandoning this same relativity for mechanical ones. What had been gained on one side had been lost on the other. Inasmuch as the laws of mechanics would certainly have to remain invariant to the Lorentz-Einstein transformations if absolute velocity were to remain meaningless, and inasmuch as the classical laws of mechanics did not satisfy this condition, there was only one way out of the difficulty. It was necessary to assume that the classical laws were incorrect.

Accordingly, the problem confronting Einstein was to formulate new laws of mechanics which would be invariant under the Lorentz-Einstein transformations, and which at the same time would tend to coincide with the classical mechanical laws when low velocities were considered. This last restriction stemmed from the fact that for low velocities the classical laws of mechanics were known to be very approximately correct. Einstein obtained the revised mechanical laws; and it is with these laws that we shall now be concerned.

The revised laws were found to entail the relativity of mass. By this we mean that the mass of a given body could no longer be regarded as an invariant, independent of the body's relative motion. On the contrary, it was now found to increase with the relative velocity of the body, becoming infinite when the velocity of light was reached. In the same way the relativity of force was established, and classical Newtonian mechanics, with its invariance of mass and force, was proved to be incorrect, and acceptable only as a first approximation for slow velocities.

It also followed that no material body could ever move with a speed greater than or even equal to the speed of light, since it would require a force of infinite magnitude to give it this critical velocity. A number of paradoxes associated with the theory are due to the critic's initial assumption that the observer may be moving with the speed of light or with still greater speed. But all these paradoxes may be dismissed at the outset, since no observer could ever move with such a speed.

Further important consequences of the revised mechanical laws concern the identification of mass and energy. When the mathematical expression of the mass of a body in relative motion was considered, and when appropriate units were used, it was found that the mass was composed of two parts: the first part represented the mass of the body at rest, called the *rest-mass;* and the second part was given by. a mathematical expresssion which for low velocities was seen to approximate to the classical kinetic energy $\frac{1}{2}mv^2$ (*i.e.* energy of motion) of the body. This second part of the mass represented the relativistic expression of the energy of motion, a refinement of the classical definition of kinetic energy. Thus the mass m of a body in relative motion exceeded its mass at rest, m_0, by reason of the addition of the generalized kinetic energy, with the result that the distinction between mass and energy became obscured. Calling m_0 the rest-mass and m the total mass, the results just stated may be expressed by the equation

$$m = m_0 + \text{relativistic kinetic energy.}$$

Since part of the mass of a body when in motion was a manifestation of its kinetic energy, it appeared probable that the entire mass, the rest-mass as well as the added mass, should be identified with energy.* In other words, mass and energy are one and the same. Now the rest-mass cannot be identified with energy of motion, but we may identify it with a kind of potential or latent energy, called *bound energy,* expressing, as it were, the existence of matter itself. We may also say that the bound energy measures the work expended by the Creator when matter was formed.

Suppose then a particle of matter, whether at rest or in motion, were to be completely annihilated. Material mass (m_0 or m, according to whether the matter was at rest or in motion) would then be converted into non-material mass, or energy, called *loose energy,* or the energy of radiation. When ordinary C.G.S units are used, we find that the energy released by

* Einstein had previously shown that electromagnetic energy (*e.g.* radiation) possesses mass, so that mass and energy had already been identified on a prior occasion.

matter would be given by mc^2 i.e. by the mass m multiplied by the square of the velocity of light. Thus the annihilation of 1 gram of matter (say, at rest), i.e. $m_0 = 1$ gram, would release $c^2 = 9.10^{10}$ ergs of radiated energy. If the matter were in motion, the energy released would be greater on account of our having to replace m_0 by m. From this formula it is easy to realise the tremendous liberation of energy and the dire consequences that would result from the complete disintegration of even a small parcel of matter.

Another consequence of these discoveries was that any variation in the internal energy of a body maintained at rest, such as would result from compressing it, heating it or electrifying it, would be accompanied by an increase in its mass at rest. This discovery enabled us to account for a curious discrepancy in the comparative masses of the helium and of the hydrogen atom, a mystery which had never been accounted for by classical science.

To illustrate: The nucleus of the hydrogen atom is composed of one proton; and the nucleus of the helium atom was believed to contain four protons and two electrons. Today, we believe that the two electrons combine with two of the protons to form two neutrons, so that the helium nucleus is formed of two protons and two neutrons. From our present standpoint, however, this change is of minor importance, because in either case the helium nucleus appears to represent a combination of four hydrogen nuclei. It would, therefore, be natural to expect the helium nucleus to be four times as massive as that of hydrogen. Now accurate measurements proved that this surmise was wrong, for there is a slight deficiency in mass, i.e. the mass of the helium nucleus is less than four times that of the hydrogen nucleus, the deficiency being about 1%. Classical science was unable to account for the discrepancy, but the theory of relativity by its identification of mass and energy furnished an obvious solution to the mystery: Thus suppose four hydrogen nuclei were to combine, giving rise to a helium nucleus; the fact that a decrease in mass would result would imply that the combination was accompanied by a corresponding liberation of energy, equal in magnitude to c^2 times the mass that had vanished.

The foregoing consequence of the relativity theory enabled astrophysicists to clarify a second mystery, viz, to account for the high temperatures of the stars maintained through billions of years in spite of incessant radiation in a frigid space. Calculation showed that the enormous amount of energy that would have to be liberated by a star could be accounted for if we assumed that helium was constantly being formed in the star at the expense of hydrogen. The actual transformation of hydrogen into helium is believed, however, to be indirect; intermediary atoms, namely carbon and nitrogen, are involved in the transformaton, and in the process two protons lose their charges and become neutrons. Nevertheless, the ultimate result is the same: hydrogen vanishes, helium is formed, mass is destroyed, and a tremendous amount of energy is radiated. When all the hydrogen has been converted into helium, the star gradually shrinks and eventually cools down.

Since light is a form of energy, light should possess mass and momentum and exert a pressure over bodies on which it impinges. This pressure, indeed, was not unknown, since it resulted from Maxwell's equations and had been verrified experimentally by Lebedew and by Nichols and Hull. As a matter of fact it was to this pressure of light that the apparent repulsion of comets' tails away from the sun was supposed to be due. Nevertheless, though not a new discovery, the fact that Einstein's theory entailed momentum for light constituted one of the many indrect confirmations of the theory.

An objection which has often been raised is the following: Since energy moving with the velocity of light must possess an infinite mass, we should be crushed under the tremendous impact of light rays falling on our bodies. In an elementary book of this sort we cannot go into mathematical details, but we may say that the argument is fallacious. It is only ponderable matter constituted by bound energy which would become infinite in mass when moving with the velocity of light. Light rays represent another type of energy, known as free or loose energy, and calculation shows that the momentum produced by this form of energy moving with the velocity of light is not infinite.

Of course the theory of relativity is not one of pure mathematics. Einstein was not solely interested in juggling with equations and laws to make them fit into his scheme. The final verdict of the correctness of a theory of mathematical physics must always rest with the experimenter, and it remained to be seen whether the Einsteinian equations of mechanics corresponded more accurately to facts than did the classical ones. We may state that the most precise experiments have proved the correctness of the Einsteinian laws of mechanics and that Bucherer's experiment proving the increase in mass of an electron in rapid motion is a case in point.*

It should be clear by now that very important differences distinguish the theory of Einstein from that of Lorentz. Lorentz also had deduced from his theory that the mass of an electron should increase and grow infinite when its speed neared that of light; but the speed in question was the speed of the electron through the stagnant ether; whereas in Einstein's theory it is merely the speed with respect to the observer. According to Lorentz, the increase in mass of the moving electron was due to its deformation or FitzGerald contraction. The contraction modified the lay of the electromagnetic field round the electron; and it was from this modification that the increase in mass observed by Bucherer

---

* A further verification was afforded when the mechanism of Bohr's atom was studied by Sommerfeld. In Bohr's atom the electrons revolve round the central nucleus in certain stable orbits. Sommerfeld, by taking into consideration the variations in mass of the electrons due to their motions as necessitated by Einstein's theory, proved that in place of the sharp spectral lines which had always been observed it was legitimate to expect that more accurate observation would prove these sharp lines to be bundles of very fine ones closely huddled together. These anticipations were confirmed by experiment, though Sommerfeld's original theory was subsequently supplemented by the hypothesis of the spinning electron.

was assumed to arise. In Einstein's theory, however, the increase in mass is absolutely general and need not be ascribed to the electromagnetic field of the electron in motion. An ordinary unelectrified lump of matter like a grain of sand would have increased in mass in exactly the same proportion; and no knowledge of the microscopic constitution of matter is necessary in order to predict these effects, which result directly from the space and time transformations themselves.

Furthermore, the fact that this increase in mass of matter in motion is now due to relative motion and not to motion through the stagnant ether, as in Lorentz's theory, changes the entire outlook considerably. According to Lorentz, the electron really increased in mass, since its motion through the ether remained a reality. According to Einstein, the electron increases in mass only in so far as it is in relative motion with respect to the observer. Were the observer to be attached to the flying electron no increase in mass would exist; it would be the electron left behind which would now appear to have suffered the increase. Thus mass follows distance, duration and electromagnetic field in being a relative having no definite magnitude of itself and being essentially dependent on the conditions of observation.

Owing to the general validity of the Lorentz-Einstein transformations, it becomes permissible to apply them to all manner of phenomena. In this way it was found that temperature, pressure and many other physical magnitudes turned out to be relatives. On the other hand, entropy, electric charge and the velocity of light *in vacuo* were absolutes transcending the observer's motion. Later on we shall see that a number of other entities are found to be absolutes, the most important of which is that abstract mathematical quantity called the *Einsteinian interval,* which plays so important a part in the fabric of the new objective world of science, the world of four-dimensional space-time.

CHAPTER XV

IN the preceding chapters we endeavoured to show how Einstein's fundamental premises—the relativity of Galilean motion through space or the ether and the postulate of the invariant velocity of light—had been posited as a result of induction from ultra-precise experiment. Once these postulates are accepted, all the strange consequences of the special theory are obtained deductively by elementary mathematical reasoning.

In view of the extreme importance of the problem let us recall that the postulate of the invariant velocity of light assumes that a ray of light, *in vacuo,* will pass through any Galilean frame whatever with the same invariant speed c in all directions, provided this speed be measured by the observer at rest in the frame, *and not by some other observer in some other frame, whether Galilean or accelerated.** And here we must recall that classical science, believing in the absoluteness of time, assumed that events which occurred "now" anywhere throughout space would continue to occur in the same simultaneous fashion regardless of the motion of the observer. This belief was equivalent to assuming that a propagation moving with infinite speed with respect to one observer would also advance with infinite speed when measured by any other observer. In other words, an infinite speed was an absolute; it constituted the invariant speed of classical science. No such physical propagation was known to science, though it was suspected that gravitation might be of this type. At any rate, classical science accepted the existence of this infinite invariant velocity in a conceptual way. This implied the existence of the Galilean transformations, hence of the ordinary laws for the addition of velocities. Inasmuch as all measurements with rigid rods, and with clocks such as vibrating atoms, appeared to bear out the anticipations of classical science, at least so far as crude observation could detect, the classical stand appeared to be vindicated *a posteriori* even though the hypothesis of absolute time presented no *a priori* justification.

Now, in the theory of relativity, Einstein asks us to agree that a certain finite invariant velocity which turns out to be that of light *in vacuo*

---

* Whenever we refer to "another Galilean frame," we invariably mean one in motion with respect to the first. Were it not in a state of relative motion, it would constitute the same frame.

must be considered invariant for all Galilean observers.*   We are thus led, as explained previously, to the Lorentz-Einstein transformations, an immediate consequence of which is that molar matter can never move with a velocity greater than or even equal to that of light.   This in itself is sufficient to exclude our taking into consideration observers moving with any such speeds.

But suppose that for argument's sake we wish to consider, even though it be in a purely conceptual way, a velocity greater than light—in particular, a velocity which would constitute infinite speed for any definite Galilean observer.   How would a velocity of this sort manifest itself to some other Galilean observer in motion with respect to the first?— The sole point it will be necessary to stress is that this erstwhile infinite velocity could never be invariant without coming into conflict with the invariance of the velocity of 186,000 miles per second.   This becomes obvious when we consider that, according to classical science, if we added any speed—say, the speed of light c — to the infinite speed, we should obtain the infinite speed; whereas, in Einstein's theory, we should obtain the speed of light.   From this it follows that the premises of classical science and of relativity are incompatible.   Inasmuch as it is Einstein's theory that we propose to discuss, we shall abandon the classical belief in the invariance of the infinite velocity.

Henceforth the invariant velocity of the universe is to be c, or 186,000 miles per second.   As for an infinite velocity, this will now become a relative, infinite for one observer, finite for another.   Now it is obvious at first sight that if our space and time measurements were such as classical science believed them to be, it would be impossible for a ray of light to pass us with the same speed regardless of whether we were rushing towards it or fleeing away from it.   A simple mathematical calculation shows us, however, that we can make our results of measurement compatible with the postulate of invariance provided we recognise that our space and time measurements are slightly different from what classical science had assumed.   This is purely a mathematical problem and can be solved by mathematical means.   It leads us, of course, to the Lorentz-Einstein transformations; and from these transformations it is easy to see that rods in relative motion must be shortened, durations of phenomena extended, and the simultaneity of spatially separated events disrupted.

Einstein's premises being consistent (as can be proved mathematically), no question can arise as to the consistency of Einstein's conclusions; premises and conclusions must all stand or fall together.   The major problem, then, is to analyse the legitimacy of the theory from the standpoint of its ability to portray the real world of physical phenomena. Some lay critics, such as Bergson, alarmed by the revolutionary nature of Einstein's discoveries and animated by a desire to retain their classical belief in the absolute nature of simultaneity remaining the same for all

* We may mention that the physical existence of a finite invariant velocity is by no means impossible. It is acceptable mathematically, and the only question that we shall have to consider is whether it corresponds to physical reality.

observers, have endeavoured to uphold this intermediary view. They have maintained that from a purely mathematical standpoint the systems of measurement of relativity might confer greater simplicity on the purely mathematical treatment of the equations of electrodynamics; but that in no case was it permissible to assume that the anticipations of the theory could ever correspond to what we should actually measure with rigid rods and ordinary clocks. We may dismiss this view summarily. Were it correct, none of the physical anticipations of Einstein's theory would have been verified by physical experiments.

Furthermore, a theory such as Einstein's is one of mathematical physics, not one of pure mathematics. It is not merely necessary that the deductions of the theory be mathematically consistent with the premises. They must also lead us to anticipations which will be verified by experiment. If they failed in this respect, they would still portray the workings of a possible rational world; but the theory itself, yielding erroneous physical previsions, could never allow us to foresee and to foretell, and no scientist would ever dream of upholding it. In short, we must recognise that if Einstein's premises are accepted, we must assume that measurements conducted with natural rods and natural clocks would yield the previsions of the theory. We must also remember that by natural clocks *we do not mean clocks arbitrarily adjusted.* We mean clocks such as vibrating atoms, or, more generally, all periodic physical processes evolving in a state of isolation. Likewise, by natural rods we mean our ordinary material rods maintained in the same conditions of temperature and pressure.

The type of measuring instruments is thus exactly the same as in classical science; and all that relativity tells us is that if our classical measurements had been pursued with greater precision, we should have realised long ago that the classical anticipations were belied by experiment. There is no use in pursuing the study of the theory any farther unless we succeed in understanding its physical significance. So far as the mathematician is concerned, with the Lorentz-Einstein transformations issuing as they do from ultra-precise experiment, their mathematical consequences must be deemed to correspond to the workings of the real world; and these consequences indicate clearly that length, duration and simultaneity must be relatives in a real physical sense. However, in view of the importance of the new ideas, we may consider the problem in an indirect way and thereby throw additional light on the concrete significance of these transformations.

Accordingly we shall have to discuss what the physicist means by an "event." An event is exemplified by a something happening in a certain region of space at a certain moment of time. It matters not what actually happens. It may be a thought flashing through our minds, a bird bursting into song, or two billiard balls colliding; what is essential is a specification of position in space and in time.

When we try to think of an event, we are compelled to credit it with more or less extension in space, and more or less duration in time. But in all problems where accuracy of localisation is demanded, we must conceive of an event as ideally restricted to an instantaneous point in space; or, if we prefer, to a mathematical point in space at a mathematical instant in time. We then get what has been called by Minkowski a **Point-Event**. This procedure of restricting events ideally is of course purely a mathematical abstraction and is in every way similar to the procedure followed by the geometrician when he speaks of points without dimensions, lines without width, or surfaces without thickness.

Even in ordinary life we are obliged to follow this same method of restricting dimensions ideally, in order to render a precise localisation possible. For instance, when we say we are so many miles from New York our assertion is extremely vague. What part of New York are we referring to? Even when we specify a definite building in the city, it could always be asked: "What part of the building?" We see, then, that to obtain any degree of precision we must select a definite point, and our statement that we are fifty miles from New York is equivalent to regarding New York as reduced to a point. Of course we know full well that a city is not reduced to a point; but arguments of this sort are irrelevant to the problem of localisation with which we are faced.

It is the same for time. An event lasts, it manifests duration; but in order to obtain precision we must conceive of it as ideally restricted. We thus obtain an instantaneous event occurring at a definite point of our space.

We may say, then, that a billiard ball, subsisting motionless at a point of the space of our frame, constitutes a continuous succession of point-events all located at the same point of our space (where the point-event itself would be represented by the instantaneous existence of the ball assumed to be reduced in size to a spatial point). Likewise, a billiard ball moving through the space of our frame would be represented by a succession of points in space at a succession of instants in time; hence by a chain of point-events.

It is not absolutely necessary to conceive of an event as something objective, actually occurring or existing. We may conceive of events and point-events as empty. Thus, a point of our space a million miles away in a definite direction at six o'clock constitutes a point-event. In the same way it is not absolutely necessary to designate a point on our sheet of paper by a pencil mark; for we can define its position by numbers. In physics, however, we are more generally interested in discussing events that are observable; so although in theory a point-event is merely the localisation of an abstract entity at a point of space at an instant of time, in practice the majority of events to which we shall refer will serve to localise observable existents. For similar reasons we usually tint the colourless liquid in the thermometer so as to render it observable; though by so doing, nothing is changed, of course, in the reading of the instrument.

Let us now pass to the problem of determining the simultaneity of events. As a fact of common experience it is possible to localise our various thoughts and sense impressions in the time-stream of our individual consciousness; and the judgment that two sensations are simultaneous or follow in succession is essentially *a priori,* in that it cannot be analysed further. Science, however, could never have existed had we restricted ourselves to registering sense impressions without attempting to co-ordinate them and account for them. To these attempts of co-ordination we are indebted for the conception of the outside universe of space and time (or space-time according to the modern view)—the common objective world of science in which events occur. It is objective in that we are enabled to conceive of it as presenting the same relationships of structure for all men, hence as pre-existing as a concrete existent to the mind that discovers it and explores it bit by bit. Whether this objective universe represents a reality in the metaphysical sense (a reality which has merely been discovered by us) or whether, as the majority of scientists would assert, it reduces in the last analysis to a mere mental construct, the product of a synthesis of the mind, is a problem that need not detain us in the present chapter. In either case the physicist may speculate on this objective universe as though it were a reality pre-existing to the observer who discovers it.

Commonplace reason, or rudimentary science, was thus led to conceive of a common external objective universe in which external events occurred, occupying definite positions in space and in time. These external events, such as the explosion of a barrel of gunpowder, were assumed to be the causes responsible for our immediate sense perceptions of light and of sound. Until some three hundred years ago it was thought permissible to identify the instant of time at which an external event occurred with the precise psychological instant at which the event impressed itself on our consciousness in the form of a sound or of a visual sensation. But this very crude view led to inconsistencies when it was realised that an explosion which occurred at our sides and which appeared to us as a simultaneous occurrence of light and sound impressions would have appeared as a sequence of two separate events, a luminous one followed by a sonorous one, had we been situated elsewhere.

If we were to retain our belief in the possibility of conceiving of a common objective universe in which a single event remained a single event regardless of the position of the observer, it became imperative to account for these discrepancies by assuming that sound transmission from place to place was not instantaneous. It was still thought, however, that light propagation was instantaneous, so that at least the instants of our visual perception of external events could be identified with the instants of their occurrence in the objective world. For example, it was usual to claim three hundred years ago that when we beheld a sudden red brightness in a star we were witnessing a stellar catastrophe occurring *there-now.* It appeared, then, as though men were gifted with the same *a priori* intuitional understanding of the simultaneity or succession of events throughout space which experience had shown to exist in the case of sense impressions located in the individual streams of consciousness. But, as Eddington points out, "This crude belief was disproved in 1675 by Römer's celebrated discussion of the eclipses of Jupiter's satellites; and we are no longer permitted to locate external events in the instant of our visual perception of them." Following Römer's discovery, scientists were compelled to discriminate between the instant an external event occurred and the instant we became cognisant of its occurrence. This discovery led them to recognise that pure intuition could yield us no reliable information on the order of succession and on the simultaneity of external events throughout space. For instance, when we suddenly perceive some stellar conflagration through the telescope, we have no means of deciding by direct intuition whether the stellar catastrophe was simultaneous with the signing of the Armistice, the battle of Waterloo or the building of the Pyramids. In the same way, were we to perceive two stellar catastrophes, it would be impossible for us to decide *a priori* which had preceded the other.

Now it is the problem of locating external events in a common space and time, or a common space-time according to the modern view, that is the one of paramount importance for the student of nature. Direct intuition failing him, if duration and simultaneity as between external events in the outside world were to have any meaning at all, they could be defined only as a result of some convention similar to that selected for the measurement of space with rigid material bodies. The aim of physics was therefore to decide on some conventional means of determining the simultaneity and succession of external events *in a manner compatible with the existence of a common objective universe.*

The most obvious method of arriving at some consistent determination of the simultaneity of two external events was to appeal to physical propagations, such as those of light or sound waves, by taking into consideration the times these propagations required to reach us.

Of course, in everyday life, the distinctions the scientist is compelled to consider are very often of scant importance. When, for instance, we bark both our shins and are conscious of a psychological simultaneity of painful impressions, we are perfectly justified in practice in assuming that we have barked our shins simultaneously. It is only when we subject an example of this sort to the rigorous logic of scientific thought that the crude view becomes untenable in theory though of course still permissible as an approximation.

The fact is that the barking of each of our two shins constitutes an external event,* and our awareness of each of the two pains that ensue constitutes in turn a psychological event. Thus we have to take into consideration two external events and two psychological events. We have no more right to muddle up these two species of events than we have to confuse the firing of a shot with the hitting of the target. In any case, a certain period of time, however short, must elapse between the instant

* By an external event we do not necessarily mean an event occurring outside our body. We mean one that does not reduce to a mere awareness of consciousness. A sudden pain in our toe would constitute an external event in exactly the same measure as would the explosion of a barrel of gunpowder a mile distant.

we bark one of our shins and the instant we become aware of the pain. Physiologists who have measured the speed of nerve transmissions could tell us exactly what the lapse of time would be, as measured by a clock. Unless we have reason to assert that the nerve transmissions proceed at the same speed along both legs, and unless the bruised spots on our two shins are situated at the same distance from the seat of our consciousness, our intuitional sense of the psychological simultaneity of the two sensations of pain does not permit us to decide on the simultaneity of the two external events (the barkings of our shins against some object). In practice, however, as we have mentioned, we may neglect all these subtleties; for as our psychological awareness of simultaneity is after all extremely crude, there would be no advantage in submitting our results to corrections of comparatively insignificant importance.

Nevertheless, although the differentiation we have established is of no practical interest in the case we have discussed, it possesses a vast theoretical importance, as will be seen from the following illustration: If we were to confuse the external events with our awareness of them, we might argue that a human giant with legs extending to the stars and touching, for an instant, the sun with his left knee and Sirius with his right foot, would be aware by direct intuition of the simultaneity of these events; since all he would have to do would be to register the simultaneity of his two conscious impressions—the two burns. But the argument would be faulty; for in a case of this sort, where the external events occurred at such vast distances from the seat of his consciousness, the giant could no longer afford to neglect possible variations in speed of the two nerve propagations and the difference in distance of the two points on his legs at which the external events had occurred. The only reason the popular mind retains a lingering belief in some vague intuitional understanding of the simultaneity of external events, however distant, is that when the events are near us their simultaneous perception justifies our belief in their simultaneous occurrence. We then extrapolate these results unconsciously from place to place, and fall into the erroneous belief that we possess this same intuitional understanding of the simultaneity of external events throughout space.

The significance of the preceding discoveries can be summarised as follows: When considering the simultaneity or order of succession of events, we must differentiate sharply between events which reduce to our awareness of sense impressions, and those which refer to external occurrences. Our recognition of the simultaneity or order of succession of sense impressions or of thoughts flashing through our mind is fundamental in that it cannot be analysed further. For this reason a physical theory cannot be called upon to give a physical definition of judgments which it recognises as *a priori*. In contradistinction to the temporal position of sense impressions, we have to consider the simultaneity or order of succession of external events. In this case direct recognition fails us. All we can do is to note the instants at which we became conscious of these external events, and then by appealing to physical propagations deduce therefrom the instants at which the events occurred in the outside world.   In order to avoid any confusion we shall therefore refer to the simultaneity of external events throughout space as *physical simultaneity*, reserving the name *psychological simultaneity* for our awareness of the simultaneity of two sense impressions.

Now all the conclusions we have reached up to this point are quite irrelevant to Einstein's theory proper.   They embody the discoveries of classical science and, while they still hold in the relativity theory, we shall see that Einstein's contributions to the problem of simultaneity lie in another direction entirely.   If these points are understood we may pass to a definite example and investigate how the physicist will proceed when he wishes to determine whether or not two external events occurring in different places are simultaneous in the physical sense.

Suppose that two instantaneous flashes of light occur at two points, A and B, on the road; and that we wish to discover whether or not the two flashes were produced simultaneously (always in the physical sense). The mere fact that we might perceive the two flashes simultaneously or in succession would not of itself permit us to settle the question, for the order of our perceptions would be found to vary with our position (of course only to a very small degree in the case of visual perceptions). But suppose now that we stood at the midpoint C between A and B, as measured with our rigid rods.   If we perceived the two flashes simultaneously (in the psychological sense), should we be justified in asserting that the two flashes had occurred simultaneously in the physical sense? Classical science answered no; for we were not at all certain that the wave of light was propagated from B to C with the same speed as the wave propagated in the opposite direction from A to C.   Classical science assumed that the unknown velocity of the ether drift caused by the earth's absolute velocity would interfere with our determinations, so that nothing definite could be said.

But the entire situation changes as soon as we accept Einstein's principle of relativity and his postulate of the invariant velocity of light. We are then in a position to assert that the ether drift, whether it exists or not, can have no physical significance; and that we are therefore justified in assuming that the rays of light are propagated with the same speed in all directions.   If this is the case, we have a right to conclude that the simultaneity of our perceptions of the two light flashes (in the psychological sense) ensures the simultaneous occurrence (in the physical sense) of the two external events, namely, the two flashes on the road.

Now, in the particular example we have considered, the observer was able to station himself at a midpoint between the two points A and B. But situations may arise where a procedure of this sort would be impossible.   Suppose, for instance, that we wish to ascertain whether two events occurring one on the sun and the other on the earth by our sides are simultaneous or not.   How shall we proceed?

We will note on our watch the precise instants at which we perceive both events. Then we will take into consideration the time required for the light message from the solar event to reach us; we will deduct this duration from the instant at which the solar event was perceived, and we thus obtain the instant at which the event actually occurred on the sun. If this instant of time is identical with the psychological instant at which the earth event at our sides was perceived, we will assert that the two events occurred simultaneously. And now let us notice that by subtracting the time that light has taken to reach us from the solar event and by noting the instant thus defined, we are in effect re-establishing matters as they would have presented themselves had the news of the solar event been propagated to us through our frame instantaneously with infinite speed, so that we may present our definition of simultaneity in the following conceptual way:

We will say that if a perturbation (say, a telepathic one) is propagated with infinite speed from the surface of the sun, coincidently with the occurrence of the solar event and carrying the news of the solar event, and if this perturbation reaches us on earth at the precise psychological instant at which the earth event is perceived at our sides, then the two events will be simultaneous. Likewise, always considering the hypothetical case of instantaneous transmission, if the solar perturbation reaches us prior or subsequent to our perception of the earth event occurring at our sides, we will conclude that the solar event has preceded or followed the earth event. Of course this definition by means of a propagation moving through our frame with infinite speed is purely conceptual, since no such type of propagation is known to physicists; but as it would lead to exactly the same determination of simultaneity, we may retain it provisionally, as it will facilitate our understanding of the nature of Einstein's discoveries.

Now, according to classical science, simultaneity having been defined in this conceptual way, any other observer, regardless of his motion, would have agreed that the perturbation had travelled from sun to earth with infinite speed, *i.e.*, in zero time; so that in all cases the two events could be regarded as simultaneous in an absolute sense. By this we mean that they would be simultaneous not merely from the local point of view of the observer on earth, but also for all observers. In other words, physical simultaneity would be absolute; likewise, the stretch of physical time separating two successive pairs of simultaneous events occurring in different regions of space would be absolute; in other words, physical time itself would be absolute. This indeed was the belief of classical science.

But if now we view the same problem in the light of Einstein's theory, and if we remember that an infinite speed is a relative depending on the frame of reference or relative motion of the observer, we find that these opinions are no longer tenable. As before, the observer on the earth will be perfectly justified in believing that the two events (the solar one and the one occurring at his side) are simultaneous; but, on the other hand, another observer in relative motion would discover that the earth observer was wrong, for as referred to his own frame, the perturbation would not have spanned the sun-earth distance in zero time with infinite speed. According to him, therefore, the two events would not be physically simultaneous.

In short, *physical simultaneity and duration, which were considered absolutes by classical science, must now be considered relatives having no absolute, universal significance,* since they vary when computed in frames moving at various relative speeds.

But it is most important to realise that this relativity of simultaneity holds only for events occurring in different places and does not apply to events occurring in the same place. The reason for this distinction is easily understood. Suppose that two instantaneous events occur at the same point and at the same time in our frame of reference. The events, being simultaneous and copunctual in our frame, are said to constitute a **coincidence of events.** If now we change our frame of reference, will the two events as referred to this new frame also be simultaneous and copunctual? Obviously yes. The principle of sufficient reason itself should satisfy us on this score, since whatever disruption in copunctuality or simultaneity a change of frame might bring about, as there is nothing to distinguish one event from the other, either in position or in time, there is no reason why one event should precede rather than follow the other, either in position or in time, when the frame is changed. A concrete illustration would be given by a collision between two automobiles. Here we have two bodies situated at the same point of the road (or thereabouts) at the same instant of time; hence the observer on the road is witnessing a coincidence of events in the road-frame. Obviously, were he to view these happenings from some passing train, the collision would occur just the same; hence, so far as the train-reference-system was concerned, the two events would continue to constitute a coincidence. From this we see that, in contradistinction to the simultaneity of events occurring in different places, which is essentially a relative depending on the choice of our frame of reference, a coincidence of events is an absolute, that is, remains a coincidence or a simultaneity in all frames of reference.

We may note that in a world in which coincidences were relative, that is, in a world in which two cars would appear to collide and all the passengers killed when viewed from the road whereas no such collision would take place when viewed from the train, science would be quite impossible; for the opinions of the various observers would be relative to the point of being inconsistent. In the squirming world of relativity something at least must be absolute, and one of these absolutes is found to reside in the coincidences of events.

We may contrast the teachings of classical science and of relativity in the following way: Just as classical science recognised that there was no physical significance in speaking of the same point of space at different times until, by selecting a frame of reference, we had objectivised, as it were, the space we were discussing, so now relativity compels us to add: "There is no meaning in speaking of the same instant of time in different places until we have objectivised time, as it were, by specifying our frame of reference." In both sciences, however, the classical and the relativistic, the coincidence of events remains an absolute, transcending the choice of a frame of reference. From a philosophical point of view, this discovery of the relativity of simultaneity marks a date of the same momentous importance as did the discovery of the Copernican system in astronomy.*

Having established the relative nature of physical simultaneity, we find it an easy matter to rediscover that other consequence of the theory, namely, the contraction of length. Consider, for example, an observer at rest on a track observing a train also at rest. What is the length of the train as referred to the observer's frame, namely, to the earth? Obviously it is the difference in his distance from the engine and from the rear car. But if now we consider the more general case where the train or the observer with his reference frame is in relative motion (either choice comes to the same thing, so long as the motions are Galilean), it becomes imperative to state that the measurements must be performed *at the same instant of time*. It would be absurd to measure our distance from the rear car at one o'clock, then our distance from the engine at two o'clock; for we might find that the train had a length of over sixty miles owing to its displacement during the interval. But whereas in classical science the significance of the same time in two different places was absolute, the same for all observers, in relativity it becomes indeterminate. According to the relative motion of the observer, different simultaneity determinations will be obtained, and as a result the length of the same train will vary in value. This is what is meant by the relativity of length. Calculation proves that the greater the relative velocity, the shorter will the train measure out.

* The relativity of simultaneity is a most revolutionary concept, as will be seen from the following illustration:

Consider two observers, one on a train moving uniformly along a straight line, the other on the embankment. At the precise instant these two observers pass each other at a point P, a flash of light is produced at the point P. The light wave produced by this instantaneous flash will present the shape of an expanding sphere. Since the invariant velocity of light holds equally for either observer, we must assume that either observer will find himself at all times situated at the centre of the expanding sphere.

Our first reaction might be to say: "What nonsense! How can different people, travelling apart, all be at the centre of the same sphere?" Our objection, however, would be unjustified.

The fact is that the spherical surface is constantly expanding, so that the points which fix its position must be determined at the same instant of time; they must be determined simultaneously. And this is where the indeterminateness arises. The same instant of time for all the points of the surface has not the same significance for the various observers; hence each observer is in reality talking of a different instantaneous surface.

In short, classical science recognised that apart from our intuitional understanding of the succession or simultaneity of the awareness of two sense impressions, such as two impressions of pain for instance (*i.e.*, the *here-now* type of simultaneity), the succession or simultaneity of events could be determined only indirectly by physical means. And even so, there always remained a certain indeterminateness owing to our ignorance of the velocity of the ether drift. Assuming, however, a spread of simultaneous events to have been determined by any given observer, it was thought that these same events would remain simultaneous for any other observer. In other words, time and simultaneity were absolute.

Einstein's theory compels us to revise the classical views. It still permits us to retain our belief in the absolute nature of a coincidence of events, that is, of events which are copunctual and which occur at the same time in any given frame. But it denies the absoluteness of the simultaneity of spatially separated events. Thus, if two observers pass each other, the events which are simultaneous with one another and with this passage according to one of the observers would be recognised as unfolding themselves in succession for the other observer. It follows that a simultaneity of events, or a sameness of time throughout space, becomes an indeterminate concept until an observer or a frame of reference has been specified; in much the same way, the weight of a given brick situated anywhere in space will be indeterminate until such time as the distribution of the surrounding masses has been decided upon. So far as the special theory is concerned, this is the sole difference between classical science and relativity.

Later, we shall see that in the general theory, where large masses of matter are in relative motion, the entire concept of the simultaneity of external events becomes obscured, for space and time determinations over extended areas become impossible. In the general theory, therefore, relativity breaks away completely from classical science. And here we may mention a number of criticisms directed against Einstein's selection of optical signals for the purpose of defining simultaneity.

Thus, Dr. Whitehead writes: "The very meaning of simultaneity is made to depend on light signals. There are blind people and dark cloudy nights and neither blind people nor people in the dark are deficient in a sense of simultaneity. They know quite well what it means to bark both their shins at the same instant." *

In this illustration, Dr. Whitehead, by referring somewhat loosely to a *sense of simultaneity,* is confusing the two species of simultaneity which physicists have found it imperative to treat separately. We may recall that these two types were exemplified by a simultaneity in our awareness of two impressions (psychological simultaneity) and by a simultaneity in the occurrence of two external events (physical simultaneity). Of course, in the illustration selected by Whitehead, all the events considered, namely, our awareness of the two pains and the actual coming into contact of our shins with the rail, are so close together

* "The Principles of Natural Knowledge."

both in space and in time that for all practical purposes the confusion would be legitimate. Nevertheless, inasmuch as a confused and obscure illustration is liable to generate confused conclusions, it is necessary to clear up the matter from the start and specify the type of simultaneity we propose to discuss.

First let us assume, for argument's sake, that it is the psychological type of simultaneity which is at stake. The criticism would then amount to this: Einstein has sought to render our immediate awareness of the simultaneity of two sense impressions contingent on the results of optical measurements executed *a posteriori*. In other words, if we should ask our friend whether he had sensed two sounds simultaneously or in sequence, he would answer: "I cannot tell you before I have looked up Einstein's rules for simultaneity determinations and performed certain optical measurements." It would follow that were no light rays obtainable or were our friend blind, he would never be able to give us an answer. Einstein would thus have made our immediate recognition of a simultaneity or of a sequence of sensations contingent on the accidental circumstance that we were gifted with eyesight. This seems to be Dr. Whitehead's contention, as would appear from his reference to "blind people" in his criticism.

It seems scarcely necessary to state that neither Einstein nor any other scientist has ever attempted to defend so paradoxical an assertion. Quite the reverse. Both classical and relativistic science are fully agreed that our ability to recognise the simultaneity of two impressions (be they visual, tactual or auditory) is in the nature of a fundamental recognition which cannot be analysed further. We *know* that we feel two pains at the same instant, or that two thoughts flash through our mind simultaneously, and that is all there is to be said about it. Any discussion as to *why* we know or *how* we know does not concern the physicist. It is a problem which must be left to the brain specialist or psychologist. So far as the physicist is concerned, it constitutes a form of *a priori* knowledge over which it is useless to argue. On no account, then, can it be claimed that the theory of relativity professes to interpret this psychological species of simultaneity in terms of optical measurements or, for that matter, of physical measurements of any kind.

But there is another way to view the problem. We might claim that though Einstein has never sought to render psychological simultaneity contingent on the performance of optical measurements, yet, on the other hand, his theory fails to take this fundamental type of simultaneity into account.

But any one at all conversant with physical science would perceive the fallacy of this contention at first blush. He would realise that physical science would be quite impossible were we to disregard the data furnished us by our fundamental recognitions. Consider, for instance, the manner in which Einstein defines the simultaneity of external events, e.g., of two flashes occurring on the embankment. He tells us that this simultaneity will be realised if the observer standing at a midpoint between the two flashes *becomes aware of the two luminous impressions simultaneously*. Obviously, for this definition to convey any meaning, it is essential that we credit the observer with the ability to recognise the simultaneity of these two luminous impressions. Had this fact been ignored or denied by the theory, Einstein would never have suggested his definition.

Summarising, we may say that in relativity, as in classical science, the sophisticated rules given for the determination of a simultaneity of external events presuppose that we possess the innate ability to differentiate between a simultaneity and a succession of sense impressions. These fundamental recognitions then serve as a basis for future discovery; and the sophisticated rules are used merely to allow us to pass from the recognition of a simultaneity which is *a priori* to those simultaneities which cannot be determined *a priori*. As for the rules themselves, those suggested by relativity constitute but a refinement of the ones entertained by classical science. Aside from this purely quantitative difference, no change in method is involved.

It is therefore obvious that Dr. Whitehead's criticism, when interpreted as referring to psychological simultaneity, appears to be totally unwarranted. Let us see whether it would possess any greater merit when interpreted as referring to a simultaneity of external events. The criticism would then have the following significance: Our recognition of the simultaneity of external events is fundamental. We know that two events, one on the sun and the other on the earth, are simultaneous, when we perceive them simultaneously. Physical measurements are irrelevant to the problem. Einstein's theory fails to recognise this fact.

Now there is no doubt that the theory of relativity does fail to recognise the type of simultaneity presented in this last interpretation of Dr. Whitehead's criticism. Hence the man who might have this understanding in mind would be perfectly justified in dissenting from a theory which would characterise his conception of simultaneity as altogether indefensible. But, for that matter, so would classical science. Both sciences refuse to agree that we possess any *a priori* means of determining a simultaneity of external events. Both sciences claim that in all such cases we are compelled to appeal to physical measurements of one kind or another. In other words, both sciences refuse to accept that bizarre doctrine known as "neo-realism" which consists in confusing the event that occurs with the event that we sense.

According to neo-realism as championed by Whitehead, and by a number of other philosophers, vision is a matter of direct apprehension.* Events would thus be simultaneous with the instants of their perception.

---

* In the present chapter we are classing under the common name of "neo-realism" all those metaphysical doctrines which agree in considering perception, especially visual perception, a matter of direct apprehension. This classification may not be in accord with that of the metaphysicians. But as, from the standpoint of science, all doctrines that hold to the theory of the direct apprehension of reality are equally objectionable, we shall not trouble to differentiate between them.

In the same spirit, the neo-realist speaks of events occurring behind a mirror, or of the space he sees behind the mirror, or of the convergence of the rails on the embankment. No distinction in terminology is made between the status of such perspectives and that of the physical entities existing in the objective world of science. In ancient times Æsop warned his countrymen against the dangers of neo-realism when he composed his fable of the dog mistaking the bone for its reflection in the water. Indeed, for the plain man there appears to be no difference between the attitude of the neo-realist and that of the unsophisticated infant or primitive savage who views a mirror for the first time.

In order to extricate himself from the paradoxes in which he finds himself involved, the neo-realist is finally compelled to agree that the objects behind the mirror, etc., are not quite the same things as physical objects. So he calls them *sensa* or *sense-objects*. When requested to justify his belief in these mysterious, shadowy existents, he tells us that sensa and sense-objects certainly exist, for otherwise how could a round coin ever be seen as an ellipse? Obviously (so he maintains) in addition to the round physical object, there must be an "elliptical sensum" present, and it is this that we are sensing.

It is scarcely necessary to say that no scientist could be in sympathy with such loose arguments, and accept as sound this arbitrary materialisation of qualities. Forms and colour sensations are all that we are conscious of visually. It is only by inference, as a result of a co-ordination of sensations, that we are finally led to believe in the existence of a round coin in the objective world. After we have reached this stage of knowledge it is a simple matter to account for the elliptical impression we receive of the coin, since an image of this shape can be proved to have been formed on the retina. An application of the laws of geometrical optics, and a slight knowledge of the eye's structure, suffice; and the "sensa" and "sense-objects" of the neo-realist can be discarded as perfectly useless hypotheses.

A more detailed study, which we shall spare the reader, would convince us that neo-realism, when taken literally, rests on a confusion between inferences and direct recognitions, and leads to contradictions when an attempt is made to conceive of an objective world transcending our private views. On the other hand, if taken with a grain of salt, mitigated, as occasion demands, by such palliatives as "pseudo" or "peculiar sort of" or "not quite the same as," it resolves itself into a mere play on words, into a giving of the same name to things which even the neo-realist would concede to be essentially different.

Physical science, it must be remembered, deals with a common objective world. This world is not apprehended directly, but issues from a long, laborious, ever-widening synthesis of private views. Primarily, all we are aware of is a complex of sensations. Even the crude knowledge of the savage is therefore itself the result of a highly complex synthesis. But science must pursue the synthesis still farther, taking the results of previous syntheses as the· elements of its more sophisticated ones.

And so what was formerly the synthetic view becomes the private one when we rise to a higher plane. Thus, if we consider the aspect of the earth's surface, viewed from any one spot, our private view would suggest that the earth was flat. Only a synthesis of a number of private views and experiences can lead us to assert that the earth is round; and science accordingly speaks of the rotundity of our planet and ignores its flatness. To put it in more technical language, the search for the objective world is similar to that of the search for the invariants or invariant equations of a group of transformations. In this illustration, each transformation corresponds to the passage from one private perspective to another, whereas the invariants correspond to the entities of the objective world with which science deals. The discovery of the space-time universe, to be discussed in a future chapter, illustrates the procedure.

To be sure, in many instances the private view may be of greater interest than the collective one. This will be the case if an artist wishes to paint a landscape on his canvas. He will, of course, draw distant objects on a smaller scale and make the rails converge. But, in contra-distinction to the artist, the scientist is not particularly interested in the private view except as a means to an end. His ultimate goal is the common objective world.

The only charge that might be directed against the scientific attitude (which in this case is also that of the man in the street) is that this objective world resolves itself into a series of inferences, a mere mental construct, hence cannot possess the same measure of reality as the impressions we experience directly. The majority of scientists would grant the justice of this charge if by "what we experience directly" we mean our awareness of sensations. But they would strenuously oppose the claims of the neo-realist who endeavours to extend the meaning of these words so as to embrace the existence of his sensa and sense-objects; these represent nothing but inferences which he has drawn from his awareness of sensations. And this is precisely the error into which the neo-realist falls when, on perceiving a red flash in the sky, he talks of a star bursting into prominence "there-now." If he wishes to avoid inferences, he should confine himself to saying: "I experience 'here-now' a luminous impression on a dark background." For all he knows, the luminous impression may be due to an aviator's light, or to a firefly, or to a purely subjective hallucination.

At any rate, the point we wish to stress is that while it is correct to say that the objective world is a synthesis of private views or perceptions, we should not forget to add that the private views are themselves the product of a synthesis of sensations. But inasmuch as science must deal with the general, and not merely with the particular, and inasmuch as it is the common objective world that renders this general knowledge possible, it will be this world that the scientist will identify with the world of reality. Henceforth the private views, though just as real, will be treated as its perspectives. We must recognise, therefore, that the common objective world, whether such a thing *exists* or is a mere convenient fiction, is indispensable to science, if in no other capacity than as a working hypothesis. And neo-realism, as we have said, *when pursued consistently,* renders a collective view impossible. Furthermore, from the standpoint of science, whether it be classical physics, relativity, physiology or psychology, neo-realism is untenable, for under no circumstances can vision and our knowledge of nature in general be considered a matter of direct apprehension. In much the same way, we know that 7 times 8 is 56, and we obtain this result immediately, without reflection and without having to count on our fingers. Were it not for the fact that the neo-realist recollects the difficulty he experienced in learning the multiplication table, he would presumably be telling us that our knowledge of the properties of numbers was also a matter of direct recognition. From the standpoint of science, there is very little difference between the alleged immediacy of perception and our knowledge of the product of 7 times 8. Neither of these two forms of knowledge can be viewed in the light of direct recognitions. At any rate, regardless of our attitude towards neo-realism, in scientific discussions we must define the meaning of our words clearly and cease to apply the same word, "simultaneity" to a simultaneity in the scientific sense, as holding for the objective world, and to a simultaneity in the neo-realistic sense, as assumed to be apprehended directly. It is remarkable that Dr. White head, who has coined such a profusion of new words when the necessity was scarcely apparent, should have missed this splendid opportunity when it happened to be imperative.*

However, it is not our intention to discuss neo-realism further, for the doctrine is of no interest to science. All we wish to point out is that in any scientific theory, whether it be due to a Newton, a Maxwell or an Einstein, we can make up our minds from the start that the author will deal with the objective world of science, treating the private views as mere perspectives, and will frame his concepts and definitions accordingly. It would appear, then, that if Dr. Whitehead wishes to defend a doctrine which stands in conflict with the views of every scientist from Newton to Einstein, it would have been conducive to greater clarity had he restricted his attacks to classical science. For when we attack a new theory such as Einstein's, which has introduced such a wealth of

---

* By giving the same name to these two conflicting types of simultaneity, Dr. Whitehead soon falls a victim to his own terminology. Thus, in his book, "The Theory of Relativity," he informs us that he accepts Einstein's discovery of the relativity of simultaneity. But in so doing he does not appear to notice that the type of simultaneity which Einstein has proved to be relative to motion is the scientific type; and it has nothing in common with the neo-realistic type. This latter species is never relative to motion, but solely to position. Hence, without any justification whatsoever, Whitehead has suddenly extended to his own understanding of simultaneity a characteristic which belongs to the type he is attacking. But owing to the same name having been given to the two types, the non-scientific reader is apt to overlook the confusion, and in this way be led to a very erroneous understanding of the nature of Einstein's discoveries.

revolutionary ideas into our understanding of nature, it is to be presumed that the reasons for our dissent arise from those ideas and not from the classical ones which the theory has left intact. Thus, if it were to be proved that matter was continuous and not atomic, we should not hold Einstein's theory responsible for our past belief in the atomicity of matter. Rather would we challenge the experiments and theories of those classical physicists who thought they had established atomicity. On the other hand, if we objected to the view that the simultaneity of external events was relative to the observer's motion, claiming it to be absolute, then of course we should be justified in criticising Einstein's theory proper, since the relativity of simultaneity is peculiar to it.

To be sure, Dr. Whitehead appears to be under the impression that he is merely attacking the new revolutionary ideas introduced by Einstein. Thus, he tells us that he is merely defending "the old-fashioned belief in the fundamental character of simultaneity," but one may well wonder to what period of human thought he is referring. Classical science, some three hundred years ago, had recognised the error of this confusion between external events and the instants at which they were perceived, precisely because this attitude was found to lead to results in utter conflict with experience.

As Weyl expresses it, referring to Römer's discovery in 1675: "The discovery that light is propagated with a finite velocity gave the death-blow to the natural view that things exist simultaneously with their perception." Needless to say, that which applies to luminous transmissions applies in exactly the same measure to the propagation of nerve excitation through the body and to all other species of transmissions.

The general conclusions we have reached from a discussion of simultaneity in different places apply without great change when we consider a number of other concepts which in a dim way we half feel and half comprehend. Thus, consider the case of temperature. Why does the scientist have recourse to thermometers rather than trust to the immediate disclosures of his senses? Merely because the thermometer furnishes an objective definition of the equality of two temperatures, and because it can discern differences which our senses would be unable to detect. The result is that when his thermometer indicates differences in temperature of several hundred degrees centigrade—say, temperatures of $-200°$ and $+400°$—he maintains that such a difference exists, even though to our senses a scalding sensation would result in either case. But, however imprecise and unreliable our direct appreciation of temperature may be, our direct recognition of a simultaneity of spatially separated events is vaguer still. For whereas men may have no great difficulty in agreeing that one room is warmer than another, even without the aid of a thermometer, they will express the most conflicting opinions when asked which of two explosions occurred first. Their answers, unless governed by the results of physical measurements, will vary with their relative positions.

Now, because the scientist measures temperature by the thermometer, no one would accuse him of wishing to imply that the concept of temperature would cease to exist were his thermometer to fall to the ground and be broken. Why, then, should the fact that we determine simultaneity in different places by means of light signals imply that we are making the concept of simultaneity dependent on the behaviour of light, and that simultaneity could have no meaning for blind people?

The sole purpose of all these measurements is to render precise and to clarify concepts which we understand only in a dim, indefinite way. In the majority of cases these concepts have arisen from crude experience. There are instances, however, where they have arisen in a more sophisticated manner. Suppose, for example, that our bodies were insensible to heat or cold. The behaviour of our thermometers would still suggest an abstract concept which we might call "temperature." In fact, even were no thermometers at our disposal, the concept of temperature would be introduced eventually into physics for theoretical reasons. "Entropy" affords us an illustration of a concept introduced in this way. Our senses cannot detect it and our instruments do not record it directly. Nevertheless, it has imposed itself on science. A difference in entropy can be measured indirectly, just as temperature, duration and space can be measured, but always by physical means; and this is indeed the only way we can increase our understanding of these concepts, and agree on what we are talking about. Incidentally, we may recall that it was through the medium of physical measurements that Einstein was able to demonstrate the relativity of simultaneity, which our so-called direct apprehension of nature had never even led us to suspect. Hence, to accept the relativity of simultaneity, which issues from highly sophisticated physical measurements, and then to cling to the theory of a direct apprehension of nature, is a course which, to any logical mind, must appear to be the height of inconsistency.

We have now to consider a second criticism also advanced by Dr. Whitehead. In it he appears to have abandoned momentarily the doctrine of direct apprehension, recognising (presumably for the sake of argument) that a determination of simultaneity at spatially separated points may require the introduction of physical propagations, but he objects to Einstein's exclusive appeal to optical transmissions for this purpose. Thus, he tells us:

"Also there are other physical messages from place to place; there is the transmission of material bodies, the transmission of sound, the transmission of waves and ripples on the surface of water, the transmission of nerve excitation through the body, and innumerable other forms which enter into habitual experience. The transmission of light is only one from among many."

Inasmuch as these arguments are advanced in a spirit of criticism, we must presume that according to Dr. Whitehead, had transmissions other than optical ones been considered, determinations differing from Einstein's would have ensued.* But a contention of this sort would be fundamentally untrue. Einstein's determinations would have yielded exactly the same results had sound transmissions been substituted for optical ones; and the only reason optical ones are continually referred to is because they permit observations of greater accuracy. We shall now consider the reasons for these statements.

The problem of time determination may be divided into two parts: First, we wish to define equal successive durations at the same point of space, for instance, at the point where we happen to be standing. Secondly, we wish to co-ordinate our time reckonings with those computed by other observers situated at different points of space.

As we have already considered the problem of time-congruence at a point, we will not dwell on it unduly. Suffice it to recall that our direct intuition of time-congruence is far too vague and uncertain to be of any use in scientific investigation. Accordingly, throughout the course of history we find men relying on physical processes of one sort or another, the burning of candles, sand clocks, mechanical clocks, rotation of the earth, vibrations of atoms, etc. Newton gave a theoretical definition of time-congruence when he formulated the law of inertia, according to which a perfectly free body described equal distances in equal times. According to this definition, by measuring equal distances along the body's path, we were enabled to mark out successive equal durations. But in common practice it was more convenient to appeal to chronometers regulated ultimately by the earth's rotation. We may now pass to the problem of synchronisation in classical science.

Let us assume that we are in possession of a number of chronometers which, when placed side by side, beat out their hours, minutes and seconds in perfect unison, hence advance at the same rate. Suppose now that having drawn a circle of, say, ten miles' radius, we maintain one chronometer at the centre and carry the remaining chronometers to various points distributed on the circumference of the circle. If some one were to play a prank and displace the hands of the chronometers located on the circumference, we should be faced with the problem of re-establishing synchronism. Classical science suggested various methods for obtaining this result.

First, we might transport our centre chronometer to each of the

* The majority of Dr. Whitehead's views are unfortunately couched in such loose and obscure terms that it is usually possible to place a variety of conflicting interpretations upon them. As he himself candidly admits in his book, "The Principles of Natural Knowledge": "The whole of Part II, *i.e.*, Chapters V to VII, suffers from a vagueness of expression due to the fact that the implications of my ideas had not shaped themselves with sufficient emphasis in my mind."

Obviously, if the author of a book is not quite clear as to what he means, there is always some danger in settling the matter for him. Accordingly, we should have omitted any reference to Dr. Whitehead's writings had it not been that some of his criticisms (though totally erroneous in our opinion) are often of considerable interest, leading as they do to a novel vision of things. At any rate, in any further reference to Whitehead, we shall view his criticisms in the light of type-objections, regardless of whether or not, as interpreted by us, they depict his own personal views.

circumference chronometers in turn, and thus re-establish synchronism. But this method was open to the objection that by displacing a chronometer back and forth we might in some way disturb its working.

A second method was to appeal to the earth's rotation on its axis. By ascribing a value of 24 hours to the complete rotation of our planet, we could divide its surface with equally distanced meridians, and then assert that the passage of the same star from one meridian to the next would always take the same time. Suppose, then, for argument's sake, that the earth's surface had been divided by 24 half meridians, all equally spaced. Then if a star passed a certain meridian at A at a certain time t, it would pass the contiguous meridian B one hour later. In order to synchronise two clocks situated on these two meridians A and B, all that would be necessary would be to set the first clock A at "zero" hour as the star passed its meridian A, then set the second clock B at the hour "one" as the star passed over the second meridian B. The clocks would then be synchronised, so that an event occurring by the first clock as its hands marked a time t, would be simultaneous with an event occurring by the second clock as its hands also marked t. But this method again was open to certain objections which it is unnecessary to recall since we have discussed them in a previous chapter. They deal with the retarding influence of the tides and with the earth's "breathing." To be sure, these objections are of a hair-splitting nature, but they are theoretically valid just the same since they do not reduce to mere errors of observation.

A third method of synchronisation considered by classical science was obtained by appealing to propagations, which would serve to establish communications between the spatially separated clocks. The propagation of sound waves suggested itself in this capacity. According to this method we should operate as follows: At noon, sharp, as marked by our chronometer at the centre, we should fire a gun. All the observers standing by their clocks on the circumference would then be instructed to set their clocks, the instant they heard the report, at the hour "noon plus the time required by the sound to progress from centre to circumference." Given the velocity of sound in air, and given the distance from centre to circumference, this lapse of time could be computed, and the clocks synchronised. But here again we should have to make sure that no wind was blowing; also, even in the case of a stagnant atmosphere, we should have to assume that the speed of sound was constant and remained the same in all directions, i.e., that its propagation was isotropic. We knew that this was the case to a first approximation, but no highly refined experiment had ever established the fact.

Finally, for sound waves we might substitute light signals or radio signals. This was the method most generally employed by classical science; it permitted greater accuracy than sound signals and was of wider applicability than determinations in terms of the earth's rotation. When, for instance, the astronomer wished to determine the present position of a comet, the optical method was the only one possible.

Accordingly, he was compelled to take into consideration the time required for the comet's light to reach him. In other words, simultaneity was determined by optical signals. In much the same way, the Eiffel Tower in Paris sends out a radio message at noon every day, and ships at sea can regulate their clocks accordingly. (Needless to say, all this has nothing to do with the Einstein theory.) Nevertheless, while recognising that a determination of simultaneity via optical signals was the most precise method possible, classical science had certain misgivings as to its theoretical accuracy. The effect of the ether drift, generated by the earth's motion, was assumed to produce an anisotropy in the propagation of light, much as the wind would do in the case of sound, causing it to proceed more rapidly in one direction than in another. Inasmuch as the velocity of this ether drift was unknown, there was no means of introducing the necessary corrections.

Now, what Einstein urges us to recognise, as a result of the negative experiments in electromagnetics, is that these misgivings of classical science were unfounded. Whether it exist at all or whether it be a myth, the ether drift exerts no effect on the propagation of light. The space of a Galilean frame is isotropic so far as the propagation of light is concerned; that is to say, light proceeds with the same speed in every direction. These results having been established by the most accurate experiments known to science, there is every reason to accept them until such time at least as their validity is disproved by further experiments still more precise in nature. Under the circumstances, the optical method of fixing simultaneity, which was already the one followed prior to the theory of relativity, becomes, in view of Einstein's discovery, the most perfect method at the command of science.

From this rather long discussion we see that in the determination of simultaneity Einstein's theory does not suggest any essentially new method. With either science, the measure of time and the determination of simultaneity in different places can be arrived at only by an appeal to physical processes of one sort or another. If we wish to discard physical determinations, and define things in the abstract, the concept of simultaneity in different places degenerates into a mere verbal expression. The following passage quoted from Einstein, in his Princeton lectures, may clarify the matter:

"The theory of relativity is often criticised for giving, without justification, a central theoretical rôle to the propagation of light, in that it founds the concept of time upon the law of propagation of light. The situation, however, is somewhat as follows: In order to give physical significance to the concept of time, processes of some kind are required which enable relations to be established *between different places. It is immaterial what kind of processes one chooses* for such a definition of time.*

"It is advantageous, however, for the theory, to choose only those processes concerning which we know something certain. This holds for

* Italics ours.

the propagation of light *in vacuo* in a higher degree than for any other process which could be considered, thanks to the investigations of Maxwell and H. A. Lorentz.

"From all these considerations, space and time data have a physically real, and not a mere fictitious significance."

Let us return to Whitehead's criticisms. It must be admitted that from the standpoint of experiment and observation, optical determinations are amenable to greater accuracy than all other methods of observation based on the propagation of sound, etc. Not only are they easier to observe, but they are also less contingent on variations in surrounding conditions (*e.g.*, velocity of wind). Now, as we have·remarked previously, when the critic attacks the theory of relativity on the ground that it defines simultaneity by optical methods, it must be presumed that he attributes the strangeness of Einstein's theory to its exclusive use of light signals. Were this opinion correct, it would certainly appear rather arbitrary to found a theory of the universe on one particular choice of time determinations, when other methods would have yielded our classical results. Why, indeed, define time in terms of light rather than of sound or moving bodies? But, as mentioned elsewhere, the opinion *is not correct,* for regardless of the type of propagation to which we appeal, exactly the same determinations of simultaneity would ensue under ideal conditions of observation. Every experimental test which has verified the anticipations of the theory proves this last statement indirectly.

In order to make this fundamental point clearer, let us consider the following concrete example: Suppose we stand at a midpoint between two guns. According to Einstein's optical criterion, the two guns would have been fired simultaneously were we to perceive the two flashes at the same instant of time. Classical science would have dissented from this view on the ground that the ether drift would have vitiated our results in a minute degree; but even classical science would have granted that had the physical effect of the ether drift been proved inexistent, there was nothing further to dissent from in Einstein's definition. If, then, with Einstein, we assume that the negative experiments have proved the irrelevance of the ether drift, both classical science and relativity would recognise that a simultaneity in the perception of the two flashes would have entailed a simultaneity in the hearing of the two reports (assuming no wind was blowing). Also, both bullets would pass us at the same instant (assuming the guns, charges and bullets were identical).

It follows from this that Einstein's theory, since it has started out by claiming the effect of the ether drift to be nil, must inevitably lead to the conclusion that however we elect to define simultaneity (whether by the optical, acoustic or bullet method), exactly the same simultaneity determinations will ensue. Dark, cloudy nights or the accidental circumstance that we are gifted with eyesight have therefore nothing to do with the matter; and blind men would not be debarred, since they could always synchronise their clocks by means of sound signals.

We see, then, that the strangeness of Einstein's theory arises not from his use of optical signals, but from his fundamental premise that the ether drift is incapable of affecting the course of physical phenomena. Those who wish to dissent from the theory should therefore concentrate their attention on the ether-drift problem.   Einstein's preference for optical signals is a mere side issue which could perfectly well be discarded in favour of sound transmission were it not for reasons of practical convenience and accuracy.   No change in the theory would ensue on this score.   We have established this point in the case of simultaneity, but in view of its importance we may consider the problem in a more general way.

As we have noted, once the irrelevance of the ether drift is admitted and the equations of electromagnetics accepted as accurate, we are led to those space and time transformations known as the Lorentz-Einstein transformations.   When we assert that duration might just as well have been defined in terms of moving bodies or sound transmissions (under ideal conditions of observation), our statement is equivalent to asserting that the Lorentz-Einstein transformations, though obtained from a study of electromagnetic and optical phenomena, are yet of universal applicability to all manner of phenomena.   Here the critic may object that this is a gratuitous assumption.   Maybe it is, but we must understand that were its validity to be contested, were the classical transformations to hold for mechanics and acoustics, while the Lorentz-Einstein ones were valid for electromagnetic processes, we should be faced with insufferable contradictions.   Thus, according to the Lorentz-Einstein transformations, as applied to electromagnetics, a ray of light proceeding *in vacuo* will pass any two different Galilean observers with the same speed c.   But according to the classical transformations, as applied to mechanics, a body which is moving with the velocity c with respect to one observer will be moving with a different velocity relatively to the second observer. If, then, a body were moving along with the crest of a light wave in the opinion of one observer, it would either outstrip the wave or fall behind it in the opinion of the second observer.   A coincidence for one would no longer be a coincidence for the other.   Coincidence would thus cease to be absolute, and it is doubtful whether any science could exist, since all unity in nature would be lacking.   Furthermore, the coexistence of two conflicting species of space and time transformations appears very improbable when we remember that it is impossible to establish a clear-cut distinction between mechanics and electrodynamics, since matter seems to be in large part an electromagnetic phenomenon.

Hence it is only natural that the scientist should assume that the Lorentz-Einstein transformations, though deduced from electromagnetic or optical processes, should yet hold in all domains of physical science, even in the interior of the atom.   It is true that these transformations were obtained from a study of optical and electromagnetic phenomena, not from acoustics or mechanics; but this is merely because experiments as sensitive as those of Michelson and Fizeau could scarcely have been performed with sufficient accuracy had sound waves, nerve excitations or ripples on the water been contemplated. And yet, had Fizeau's experiment been performed with sound waves, there is every reason to suppose that the curious composition of velocities would have been observed.

The arguments thus far presented show that the space and time determinations derived from electromagnetics and optics are in all probability also those holding for the entire physical universe. Such is indeed Einstein's attitude when he unhesitatingly applies the relativistic transformations to all manner of phenomena. But it is precisely because as a result of this application he has anticipated phenomena at variance with those deduced from the classical transformations, and it is precisely because his anticipations have been verified by highly accurate experiments, that his assumption of the universal validity of the relativistic transformations must be considered justified.

On no account, therefore, may we regard time and simultaneity, when defined by optical signals, as constituting a special kind of optical or electromagnetic time and simultaneity, at variance with equally justified determinations issuing from mechanical processes or sound transmissions.

In short, if we wish to co-ordinate natural phenomena with simplicity, we must thus conceive the various time-flows and simultaneities defined by Einstein, to represent the times imposed upon us, not by the caprice of a mathematician, not by optical transmissions alone, but by the entire physical world. The times of relativity are therefore those which will govern all physical processes, not merely optical ones; and for this reason we must regard them as representing the real time-flows of the universe.

An understanding of this point is essential. For instance, in the general theory we shall see that Einstein's law of gravitation necessitates a slowing down of time in the proximity of large masses of matter. If we accept the general significance of this fact as affecting all physical phenomena, it will entail the slowing down in the rate of vibration of an atom located in a gravitational field. This is the celebrated *Einstein shift-effect*, which has since been verified. Before it was detected, however, and while it was still considered doubtful, some thinkers had argued that it might perfectly well prove to be non-existent on the ground that the atom might not act like a perfect clock and might fail to adjust itself to the local rate of time-flow. Einstein, however, refused to accept this means of escape and stated that the entire theory would have to be rejected if the effect were not verified, for the times of relativity should be of universal significance or else worthless.

We may conclude by saying that it appears unjustified to refer to Einstein's time and simultaneity determinations as depending on the behaviour of light. They depend on the processes of nature of which the behaviour of light is but a particular illustration. We must therefore amend Dr. Whitehead's criticism of relativity, as making the very meaning of simultaneity depend on light signals, by replacing the words "light signals" by the words "physical processes." But when amended in this way, the criticism loses all its force. For we may speak of simultaneity throughout space, of an instantaneous space, as much as we please, but until we know enough about these evasive concepts to be able to distinguish events that are simultaneous from those which are not, we cannot claim to have any definite idea of what we are talking about. On this score, both classical science and relativity are in perfect agreement.

CHAPTER XVI

PRACTICAL CONGRUENCE IN RELATIVITY

In a preceding chapter we mentioned certain of the most important aspects of the problem of physical space. We saw that the concept of spatial equality or congruence was deemed to have arisen from the facts of experience. Certain objects appeared to maintain the same visual aspect wherever displaced, provided we modified our own positions as observers in an appropriate way. But such fundamental recognitions were too vague to be of any use to science; hence congruent bodies were defined as those which, when maintained at constant temperature and pressure, coincided when placed side by side.

Congruence, as thus defined, involved physical measurements with material bodies; and, as Poincaré remarked, all we could ever discover in this way would reduce to the laws of configuration of solid bodies, space itself transcending our experiments, since the bodies might behave one way or another in the same space. Poincaré's attitude drives us to complete agnosticism so far as the geometry of space is concerned. If physical measurements are denied us, there is no means of solving the problem of space, for we have no *a priori* means of deciding that the structure of space is this or that. Logical arguments are of no avail, for they do not lead us to any definite solution, only to a variety of possibilities. We are thus thrown back on Poincaré's main contention, *i.e.*, "space is amorphous." In it we can define congruence in any way we please (theoretical congruence), although for reasons of practical convenience it is necessary to be guided by the properties of so-called rigid bodies. Thus we obtain a physical definition of practical congruence, which permits us to determine the geometry that for all practical purposes is to be called the geometry of real physical space.

When we consider Poincaré's arguments, there is very little to be said against them. Nevertheless, the physicist must concentrate his attention on the space with which he can cope, with the one that can be explored by physical methods. This attitude defended by Einstein (which was also the attitude of Gauss and Riemann) leads us to the space whose geometry is defined by the paths of light rays, by the laws of nature when expressed in their simplest form, and again by the numerical results of measurement obtained with material rods maintained under constant conditions of pressure and temperature. It is with this space that we shall be concerned.

But before proceeding, let us make it quite plain that according to the present view the office of our rods is to reveal the pre-existing structure of space much as a thermometer reveals the pre-existing differences of temperature throughout the house. On no account, therefore, may it be claimed that the rods are assumed to create space any more so than the thermometer creates temperature or the weather glass creates the weather. The rods permit us to explore space bcause they adjust themselves or mould themselves to its structure, or metrical field. The origin of this structure is still very mysterious. For Riemann it was created by matter; but by matter he did not mean the matter of our rod: he meant the total aggregate matter of the entire universe. Einstein's cylindrical universe confirms Riemann's views. At all events, regardless of the mysterious problem of the ultimate origin of the metrical field, regardless of whether this structure of space exist *per se* as a characteristic of physical space or whether it be generated by the masses of the cosmos, in either case it pre-exists to our local exploration with rods, whose masses are far too insignificant to modify it to any perceptible degree.

Also let us recall that when we discuss space, we must specify it by referring it to a frame of reference, in particular to a Galilean frame (one in which no forces of inertia are experienced). We then proceed to consider the laws of disposition of our congruent bodies in this frame, and the nature of the geometrical results obtained will define the geometry of space. As we know, Euclidean geometry is the outcome (at least to a first approximation).

Now classical science had never thought it necessary to stipulate that our rods should stand at rest in our frame, but in view of the disclosures of relativity we see that it is essential to specify this condition. For two rods whose extremities would coincide when at relative rest would cease to coincide when in relative motion, so that a square traced on the floor of our frame would turn out to be an elongated rectangle when measured by a rod in relative motion. To be sure, similar results would have been expected even in classical science, had it been assumed that the moving rod was compressed or distorted owing to its motion through the ether. In a case of this sort, the modified readings of the rod would not have involved space and its geometry, any more than when by heating or crushing a rod we alter its readings. The blame for the discrepancy would have been placed solely on the disturbing conditions affecting the rod.

Under Lorentz's theory this interpretation could be entertained, since the FitzGerald contraction would have been explained in terms of the pressure generated by the rod's motion through the stagnant ether. But under the theory of relativity this interpretation can no longer be defended. For when relative velocity exists between A and B (*not relative acceleration*), there is no sense in enquiring whether it is A or B which is in motion. This, indeed, constitutes the essence of the special principle of relativity. Hence, when we consider the moving rod, we can ascribe no absolute significance to its motion. Whatever motion exists between frame and rod might with equal justification be attributed to the motion of the frame gliding beneath a motionless rod, there being no absolute term of comparison like the stagnant Lorentzian ether to decide the issue. And so we cannot claim that the so-called moving rod has suffered a physical change by reason of its motion, a change rendering it unfit for measurement purposes.

And yet, if we retain all our rods as equally valid, whether in motion or at rest, we obtain conflicting results which are such as to deprive the geometry of space of all significance. When we analyse this anomalous situation we find that the following facts stand out clearly: Rods remain unmodified whether in motion or at rest, and yet their length varies. The truth is that length is not a definite characteristic of a body; it appears as a shadow of a something else projected into the space of our frame. When the rod is at rest in our frame, its length, or the shadow, presents a maximum; it is as though the something and its shadow coincided. But when relative motion is present, the something becomes tilted—tilted along a fourth dimension out of the space of our frame; and its shadow is shortened in proportion. Yet this rod which is moving in our frame is at rest in some other frame, and in this other frame, therefore, the something again coincides with its shadow. And so we must assume that the various Galilean frames and the spaces they serve to define are variously tilted, rotated in a fourth dimension with respect to one another.

There is then no longer one all-embracing Euclidean space which may be defined with respect to one Galilean frame or another. In its place we must conceive of an indefinite member of Euclidean spaces variously tilted, yet fused together.* To each separate Galilean frame one of these spaces will correspond, so that when we change our motion we are also changing our space.

The situation is analogous to that which arises when we consider various verticals drawn to the earth's surface.

John, in London, will assert that his vertical is truly vertical and that Peter's at New York is slanting, and Peter will return the compliment. Owing to these discrepancies, we cannot regard the two verticals as defining one same direction, though they both may be truly vertical for John and Peter respectively. Rather must we say that there exist an indefinite number of different vertical directions, no one of which is more truly vertical in any absolute sense than any other. Now replace the concept of vertical by that of three-dimensional space, and we have a picture of the spaces of relativity.

Very similar conclusions apply to time. Thus, consider two observers A and B in their respectively moving frames. Each observer holds a watch in his hand, and these watches, when at rest with respect to each

---

* For the present we are confining ourselves to the consideration of Galilean frames; the problem becomes highly complicated when we study the peculiarities arising from a choice of accelerated frames. In this latter case, a non-Euclideanism, or curvature of space, appears with the presence of acceleration (as also of gravitation), so that we cannot even refer to a multiplicity of superposed Euclidean spaces. Also, space and time become so hopelessly confused that we are compelled to express ourselves in terms of four-dimensional space-time, where everything becomes clear once more.

other, beat the same time; they define congruent stretches of time. The classical idea of practical time-congruence, founded on approximate observations, led us to maintain that these conditions would still endure regardless of the positions of the clocks in space *and regardless of their relative motions.*

But it now appears that the time-stretches defined by the ticking of a clock at rest in the frame of reference A would appear to an observer B in relative motion to be longer than they would appear to A. The result is quite general and applies to all processes. Thus, the duration of the revolution of a top will be possessed of varying values according to the relative motion of the observer. Arguments similar to those we elaborated in the case of space go to show that time is no longer unique. A multiplicity of different time directions, variously tilted, must coexist, each one referring to some particular frame of reference, hence to some particular observer. As a result, the simultaneity of two events occurring at different points of space can no longer be absolute, since simultaneity depends essentially on the time which we select, and therefore on our frame of reference.

Just as each frame had its own space, so also would it now have its own time and its own definition of simultaneity. We may summarise these various discoveries by stating that there can exist no universal definition of practical congruence either for space or for time; there can exist, therefore, no geometry for a universal space or for a universal time.

The philosophical consequence of these new points of view is to deprive universal space and time of the objective significance with which they were formerly credited. Henceforth, in the words of Minkowski, "space and time, by themselves, sink to the position of mere shadows."

Thus, consider a yardstick. If this same yardstick is measured by various observers passing it with different velocities, it will be possessed of different lengths. The space this rod occupies can then be credited with no particular magnitude, since this magnitude itself depends as much on the point of view of the observer as on the characteristics of the rod. This is what is meant by the relativity of distance. We must not confuse this discovery of Einstein's with the fundamental relativity of all distance in conceptual space. For while classical science recognised that there was no such thing as absolute distance in conceptual space, yet it believed that in the real space of the physicist the distance between two given points in a frame of reference could be defined unambiguously for all observers, by means of rigid rods. It is this last opinion which is shattered by Einstein's discoveries.

It is the same for duration when we remember that the duration of any phenomenon will be possessed of varying values according to the relative motion of the observer; and it is this fact which is expressed by the statement that time or duration is relative.

Thus, neither space nor time by itself can exist in the real phenomenal world which the physicist explores. Space and time appear

as mere modes of perception, mere relations, varying and changing according to the conditions of relative motion existing between the . observer
and the observed. With the disappearance of the objectivity of space
and time regarded as absolute and universal forms of our perception, the
objectivity of the entire physical world seemed to sink into a mist. If,
then, any kind of objectivity was to be retained, it was necessary to
effect a synthesis of all the individual points of view of all the different
observers, to mould into one sole representation the multiplicity of individual spaces and the multiplicity of individual durations.

Could this result be accomplished, the relativity of practical congruence for both space and time would give way to some universal definition of practical congruence holding for all observers regardless of their
relative motion, and connoting, therefore, the existence of some absolute
continuum no longer a mere shadow. In place of the individual point
of view varying with the relative motions of the observers, we should
obtain an impersonal and hence common objective understanding of
nature.

The achievement of this supreme synthesis of the points of view of
all observers was accomplished by Minkowski in 1908. He succeeded in
obtaining an invariant definition of congruence by combining any given
observer's definitions of practical congruence for space and for time,
and by showing that this combination possessed an invariant value holding equally for all other observers. Of course, the type of congruence
obtained was one neither of space nor of time. It was a combination
of both. But its existence showed us that, transcending space and time,
there existed an impersonal and fundamental four-dimensional metrical
continuum which Minkowski called space-time. The definition of congruence obtained by him for this mysterious continuum proved it to be
Euclidean (more properly, semi-Euclidean).

The elusiveness of space and time, or rather the ambiguity of these
concepts, depending as they do on the observer who measures and senses
them, is replaced by the common objectivity of this fundamental continuum, which is the same for all and transcends the particular conditions
of motion of the observer. This continuum, though of itself neither space
nor time, yet pertains to both in that the observer is able to carve it up
into that particular space and that particular time which are characteristic of the frame of reference in which he is stationed, and which constitute the space and time in which he lives and experiments.

From now on, the real extension of the world has a fourfold order,
and the geometry of relativity is necessarily four-dimensional. What we
call the space of our Galilean frame at an instant, is but a cross-section
of this four-dimensional space-time world; and what we call the time
of our frame lies along a perpendicular to this section. As we pass from
one Galilean frame to another, we change our space and we change our
time. Our new instantaneous space, though still a three-dimensional
cross-section of the four-dimensional space-time world, is tilted with
respect to our former cross-section; and the same tilting ensues for the

perpendicular direction called time. In fact, the situation presents a striking analogy with the various directions of the verticals to the earth's surface. They, also, manifest different directions because the various two-dimensional portions of the earth's surface are variously tilted in three-dimensional space. The great change in our understanding of the spatio-temporal background, the great difference between space-time and the separate space and time of classical science, resides, therefore, in this tilting process which accompanies a change of relative motion.

In order to make these points clearer, let us understand that when we speak of the instantaneous spaces of two observers in relative motion as being tilted one with respect to the other, hence as lying outside of each other (except for their common two-dimensional cross-section), we do not mean that the space of one is some incomprehensible entity in the opinion of the other. The tilted instantaneous spaces will enter into our perception, but not as spaces alone: they will appear as a succession of two-dimensional surfaces moving with uniform speed.

All these spaces are equally justified. No one of them stands out more prominently than any other on the ground of symmetry with respect to the four-dimensional world. Only when we specify the frame of the observer will one particular space and one particular duration be allotted. Hence we may say that practical congruence exists for space and for time, as in classical science, provided the Galilean observer effects his measurements with rods and periodic mechanisms, such as clocks, which do not move about in his Galilean frame while performing measurements. There is thus a perfectly definite physical meaning in stating that the distances between two point-pairs at rest in the frame are congruent or unequal; but we must specify that this distance is measured according to the standards of the frame; and the same applies to time-stretches. Likewise, as in classical science, the Galilean observer will discover upon measurement with his congruent rods that his space is Euclidean. It is only when the rods and clocks are in relative motion or when, while at rest in our frame, the frame happens to be accelerated or submitted to gravitational action, that they cease to measure congruent stretches. It is only, therefore, when we reason in an impersonal way, without specifying any particular frame, only when we reason from the standpoint of all possible observers, whatever be their motion, that space and time fade away into shadows; and it is only then that we are compelled, whether we like it or not, to reason in terms of the common objective world of relativity, that is, in terms of four-dimensional space-time.

CHAPTER XVII

THE MATHEMATICAL EXPRESSION OF EINSTEIN'S FUNDAMENTAL PREMISES

CONSIDER a Galilean system, and two points A and B in this system. If a ray of light is propagated from A to B, the postulate of the invariant velocity of light demands that the distance AB measured by *us* in the frame, divided by the duration required (according to the time of our frame) for a light wave to cover this distance, be always equal to c, the velocity of light. If, now, we had viewed the same phenomenon from some other Galilean system, the points A and B of the first system at the instants when the ray of light passed them would have corresponded to some other points A' and B' of our second system. But, just as in the first case, we should have been able to measure the velocity of light between the two points A' and B' of our second system. The principle of the invariant velocity of light states that in whatever Galilean system we might have operated, the measured velocity of light *in vacuo* would always be the same.

Of course this velocity may be positive or negative, according to whether the light ray is directed to the right or to the left; but we can obviate this ambiguity of sign by considering squared values.

The mathematical translation of this principle of physics yields us the following equation, which must remain invariably zero in value for all Galilean frames:

$$dx^2 + dy^2 + dz^2 - c^2 dt^2 = 0 \quad \text{(using differentials)}$$

When we pass from one Galilean frame to another, dx, dy, dz, dt may vary in value; but the variations must be so connected that the sum total of the above mathematical expression remains invariably zero. Such is the mathematical condition which expresses the principle of the invariant velocity of light.

Now it is our object to determine exactly in what measure these different magnitudes dx, dy, dz, dt must vary in value when we pass from one Galilean frame to another, moving with a constant velocity v with respect to the first, if the previously written mathematical condition of invariance is to remain satisfied. From a purely mathematical standpoint problems of this type form part of a branch of mathematics known as the theory of invariants. Such problems had been studied many years before, and it was known that the relations between the variables dx, dy, dz, dt, or, what comes to the same thing, the transformations to which it was necessary to subject these variables (in order to satisfy the condition of invariance set forth above), were given by a wide group of transformations known as *conformal transformations*.*

But when, in addition, the relativity of velocity is taken into consideration it is seen that *conformal transformations* are far too general. We must restrict them; and when the required restrictions are imposed we find that the rules of transformation according to which the space and time co-ordinates of one Galilean observer are connected with those of another depend in a very simple way on the relative velocity v existing between the two systems. These rules of transformation are given by the **Lorentz-Einstein** transformations.†

Now these transformations are, as we have said, more restricted than the conformal transformations; and this lesser generality of the Lorentz-Einstein transformations has, as a consequence, the further restriction of the conditions of invariance of the mathematical expression mentioned previously. Not only will this expression have a zero value for all Galilean frames when it has a zero value for one particular Galilean frame, but in addition, if it does not happen to have a zero value in one frame but has some definite non-vanishing numerical value, it will still maintain this same definite non-vanishing value in all other Galilean frames. In other words, Einstein's premises are represented mathematically by the invariance of the total value of $dx^2 + dy^2 + dz^2 - c^2 dt^2$ for all Galilean frames, regardless of whether this value happens to be zero or non-vanishing.

The deep significance of this condition of invariance was first noted by Minkowski, and it led, as we shall explain in the next chapter, to the discovery of four-dimensional space-time.

---

* Conformal transformations are those which vary the shape of lines while leaving the values of their angles of intersection unaltered. They are of wide use in maps, *e.g.*, in Mercator's projection or in the stereographic projection.

$$\dagger \; dx' = \frac{1}{\sqrt{1 - \dfrac{v^2}{c^2}}}(dx - vdt) \; ; \; dt' = \frac{1}{\sqrt{1 - \dfrac{v^2}{c^2}}}\left(dt - \frac{vdx}{c^2}\right)$$

the direction of motion being situated along the Ox axis.

CHAPTER XVIII

THE discovery of this invariant

$$dx^2 + dy^2 + dz^2 - c^2 dt^2$$

whose value we shall designate by ds^2, marks a date of immense importance in the history of natural philosophy. The fact is that with Einstein's discoveries such familiar absolutes as lengths, durations and simultaneities were all found to squirm and vary in magnitude when we passed from one Galilean system to another; that is, when we changed the constant magnitude of the relative velocity existing between ourselves as observers and the events observed. On the other hand, here at last was an invariant magnitude ds^2, representing the square of the spatial distance covered by a body in any Galilean frame, minus c^2 times the square of the duration required for this performance (the duration being measured, of course, by the standard of time of the same frame). It mattered not whether we were situated in this frame or in that one; in every case, if ds^2 had a definite value when referred to one frame, it still maintained the same value when referred to any other frame.

Obviously, we were in the presence of something which, contrary to a distance in space or a duration in time, transcended the idiosyncrasies of our variable points of view. This was the first inkling we had in Einstein's theory of the existence of a common absolute world underlying the relativity of physical space and time.

Minkowski immediately recognised in the mathematical form of this invariant the expression of the square of a distance in a four-dimensional continuum. This distance was termed the **Einsteinian interval,** or, more simply, the **interval.** The invariance of all such distances implied the absolute character of the metric relations of this four-dimensional continuum, regardless of our motion, and thereby implied the absolute nature of the continuum itself. The continuum was neither space nor time, but it pertained to both, since a distance between two of its points could be split up into space and time distances in various ways, just as a distance in ordinary space can be split up into length, breadth and height, also in various ways. For these reasons it was called **Space-Time,** and the interval thus became synonymous with the distance between two points in space-time.

And here certain aspects of the problem must be noted. In the first place, it may appear strange that measurements with clocks can be co-ordinated with measurements with rods or scales. This difficulty, however, need not arrest us; for although dt is a time which can only be measured with a clock, yet cdt, being the product of a velocity by a time, is a spatial length since it represents the spatial distance covered by light in the time dt. For this reason we may consider our four-dimensional continuum to possess the qualifications of an extensional space.

Our next problem is to determine the nature of the geometry of this mysterious continuum. The mere fact that it has been possible to write the expression of this square of the interval with squared magnitudes, such as dx^2, dy^2, etc., without having recourse to terms such as $dx\,dy$, $dx\,dt$, etc., would of itself imply that the continuum was flat and free from all trace of non-Euclideanism or curvature. Furthermore, it implies that the mesh-system to which we refer this distance is one of straight lines, and is not curvilinear (see Chapter VII).

The only flat continuum we have studied so far is the Euclidean one, but in the present case we see that though flat, our continuum is not strictly Euclidean owing to the minus sign preceding cdt; or, again, owing to the fact that $g_{44} = -1$, instead of $g_{44} = +1$, as in a four-dimensional Euclidean space. Yet, because of the constancy in the values of the g_{ik}'s throughout the mesh-system, hence because of the flatness or absence of non-Euclideanism or curvature of the continuum, the analogy with a Euclidean continuum is very great, and for this reason it is called semi-Euclidean.*

We say, therefore, that space-time is a four-dimensional semi-Euclidean continuum, and that it is differentiated from a truly Euclidean four-dimensional one solely because it has one imaginary dimension and three positive ones in place of four positive ones. In this continuum, time represents the so-called imaginary dimension, and the three dimensions of space represent the three positive ones; though it must be remembered that all we wish to imply by this statement is that there exists a difference between the dimensions of space and that of time. We should be equally justified in calling time a positive dimension, and the three dimensions of space imaginary dimensions.

Now we have also seen that the mathematical form of the expression which gives the square of the distance depends not only on the geometry of the continuum, but also on the system of co-ordinates we may have adopted. In the present case the form of the mathematical expression tells us that our co-ordinate systems or mesh-systems are necessarily Cartesian, that is, are constituted by mesh-systems of four-dimensional cubes. Hence we see that all our different Galilean mesh-systems, when taken as frames of reference, correspond to differently orientated Cartesian mesh-systems in four-dimensional space-time.†

---

* This type of geometry is sometimes called *flat-hyperbolic;* but as the appellation "hyperbolic" is also attributed to Lobatchewski's geometry, it is apt to be misleading. Hence it is preferable to use the expression "semi-Euclidean."

† Minkowski demonstrated the significance of the expression for ds^2 by taking a new variable $T = ict$, where i stands for $\sqrt{-1}$. With this change, ds^2 can be written:

$$ds^2 = dx^2 + dy^2 + dz^2 + dT^2,$$

In short, space-time must be regarded as the fundamental continuum of the universe. Of itself it is neither space nor time. Any Galilean observer splits it up into four directions by means of his particular Cartesian mesh-system; one of these directions corresponds to his time measurements, while the three others correspond to his space measurements.

Henceforth an instantaneous event occurring at a certain point of our Galilean system and at a definite instant of our time will be represented by one definite point in space-time, a point in four-dimensional space-time being called a **Point-Event.** Space-time itself appears then as a continuum of point-events, just as ordinary space was considered to be a continuum of points, and time a continuum of instants. Of course point-events, like points and instants, are mere abstractions; for a point without extension and an instant without duration are mathematical fictions introduced for purposes of definiteness; they are mere abstractions from experience.

Now just as an instantaneous event is represented by a point-event, so a prolonged event, such as a body having a prolonged existence and occupying therefore the same point or successive points in the space of our frame at successive instants of time, will be represented by a continuous line called a **World-Line.** If this line is straight, the body is at rest in our Galilean system, or else is animated with some constant motion along a straight line. If the world-line is curved, the motion of the body is accelerated or non-rectilinear in our Galilean system; hence it is accelerated or curvilinear in space (owing to the absolute character of acceleration). In this way all phenomena reducible to motions of particles are represented by absolute drawings in space-time.

According to the Galilean system of reference we may adopt, we divide space-time into space and time in one way or in another. Hence the resolution of the drawings of phenomena into space and time com-

which is the expression of the square of a distance in a four-dimensional Euclidean space when a Cartesian co-ordinate system is taken. Since this expression is to remain unmodified in value and form in all Galilean frames, we must conclude that in a space-time representation a passage from one Galilean frame to another is given by a rotation of our four-dimensional Cartesian space-time mesh-system. Now rotation constitutes a change of mesh-system to the same extent as would a deformation of the mesh-system, and all changes of this sort entail a variation in the co-ordinates of the points of the continuum. In other words, they correspond to mathematical transformations. The transformations which accompany a rotation of a Cartesian co-ordinate system are of a particularly simple nature; they are called "orthogonal transformations." It follows that if we write out the orthogonal transformations for Minkowski's four-dimensional Euclidean space-time, we should obtain *ipso facto* the celebrated Lorentz-Einstein transformations which represent the passage from one Galilean system to another. This fact is easily verified. If, now, we recall that Minkowski had got rid of the minus sign in the expression for ds^2 by writing T in place of ict, we obtain the following result: Two Galilean systems moving with a relative velocity v are represented by two space-time Cartesian co-ordinate systems differing in orientation by an imaginary angle θ, where θ is connected with v by the formula $\tan \theta = \dfrac{iv}{c}$.

ponents is effected in various ways. We thus obtain variations in the spatio-temporal appearances of phenomena. Notwithstanding these variations, dependent on our relative motion, the absolute space-time drawings themselves, with their intersections, remain unchanged, and a direct study of the absolute world would consist in a study of these absolute drawings without reference to their varying spatio-temporal appearances.

It should be noted that inasmuch as space-time appears as an absolute continuum, existing independently of our measurements, there is no reason to limit ourselves to Cartesian mesh-systems when we wish to explore it. The mesh-system is, as we know, a mere mathematical device facilitating our investigations; and from a purely mathematical point of view we may always with equal justification split up a given continuum with one type of mesh-system or another. However, it must be remembered that in the case of space-time, when we wish to express the results of our measurements in terms of the space and time proper to our frame of reference, the mesh-system we must adopt is no longer arbitrary.

All Galilean observers, as we have seen, must split up space-time with their particular Cartesian mesh-systems. Certain curvilinear mesh-systems will correspond with more or less precision to the partitioning of space-time by accelerated or rotating observers. But in the most general case, arbitrary curvilinear mesh-systems do not correspond to the space and time partitions of any possible observer, though this fact does not detract from their utility.

So far as the essential characteristics of space-time itself are concerned, these Gaussian mesh-systems are just as legitimate as the Cartesian ones. As a matter of fact, when space-time becomes curved, owing to the presence of gravitation, Cartesian co-ordinates cannot be realised, being incompatible with the curvature of the continuum.

Yet, regardless of the mesh-system we use, certain general characteristics of our space-time drawings remain absolute. Thus, we have seen that the interval between two space-time points has a value which is invariant to a change of mesh-system. While it is true that a world-line which measures out as straight from a Cartesian system may appear curved when referred to a curvilinear or Gaussian one, yet, on the other hand, intersections or non-intersections of world-lines are absolute and are in no wise affected by our choice of a mesh-system. It is for this reason that phenomena such as coincidences are absolute, in contrast to simultaneities of events at spatially separated points. These are relative.

Thus, if two billiard balls kiss in the observation of one man, they will continue to kiss in the observation of all other men, regardless of the relative motion of these men. The kissing of the balls constitutes a coincidence, an intersection of the world-lines of the two balls; hence it is an absolute. On the other hand, if two billiard balls hit different cushions simultaneously in the observation of one man, they will not in general hit the cushions simultaneously in the observation of a man in motion with respect to the first. The reason is that our second observer will have adopted a new mesh-system, oriented differently from the first; and in this new mesh-system the two space-time point-events represented by the instantaneous impacts of the two balls against the two cushions will no longer lie necessarily at the same distance along the new time direction.

Owing to the common use of curvilinear mesh-systems in the general theory, we must recall that in a curvilinear mesh-system the square of the interval adopts the more complicated form

$$ds^2 = g_{11}dx_1{}^2 + 2g_{12}dx_1dx_2 + 2g_{13}dx_1dx_3 + \ldots \ldots$$

expressed more concisely by

$$ds^2 = \Sigma g_{1k}dx_1 dx_k.$$

The g_{1k}'s are no longer constants, as when we had $g_{11} = g_{22} = g_{33} = +1$ and $g_{44} = -1$, with all other g_{1k}'s vanishing; they now have variable values from place to place, and as for dx_1, dx_2, dx_3, dx_4, these no longer necessarily represent space and time differences; they are mere mathematical Gaussian number differences. In spite of all this, the numerical value of ds^2, the square of the interval, remains the same regardless of the mesh-system selected.

Psychological duration, or aging, can also be represented in space-time. By psychological duration we mean the flow of time which the observer experiences; or, more precisely still, the duration that would be marked out by a clock held in his hand. Psychological duration is measured by the length of the space-time world-line followed by the observer. When we keep this point in mind the paradox of the travelling twin to be discussed in a later chapter loses much of its paradoxical appearance.

We can also understand how it comes that the discovery of space-time permits us to account for the duality in the nature of motion, relative when translationary and uniform, absolute when accelerated or rotationary. As long as we were endeavouring to account for this duality by attributing an appropriate structure to three-dimensional space, we were met by a peculiar difficulty: If we assumed space to have a flat structure, that is, one of planes and straight lines, it would be possible to understand why a free body, when set in motion with any definite velocity, should be guided along one of these straight lines or grooves in space. It would be following its natural course and there would be complete symmetry all around it; everything would run smoothly and no forces would arise. Then, if we attempted to tear the body away from its straight course, compelling it to follow a curve, we should be forcing it to violate the laws of the space-structure, tearing it away from the space-groove which it would normally follow. It was not impossible to assume that the groove would react and that antagonistic forces would arise. Thus the fact that curved or rotationary motion was accompanied

by forces of inertia, hence manifested itself as absolute, did not constitute a difficulty of an insuperable order. But a real difficulty arose when
we considered the other species of accelerated motion, *i.e.*, variable motion
along a straight line. In this case, the paths of the bodies being straight
(hence in harmony with the flat structure of space), no violation of the
laws of this spatial structure was evidenced. And yet forces of inertia
were again present. It was difficult to interpret their appearance in terms
of the space-structure.

*When, however, we substitute four-dimensional space-time for separate
space and time,* the difficulty vanishes. For now we notice that *all* accelerated motions, regardless of whether they be rectilinear or curvilinear,
are represented by curved world-lines in space-time. All these accelerated
motions violate, therefore, the flat space-time structure; and for this
reason forces of inertia will always be generated by them. On the other
hand, when we consider Galilean or uniform or translational motions, we
see that, as in the case of three-dimensional space, they will be represented
by straight lines. These motions will then stand in perfect harmony with
the flat space-time structure, and no forces of inertia will arise.

In this way, thanks to space-time, one of the outstanding difficulties
which confronted classical science, namely, the dual nature of space and of
motion, is accounted for.

CHAPTER XIX

THE IRREVERSIBILITY OF TIME

WE saw, when discussing the existence of a finite invariant velocity in nature, that the presence of this velocity was going to work havoc with our belief in absolute simultaneity. Thus, if two events take place in different places, we can no longer attribute any universal meaning to the opinion that these two events have taken place at one and the same instant of time or at two different instants. In other words, there exists no absolute clock giving us the correct time at all points of the universe. According to the motion of the observer, two events which are simultaneous for one may follow in sequence for another.

At this point we must guard against possible confusion. It might be argued: "There is nothing very new in this relativity of simultaneity and of the order of succession of our perceptions; for we encounter such examples even in classical science. For instance, if two explosions occur at different spots, by adjusting our position as observers it may be possible for us to perceive the two noises now simultaneously and now in succession." But such arguments would indicate that the revolutionary notion of the relativity of simultaneity had not been grasped.

Thus, in the example of the two explosions happening in different places on the earth's surface, while it is perfectly true that the order of succession in which we hear or see these explosions will depend on our position as observers, yet, on the other hand, if we take into consideration the time which sound or light waves require to cover the distances separating us from the two explosions, we can always decide without ambiguity whether the two explosions occurred simultaneously or in succession. It was never the order or the simultaneity of our perceptions which was considered absolute in classical science; it was solely the simultaneity or order in which these explosions or events had taken place in the outside world.

Now, according to the theory of relativity, not only is the simultaneity of our perceptions relative, but in addition, even when we take into consideration the speed with which the sound waves or light waves advance towards us, it is still impossible to decide whether the two events (regarded as existing in the outside world independently of any observer) are simultaneous or not; for our calculations would show that this simultaneity in the outside world would vary from one observer to another.

The relativity of simultaneity leads us to the kindred subject of the relativity of the order of succession of two events occurring in different places. Here it is impossible to lay down a rigid rule; for we shall see that the theory compels us to recognise that in certain cases the order of succession is indeterminate or relative to our motion, while in others it is absolute, remaining the same for all observers.

The problem is of sufficient importance to warrant a more detailed explanation. Our awareness of the passage of time constitutes one of the most fundamental facts of consciousness. Not only is time continuously passing, but it is ever flowing in the same direction. To this mystery of the unidirectional passage of time, the theory of relativity contributes no new information, so that we may discuss the problem from the standpoint of classical science. Suppose, then, that in a bottle we place two layers of powder, one white and one black. If we shake the mixture long enough, we know that the final result will be a uniform grey mixture. Now we might have anticipated this result without actually performing the experiment. For if we consider the various ways in which the particles of the powders could be distributed in the bottle, sufficient reason would urge us to assert that all distributions were equally probable. But it so happens that by far the greater number of these possible schemes of distribution would yield the appearance of a uniform grey mixture. Probability would suggest, therefore, that on shaking the bottle long enough, the uniform grey mixture would finally appear. When this homogeneous state had been obtained, only once in æons of time would the black and white separation reappear, and then but for an instant. We may express these facts by saying that the general tendency would be a passage from the heterogeneous (black and white layers) to the homogeneous (uniform grey mixture).

The example we have considered is one of extreme simplicity, but inasmuch as the arguments involved appear to be of universal validity, we may say that natural phenomena present a unidirectional sense of advance, passing from the states of lesser probability to those of greater probability. We may therefore identify the states of lesser probability with the past and those of greater probability with the future. The direction of time's passage can thus be defined physically in terms of probability considerations, entailing a mere appeal to sufficient reason and to operations of counting.

This was the definition suggested by Boltzmann (the founder of the kinetic theory of gases). According to him the universe was passing from states of lesser probability to that of maximum probability, and the direction of this passage defined that of time. As can be gathered from the preceding explanations, a unidirectional passage in the course of phenomena is to be ascribed to the fact that phenomena are irreversible, *i.e.*, that the various states are not equally probable.

It is true that there also exist types of phenomena for which all states present the same probability. In this case no privileged direction exists, and the phenomena are called *reversible*. An adiabatic transformation under ideal conditions, and the rotation of a body in the absence of friction, are illustrations of reversible processes; such phenomena would of course be incapable of defining the direction of time.

But by far the greater number of phenomena in nature are of the *irreversible* type; with these a definite direction of change is privileged. When we wish to force the phenomenon against its natural trend, work must be furnished. All phenomena entailing friction are of the irreversible sort, for whereas motion generates heat through friction, the heat cannot be used to regenerate the motion. The example of the two powders also presented us with an illustration of the irreversible type of phenomenon, since the natural evolution was from heterogeneity to homogeneity. To be sure, it would be possible to reverse the process, but only through the medium of some intelligent activity sorting out the particles and distributing them according to states of lesser probability. The action of a demon of this sort, *Maxwell's demon,* would cause the direction of the irreversible phenomena to be reversed, so that the direction of time would appear to change. Needless to say, however, Maxwell's demon is but a fiction.

In the illustration of the two powders we can readily understand the reason for the irreversibility, but it was not by this method that the principle involved was first discovered. We must go back to Carnot and to his investigations on the cycle of the steam engine in order to trace the origin of what was to become one of the most fundamental principles of science. Carnot's celebrated principle relating to the efficiency of the steam engine was shown by Clausius to be a special case of a general physical principle which may be stated thus: "Heat cannot flow unaided from a colder to a hotter body, but tends invariably to seek lower levels of temperature."

By introducing a new concept called **Entropy,** defined as the ratio of a quantity of heat to a temperature, Clausius was able to give a more general form to this principle. It then became the *Principle of Entropy,* according to which, in any irreversible change, the entropy of a system was increased; only in the case of a reversible change would it remain constant. Under no circumstances, however, would the total entropy decrease. Thus all natural processes involve an increase of entropy since none are ideally reversible.

Now when we consider the principle of entropy in its bearing on the concept of energy, we find that it implies a continual loss of potentiality in the energy of the system. The energy is thus constantly dissipated, or, more properly speaking, *degraded* into that lowest of forms, namely, disorderly molecular agitation, *i.e.,* heat; whence the name, *Principle of the Degradation of Energy,* under which it is often referred to.

The principle is of such extreme importance that it has been named the *Second Principle of Thermodynamics;* the appellation *First Principle of Thermodynamics* being reserved for the principle of the conservation of energy. It is important to note that these two principles, that of degradation and that of conservation, do not conflict. In quantity the energy is conserved; it is solely in quality, or in potentiality, or in its ability to perform work, that it is degraded.

We now come to Boltzmann's contributions. Entropy as defined by

Clausius was an exceedingly abstruse thermodynamical concept. Boltzmann, by basing his deductions on the kinetic theory of gases, succeeded in giving a more concrete representation of entropy, defining it as the logarithm of a probability. Thus, if we admit the correctness of the kinetic theory, the principle of entropy becomes one of maximum probability, losing thereby its status of absolute validity and assuming a statistical significance. The representation of entropy in terms of probability permits an easy application of the principle to the problems of molecular physics and to atomistic physics in general. We may note that it was by following this method, or rather by combining the two definitions of entropy, that Planck established his quantum radiation formula, and was thus led to his quantum theory. Needless to say, the principle holds only when exceedingly large numbers are considered. In other words, it is a principle governing the chaos, or, again, it is a statistical principle.

It will be seen that the principle of entropy, by stating that the world is ever passing from states of lesser to states of greater probability, indicates that when the state of maximum probability or entropy is reached, no further change can take place. There will then be silent immobility or death. Under the circumstances it might appear that the principle was inconsistent with the existence of an original state of minimum probability. But here it should be remembered that the principles of science do not aspire to any absolute measure of value. All we can demand is that they be satisfied in the very restricted domain which science can explore. In the case of the principle of entropy, however, there is no reason to fear any conflict with the possibility of rebirth or re-creation. For instance, when shaking the powders, we saw that a uniform grey mixture would be the outcome; but were we to go on shaking the bottle for trillions of centuries, at some time or other the black and white separation would reoccur, and the cycle begin all over again. In short, the entropy would continually increase, but once every trillion years it might suddenly decrease, and then start to increase once more. We may mention that although there is no reason to assume that vital phenomena are in any wise inconsistent with the principle of entropy, it is safer to restrict the principle to the physical world. With biological phenomena, questions of probability are far too complex to be treated with any measure of assurance in the present state of our knowledge. Only in certain special cases, such as those connected with Mendel's observations, have probability considerations been introduced with any profit.*

* It is important to understand the precise significance of the principle of entropy. Keeping in mind the fact that the principle refers to probabilities and not to absolute certainties, we may express its general formulation as follows: For all changes of a system the total entropy either is increased (irreversible changes) or else remains constant (reversible changes). At first sight it might appear as though the transformation of heat into work in the steam engine, or the withdrawal of heat from a colder body and its transfer to a warmer one, as in the case of refrigerating machines, would constitute flagrant violations of the principle. But, as a matter of fact, such is not the case. For, with the steam engine, the unnatural change exhibited by the transformation of heat into work is counterbalanced by a natural change illustrated

Now the two points we wish to stress are as follows: First, the unidirectional progress of time is imposed by common experience. Secondly, the principle of entropy is suggested both by experience and by arguments of a rational order. We do not assert thereby that the principle of entropy, even when reduced to the rank of a statistical principle, can be claimed with assurance to be of universal validity. It is far from impossible that in certain phenomena, of which as yet nothing is known, some counterbalancing activity may be at work. Nevertheless, it may be pointed out that a denial of the principle of entropy, at least as regulating those physical phenomena with which science deals, would arrest further progress.

For instance, we can readily understand in what an embarrassing situation we should be placed were a reversal of time to occur. Quite aside from such trivial examples as that of a hen becoming a chick and disappearing into an egg, or from effects becoming causes, a number of more scientific considerations must be mentioned. Phenomena would now pass from states of maximum probability to those of minimum probability; we could not merely interchange the meanings of the words "maximum" and "minimum," for these are given by purely numerical considerations, which are quite irrelevant to the direction of the time-flow. Hence phenomena would pass from the homogeneous to the heterogeneous. But whereas a homogeneous state is unique, and can be predetermined, a heterogeneous one is more or less arbitrary, and baffles prevision. Thus, in the case of the two powders, the homogeneous state is a uniform grey mixture; whereas a heterogeneous state might be a succession of alternate layers, or only two layers, or a juxtaposition of black and white cubes. In fine, prevision would become well-nigh impossible.

From this we may readily anticipate that should Einstein's theory render the direction of time's progress indeterminate, should past and future be interchangeable according to our circumstances of motion, the utmost chaos would ensue. However, we need have no fear on this score, for as we shall see, the theory of relativity leads to no reversal of causality. Let us now consider the space-time theory in greater detail.

When we consider the four-dimensional space-time continuum, where space and time are on the same footing, there is nothing to suggest either a flowing of time or a privileged direction for this flow. In order to

by the fall of heat from generator to condenser. In fine, the total entropy thus remains constant (in a perfectly reversible cycle, for instance in a Carnot cycle), or else suffers an increase. Again, with the refrigerating machine, the unnatural passage of heat from the cold to the warmer body is compensated by a natural transformation; this is represented by the transformation into heat of the mechanical work performed on the apparatus. So once again the total entropy is increased, or, at best, remains constant. Still more striking cases, where the principle would appear to be at fault, are illustrated by the endothermic reactions in chemistry. Contrary to what we should expect, they absorb heat so that low-grade heat energy is transformed into higher-grade chemical energy; but here again compensating influences are at work. We cannot dwell further on these points. The reader who is interested in studying them must consult some standard work on thermodynamics.

conform the theory to the facts of experience, it is therefore necessary to postulate that our consciousness rises along the world-line of our body through space-time, discovering the events on its course. Obviously, we might reverse the presentation by assuming that it was space-time that was moving past our consciousness; we might also claim that the very texture of space-time possessed dynamic properties urging our consciousness along time directions. But in view of the vagueness of the subject, not much is to be gained by speculating on this score.

Now when, along the world-lines which each one of us follows, a common direction from past to future has been specified, we find that an absolute past and an absolute future (though not an absolute present) are demanded by the theory of relativity. To be more explicit, we find that whereas, according to the nature of our relative motions, certain events may appear as simultaneous, or as antecedent and consequent, or as consequent and antecedent, others will invariably present themselves in the same order of temporal sequence regardless of our motion.

This division between events whose order of occurrence is absolute and those whose order is relative is obtained as follows: Whenever it is possible for a ray of light to pass from one event to a second event, leaving the first event at the instant it is produced and reaching the second event before or even at the instant at which it is produced, the time sequence of the two events cannot be reversed by relative motion; it is absolute.

On the other hand, if it is impossible for the ray of light to satisfy the preceding requirements, owing to the too great distance or the too small period of time separating the two events, the time sequence of the two events becomes indeterminate. One observer may claim that A precedes B, whereas the other may claim that B precedes A, and yet another may assert that A and B are simultaneous. If we remember that in Einstein's theory no active influence, no propagation, can be transmitted with a speed greater than light, we see that when a ray of light cannot proceed from A to B, leaving A when A is produced and reaching B when or before B is produced, it is quite impossible for us to suppose that A has influenced B. No causal connection can exist between these two events. Hence we see that when two events are causally connected, their order of occurrence will remain the same for all observers. Only when no causal connections could possibly exist can the temporal sequence of the events be reversed by an appropriate motion of the observer.

It is not because a causal connection happens to exist between two events that the order is non-reversible; the non-reversibility is due entirely to the spatio-temporal separation of the events. As for the causal connection, it may or may not exist. The only reason we mention causal connections with reference to this problem is in order to show that there is nothing in the relativity theory to suggest a reversal of causality; hence, though Einstein's theory entails the relativity of simultaneity and in certain cases that of temporal sequence, there is no danger of a cause appearing as an effect or vice versa. In particular, no relative motion of the observer could ever lead him to believe that the glass we had let fall had in reality leaped up from the floor into our hand. No danger of the black and white powders which we had shaken into a grey mixture, appearing to separate back into black and white, as a result of a change in our relative motion. In short, no danger of the principle of entropy being overthrown.

We may also add that whereas, in classical science, owing to the possible existence of influences transmitted with infinite velocity, any two events happening in space might have been conceived of as causally connected or at least as related, Einstein's theory, by requiring the velocity of light to be unsurpassable, permits us to restrict the cases of possible causal relationships and thereby to gain a better understanding of the cross influences which may be active in the universe.

CHAPTER XX

THE REALITY OF THE CONTRACTION OF LENGTHS AND OF THE LENGTHENING OF DURATIONS

LET us now pass to the question of lengths and durations. Consider a rigid rod placed on a table, and suppose that both table and rod are drawn away from us with constant speed along a straight line. The contraction of lengths implies that both rod and table will appear to us to be contracted, owing to their relative motion with respect to us.

Here we must emphasise the fact that this contraction has nothing to do with the motion of the rod through space or the ether. As we know, motion, or at least Galilean motion, through space has no meaning. It is purely a relative; so that exactly the same effects of contraction would ensue were we to conceive of the table and rod as at rest and of the observer as moving away from them. It follows that an observer attached to the rod would perceive no contraction, although his friend in motion with respect to him would assert that a contraction had taken place.

The nature of this contraction is obviously rather mysterious, and the question is to decide as to its reality. Here we are met with the difficulty of defining what we wish to convey by the word "reality"; in some respects the contraction of the rod is a physical reality, in others it is a fiction, but in no case is it an illusion.

Let us consider the problem of illusions. For instance, we experience an aggregate of visual impressions which we unhesitatingly ascribe to the presence of a table situated "there in the room." Yet all we have any right to be positive about is that certain visual impressions have been experienced; our belief in the existence of the table situated "there" results from an inference, a judgment formulated in an attempt to ascribe a cause to our visual sensations. By an association of ideas and by recalling past experience, we feel justified in claiming the existence of the table "there" as we see it. In the majority of cases our judgment will be correct, but in a certain number of instances it will be erroneous. Such would be the case were we to be viewing the reflection of a table in the mirror. An illusion is therefore an error of judgment. And what prompts us to state that our belief in the object behind the mirror is an illusion? Obviously, the bare fact that on testing the matter out in a variety of different ways it appears impossible to credit the table with any existence at the spot where we claim to see it in the common objective world of science. In much the same way we see a mirage and interpret our sensations as connoting the presence of water. Yet were a meteor to fall

208

into the water it would not appear to be extinguished by the waters of the lake. Kindred observations of an indirect order would finally compel us to recognise that our initial judgment was at fault, and that no water lay before us.

And now let us consider the FitzGerald contraction. Is it an illusion? To this question our reply must unhesitatingly be in the negative. Not only should we perceive the contraction were our eyes capable of yielding a snap-shot picture of a rapidly moving object, but in addition every conceivable physical experiment performed in our frame would indicate that the contraction was physically present. And yet we have said that the contraction differed essentially from the one contemplated by Lorentz, generated as was the latter by a real compression caused by the motion of the object through the stagnant ether. The explanation of the dilemma is as follows:

In Einstein's theory, the rod is contracted, but we must add that this contraction holds only in the space of a frame with respect to which the rod is in motion. In the space of the frame accompanying the rod in its motion, no contraction would occur. We see, then, that as there exists no privileged space, only "my space," "your space," "his space," there is no particular length to the rod, only a length "for me," "for you" and "for him." The contraction is thus embodied in a two-term relationship extending between the observer and the observed; it is a "relative." But, once again, for a given observer it is as physically real as a chair or a table. Only when we consider the contraction from an impersonal standpoint, from that of space-time, without mentioning the conditions of observation, is the contraction not an illusion, but downright meaningless; for in space-time length itself is meaningless. It is, as we see, the concept of spatial magnitude, of primary qualities, which is at stake, in a form entirely different from anything science or philosophy had ever dreamed.

Although we may be repeating ourselves unnecessarily, we cannot help but feel that the easiest way for the layman to grasp these difficult concepts will be to return to the well-known illustration of the relativity of a visual angle. When we perceive an object, say a telegraph pole, we see it under a certain angle, which we may call the visual angle. Is this visual angle real? Of course it is; we can measure it and define it physically as the angle subtended by two Euclidean straight lines extending between our eye and the two extremities of the pole. It is real but it expresses a relationship—a relationship between the positions of the observer and pole. It is certainly not inherent in the pole, since as the observer changes his position the visual angle changes in magnitude, though nothing happens to the pole in the meantime.

Visual angle in classical science is analogous to length in Einstein's theory. In the relativity theory our rod has no true length any more than the telegraph pole has any true visual angle. The whole significance of relativity is precisely to deprive length and duration of the absolute characteristics which classical science attributed to them. In short, when we are asked whether the rod has really been contracted and lost part of its length, we might counter by asking whether the telegraph pole has really been modified and lost part of its visual angle. Just as, if we omit to stipulate the position of the observer, the visual angle of the pole has no meaning and can teach us nothing about the pole, so now, in relativity, if we omit to specify the relative velocity, the length of the rod has no meaning and can teach us nothing about the rod.

All these arguments developed on the score of length apply in similar fashion to duration. Both length and duration are relatives having no absolute significance in the universe.

The critic will generally balk at these statements and be inclined to say: "What nonsense! The length of the rod is immanent in the rod and can in no wise be affected by the observer's motion or by the frame's existence. The observer might die, the frame might disappear; what has all this to do with the length of the concrete entity we call the rod? The length of the rod may *appear* longer or shorter according to its relative motion; but we must not confuse appearance and reality. Thus, a man standing 100 yards away *appears* to us shorter than when he is close up, but it would never enter our minds to say that he was really reduced in stature. Just as the height of the man is independent of our position and of our existence as observers, so is the length of the rod independent of our relative velocity."

Now we might readily concede to the critic the right to maintain that the existence and absolute characteristics of the rod are in no wise dependent on the motion or the existence of an observer. But the critic is begging the question when he assumes without further ado that these absolute characteristics are expressed by *the length* of the rod. This length enters into existence only when a definite space is specified; till then it is meaningless. And in the absence of an observer there is no such thing as a space, but merely four-dimensional space-time. The absolute characteristics of the rod thus refer to space-time, and this eliminates length as representative of a primary quality in the objective world.

With the rejection of such classical absolutes as length and duration, our ability to conceive of an objective impersonal world, independent of the presence of an observer, seems to be imperiled. The great merit of Minkowski was to show that an absolute world could nevertheless be imagined, although it was a far different world from that of classical physics. In Minkowski's world the absolute which supersedes the absolute length and duration of classical physics   is the *Einsteinian interval*. We mentioned Minkowski's discovery in Chapter XVIII, but there we were stressing its geometric significance. Here we shall be concerned with its physical significance from the standpoint of the objective world.

Thus suppose that, as measured in our Galilean frame of reference, two flashes occur at points A and B, situated at a distance l apart, and suppose the flashes are separated in time by an interval t. If we change our frame of reference, both l and t will change in value, becoming l' and t' respectively, exhibiting by their changes the relativity of length and duration.

In Minkowski's words, "Henceforth space and time by themselves are mere shadows." On the other hand, the mathematical construct $l^2 - c^2t^2$ will remain invariant, and so we shall have $l^2 - c^2t^2 = l'^2 - c^2t'^2$. It is this invariant expression, which involves both length and duration, or both space and time, which constitutes the Einsteinian interval; and the objective world which it cannotes is the world of four-dimensional space-time. The Einsteinian interval defines a distance in space-time, a distance which remains the same for all observers, just as distance alone or duration alone were mistakenly believed to remain the same for all observers in classical physics. It is true that we have considered only Galilean frames, and that accelerated frames also exist. However, the Einsteinian interval still remains an invariant as measured from all frames of reference, whether accelerated or not. In the case of accelerated frames, however, we must restrict our attention to Einsteinian intervals of infinitesimal magnitude, and then add up the intervals when finite magnitudes are involved.

CHAPTER XXI

THE inevitable rational consequences which follow with mathematical certainty when once the existence of space-time as a model of the universe is conceded, are often puzzling to beginners, who feel that these consequences are too paradoxical to be acceptable. Of course, we must distinguish between a paradox of feeling, and a paradox of logic or an inconsistency. Any inconsistencies would be fatal to a rational construction and would denote either errors of reasoning or the presence of some fundamental contradiction in our basic premises. If, therefore, when analysing any given example in Einstein's theory, we proceed to follow our deductions by adopting successively two different alternatives in our point of view, both of which would be consistent with the premises of the theory, and if we finally compare the results thus obtained and find them to be contradictory, hence inconsistent, the theory of relativity could never survive. However, it is scarcely probable that any such inconsistencies will ever be discovered by the layman, inasmuch as the theory of relativity is the work of mathematicians, for whom, more than for any one else, accurate and consistent reasoning is the indispensable condition of research.

On the other hand, though no inconsistencies or paradoxes of logic can exist in Einstein's theory, this does not mean that the conclusions of relativity may not appear strange and lead to paradoxes of feeling. But once more it is important to differentiate between strangeness or paradoxes of feeling, and inconsistency or paradoxes of reason.

The important point which must be settled, therefore, is whether our habitual notions of time, space and simultaneity are rational necessities, or whether they are merely feasible notions among a number of equally feasible ones. The question we have to decide is analogous to that which Kant proposed to solve when he stated that Euclidean geometry and three-dimensional space were the *a priori* forms of pure perception. Kant was wrong, as the discovery of non-Euclidean geometries proved without doubt. To-day Einstein has done for space, time and motion what Riemann and Lobatchewski did for the geometry of space alone. He has furnished us with a model just as rational as the model of classical science; and this model has the additional advantage of being in accord with empirical observation to a high order of precision, instead of being a mere approximate construction as was the classical structure. Indirectly, then, Einstein has proved that our innate classical conceptions of space, time, simultaneity and motion were by no means rational necessities.

Many people, however, are of the mistaken opinion that the theory of relativity leads to paradoxes of logic or inconsistencies. But when we analyse their arguments we are always confronted with the same erroneous method of reasoning. In many cases the critics seem to have a very hazy idea of the premises on which the theory is built. Others, while stating that for the sake of argument they are willing to recognise the validity of these premises, appear to forget their good intentions before reaching the end of their discussion. Inadvertently, in the very middle of their demonstrations, they relapse into the premises of classical science and, of course, are led to insufferable contradictions; whereupon they conclude that the theory of relativity leads to inconsistent results.

Let us consider, for example, the relativity of duration. A number of clocks or atoms which beat at the same rate and mark the same time when placed side by side are displaced along different routes to the same terminal point. The relativity theory then demands that when set side by side once again they shall be marking different times. The critic immediately exclaims: "This is inconsistent!" But where is the inconsistency? There would be inconsistency, indeed, had Einstein started out by positing as an axiom that time was absolute. This, in fact, is precisely what the critic has done. But then, when, as a result of this uncalled-for presupposition, he finds that the theory leads him into inconsistencies, he should not blame Einstein; for the theory is inconsistent *not with itself* as a logical structure, but solely with the critic's own presuppositiohs.

Of course, the source of all the trouble lies in the fact that a belief in the absoluteness of duration has become second nature to us; and it requires a certain effort of introspection on our part for us to realise that this belief presents no rational justification. Eventually, however, our ideas become clarified, and we concede that we have no means of figuring out *a priori* whether time will be absolute or relative. We realise that it is impossible to lay down the law at the start one way or the other, unless we wish to force nature into a preconceived mould—which may not fit.

Our sole means of settling the problem is, then, to appeal to experiment; to ultra-precise experiment in the present case, since the crude variety would not enable us to reach a decision. The experiment of the clocks, could it be performed with sufficient refinement, would constitute one of these. Obviously it would be unwise to assume that its result could be known with certainty in advance; for were men to labour under any such belief, there would be no need for them to experiment. We may, of course, question the likelihood of a certain result occurring, basing our opinion on previous experiments of a similar nature; but in the final analysis our understanding of nature will issue from an accurate observation of facts and not from *a-prioristic* metaphysics.

Unfortunately, as we mentioned previously, this experiment with clocks or atoms cannot be performed with sufficient accuracy, so that indirect methods must be considered. But it is precisely because all the indirect experiments thus far attempted have yielded results incompatible with the absoluteness of duration and distance that Einstein was compelled to reject these secular beliefs of mankind. Accordingly, we have every reason to anticipate that the clock experiment would bear out Einstein's views; in fact, were this not the case, the entire theory would collapse.

Let us abandon these generalities, and consider the clock experiment in a more vivid form. We refer to what is known as the *problem of the twins*. In this we consider twin brothers one of whom remains on earth, while the other steps on to a magic carpet and visits distant Arcturus. On his return he finds that his brother has grown old and decrepit, while he himself has preserved his youth.

This particular consequence of the theory has been one of the stumbling blocks of practically every lay writer who has devoted his time to criticising the theory of relativity. Some have stated that Einstein never upheld any such absurdity, and that the whole trouble must be ascribed to his too enthusiastic followers. Others have stated that the conclusions violated the rules of logic in that they led to inconsistencies; and that if Einstein's theory necessitated such results, it should be regarded as a gigantic hoax from beginning to end.

Now, in order to avoid any misconception, we may mention that in our example both twins hold in their hands clocks which, when placed side by side, beat out their seconds and minutes in perfect unison. Einstein's theory indicates that the clock carried by the travelling twin will be behind time when brought back to earth and compared again with the clock that has been left there, in the keeping of the stationary twin. Some critics, such as Bergson, while accepting the postulate of the invariant velocity of light, deny that this slowing down in the beatings of the moving clock is required by the theory in any real sense. It is a mere appearance, Bergson maintains; a mathematical illusion. Just as the gradual shrinking of a man's apparent size as he moves away from us would not connote any real decrease in his stature, so now, when the travelling twin returns with his clock to his brother on earth, the illusion would vanish, and both twins would recognise that their respective clocks were still marking the same time in perfect agreement. one with the other.

Now, it cannot be stressed too strongly that this interpretation of the problem is worse than incorrect. Not only is it incorrect from the standpoint of Einstein's theory, but it denotes on the part of the critic a colossal ignorance of the significance of a theory of mathematical physics. Here it must be recalled that the comparative retardation of the travelling twin's clock is an inevitable consequence of the Lorentz-Einstein transformations. To deny, therefore, that these transformations correspond to what would actually occur, to what would be detected and measured by the most accurate instruments, would be to deny that the transformations possess any physical basis. If this be the stand we wish to take, well and good; but the logical outcome of this attitude will be to deny the legitimacy of the entire theory; it would be quite impossible to retain the theory while contesting the real significance of the transformations. Once again a theory of mathematical physics is not one of pure mathematics. Its aim and its *raison d'être* are not solely to construct the rational scheme of some possible world, but to construct that particular rational scheme of the particular real world in which we live and breathe. It is for this reason that a theory of mathematical physics, in contradistinction to one of pure mathematics, is constantly subjected to the control of experiment. If, therefore, the Lorentz-Einstein transformations failed to yield the results that would be actually measured, Einstein's theory as a means of physical discovery, as a co-ordination of physical facts, would be not only useless, but entirely incorrect, and we should have to abandon it.

Furthermore, it should also be noted that every precise physical experiment that it has been possible to accomplish has verified the anticipations of the Lorentz-Einstein transformations; it is owing to this fact that the theory was not abandoned years ago. This in itself should warn us against asserting that the transformations possess no physical significance.

Thus far we have been concerned with generalities and have merely attempted to show why Bergson's solution of the problem must be discarded on first principles. Leaving aside this aspect of the question, it is instructive to consider the formal proof which he presents in support of his contentions.

He appears to realise that the entire matter hinges on the measure of physical reality which should be credited to the Lorentz-Einstein transformations. Accordingly, he attempts to prove that these transformations do not represent reality, but possess a mere fictitious mathematical significance having nothing in common with the physical world. He calls them "fantasmatical," contending that mathematicians, not being trained in philosophy, are severely handicapped when they attempt to interpret their equations, and are incapable of distinguishing mathematical fictions from physical reality. So Bergson proceeds to demonstrate the fictitious significance of the transformations.

He first serves us with a series of arguments based on the principle of the relativity of motion itself. In the very first paragraph, however, he confuses the relativity of Galilean motion with the complete relativity of all motion, Newtonian and special relativity with visual relativity, velocity with acceleration. All these errors relate to classical science itself; and it is obviously quite impossible to discuss Einstein's theory intelligently until we have acquired at least an elementary knowledge of classical mechanics. Following this philosophical demonstration, he goes on to give us a mathematical one; but all he succeeds in proving is that he has not the slightest conception of the most elementary properties of mathematical transformations.

Now, without going into mathematical details, let us make an effort to see why it is that the clock of the travelling twin must necessarily have suffered retardation. To understand the reason, we must revert to

Einstein's premises, and in particular to the postulate of the invariant velocity of light. This postulate, as we have explained on several occasions, states that all Galilean observers situated in their respective Galilean frames will discover, as a result of measurement with rigid rods, that waves of light invariably pass through their frames with the same constant speed of 300,000 kilometres per second. If this postulate be accepted, we must infer that in every Galilean frame, time can be measured by the distance in space covered by a wave of light (the distance being measured with rigid rods at rest in the frame). As a result the postulate allows us to construct a sort of optical clock. We may, for example, consider two parallel mirrors situated at a fixed distance apart of 1.5 kilometres in our Galilean frame, and we may suppose that a wave of light is reflected back and forth from one mirror to the other. These oscillations of the ray of light, covering always the same distance, require equal durations; and in the present case 100,000 such double oscillations would be seen to define a duration of one second for that Galilean frame in which the optical apparatus is situated. We must henceforth consider every Galilean frame as possessing an optical clock of this kind defining the passage of time for its own frame.

Let us now analyse the problem of the two twins; and let us suppose that *we* as observers are stationed on the earth, say, at the North Pole. To a high degree of approximation we may consider ourselves, therefore, as at rest in a non-rotating or Galilean frame. The moving observer is also situated in a Galilean frame, moving away from us with a speed which we shall assume to be $\frac{\sqrt{3}}{2} c$. *i.e.* about four-fifths that of light. He visits a distant star which *we* should recognise as being $10\sqrt{3}$ *i.e.* about 17 light years distant* from us, then reverses his velocity and returns to earth. In our estimation the absence of the traveller will have lasted 40 years.

Both *we* and the traveller carry optical clocks, of precisely the same dimensions when placed side by side; so that for either one of us, 100,000 oscillations of our respective light waves in our respective optical clocks correspond to one second of our respective times. Let us assume that the optical clock of the travelling observer is placed crosswise to his motion with respect to us, so that his light waves oscillate back and forth in a direction perpendicular to his relative motion. So far as the traveller is concerned, the light waves of his optical clock will oscillate along one same straight line 1.5 kilometres in length; and if during his entire trip back and forth he finds that his clock has oscillated N times, he will be justified in saying that according to his optical clock his trip has lasted $\frac{N}{100,000}$ seconds.

Now let us examine the problem from our own point of view; that is to say, let us refer all measurements to our own Galilean system attached to the earth. In this Galilean frame of ours the oscillating wave of light

* A light year is an astronomical unit of length defined by the distance a wave of light covers in one year.

of the traveller will not follow a to-and-fro motion along the same straight line. It will follow a zigzag line analogous to a series of *V*'s placed end to end. The total length of this path followed by the light wave must therefore be longer than when referred to the frame of the traveller; in the present numerical example it would be just twice as long. The question then arises: "How long will it take for the traveller's wave of light to describe this zigzag line back and forth *when time is measured according to the standards of our Galilean frame attached to the earth?*" The postulate of the invariant velocity of light gives us the answer immediately. Since the zigzag line followed by the light wave in our frame is just twice as long as the succession of superposed up-and-down lines which it follows in the traveller's frame, and since, whether referred to one frame or the other, the velocity of the light wave along its path must be the same (invariant velocity of light in all Galilean frames), then the immediate conclusion is that the total duration of the N oscillations will be twice as long when measured by us as when measured by the traveller.

As these N oscillations start and end with the beginning and end of the trip, we must assume that the trip, which lasted forty years in our own estimation, will have lasted twenty years according to the traveller. In other words, any registering device would show that when both optical clocks were again placed side by side, our earth clock had performed $2N$ oscillations during the trip, whereas the moving clock had performed just half that number, namely, N. Obviously, there is no illusion about the matter; for, as we have said, the oscillations of the two clocks could perfectly well be registered graphically, so that we should be dealing with a definite enumeration of registrations which any third party could verify.

It is quite easy to understand why these extraordinary results must take place. The cause is to be ascribed to the invariance of the velocity credited to any light wave in all Galilean frames. Were it not for this postulate, the mere fact that the total length of the zigzag line was longer than the up-and-down line would not connote any such difference in duration. Classical science would have assumed that since the path was longer, the velocity of the light wave along this path would have measured out as greater; so that finally both durations would have been the same, in full accord with the classical belief in the uniqueness of time. In short, the comparative retardation of the moving clock follows as an inevitable necessity as soon as we agree to accept the validity of the postulate of the invariant velocity. On no account is it permissible to accept the postulate and then deny its mathematical consequences. If, therefore, the example of the trip to the star is too great a tax on our credulity, we must reject the postulate of invariant velocity and be faced with all the difficulties which surrounded electrodynamics prior to Einstein's discoveries.

And now we may discuss another type of criticism, not quite so misplaced as that of Bergson, but still utterly erroneous. It has been argued, for instance, that all we have succeeded in proving is that if each Galilean observer measures time with the help of his optical clock, it is true that the duration of the trip will measure out differently according to the observer, but that if, in place of optical clocks, each observer had defined duration in terms of the usual mechanical watches, these strange discrepancies would never have arisen.   In other words, the so-called slowing down of time in the frame of the travelling twin would be due entirely to the slowing down, *not of time,* but of the optical clock wherewith he proposed to measure time.   Always according to this same criticism, the optical clock measured duration only in virtue of an artificial convention, and its readings could not be claimed to represent true lapses of time. Real time would flow independently of any clock; and to assert that an arbitrarily postulated clock defined true time would be about as reasonable as claiming that we should be dead were we to forget to wind up our watch.   In other words, we might adjust our clock as we pleased, accelerate its motion, retard it, and even stop it completely, but we should go on living just the same.

The only way to refute this fallacious argument is to examine once again the premises of the relativity theory.   Einstein starts with the assumption that Galilean motion through the ether is relative.   From this it follows that no observer attached to a Galilean frame can by any means whatsoever ascertain whether his frame is in a state of uniform rectilinear motion or in a state of rest (special principle of relativity). Then he lays down his famous postulate on the invariant speed of light (300,000 kilometres per second) through any Galilean frame when measured by the observer attached to the frame.   This postulate, as we have seen, enables any Galilean observer to construct an optical clock and to measure duration as referred to his clock, hence to his frame.

Now, in order to answer the criticism which charges Einstein's selection of optical clocks with being artificial, we must proceed to investigate whether the definitions of congruent durations marked out by the optical clocks are justifiable.   And, first, what do we mean by a justifiable definition of the equality of two successive durations?   Would it have been justifiable to define equal durations by the durations a body falling vertically towards the earth would require to cover equal Euclidean distances in space?   Obviously not.

We saw, when discussing congruence generally, that mathematical analysis by itself afforded us no means of exalting above all others one particular definition of congruent stretches either for space or for time; but that when we combined physics with mathematics, certain precise methods of defining congruence imposed themselves immediately in any given Galilean frame of reference.   In the case of time-congruence we found it necessary to assume that any periodic phenomenon at rest in our frame, in a complete state of isolation from all external influences, would have to be regarded as passing at regular congruent intervals of time through the same initial state.   This assumption was demanded by our belief in causality, and by a justifiable definition of time-congruence we mean to imply one which would satisfy these consequences of causality.

When, therefore, we stop to consider whether Einstein's optical clocks yield us a justifiable definition of equal successive durations, the whole question simmers down to deciding whether or not the oscillations of a ray of light in an optical clock constitute an isolated phenomenon, or whether, according to the motion of the Galilean frame in which the clock is situated, certain perturbating influences may not interfere with the state of isolation we credit to the working of the clock.

Until such time as the effect of the ether on an object moving through it had been settled, there was every reason to suppose that in the case of a Galilean frame moving through the ether, the effect of the ether wind would be to disturb the free oscillations of the light rays between the mirrors of the clock. And so the clock could not be considered isolated, and its readings would be valueless, or at least in need of correction. But Einstein's special principle of relativity has precisely for its object to inform us that the ether behaves as though it were non-existent, so that any perturbative effect which classical science had suspected was *ipso facto* eliminated. In other words, if we accept the special principle of relativity, the optical clocks may be regarded as perfectly isolated; and their readings must therefore afford us with perfectly reliable measures of congruent time-stretches, as required by the principle of causality.

It follows that if in a Galilean frame we place side by side, at rest, an optical clock, a vibrating atom, a piece of radium gradually losing its mass, and a top revolving without friction, the time determinations defined by all these phenomena will be identical, since all these are perfectly isolated from external influences. To be more precise, if a sodium atom vibrates so many times per second, as measured by the optical clock of its Galilean system, it should still vibrate exactly the same number of times per second (as measured by the clock of the frame) even if the constant velocity of the Galilean frame (with respect to some standard Galilean frame) has been changed to some other constant velocity. In short, all physical phenomena at rest in a frame would evolve at exactly the same rate in terms of the optical clock of the frame, regardless of the magnitude of the Galilean motion of the frame; since, according to the special principle of relativity, this absolute Galilean motion is entirely meaningless.

If observation, by any chance, should prove that the relative rates of vibration of the atom and the clock were to vary according to the Galilean frame in which we set them, we should have a physical means of differentiating one Galilean frame from another, hence of determining the velocity of our frame through the ether; and the special principle of relativity would have to be abandoned. Assuming, therefore, Einstein's premises to be correct, we must agree that the vibrations of the optical clock constitute as perfect a means of determining congruent intervals of time as it is possible for physicists to obtain. Finally, we see that in any Galilean frame exactly the same time-determinations would be obtained whether we made use of an optical clock, a top spinning on its axis without friction, a vibrating atom, or a mechanical clock, provided all these bodies were at rest as a whole in our Galilean frame.

Consider, then, two Galilean frames, each containing an optical clock and a top spinning without friction on its axis. If, when the two frames are brought to relative rest, side by side, the two tops are spinning at exactly the same rate, both executing N rotations per second when measured in terms of their respective clocks, then exactly the same correlation must exist when the Galilean frames are moving apart. It is therefore a matter of indifference whether the observers in the frames measure time by means of their optical clocks or by means of the rotations of their tops.

Now a planet revolving on its axis in space is nothing but a huge top; and if, for instance, we stand at the non-rotating point of a planet, say the North Pole (so as to be in a Galilean frame), we should be perfectly justified in measuring our unit durations by the successive rotations of our planet with respect to the fixed stars. Consider, then, two planets rotating at exactly the same speed when held side by side, and let us call one of these planets the earth. Regardless of whether these two planets be set flying apart or held together, the observer on either planet will be justified in measuring his unit time-stretches in terms of his own planetary rotations. Classical science assumed that since time was one and universal, both planets would always beat out exactly the same unit time through their rotations, whether they were in relative motion or at rest, so it mattered little which particular planet was selected. Under the circumstances, from motives of convenience, the earth was chosen as standard.

The novel point of Einstein's theory consists in proving that this attitude is untenable. Assuming the various planets to rotate with the same speed when their axes are at relative rest, each different observer must henceforth abide by the rotations of his own planet and not by those of the other planets when relative motion is present. Only thus can the special principle of relativity be maintained. In particular, if we assume one rotating planet to be transferred to a distant star, then returned near the rotating earth as in the example of the two twins, it would be found that, owing to the solidarity between the optical clocks and the rotations of the planets, the wandering planet would have revolved less often during its trip than would have been the case with the earth. So we see that the slowing down we have mentioned is absolutely general, and must indeed be so if the special principle of relativity is to endure.

Furthermore, we see that the great difference between classical science and relativity consists, *not* 'in having substituted an optical clock for the earth's rotation, since both these determinations of time would be identical, but in having proved that the earth's rotation or the earth's optical clock is valid only as a measure of time for an earth observer or for one at rest with respect to him.

Having established the solidarity of the evolution of all physical phenomena at rest in the same Galilean frame as an inevitable consequence of the relativity of Galilean motion, we must now enquire whether this general solidarity should be assumed to extend to vital phenomena and to our consciousness of the flight of time. Of course we cannot deny that aging is accompanied by a physiological alteration of our bodies; hence, on this ground alone, there can be no doubt that physiological aging will proceed in terms of the general solidarity of all physical phenomena. But there exists another aspect of the passage of time, and that is our consciousness of its stream.

Our awareness of the flight of time is probably the most fundamental fact of consciousness; and it remains an ever-mysterious enigma. If we credit it to certain physiological processes occurring in the brain, it must of course accompany the general evolution of physical processes; but if we wish to regard this flight of time as a vital or spiritual phenomenon different in its essence from anything physical, there can be no *a priori* reason for assuming that it would accompany the general physical processes. Of course our awareness of the flight of time is an extremely vague feeling; nevertheless, so long as we retain a conscious state, we have no difficulty in differentiating a duration of twelve hours from one of one second.

If, then, in the example of the two twins, we assume that the length of the trip and the velocity of the travelling twin have been so selected that when the twins meet again, the clock of the traveller marks an advance of only one second, whereas that of his stationary brother marks an advance of ten years, our differentiation between physiological time and our consciousness of the time-flow would lead to the following result: The travelling twin would not have aged, physiologically speaking, but he would be fully aware, in his inner consciousness, that a stretch of time far in excess of one second had elapsed during his trip. However, we cannot fail to observe that if such were the case, if there existed two rates of aging for the same man, one physiological and the other corresponding to a psychological sensing of the flow of time, the special principle of relativity, and with it Einstein's theory, would have to be abandoned.

The reason for this last statement is easy to understand: If the two species of time-flow varied independently, the rate of evolution of all processes would appear to vary according to the frame in which we might be situated—a tortoise would appear to crawl or to be galloping madly through space. Theoretically, therefore, it would be possible to detect a difference as between Galilean frames, hence to discover absolute velocity through the ether or through space. And this would be contrary to the principle of relativity.

When one places in the balance the mass of empirical evidence which lends support to the special principle of relativity, and the total lack of evidence of any sort or kind which would encourage the belief that the time sensed and lived is different from the time of physics and of physiological aging, one cannot be surprised at the total lack of interest the mathematicians of relativity must take in such vague hypotheses of certain philosophers.

Moreover, there is yet another aspect of the question. The fundamental reason for all the criticisms we have mentioned, is that deep down at the bottom of his heart the critic wishes to preserve, in spite of all, his hereditary belief in the absoluteness of time. Even when convinced that as a result of Einstein's physical discoveries any such belief is incompatible with the evolution of things physical, he endeavours to assert that absolute time must continue to endure when the sensing of the flow of time and our consciousness of being alive are contemplated. In this way he proceeds to draw a distinction between time actually lived and time as measured by our optical clock. But the critic overlooks the fact that this belief in an absolute time which he refuses to abandon is in no wise a rational necessity, any more than is the hypothesis of the plurality of times. Experiment alone can yield us a clue; and granting that physical experiment may not be capable of throwing light on the spiritual, nevertheless, if we are to be eternally ignorant of things spiritual, there is surely no reason to state *a priori* that they must be governed by absolute time or that time itself should have any meaning so far as they are concerned. Thus the trouble with the critic who wishes to understand the significance of Einstein's discoveries is that in spite of himself he remains obsessed by this unwarranted belief in the absoluteness of time. This conviction is undoubtedly hard to eradicate, and only prolonged introspection can gradually rid us of an obsession which is so thoroughly ingrained in the species. Then, and only then, will the so-called paradox of the two twins cease to be a paradox.

For all these reasons, if we accept Einstein's premises there is no alternative but to recognise that the duration marked out by an optical clock in a Galilean system will give us the same measures of duration as will any other isolated periodic physical phenomenon; and that our aging, our awareness of the flow of time, and the time we are conscious of living accompany the oscillations of the optical clock. In short, therefore, if the optical clock in our Galilean system indicates that one year has elapsed, we must have aged by one year. We shall have gone to sleep and we shall have risen from bed on 365 different occasions, and we shall have had enough time to perform all those duties which usually occupy the space of one year.

Under these circumstances, to what does the example of the two twins finally reduce? Simply to this: The departure of our twin brother and his return to earth constitute two definite events. The duration separating two events being robbed of any definite value by the theory of relativity, there is no cause to be surprised that this duration should manifest different magnitudes to different observers.

Again let us put the paradox in the following alternative form: Our twin brother covers a distance of $20\sqrt{3}$ light years (when he goes to the star and then returns), and the duration of his trip is forty years; hence his speed along the distance is $\dfrac{\sqrt{3}}{2}$ that of light. How, then, could he ever be misled into believing that he had required only twenty years to cover this total distance? Would not this belief suggest to him that his speed had been greater than that of light, and would not this fact conflict not only with relativity but also with the assertion which he would voluntarily offer, that his speed with respect to us was only four-fifths that of light?

But here, again, the criticism is occasioned by another erroneous obsession, that of absolute distance. The critic does not grasp the fact that the distance from earth to star, just like any other distance, is indeterminate and has no meaning in itself. While it is true that, so far as *we* are concerned, this distance is $10\sqrt{3}$ light years, yet in the opinion of the traveller it is only one half as much; and his opinion expresses exactly the same measure of reality as does our own. We can only repeat once more that absolute duration and distance are incompatible with Einstein's premises; and if we insist upon preserving our classical obsessions, a study of Einstein's theory will be little better than a nightmare.

Let us now pass in review a totally different order of criticism. It is contended, for example, that the principle of relativity permits us to interpret any problem of relative motion between two bodies in two different ways: by considering either one body or the other as being at rest. Then, according to the critic, since in the problem of the two twins it is the twin's motion that is the cause of the slowing down of his time, all we need do is to consider him at rest and his brother in motion, in order to reach the ridiculous conclusion that either twin is younger than his brother.

This criticism is based on false premises. The Special Principle *does not state* that when two bodies are in relative motion we may always consider either body indifferently as being at rest and the other in motion. The special principle states explicitly that a reciprocity of motion exists *only when the motions of both bodies are Galilean;* or in other words, only when both bodies remain permanently attached to Galilean frames. When a body is not permanently attached to a Galilean frame it is necessarily in a state of acceleration (according to the special theory), and so the reciprocity of motion is no longer valid.

Now in the example of the twins, the earth-twin remains fixed in a Galilean frame (if we neglect the earth's rotation and its accelerated motion along its orbit); but the second twin, on reaching the star, changes his position from one Galilean frame to another which is moving in the opposite direction, and so the second twin does not remain fixed to a given Galilean frame. It is this change in Galilean frames (accompanied as it is by acceleration) which differentiates the present problem from the usual ones studied in the special theory of relativity, and which is responsible for the absence of reciprocity. At the same time, it is this change in Galilean frames on the part of one twin and not on the part of the other, which creates a difference between the status of the twins and which prevents us from viewing indifferently the one or the other as at rest. In short, the twin that has been subjected to acceleration, and so to forces of inertia, will be the younger of the two when the twins meet again.

We need not be surprised at these very real effects produced by acceleration, for we must remember that acceleration is an absolute and not a relative like mere velocity. A body has one velocity or another, according to our point of view or choice of a frame of reference; but when a body is accelerated the choice of our frame of reference is irrelevant for the simple reason that acceleration betrays itself by physical forces and stresses which may exert palpable influences.

We must not allow ourselves to be misled by a further generalisation of Einstein's, known as the **Postulate of Equivalence,** according to which even an accelerated observer may be considered at rest. This new postulate modifies to a certain degree the position of the problem of the two twins. Whereas the special principle did not allow us to consider the accelerated wanderer as at rest, the postulate of equivalence permits us to adopt this alternative view. However, those who hope to discover by this means a logical inconsistency in the theory, will again be disappointed; for the postulate states expressly that an accelerated observer may be considered at rest *only if we assume that a field of gravitation takes the place of the field of inertial forces* to which the accelerated observer was submitted. Then, in virtue of the postulate of equivalence itself, exactly the same relative difference in agings will take place except that this difference will now be ascribed to gravitation instead of to acceleration.

It is hoped that the paradox of the two twins has been presented in a sufficient number of ways to enable the conscientious student to master the difficulty. The problem requires a certain effort of introspection; but once we have succeeded in ridding ourselves of our prejudiced ideas regarding space and time, the paradox will vanish of itself.

And now let us mention another type of so-called paradox which appears to cause considerable trouble to beginners. We refer to the Einsteinian addition of velocities. Thus, the critic begins by stating that according to the dictates of common sense $1 + 1 = 2$, so that it is only natural to assume that a velocity of one mile per second plus another velocity of one mile per second equals a velocity of two miles per second. Einstein's contention that the resultant velocity is slightly less than two miles per second is therefore an absurdity on the face of it. But if we examine these arguments carefully we shall see that they again are based on a faulty understanding of the problem.

Even in classical science it is not necessarily correct to state that velocities add up like the numbers of arithmetic. Consider, for instance, the following illustration: If, while we are standing on an embankment, a train moves away from us with a speed v and if a rifle bullet is shot from the train with a speed V, it is not always correct to state that the velocity of the bullet relatively to us on the embankment is $v + V$. This statement is true *only when* the velocities of the train and the bullet lie along the same straight line. If we neglect to specify this restriction, our statement is incorrect. In Einstein's theory the problem is very similar.

If a Galilean frame moves away from us with a speed v and if, *with respect to this frame* and in the direction of its motion, a bullet speeds with a velocity V, the resultant speed of the bullet might be $v + V$ only provided our methods of measuring a velocity, and those of the observer in the Galilean frame, were identical; that is, provided our measurements of space and time were identical. This was always tacitly assumed to be the case by classical science; but we know that in Einstein's theory the identity of our standards of measurement no longer holds, since such a state of affairs would be incompatible with the invariance of the velocity of light.

We may therefore assimilate the conditions surrounding the various space and time measurements in Einstein's theory to those which would endure in classical science if by reason of some law of nature it were impossible for the observer in the train to fire a bullet in the precise direction of the train's motion. In other words, in Einstein's theory, velocities lying in the same direction add up like algebraical numbers, provided the velocities we propose to add and subtract are computed under exactly the same conditions, that is, *by the same Galilean observer*. It is when one observer computes one velocity and a second observer, in relative motion, computes the second velocity that the classical law of addition is at fault.

Very similar to this problem of the addition of velocities is that of simultaneity. Thus, Einstein's theory implies that if two events A and B occur simultaneously for one observer, John, and if in the opinion of another observer, Peter, in motion with respect to John, the event B and another event C occur simultaneously, it is not correct to assume that A and C will be simultaneous in the opinion of either of the two observers. Here the critic assumes, without further ado, that the symbolic expression of these relationships would be as follows: $A = B;$ $B = C;$ $C \neq A$. Therefrom he infers that the theory of relativity is contrary to logic and is a manifest absurdity. And yet, if we examine these statements carefully, we shall have no difficulty in seeing that the error in logical reasoning lies not with Einstein but with the critic himself.

In order to dispel any doubt on this score, let us ask the critic what he intends to convey by the symbol $=$. Suppose, for example, that we are considering two objects, A and B, of the same size. Are we entitled to write $A = B$? The answer should be: Yes, if the symbol $=$ stands for "is of the same size as"; or No, if it stands for most other types of identity. In order to avoid confusion let us then represent the symbol by $\overset{S}{=}$ (S standing for size) and let us write $A \overset{S}{=} B$.

And now consider a third object, C. We will assume that C is of the same colour as B. Beyond that we know nothing of it. Once again,

are we entitled to write $B = C$? The answer is: Yes, if the symbol =
stands for "is of the same colour as"; No, if it stands for some other
type of equality. In order to avoid confusion let us then write our
symbol: $\overset{C}{=}$ (C standing for colour).

Our two equalities now become $A \overset{S}{=} B$; $B \overset{C}{=} C$; and obviously there
is no reason for us to infer that A should equal C either in size or colour,
since the symbols $\overset{S}{=}$ and $\overset{C}{=}$ do not convey the same meaning.

It may be assumed that even the most careless of critics would grant
the correctness of the preceding argument. But in the case of his criticism of Einstein's theory he would argue: "All this is very fine, but
in the relativistic absurdity which I am pointing out to you I am perfectly entitled to write $A = B$; $B = C$ and $C \neq A$. For in the present
case the symbol = between A and B, and between B and C, stands
for exactly the same concept, namely, that of identity in time; it does not
refer to such different concepts as 'size' in the first instance and 'colour' in
the second."

Obviously, it is easy to detect the *petitio principii* unconsciously introduced into this argument. By stating that the two symbols = must
mean the same thing since they both connote identity in time, the critic
is accepting *a priori* as self-evident that there can be only one time-
stream, or, in other words, that time is absolute. Of course, if he wishes
to introduce his own private assumptions, he may do so; but he must not
blame the theory if as a result thereof he is faced with a logical contradiction. He should realise that the contradiction arises solely as a
result of his own initial unwarranted assumption, and that were he
to withdraw this initial assumption the contradiction would automatically
disappear. Then, indeed, he would have to recognise that the two symbols = need not have the same meaning; that the first, which we may
write $\overset{J}{=}$, would represent "at the same time according to John,"
whereas the second one, $\overset{P}{=}$, would represent "at the same time according
to Peter." As a result the relations would be written:

$$A \overset{J}{=} B; \ B \overset{P}{=} C; \ A \neq C;$$

and just as in the case of colour and size, he would recognise that the
first two equalities in no wise compelled the equality of A and C.

But we may wonder why it is that the critic does not realise that all
the trouble that has arisen is of his own making. The answer appears
perfectly clear. It would be somewhat as follows:

"At no time does he realise that he has introduced an arbitrary assumption and has thereby begged the entire question. The traditional belief in
absolute time so dominates his mind that he thinks in terms of it even
when he knows that the theory of relativity is not based on any such
a priori belief."

Once again, the theory of relativity as a theory of mathematical physics may survive or may succumb. It is for experiment to give the answer. If the anticipations of the theory should not conform to experiment, if, for instance, the recent experiment of Miller were considered conclusive *as detecting velocity through the ether,* it would be impossible to save the theory, at least in its present form.* But we must not confuse the ability of the theory to portray the workings of nature with its legitimacy as a perfectly consistent and rational doctrine of thought. On this last point no doubt exists. The theory has been scrutinised by the greatest of living mathematicians and not the slightest error of mathematical reasoning, nor therefore of logic, has been found in it.

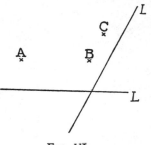

FIG. VIa

As a final illustration of this same problem of the relativity of simultaneity we may present a geometrical analogy. We have only to consider two points A and B at the same distance above a line L, and then to consider the points B and C at the same distance above another line L'. We have only to look at the figure to realise that A and C are at the same distance neither from L nor from L'. Only should the two lines L and L' coincide would they be so.

It is much the same in relativity. The straight lines L and L' correspond to the time directions of the two different observers. Classical science, assuming time to be absolute, considered the two lines coincident. Relativity has proved the error of this belief.

* Recent experiments conducted by Kennedy have shown that Miller's conclusions were erroneous, and that the null result of the original experiment was completely confirmed. (See *Proc. Nat. Acad. of Sci.,* Nov., 1926.)

PART III
THE GENERAL THEORY OF RELATIVITY

CHAPTER XXII

POTENTIALS AND FORCES

ACCORDING to classical science, force may be defined as the product of mass by acceleration. When no force is acting, the motion of a body is always Galilean; only when a force is acting will the motion of a body become accelerated.*

Now it must be noted that when a body is subjected to some definite muscular pull the magnitude of this pull or force depends solely on the muscular effort which we are willing or able to produce. On the other hand, when a body is situated in the gravitational field of the earth, the gravitational force which acts upon it and which is commonly known as the weight of the body not only depends on the mass of the earth but likewise varies with the mass of the body. In order to remedy this indeterminateness, it is usual to specify that the body on which the force is acting is one of unit mass; it is then called a test-body. Under these circumstances, we can explore a field of force by placing our test-body in successive regions of space and determining the magnitude and direction of the force which is acting on the body. When we have mapped out the magnitude and direction of the force for every point of space, we are in a position to state that we have determined the lay of the **field of force.**

In physics we are often concerned with fields of force. The gravitational field surrounding the earth constitutes only one particular illustration. We may also consider the electric field surrounding an electron, or the magnetic field surrounding the poles of a magnet, or even combinations of these two fields, such as the electromagnetic fields produced by electric currents.

When the direction and magnitude of the force are the same throughout space and do not vary with time, the field of force is called **uniform.** Uniform fields of force are very difficult to obtain; the simplest illustration we can think of would be that given by a river flowing with constant speed along a rectilinear course. The force exerted by the current would of course remain the same in magnitude and in direction at all points in the river; and for this reason the field of force produced by the current would constitute a uniform field.

---

* It might be argued that a rotating disk is an accelerated body and that, in spite of this fact, no force acts upon it. This view would, however, be incorrect. Cohesive forces are acting between the various molecules of the disk, and were it not for the existence of these forces, the various molecules would fly off along tangents and would pursue Galilean motions; the disk would cease to exist. It is the presence of the cohesive forces which compels the molecules to describe circles, that is, to pursue accelerated or non-Galilean motions.

Now, if we rest satisfied with this type of description of a field of force, we see that in the case of a uniform field, at any rate, there is no means of differentiating a region of the river or field that is upstream from one that is downstream. All parts of the river are identical. But it is obvious that this apparent identity between the different parts of the river is very superficial. It is a fact of common experience that whereas a boat will float downstream of its own accord, it will need the expenditure of a considerable amount of work to force it upstream again. Mathematicians were therefore compelled to take into consideration a new type of abstraction called a **potential.** Thanks to the introduction of this new concept, it became possible to differentiate between the various regions of the river (upstream or downstream) by stating that these regions differed in potential. The regions upstream were called the regions of higher potential, and the regions downstream were termed those of lower potential. Hence, it was possible to state, in the form of a general law, that bodies abandoned at rest would move of their own accord from regions of higher to regions of lower potential.*

So far we have been dealing with generalities, and it remains to be seen in what precise manner forces and potentials will be connected. The mathematical connection is expressed as follows:

At any point of a field of force, the force points in the direction along which the potential is decreasing the most rapidly; and the magnitude of the force is equal to the space-rate of change of the potential along the direction in question.

Potentials cannot always be associated with fields of force, but in the majority of cases encountered in physics, the concept of the potential is valid. As we have said, it is the variations in value of the potential from point to point that are associated with the existence of forces. When this variation is uniform in one direction and no variation exists in directions at right angles, we are in the presence of a uniform field of force. When the variation is irregular, we have a non-uniform field of force. When the potential has the same constant value throughout space, so that it does not change in value from place to place, no forces can be present; there is then no field of force.

From this we see that a certain degree of indeterminateness surrounds the precise numerical value of a potential, since it is only its variation from place to place, and not its absolute value, that defines the force. This indeterminateness can be removed in the case of a non-uniform field of force by stipulating that in those regions of the field where the force vanishes (such as would be the case with the gravitational force at an infinite distance from matter), the potential itself vanishes. A definite value having been attributed to the potential in one part of the field, we can determine its precise value from place to place in all other parts of the field.

---

* This statement bears a certain resemblance to the principle of entropy.

Let us now give a few concrete illustrations of fields of force and of potentials in classical science. We have already mentioned the field of force surrounding matter. This was the gravitational or Newtonian field, and the potential at every point derived therefrom was called the **Newtonian Potential** at the point. The distribution of the field of force was known in a perfectly definite manner when the potential distribution was known. In fact Newton's law, which tells us how the field of force is distributed around matter, can also be expressed in an equivalent form by **Laplace's Equation,** in which it is the potential distribution, and no longer the force distribution, that is described.

Let us now consider another type of field of force known as the inertial field. In perfectly free space, far from matter, there exists no field of force so long as we refer our observations to a Galilean frame. But if now we step into an accelerated or a rotating frame, we shall experience the effects of a field of force which we will ascribe to forces of inertia. Not only our own human bodies, but, in addition, all free bodies, will be subjected to the actions of these forces. Thus, consider a disk that is rotating, and a ball rolling without friction on the disk's surface. If the ball is sent from the centre of the disk to its periphery it will of course follow a straight line at constant speed with respect to the earth or to any other Galilean frame. But then, with respect to the rotating disk itself, its course can no longer be rectilinear and uniform. Instead of following one of the radii of the rotating disk, it will follow a curve as though it had been pulled sideways by some force.

This result is general. If *we ourselves* tried to advance along one of the radii of the rotating disk we should experience a real physical force pulling us sideways. To all intents and purposes, when we referred events to the rotating disk and not to the non-rotating earth (the rotation of the earth being so slow that we may neglect it as a first approximation), the physical existence of a field of force would have to be taken into consideration. The precise type of force we have mentioned is called the **Coriolis force;** in addition, there also exists another type of force, the better-known **centrifugal force.** Both these types of force are called forces of inertia. It is their ensemble which constitutes the field of inertial force existing in a rotating frame.

If, in place of a rotating frame, we had considered an accelerated railway compartment, the field of inertial force would have been disposed in a different manner. In the case of a train moving with constant acceleration along a straight line, the field of inertial force generated would have been of the uniform variety, disposed longitudinally through the train.

All these different illustrations show us that whereas, in a Galilean frame, no field of force exists, yet a field springs into existence automatically as soon as we place ourselves in any accelerated frame. It is for this reason that accelerated frames can be distinguished physically from Galilean frames; and it is owing to this generation of physical fields of force that accelerated motion must be regarded as absolute, whereas·

velocity, giving rise to no such fields, yields us no means of distinguishing one velocity from another, hence is relative.

Now we have seen that fields of force are accompanied by potential distributions. Hence it follows that, whereas, with respect to a Galilean frame, the inertial potential is zero at every point, or at least maintains some constant value, in the case of accelerated frames this inertial potential must vary from place to place in the frame. The magnitude of the force at any point of an accelerated frame is given, therefore, by a mathematical expression involving the variations in value of the potential in the neighbourhood of the point considered.

All we have explained so far pertains to classical science, but it appeared necessary to mention these results briefly before proceeding to a more systematic study of Einstein's theory. We shall now see how all these results find a natural place in the space-time theory.

Consider a Galilean frame and an object moving freely in this frame. Its path will of course be rectilinear and its speed uniform, hence it will possess no acceleration, in accordance with Newton's law of inertia. If we interpret these results in terms of space-time, we see that the world-line of the body is a straight line in space-time as referred to our mesh-system of equal four-dimensional cubes. If we now examine the motion of this same free body from some definite accelerated frame, all we have to do is to change our four-dimensional space-time mesh-system in an appropriate way. This change of mesh-system does not affect the world-line of the body, since this line remains a straight line or geodesic through flat space-time; but, on the other hand, it will certainly alter the appearance of the straight line when we refer our measurements to our curvilinear mesh-system. The erstwhile straight world-line will now appear curved in a definite way, and its precise mathematical equation as referred to our new curved mesh-system can be obtained without difficulty.

The physical significance of this apparent curvature of the world-line of the body will be that the body will now appear to possess acceleration with respect to our new non-Galilean frame. The precise magnitude of this acceleration will therefore be given by the apparent curvature of the world-line followed by the body, and this curvature is of course expressed implicitly by the equation of the world-line when referred to our curvilinear mesh-system. But for a given test-body the acceleration is related to the force acting on the body. From this it follows that the mathematical expression of the force can be obtained immediately.

This mathematical expression of the curved world-line, hence of the force of inertia at any particular point, is seen to be built up with the variations in value of the g_{ik}'s around this point. Were the g_{ik}'s to remain constant in value throughout, as would be the case in a Galilean frame, this mathematical expression would vanish in value. It follows that the g_{ik}'s, of which this expression of the force is built, must correspond to the potentials of inertia. We have thus discovered the physical significance of the g_{1k}'s of space-time; they define potentials. We see, indeed, that this identification is legitimate in every respect. Thus, in a Galilean frame, there are no forces of inertia, so that the potentials of inertia must be constants; and we know that in a mesh-system of equal four-dimensional cubes, which corresponds to a Galilean system, the g_{1k}'s are all constants and are given by $g_{11} = g_{22} = g_{33} = +1$; $g_{44} = -1$, all other g_{1k}'s being zero. Again, in an accelerated frame, a field of inertial forces appears; hence the potential must vary from place to place; and we know that in a curvilinear mesh-system (corresponding to an accelerated frame) the g_{1k}'s lose their constant values and vary from place to place.

So far the reason for the existence of forces of inertia has been made apparent. They arise owing to the uneven spread of g_{1k} numbers which accompanies all curvilinear mesh-systems (accelerated frames).* When this occurs, the mathematical expression of the force of inertia assumes a definite numerical value at each point, whereas, when the g_{1k}'s are constants, as in a Cartesian mesh-system, this expression of the force maintains a zero value.

From this we see that forces of inertia arise from an attempt on our part to cut up space-time with curvilinear mesh-systems instead of Cartesian ones, just as they arise from our substitution of accelerated frames for Galilean ones. We cannot help but feel, however, that having proceeded thus far, a further generalisation is required. To be more explicit, it should be understood that the scheme of physics we have developed has compelled us to attribute a fundamental rôle to space-time. But Newton's great law of universal attraction is expressed in terms of the separate space and time of classical science. If space-time is indeed as fundamental as Einstein has led us to believe, it appears incredible that Newton's law should remain outside its scope. Yet this it certainly does, for if Newton's law were a space-time law, it would preserve the same form in spite of any change in our space-time mesh-system. And this it fails to do.

Einstein's next attempt was therefore to weld Newton's law into the general fabric, a result he achieved about 1914. The mathematical generalisation which would allow this result to be obtained seems almost obvious to-day, since (as will be seen in Chapter XXIV) it reduces to assuming that in the neighbourhood of matter, four-dimensional space-time loses its flatness and becomes non-Euclidean or curved. However, it appears to have been through the medium of physical observation that Einstein was led to his superb generalisation, so we shall proceed to follow the historical order by explaining the significance of his postulate of equivalence.

---

* Note how the theory of relativity is establishing the fusion of the mathematical (*i.e.*, the g_{1k}-distribution) with the physical (*i.e.*, the forces of inertia).

CHAPTER XXIII

THE POSTULATE OF EQUIVALENCE

EINSTEIN'S **Postulate of Equivalence** consists essentially in an identification of forces of gravitation and forces of inertia. This identification he considered permissible because of certain well-known empirical facts which we shall discuss presently.

Let us first recall that a field of force is exemplified by a region of space, at each and every point of which a definite force would be found to be acting on a test-body with which the field might be explored. Several different types of fields were known to classical science. Electric fields acted on electrified bodies, magnetic fields acted on magnets, and both inertial and gravitational fields acted on material bodies in general, whether electrified or not. For the present we shall be concerned solely with the inertial and gravitational fields.

These two species of fields of force were regarded by classical science as of a totally different nature. There appeared to be very good reasons for this distinction. Suppose, for instance, that a train is slowing down. If we are standing in the train, we shall feel a force pulling us towards the engine and it may require a certain effort on our part to resist its pull. As such, the force is obviously real, in that it is experienced. But suppose now that an observer on the embankment views these same happenings. He will argue as follows: No force is pulling the passenger; but as the train is slowing down and as the passenger's body tends to maintain a constant velocity along a straight line, in conformity with the laws of motion (*law of inertia*), the net result is that he will overtake the engine unless he holds on to the seat. Thus we see that according to whether we judge these same happenings from the standpoint of the train or of the embankment, the force exists or becomes a fiction.

Similar conclusions would apply were we to consider those other species of inertial forces called centrifugal and Coriolis forces. From the standpoint of an observer who selects a rotating platform as frame of reference (just as prior to Copernicus men were wont to take the earth as frame of reference), centrifugal and Coriolis forces are real. But the moment we refer events to a Galilean frame, these forces become fictitious, and can be interpreted in terms of the law of inertia, just as the force in the train became fictitious when events were viewed from the embankment.

Contrast this situation with that of a stone falling towards the ground. Here, we may refer events to any frame we please, but in any case the law of inertia can never be made to account for the stone's motion towards the earth. The situation appears to be entirely different from that

236

of the passenger in the train. And so it follows that we are compelled to conceive of a real force, *a force of gravitation* pulling the stone towards the centre of the earth. Thus it would appear that gravitational force, in contradistinction to inertial force, could not be ascribed to mere conditions of observation. It was assumed, therefore, by classical science that there existed around a body like the earth a real absolute field of force distributed radially through space, whereas in the accelerated enclosure the forces were merely fictitious or relative.

There is yet another way of bringing out the difference between forces of inertia and forces of gravitation. Forces of inertia are generated by motion, whereas forces of gravitation are generated by matter. Thus, if we arrest the rotation of a disk, the observer, fixed to the disk, will note that the forces of inertia have vanished. But to remove the force of gravitation, it would be necessary to annihilate the entire mass of the earth, and this would constitute an operation of an entirely different nature.

Now, in every type of field of force, the magnitude of the pull which is exerted on a body susceptible to the influence of the field is governed by a certain definite characteristic of the body. Thus, in a given electric field, the magnitude of the pull exerted on a body at a given point of the field is proportional to the electric charge of the body. Likewise, in an inertial field, the magnitude of the pull is proportional to what is known as the **inertial mass** of the body, and in a gravitational field the pull is proportional to the **gravitational mass** of the body. Every material body was assumed, therefore, to possess two types of masses, the inertial and the gravitational, corresponding to the two types of fields to which the body would react.

A concrete illustration of the difference between inertial and gravitational mass is given by the following example: Consider a billiard ball at rest on a table. We should discover that the more massive the ball, the greater would be the effort necessary to set it in motion, as also to arrest its motion once started. The type of mass with which we should here be concerned would be inertial mass; and we may say that it is inertial mass which opposes a departure from rectilinear uniform motion or rest, hence which opposes acceleration. If, now, we were to lift the ball from the table and hold it at arm's length, the effort we should have to exert to prevent our arm and ball from falling would again vary with the mass of the ball. This time, however, the mass whose effects we were seeking to resist would be the gravitational mass, *i.e.*, the mass which is responsible for weight. It is this same gravitational mass which enters into the expression of Newton's law of attraction.

There is no *a priori* reason why any connection should exist between these two different types of masses. Thus, if a billiard ball is found to weigh twice as much as another, there is no logical necessity to anticipate that its inertial mass will also be twice that of the other. Yet it so happens that the most refined physical experiments have invariably proved the strictest proportionality between these two types of masses;

so that by choosing our units suitably it was always possible to represent both the inertial and the gravitational mass of a given body by the same number. Classical science had accepted the equality of the two types of masses as an empirical fact, but had found itself incapable of suggesting any theoretical justification for it. It appeared to be due to some miraculous coincidence.

This equality in the numerical values of the two types of masses will entail important consequences. For suppose, indeed, that the equality did not hold. This would imply that two billiard balls might present exactly the same gravitational mass, yet differ in the value of their respective inertial masses. If, then, the two balls were to be released simultaneously and allowed to fall from the same height towards the earth, their gravitational masses being equal, both balls would be subjected to exactly the same pull by the earth. But their inertial masses being assumed unequal, the ball whose inertial mass was greater would oppose the acceleration of the fall more strenuously than would be the case with the ball of lesser inertial mass. As a result, the balls would not reach the earth's surface simultaneously. On the other hand, if the two types of masses are identical, we may be assured that any two objects released from the same point will reach the earth's surface at the same instant (provided we operate *in vacuo* so that the resistance of the air does not interfere with their fall). It was precisely because experiment had demonstrated the existence of the same rate of fall for all bodies released from the same height, that Galileo was led to recognise the equality of the two masses. It may be mentioned, however, that ultra-precise experiments (notably with Eötvös' torsion balance) have since verified these results with extreme accuracy.*

* The statement that all bodies fall with the same motion *in vacuo* is correct, but, unless properly understood, is apt to lead to erroneous conclusions.

What the statement asserts is that any body, regardless of its constitution or mass, will fall in exactly the same way *through a given gravitational field*. If, however, the falling body causes a modification in the distribution of the field, then it is obvious that various falling bodies will no longer be situated in the same gravitational field; and there is no reason to assert that they will all fall in exactly the same way. For instance, if a relatively small mass—say, a billiard ball—then a relatively large one—say, the moon—were to be released in succession from some very distant point and allowed to fall towards the earth, it is undoubtedly correct to state that the duration of fall would be greater for the billiard ball than for the moon. It is easy to understand why this discrepancy would arise. In either case, owing to the mutual gravitational action, the earth would also be moving towards the falling body (moon or billiard ball), so that the point of collision would be some intermediate point, namely, the centre of gravity of the system earth-moon or earth-billiard ball. But in the case of the billiard ball, owing to its relatively insignificant mass, this centre of gravity, or point of collision, would be practically identical with the earth's centre. This implies that the earth would scarcely move at all towards the billiard ball, whereas it would move an appreciable distance towards the moon. The motion of the earth would thus shorten considerably the distance through which the moon would have to fall, whereas the billiard ball would have to fall through the entire distance.

If we wish so to modify the conditions of the problem as to re-establish the perfect identity in the rate of fall of the moon and billiard ball, we must so

We see, therefore, that the equality of the two types of masses allows us to anticipate that an observer situated in a falling elevator would have no weight. He might be standing on a weighing machine, but the needle would always register zero. The fact is that both observer and weighing balance would be falling with the same motion towards the earth; hence the observer's feet would never press against the balance.

These anticipations may appear to be self-evident; but it is important to note that they are in no wise self-evident until we have recognised the perfect equality of the two types of masses. If such an equality did not exist, the observer in the elevator might still press against the scales of the weighing machine or might hit the ceiling of the elevator, and it would no longer be correct to anticipate that elevator, scales and observer would all fall with exactly the same velocity and acceleration.

We may illustrate the example of the falling elevator in a more general way by considering an observer situated in the interior of a hollow projectile describing its parabola under the influence of the earth's attraction. The observer would detect no manifestation of weight in his enclosure. He might tip a glass full of water upside down: the water would not spill. If he threw a handful of pebbles into the air, each pebble would follow a straight course with rectilinear motion through the projectile, just as would be the case in a Galilean frame. In short, no mechanical experiment could ever reveal the presence of the earth's field of gravitation in the interior of any frame of reference moving freely under the action of this field of gravitation. Experiments conducted inside such falling bodies are very difficult to perform, but since the results we have mentioned are the immediate consequences of the equality of the two types of masses, which highly refined experiment has established, we are able to anticipate indirectly the results of such experiments.

arrange matters that the earth is unable to fall towards the body it is attracting. If, for example, it were possible to nail the earth to the Galilean frame in which earth, moon and billiard ball were originally at rest, and if this Galilean frame could be made to remain Galilean, i.e., unaccelerated, then the previous experiment attempted first with the moon, then with the billiard ball (the moon being removed entirely), would reveal exactly the same rate of fall for the two bodies. For now, indeed, the modifications in the distance covered, and in the nature of the field brought about by, the displacement of the earth, would be non-existent.

A further case which presents a theoretical interest in Einstein's discussions is afforded by what is known as a uniform field, that is to say, a field in which the gravitational force is the same in intensity and direction throughout space. A field of this sort would be generated by an infinitely extended sheet of matter of uniform density. Owing to the infinite mass a sheet of this sort would possess, its acceleration towards the falling body would be nil; all bodies would then fall with exactly the same constant acceleration towards the sheet. The reason uniform fields present a theoretical interest is because the field of force generated in an enclosure moving with constant acceleration is precisely of this type. When, therefore, Einstein identifies the field of force enduring in an accelerated enclosure with a gravitational field, we must remember that the distribution of matter which would be necessary to produce the same field is that of an infinitely extended sheet. Only to a first approximation can a finite mass of matter, like the earth, be deemed to generate a field of this kind.

So far we have considered the cancellation of a field of gravitation, but we may also consider the creation of a field of force in empty space far from matter. Thus, if a hollow chest be subjected to a variable acceleration through empty space, a variable field of force (inertial force) will be experienced by an observer in the chest's interior. If now we consider a second chest at rest in a varying gravitational field distributed throughout the chest in exactly the same way as were the forces of inertia and with the same intensity, the equality of the two masses proves that mechanical experiments conducted in either of the two enclosures would yield exactly the same results, and hence that bodies would fall towards the floor of the enclosure in exactly the same way. And so it would be impossible, *on the basis of mechanical experiments* conducted in the interior of the enclosure, to decide whether we were in a field of inertial force generated by acceleration or in a field of gravitational force generated by matter. These are the conclusions which the equality of the two masses urges upon us. Inasmuch as the equality of the two types of masses constitutes a fact established by experiment, we may also regard as established empirically the inability of mechanical experiments to differentiate between a field of inertia and one of gravitation.*

Now, up to this point, all the facts we have mentioned were perfectly well known to classical science. It is solely in the interpretation of these facts that Einstein suggests highly original views. Let us first examine what classical science had to say. Its arguments ran somewhat as follows: Our inability to differentiate between a field of inertial forces generated by acceleration and one of gravitation generated by matter is due solely to a miraculous coincidence, that of the equality of the two masses. Again, when, in the elevator falling in the earth's gravitational field, we experience no feeling of weight, it is not that the gravitational field has vanished; it is solely that the field of gravity has been counteracted by an equal and opposite field of inertia generated by the elevator's accelerated motion. The equality in the intensities of these two opposing fields is due once more to the equality of the two masses. Thus, according to classical science, there existed a very decided difference between the two types of fields of force, and it was always assumed that this difference would be detected when we performed non-mechanical experiments such as optical or electromagnetic ones.

It is this classical interpretation which Einstein challenges. He maintains that the field of force generated in an enclosure is of exactly the same nature regardless of whether it has been generated by acceleration or by gravitation. Also, in the case of the falling elevator, we should not say that the field of gravitation had been counteracted; we should say that it had *vanished*. Thus interpreted, the miraculous equality of the two types of masses is no miracle at all. There are no two types of masses which react to two types of forces. Since there is but one type of force, there is but one type of mass, to which classical science had erroneously given two different names, and then marvelled at the fact that whatever

* Subject to certain niceties which will be mentioned presently.

name was given, its value remained the same. In this way Einstein succeeded in accounting for a miraculous equality which had baffled classical science.

But, of course, Einstein's idea was susceptible of experimental proof or disproof. For if we assume that the two types of fields of force are all one, we must assume that not alone mechanical experiments, but also *all manner of experiments*, will fail to detect the slightest difference. This is indeed Einstein's attitude, and he upholds it in the **postulate of equivalence.**

This postulate states that there can exist no difference in the conditions prevailing in the interior of an enclosure, regardless of whether the field of force experienced be generated by acceleration or by gravitation.

Now, when all is said and done, it might appear that when Einstein stated his postulate of equivalence he was taking a leap in the dark. The equality of the two masses proved that so far as mechanical experiments were concerned, conditions existing in a falling elevator would be identical with those enduring in an enclosure floating in free space far from matter. These anticipations were based on experience. But what justification was there for generalising these results and extending them without hesitation to all manner of non-mechanical experiments? At first sight the generalisation might appear somewhat hasty, and most certainly such generalisations should not be encouraged at the hands of those who do not possess any profound scientific knowledge. But with Einstein the case is different. On this occasion, as on many others, his generalisations are those of a man who seems to be able to read the secrets of nature as no one else has ever done before. It is not his knowledge of mathematics or of physics that causes admiration; it is his insight into the philosophy of nature, which is stupendous. As a matter of fact, a similar generalisation on his part was encountered in the special theory. There he generalised from electro-dynamics to mechanics; here it is the reverse. But, in either case, a desire to find unity in nature appears to have been his guiding motive.

There is a great analogy between Einstein's interpretation of the problem we are discussing and his explanation of Michelson's experiment. In the latter, we remember that Galilean motion through the ether appeared to generate no ether drift. Lorentz explained this absence of an ether drift by assuming that a physical drift was really present, but that it was automatically cancelled in a miraculous way by the trickiness of nature working through the compensating influences of a real physical contraction of bodies moving through the ether and of a real physical slowing down of the duration of events. Einstein, as we have seen, preferred to assume that if no ether drift could be detected, it was because no ether drift was present. Classical science demanded the ether drift and attempted to explain why in spite of its expectations none could be detected. Einstein, by refusing to patch up classical science so as to explain the miraculous, was led to declare that classical science was wrong in attaching any credence to the existence of an ether drift, and that its belief in this ether drift was attributable to mistaken premises.

If we now pass to the problem of gravitation we see that the situation is very similar. Gravitational force vanishes in a falling elevator; this is the physical fact. Classical science took the stand that the force did not really vanish, but that it was compensated. Einstein prefers to assume that if the gravitational force *appears* to vanish, it is because it really *does* vanish.

With these new views, the force of gravitation acting at a point loses its attributes of absoluteness. Just like a force of inertia, a force of gravitation betrays a relationship existing between the frame of reference selected and the surrounding conditions of the world. Just as a force of inertia can be annulled by changing the motion of our frame, so now can a force of gravitation be annulled. This does not mean that a force of gravitation is unreal. It is perfectly real, since it can be detected and measured. But it is no longer an absolute; it is a relative, like a force of inertia, for its value varies with our choice of a frame of reference. Henceforth no intrinsic difference exists between a force of gravitation and a force of inertia; these forces are of the same essence. When, therefore, we are in the presence of a force acting on mass, we may refer to it as a force without specifying whether it has been generated by the acceleration of our frame (field of inertia) or by the proximity of matter (field of gravitation).*

Einstein maintains the appellation of a gravitational field of force. But he applies it indifferently to fields which classical science would have considered due to the acceleration of the frame, and to the fields which classical science would have recognised as generated by the proximity of matter. By classing both types of fields as gravitational, his object is to differentiate these identical types of fields from fields of other types, such as fields of electric, magnetic or electromagnetic forces.

However, it is well to caution the beginner against a misunderstanding which might cause him some trouble. In classical science the word "gravitational" was always associated with the attraction caused by matter. In Einstein's theory, when we identify an inertial field with a gravitational one and call both these species of fields gravitational, it is not meant to imply that an exact replica of the inertial field could be reproduced by disposing matter in a suitable way with respect to our frame of reference. Inversely, a field of force produced by matter cannot be duplicated in every detail by communicating a suitable accelerated motion to our frame of reference in free space far from matter. In spite of the complete identification of forces of gravitation and forces of inertia, there still exists a difference in the spatial distribution of the field of force, according to whether it is produced by the proximity of matter or by the acceleration of our frame. In consequence, although all fields of force may be called gravitational fields, regardless of whether they be generated by matter or

* To obviate any confusion .with electromagnetic forces, we are considering only forces which act on uncharged bodies.

by acceleration, we must remember that the actual lay of these fields through space will vary with their origin.

A few precise illustrations will make this point clearer. Consider a hollow chest pulled by some unseen hand through interstellar space far from matter. If we suppose that the chest is rising vertically with constant acceleration, we know that as a result of this constant acceleration a uniform field of inertial force will be present in the interior of the chest. The postulate of equivalence consists in asserting that the observer in the interior of the chest might with equal justification consider the chest to be unaccelerated or at rest in space, while the field of force he perceives would be assimilated to a field of gravitation. All that it is meant here to imply is that the physical nature of fields of gravity and of inertia is one and the same. It is *not* meant to imply that any actual distribution of matter could ever produce a perfectly uniform field of force in the chest's interior such as existed under the influence of constant acceleration. We know, indeed, that a distribution of matter under the chest would produce a somewhat similar field of force, but the field would be non-uniform, the magnitude of the force being smaller at the top than at the base of the chest. Thus, although the physical nature of both types of fields would be the same, the precise distribution of these fields through space would be different.

These, at least, are the conclusions by which we must abide at the present stage of the theory. We shall see that when we come to consider the universe as a whole, there may be grounds for modifying our opinions. But at the present stage of the discussion we must admit that fields of force produced by matter can never be distributed in exactly the same way as fields produced by acceleration, and vice versa.

Another example is afforded by the field of force generated on a rotating disk. This field of force can be split up for purposes of analysis into two separate fields: a centrifugal field constituted by forces directed away from the centre of the disk, and a Coriolis field pulling bodies sideways as they approach or move away from the centre. Now, no distribution of matter around the disk can be conceived of at this stage which would produce a field of gravitation on the disk distributed in exactly the same way as this field of inertia. Nevertheless, the inertial field on the disk can be called a field of gravitation for the reasons previously set forth in this chapter.

Again, just as matter cannot produce a field of force distributed in exactly the same way as a field generated by acceleration, so also is it impossible to produce by the sole virtue of acceleration a field of force distributed in exactly the same way as the gravitational field around matter. From this it follows that it is impossible, by merely selecting some suitably accelerated frame, to cancel in its entirety a field of force produced by matter. This statement may appear to be in conflict with the example of the falling elevator, in which it was explained that owing to the motion of the elevator the field of force produced by the earth vanished in its interior. However, there is no conflict between the two statements, if we are careful to express ourselves with precision.

In the falling elevator no field of force is experienced in the interior of the elevator, *provided the extension of the elevator is small in comparison with the size of the earth.* If the elevator were very large, if, for instance, it were a gigantic box completely surrounding the earth, no conceivable motion of the box could ever cause the gravitational field of the earth to vanish everywhere in the box's interior. Suppose, for instance, that the roof of the box were accelerated towards the North Pole, its floor would then be moving with accelerated motion away from the South Pole. Hence, whereas an observer attached to the roof of the box would perceive no force of gravitation, an observer attached to the floor would experience a force of doubled intensity. In other words, it would be impossible to banish the force of gravitation in all parts of the box. Only when the box or elevator is small in comparison with the earth, so that within its confines the field produced by our planet is practically uniform, is it possible to cancel the earth's field in the whole interior of the box. Even then, however, this cancellation would be rigorous only at a point, and would decrease in thoroughness as we moved away from this point.

It follows that a falling elevator cannot be assimilated in all rigour to a Galilean frame, since in a Galilean frame, wherever we might be stationed, no forces would be experienced. Nevertheless, as in practice, the frames we consider are not supposed to extend for great distances in all directions, conditions enduring in a falling elevator would be identical to a very high order of precision with those existing in a Galilean frame. For this reason, rigid frames of small dimensions falling freely in a gravitational field are termed *semi-Galilean.*

If we summarise the conclusions reached in this chapter, we may say that the postulate of equivalence has allowed us to identify forces of inertia with forces of gravitation. But this identification applies solely to the nature of the forces, not to their spatial distribution. And so it is not correct to say that it would be impossible for us to ascertain whether the field of force experienced in our enclosure was due to the acceleration of the enclosure or to the proximity of gravitating masses. Even without peering out and discovering whether large masses were present, we could always, at least in theory, by a mere exploration of the field distribution, ascertain the true conditions.

We see, then, that whereas velocity was relative in that it was quite impossible for us to ascertain the absolute velocity of our enclosure, acceleration still remains absolute in spite of the postulate, since (theoretically at least) it can always be detected and its effects separated from those due to matter. In fact, as we shall see later, the complete relativity of all motion can be established only if the universe is finite. Under the circumstances, we must be careful not to overestimate the philosophical significance of the postulate in its bearing on the relativity of motion.

Now it might be thought that owing to this fundamental difference in the spatial distribution of fields of force (inertial and gravitational), the postulate would not be of much use in its physical applications. But this view would be erroneous. From a purely qualitative standpoint the postulate permits us to assert that any phenomenon whose behaviour should be affected by the acceleration of the enclosure, must also be sensitive to the presence of a gravitational field due to matter. Inasmuch as it is often easy to see that the acceleration of our frame of reference must inevitably modify the observed behaviour of a phenomenon, we are able to infer therefrom that the same phenomenon will also be affected by a gravitational field generated by matter.

It was by following this method that Einstein was able to anticipate a number of gravitational effects which classical science had never even suspected. Chief among these are the bending of a ray of light in a gravitational field, and the *Einstein-shift effect,* since observed on the companion of Sirius.

We may understand without difficulty how the bending of a ray of light was anticipated. For consider an enclosure floating in empty space far from matter. A wave of light enters by a crack in the wall and travels through the enclosure along a line parallel to the floor. But if now the enclosure be uniformly accelerated, the floor advances with accelerated motion across the ray. As referred to the enclosure, the ray of light will thus assume a bent path, just as the course of a bullet would appear bent under similar circumstances.* Then the postulate of equivalence allows us to assert that exactly the same results would have ensued had the enclosure been at rest in a uniform gravitational field; so that regardless of whether or not *uniform gravitational fields* generated by matter can exist, we are led to the general conclusion that a gravitational field will bend the course of light waves and modify their velocity.

It is this qualitative discovery that is of importance and that constitutes a fact which was unknown to classical science. What the precise course of a light ray will be in a non-uniform gravitational field generated by matter, is a question of another order.

The postulate, however, enables us to answer this question, provided we know how the gravitational field is spatially distributed around matter. We may understand this point as follows: Owing to its gradual variations from place to place, even a non-uniform field may be regarded as uni-

* Here the reader may well question our right to argue as though velocities combined in the classical way, when the whole significance of the special relativity theory has been to deny the validity of the classical transformations. In point of fact the objection would be legitimate; and in all rigour the Einstein-Lorentz transformations should be applied for each successive instantaneous velocity of the enclosure. But it so happens that when the Einstein-Lorentz transformations *are* applied to a transverse beam of light, it is found that for low velocities of the enclosure the bending is practically the same as it would have turned out to be had we followed the classical rule of composition. In other words, had the motion of the enclosure been uniform, the transverse ray of light as measured in the enclosure would have been inclined much as in classical science. Indeed, were it not that the relativity transformations entailed a variation in the slant of a ray of light moving transversally, the theory would be incompatible with the well-known phenomenon of astronomical aberration.

form if we restrict our attention to very small regions of space.* It follows that we can decompose a non-uniform field into a series of contiguous volumes of space, each one of which may be assimilated to the spaces in enclosures possessing appropriate uniform accelerations. If, therefore, the spatial distribution of a non-uniform gravitational field is given, we may deduce (according to the methods of differential geometry) the path of a ray of light from place to place over contiguous infinitesimal distances. In this way, assuming Newton's law to be correct, Einstein deduced the curvature of a ray of light in the sun's gravitational field. He found that for a ray grazing the sun's limb the deflection would be 0".87, just one-half of what he was to establish subsequently, after he had recognised that Newton's law could not be correct.

In other cases, we are not concerned with the study of a phenomenon extending over large areas; its behaviour in a small area, say in our room, suffices. In this case, of course, the earth's gravitational field may always be regarded as uniform; so that by considering the effect arising from the introduction of a constant acceleration of suitable magnitude, the action of gravitation on our phenomenon can be anticipated.

In a general way, the postulate of equivalence led Einstein to the following conclusions: Inertial mass and gravitational mass being one and the same thing, all forms of existence which possess inertial mass must also manifest weight in a gravitational field. Hence, as all forms of energy possess inertial mass, we see that energy has weight. A hot brick weighs more than a cold one, an electrified body more than a neutral one; possibly, also, a living animal more than a dead one, and so on. Conversely, all forms of energy must develop a gravitational field.

In the following pages we shall show that the postulate of equivalence, which in the present chapter was derived as a generalisation from the equality of the two masses, can also be deduced in a purely rational way from the existence of space-time. We shall thus be in a position to understand how all these discoveries dealing with forces and gravitation can be woven into the common space-time fabric.

* We are assuming that the field does not vary with time. When this is not the case, we must specify that our observations must also be conducted over a very short duration of time.

CHAPTER XXIV

THE INCLUSION OF GRAVITATION IN THE MODEL OF SPACE-TIME

In the preceding chapter the following results were established:

1. No essential difference exists between a field of gravitation produced by matter and a field of inertia produced by the acceleration of our frame of reference (postulate of equivalence).

2. Although these two types of fields of force are essentially of the same nature, yet the actual spread of the gravitational field around matter is never quite the same as that of the field generated by the acceleration of a frame in free space far from matter. As a result of this feature, whereas it is always possible to cancel a field of force entirely by changing the motion of our frame in a suitable way when the field contemplated exists in regions remote from matter, on the other hand, in the neighbourhood of matter, the field of force can be modified, and even cancelled in a small region, but never cancelled over any wide extension. It is like a protuberance which can be flattened out in one place, only to reappear with increased intensity elsewhere.

These are the results which we must weave into the general model of space-time, unless we reject Einstein's postulate of equivalence, and maintain that fields of gravitation produced by matter have no more in common with fields of inertia than they have with electromagnetic fields, for instance. The postulate of equivalence, as we have seen, appears to be justified owing to the empirical fact that the two types of masses are identical. But we now propose to show that quite independently of this empirical fact, even if the postulate of equivalence had never been stated in a specific manner, a natural mathematical generalisation of the geometry of space-time would have led us indirectly to exactly the same identification of forces of inertia and gravitation. Accordingly, let us recall briefly the results we arrived at in Chapter XXIII when we examined the origin of forces of inertia in our space-time model.

We remember that space-time was assumed to be perfectly flat, so that its geodesics were Euclidean straight lines when referred to Cartesian mesh-systems of equal four-dimensional cubes. These Cartesian mesh-systems, splitting up space-time into mutually perpendicular rectilinear directions, constituted the geometrical representations of our methods of breaking up space-time into space and time for purposes of referring our measurements to Galilean systems of reference. In these Cartesian mesh-systems the g_{ik}'s of space-time maintained constant values throughout, and the numerical values of these g_{ik}'s represented the potentials of the field of inertial force. These potentials being constants, remaining unchanged in value from point to point in our Cartesian systems, no field of force could be present; and this was in full accord with what we know of Galilean systems, since these systems are characterised by the total absence of all such fields. Furthermore, as all free bodies and light waves followed geodesics in space-time according to the requirements of Least Action, and as these geodesics when referred to Cartesian systems were straight lines, free bodies and light waves would always appear to follow straight lines with constant speeds when observed from Galilean frames. This was in accord with Newton's law of inertia.

Subsequently we saw that accelerated frames moving in interstellar space would be represented by curvilinear or Gaussian mesh-systems. In these systems the values of the g_{ik}'s of space-time were no longer constant, but varied from place to place. This variation in the values of the g_{ik}'s connoted the variation in value of the potentials; and as a result the mathematical expression of a force (given as it was by relations between these variations) no longer vanished. These curvilinear mesh-systems represented, therefore, systems in which fields of force would be present. This again was in perfect accord with what we knew of accelerated or non-Galilean frames. Further, as referred to these curvilinear mesh-systems, the geodesics of space-time would appear to be curved. The physical significance of this fact would be to imply that the motion of free bodies and light waves would appear to be accelerated when viewed from accelerated frames. Once more the complete accord with experiment is apparent.

Now let us proceed to the natural generalisation of these results. So far, we have assumed space-time to be flat. Indeed, we could not escape this assumption if we were to admit that Galilean frames, free from all fields of force, could exist in space far from matter, and that as observed from these frames the geodesics of space-time would be Euclidean straight lines. Only thus would Newton's law of the motion of free bodies in interstellar space be verified. But suppose now that, for some reason or other, space-time were to develop non-Euclideanism or curvature, and so depart from its flatness. How would such a condition manifest itself physically?

In the first place, it would be impossible to split up this curved space-time into the straight directions which define a Cartesian mesh-system. Physically speaking, this would imply that in those regions where space-time was curved no Galilean systems could exist. We might, of course, conceive of mesh-systems less curved than others; but even the straightest of these systems would at best be only approximately Cartesian within a restricted region. As we considered parts of the mesh-system farther and farther removed from this special region, the meshes would become more and more widely spaced or more and more closely gathered together.

An elementary illustration of this would be given by considering a curved surface of two dimensions, such as that of a sphere. On a sphere we can trace an infinite variety of mesh-systems, but none of these mesh-systems is made up of equal Euclidean squares. The nearest approach to a network of squares would be afforded by a mesh-system of meridians and parallels. But even a mesh-system of this sort could be likened to a network of equal squares only in the immediate vicinity of the equator. As we examined our mesh-system at regions farther and farther removed from the equator, we should notice that the meridians had a tendency to run closer and closer together, so that the quadrilaterals bounded by the meridians and parallels would depart more and more in shape from those of Euclidean squares. In the general case, where we consider an arbitrarily curved surface and no longer a uniformly curved one, such as a sphere, it could be shown that the straightest type of mesh-system compatible with the surface was assimilable to one of squares only around a point, and not, as in the case of a sphere, along a line (such as the equator). The point could be selected arbitrarily, and our pseudo-straight mesh-system drawn around this point accordingly.

We see, therefore, that if space-time is unevenly curved or unevenly non-Euclidean, the nearest approach to a Cartesian mesh-system around any given point could be approximately Cartesian only in a restricted region around this point. As we moved away from this point in our mesh-system, its curvilinear characteristics would become more and more pronounced. So far as the g_{ik}'s of space-time are concerned, this would imply that in a region where curvature was present these g_{ik}'s could maintain constant values only in the more or less immediate neighbourhood of a point. Expressed in terms of fields of force, this assertion is equivalent to stating that in a region of space-time curvature it would be quite impossible to rid ourselves of a field of force throughout space. We might annul it at the point and in its immediate neighbourhood, but the field of force would reappear for more distant regions.

It is easy to see that these physical conditions are precisely those which we have found to exist in a gravitational field produced by matter. The straightest type of mesh-system around a point (namely, a semi-Cartesian mesh-system) would correspond to that of the observer in the falling elevator or in the hollow projectile, since for an observer in a frame of this type the force of gravitation disappears in the immediate interior of his frame, but reappears at more distant points. Further, space-time being curved, its geodesics would no longer be Euclidean straight lines. They would still appear approximately straight when referred to a semi-Cartesian mesh-system in the immediate neighbourhood of the point for which the mesh-system was approximately one of equal four-dimensional cubes; but their curvature would become more and more apparent when we surveyed them from this same mesh-system in more distant regions. The physical interpretation of this geometrical fact would be afforded by the seemingly uniform and rectilinear motion of free bodies and rays of light in the immediate neighbourhood of the falling elevator when viewed by an observer situated in this falling frame, and by the apparent acceleration of these bodies in regions more distant from the falling elevator.

It is therefore obvious that the existence of a non-Euclideanism of space-time around matter would betray itself physically in exactly the same way as a field of gravitation produced by matter.* Assuming this generalisation to be correct, we see that no essential difference can exist between a field of gravitation and one of inertia. Both types of field arise from a relationship between the mesh-system adopted by the observer, and the fundamental structure of the space-time continuum. The only difference that exists is that in interstellar space far from matter, space-time being flat, it is always possible to select a Cartesian mesh-system corresponding to a Galilean system in which no field of force is present, whereas in the neighbourhood of matter, owing to the inherent curvature of space-time, the nearest approach to Galilean frames is given by semi-Cartesian or semi-Galilean frames (falling elevator). In such frames the field of force cannot vanish throughout the frame at more or less distant points.

It follows that these same g_{ik}'s whose values in a given mesh-system represented the potentials of the field of inertia can now with equal justification be considered to yield the potentials of the field of gravitation. The mathematical expression comprising the variations of the g_{ik}'s from point to point, which was formerly proved to represent the force of inertia, can now with equal justification be deemed to represent the force of gravitation. As was the case with the force of inertia, this mathematical expression of the force, depending as it does on the variations in the values of the g_{ik}'s from point to point, varies with our choice of a mesh-system; in exactly the same way, the magnitude of a gravitational force in our frame of reference varies with the acceleration of our frame (as was established when we discussed the postulate of equivalence). Furthermore, the value of the mathematical expression of this force will vanish completely throughout a mesh-system when this mesh-system is Cartesian, whereas it will vanish only around a definite point when, owing to the curvature of space-time, a Cartesian mesh-system is impossible, so that we have to be content with a semi-Cartesian one. We see that the analogy is complete in all respects.

Here it must be noticed that, identifying as we are the g_{ik}'s in a given mesh-system with the gravitational potentials in the frame corresponding to this mesh-system, we find ourselves in the presence of as many different gravitational potentials as there are separate g_{ik}'s. That is, we have ten different potentials. Newton knew of only one of these, the Newtonian potential, which turns out to be g_{44}—the one directed along the time-axis of the mesh-system. It is the presence of these other potentials ignored by Newton and discovered by Einstein which is responsible for the major differences distinguishing Newton's law of planetary motions from that of Einstein. We see, therefore, that this natural generalisation of the geometry of space-time, by compelling us to recognise forces of gravitation and inertia as of the same nature, leads us to exactly the same conclusions as were obtained by Einstein when he started from the postulate of equivalence. So although we laid stress on the postulate of equivalence

* Qualitatively at least. The precise quantitative justification will be furnished later.

as a separate discovery in order to abide by the historical sequence of Einstein's investigations, we see that when we view the problem from a loftier standpoint the postulate does not add anything to what would normally have been discovered, or at least suspected, as a result of the more philosophical generalisation we have discussed.

Of course, if the postulate of equivalence should prove to be unjustified, as it would were experiment to detect a difference between inertial and gravitational mass, the generalisation would also be unjustified and gravitation could never be ascribed to a curvature of space-time. However, as the most refined experiments have invariably disclosed the complete identity of the two types of masses, the space-time generalisation appears to be in order.

CHAPTER XXV

TENSORS AND THE LAWS OF NATURE

WE have now reached a point where further progress would appear to require at least a few elementary notions on tensors and the absolute calculus. We shall endeavour to present these matters in as clear a way as possible by considering, first of all, tensors from the standpoint of pure mathematics, without any reference to their applications in Einstein's theory.

Let us suppose that we are dealing with the ordinary three-dimensional Euclidean space of classical science. We remember that it is often convenient to refer the positions of objects situated in space to some definite system of reference or system of co-ordinates. This procedure is equivalent to drawing an imaginary three-dimensional mesh-system with respect to which we locate the points of space. The precise shape of this mesh-system is arbitrary. It may split up space into equal cubes (in which case it is called Cartesian), or again it may be curvilinear and split up space into irregularly shaped volumes. In all cases, however, we shall assume that these mesh-systems remain fixed and do not vary in shape or orientation once they have been drawn.

Now, though we have stated that the precise shape of the co-ordinate system, or mesh-system, is arbitrary, there are mathematical reasons which render the use of Cartesian systems especially convenient. So long as we are dealing with Euclidean space, in which, as we know, Cartesian mesh-systems are always permissible, we need not be concerned with Gaussian co-ordinate systems. It was for this reason that classical science, when dealing with physical problems, confined itself to Cartesian co-ordinates; and by a change of co-ordinate system it meant a rotation or a displacement of the Cartesian frame of reference from one position to another.

Nevertheless, we always had to view the possibility of having to investigate problems in which Cartesian co-ordinate systems would be impracticable, owing to the curvature or non-Euclideanism of the continuum. Accordingly, even before the advent of the theory of relativity, mathematicians had endeavoured to perfect a form of calculus which would free them from all dependence on any particular type of co-ordinate system. This task was accomplished chiefly by Levi-Civita and Ricci, and the mathematical instrument evolved was called the *absolute calculus*. And so, when in the general theory Einstein was led to his conception of a curved or non-Euclidean space-time, in which Cartesian co-ordinate systems were impossible, the mathematical instrument he required was at hand. All he had to do was to put it to practical use.

Although, in the Euclidean space of classical physics which we now propose to discuss, there was no need to consider Gaussian or curvilinear co-ordinate systems, yet, in view of the greater generality of the method, we shall treat the subject matter of this chapter from the standpoint of all systems of co-ordinates, whether Gaussian or Cartesian. The conclusions we shall reach must, however, be construed as holding for all manner of co-ordinate systems. If these preliminary points are understood, we may proceed to a more detailed discussion of the problem.

Let us consider a straight rod situated in this space. If we agree on a unit of measurement, we can assign some definite length to the rod. Obviously, this length, whatever it may be, will remain unaffected by a change in our co-ordinate system in space. Magnitudes of this sort, which transcend in every respect our choice of a co-ordinate system, are known as **scalars, numbers** or **invariants.**

But suppose now that we wish to determine the exact orientation of our rod. Orientation, in contrast to length, has no meaning until some mesh-system has been prescribed; for we can define a direction in empty space only by referring it to some accepted system of standard directions at each point. The meshes of our mesh-system at each point will, of course, define such standard directions for us. When, therefore, we attempt to determine the orientation of the rod, our procedure will be to obtain what are known as the three components of the rod with respect to the mesh-system considered. If our mesh-system is Cartesian, the components will be the three projections of the rod on the three straight lines of the Cartesian system which pass through the origin of the rod. In the general case of a Gaussian, or curvilinear, mesh-system, the definition of the three components is less simple. In any case, however, the important point to remember is that in any mesh-system there will be three components, (which we may view as three numbers) and these components will serve to determine the orientation and the length of the rod.

From this example we see that a considerable difference exists between the directed magnitude we are here considering and the invariant one which was given by the mere length of the rod irrespective of its orientation. Whereas the invariant magnitude possessed no components, the directed one has as many components as the space has dimensions, three in the present case. A directed magnitude of this sort is termed a **vector.**

Suppose now that we change our system of reference, rotate it slightly, for instance. We shall thereby obtain three new components for our vector in the new co-ordinate system, and these new components will differ in magnitude from our former ones. However, the change in the value of the components for a definite change of frame of reference is not arbitrary, but will be submitted to well-defined mathematical rules.* This fact enables us to discover a criterion of objectivity.

---

* Thus, calling *X, Y, Z the components of the vector in the first mesh-system*

For, suppose that in a first co-ordinate system certain components, X, Y, Z, are revealed by measurement, and that in a second co-ordinate system new components X', Y', Z' are obtained. If under these circumstances we were to find that the two trios of components X, Y, Z and X', Y', Z' were *not* linked together by the stringent mathematical rules we have referred to, we should have to recognise that we could not have been contemplating the same vector when we passed from the first co-ordinate system to the second. Conversely, if the mathematical rules were found to be satisfied by the components, we should realise that we were indeed dealing with the same vector. In this way the vector would no longer be some indeterminate magnitude. It would be a definite and precise one to which we might attribute an objective existence transcending our choice of a mesh-system of reference. The components themselves have no absolute existence; they do not transcend our mesh-system. They represent but partial aspects of the vector, mere modes or shadows, varying, as they do, with our system of reference. But to the vector itself an objective existence can be conceded.

Now, mathematical investigation soon showed that there existed two different types of vectors, distinguished by two different rules of mathematical transformation when we passed from one co-ordinate system to another. As these two rules of mathematical transformation, known as the **contravariant** and the **covariant** rule, respectively, appeared to be the only consistent ones, it was recognised that there existed only these two different types of vectors, the so-called **contravariant vector** and the **covariant vector**. In a certain sense this distinction is somewhat artificial, for we may always regard a given vector as contravariant or covariant, as we see fit. All that we mean to imply by this terminology is that we intend to subject the components of the vector to the contravariant or the covariant rule of transformation when we change our co-ordinate system.

So far we have considered two types of magnitudes, invariants or scalars, and vectors. Mathematicians discovered, however, that there existed a vast array of other types of magnitudes named **tensors,*** whose components likewise obeyed rigid rules of mathematical transformation when we passed from one co-ordinate system to another, and which for this reason could be credited, just as truly as vectors and invariants, with an objective existence transcending our mesh-system. Tensors differed from vectors owing to the greater number of their components. Whereas a vector possessed only as many components as the space had dimensions, a

and X', Y', Z' in the second, the components X', Y', Z' will be connected with the components X, Y, Z in the following way:

$$X' = aX + bY + cZ$$
$$Y' = dX + eY + fZ$$
$$Z' = gX + hY + iZ$$

where a, b, c, d, etc., are defined by the nature of the change to which our co-ordinate system has been subjected.

* Not to be confused with the tensor of a quaternion.

tensor could have a much greater number. According to the number of their components in a space of a given number of dimensions, tensors were grouped into tensors of the second order, of the third order, of the nth order.

In a three-dimensional space a tensor of the second order has nine components instead of three, as has a vector. A tensor of the third order has twenty-seven, and so on. Just as in the case of vectors, the components of a tensor situated at a point in space vary in value with the choice of our mesh-system, and, again, just as with vectors, the values of the components in one mesh-system are connected with their values in another mesh-system by rigid mathematical rules of transformation. In the present case, however, these rules of transformation are more complex. They are seen to be generalisations of the rules regulating the transformations of vector components. As an immediate sequel, tensors appear to be mere generalisations of the older known vectors.

In particular, *a vector was found to be nothing else than a tensor of the first order*, while *an invariant appeared to be a tensor of zero order*. The general name "tensor" expresses, therefore, all those types of mathematical magnitudes which possess an objective significance transcending our system of co-ordinates, although it will often be convenient to reserve the name "tensor" for tensors of the second and higher orders, retaining thereby the classical appellations "vector" and "invariant" for tensors of lower order.

Now we shall see that the distinction we have made between the two different types of vectors (contravariant and covariant) will extend automatically to tensors. Let us consider, for instance, a tensor of the second order. When we change our co-ordinate system, the nine components of this tensor will vary, and to obtain their values in the new co-ordinate system we apply a rule of transformation which is, so to speak, the vector rule taken twice in succession. As there are two vector rules, one contravariant and one covariant, we get, in the case of tensors of the second order, a twice contravariant rule, a twice covariant one or a mixed rule both contravariant and covariant. In this way we are led to distinguish three different types of second-order tensors: the twice contravariant, the twice covariant and the mixed variety.

Here it must be stated that the two rules, the contravariant and the covariant, are in certain respects the antitheses of each other. What one rule does the other undoes. If, therefore, we combine any vector considered as contravariant with any vector considered as covariant, the result will be similar to what happens when a squirrel runs in a drum which is rotating backwards. The forward motion of the squirrel is counteracted by the backward motion of the drum, so that finally the squirrel does not move at all. The result of combining a contravariant and a covariant vector will be, therefore, to obtain a scalar or invariant. In the same way, a covariant tensor of the second order combined with two contravariant vectors, or a covariant tensor combined with a contravariant one of the same order, will each yield an invariant.

Let us now pass to one of the most important characteristics of tensors and vectors. The fact that the components of all contravariant vectors are submitted to the same rules of change when we pass from one co-ordinate system to another, proves that if two contravariant vectors situated at the same point of space are equal, that is, if the components of the one are equal to the components of the other in our co-ordinate system, this equality will inevitably endure in any other co-ordinate system. In other words, the equality of two vectors at a point of space constitutes an equality which a change in our co-ordinate system can never destroy; the equality is thus an absolute. Similar conclusions apply to the equality of two tensors at the same point of space.

It must be noted, however, that this equality endures only because the components of either vector or of either tensor on the two sides of the equation are submitted to the same rules of transformation when our co-ordinate system is changed. This implies that we are dealing with vectors of the same nature (covariant or contravariant), or with tensors of the same nature and order. If these conditions are satisfied, we see that equations between vectors or tensors often called **vector equations** and **tensor equations,** exhibit the remarkable property of remaining unaffected by a change of mesh-system.

In a certain sense, then, vector equations and tensor equations constitute invariant or absolute equations. This last appellation would, in fact, be perfectly justified were we to argue in terms of the absolute vectors and tensors which exist independently of our mesh-system. But in practice it is very difficult to dispense with a mesh-system. As a result, what we come into contact with—what we measure in reality—is not the vector or the tensor itself, which is absolute, but its components in the mesh-system that we have selected. These components are relative, since they change in value when we change our mesh-system. True, these changes being the same on either side of the equation, the equality endures regardless of our choice of mesh-system. But, on the other hand, a definite numerical change has taken place in the value of either side of our equation. The form of the equation as a whole has remained unaltered, but the contents of the two sides has changed.

Let us explain this point more fully. Consider two vectors which are situated at the same point and which are equal to each other. In three-dimensional space, this equality is expressed by the respective equality of the three components of the first vector with the three components of the second vector in our co-ordinate system; hence we have three equations, such as

$$X_1 = X_2; \quad Y_1 = Y_2; \quad Z_1 = Z_2.$$

If, now, we change our mesh-system, the components of either vector change in exactly the same way, so that our equations subsist. In the new mesh-system, however, X_1 becomes X'_1 and X_2 becomes X'_2, etc., so we now have

$$X'_1 = X'_2; \quad Y'_1 = Y'_2; \quad Z'_1 = Z'_2.$$

Obviously, although the equations subsist, yet the expressions on either side of our three equations have varied; for X'_1 is different from X_1, and so on. Thus we see that it would not be quite correct to state that our equations remained completely unchanged when we passed from one mesh-system to another.

For these reasons it is customary to speak of the **covariance** rather than of the **invariance** of vector or tensor equations. The word "co-variance" expresses the fact that both sides have varied in exactly the same way. The appellation "invariance" would tend to make us believe that neither side had varied at all, which would, of course, be incorrect. In short, we must remember that it is the relationship of equality between the two sides, and not the individual terms themselves, that remain unaffected by a change of mesh-system.*

Now, there exists still another type of vector and of tensor equation. If, for instance, a vector or a tensor vanishes, this means that all the components vanish individually. A simple glance at the rules of transformation shows that in this case all the components will continue to vanish in any co-ordinate system, so that the equating of a tensor, a vector or an invariant to zero at a point of space will endure in all co-ordinate systems if it endures in any particular system.

Summarising, we find that there exist two types of covariant equations:

The first type is given by the equality of two vectors (the vectors being both covariant or both contravariant) or, more generally, by the equality of two tensors, these tensors being of the same order and same nature (*i.e.*, covariant, mixed or contravariant). A limiting case would be given by the equality of two invariants; we should then have an invariant equation.

The second type is given by the vanishing of a vector or a tensor at a point in space. A limiting case would be given by the vanishing of an invariant.

All these deductions obtained mathematically are in a certain sense self-evident physically when we recall that a tensor, a vector and an invariant represent magnitudes to which an objective existence may be conceded. It is obvious, indeed, that if two such magnitudes are equal to each other at a point in space, a mere change of our co-ordinate system can never disturb this intrinsic equality. In the same way, if a concrete entity, such as a tensor, has vanished at a point in space, a mere change of our co-ordinate system is incapable of bringing it into existence. Once again, we realise the strict dependence which exists between physical representations of this sort and the fact that invariants, vectors and tensors must be recognised to have an objective existence transcending the particular point of view of our co-ordinate system.

If these mathematical considerations have been grasped, we may pass to physical applications of these purely mathematical discoveries. We

---

* A distinction of this sort does not apply, of course, to the equality of two invariants; for, as we have seen, a change of mesh-system can produce no effect on the value of an invariant, seeing that an invariant has no components.

will restrict ourselves for the present to the three-dimensional Euclidean space of classical science; hence by a change of co-ordinate system we again mean a mere change in shape, position and orientation of our mesh-system. *We do not mean the passage from one system to another in relative motion thereto.*

In classical science we often come upon invariants and vectors. Examples of invariants are given by the mass and the density of mass of a body. Vectors are illustrated by a velocity, an acceleration, a force, a displacement; and we may say that in a general way a vector can be represented by an arrow. The simplest physical illustration of a tensor at each point of space is afforded by the state of compression and tension at each point of an elastic medium which has been distorted.

It is impossible to express this state of tension and compression at a point by means of three components in our three-dimensional space. Owing to lateral wrenching effects at each point, nine components are necessary; and we realise that we can no longer appeal to a vector. We are now in the presence of a tensor. The physical origin of the appellation "tensor," coined by Willard Gibbs, thus becomes apparent. In the present case the tensor is a second-order tensor, and, furthermore, it is what is known as a **symmetrical tensor**. The characteristics of a symmetrical tensor are defined by the identity of certain of its components two by two; this reduces the number of independent components of a second-order tensor in three-dimensional space to six in place of nine.

It can be proved that the g_{1k} numbers which we have often mentioned constitute symmetrical second-order covariant tensors. The tensor nature of the g_{1k}'s means nothing else than that the components of the g_{1k}'s in our co-ordinate system vary according to the rules of tensors when we pass from one co-ordinate system to another. The g_{1k}'s being second-order covariant tensors, we may infer that by combining them with two displacements, or vectors, of opposite nature (*i.e.*, contravariant vectors), we shall obtain an invariant. It is for this reason that $\sum g_{1k} dx_1 dx_k$ yields an invariant,* which, as we know, is nothing but the square of the length ds.

In three-dimensional space there are nine of the g_{1k} components, but, the tensor g_{1k} being symmetrical, this number is reduced to six. Thus $g_{12} = g_{21}$; $g_{13} = g_{31}$; $g_{23} = g_{32}$. For this reason we may in the case of three-dimensional space write out ds^2 as follows:

$$ds^2 = g_{11} dx^2_1 + 2g_{12} dx_1 dx_2 + 2g_{13} dx_1 dx_3 + g_{22} dx^2_2 + \ldots \ldots ,$$

instead of

$$g_{11} dx^2_1 + g_{12} dx_1 dx_2 + g_{21} dx_2 dx_1 + \ldots \ldots .\dagger$$

---

* The tensor g_{1k} being twice covariant, and the vector dx_1, repeated twice in the formula, being contravariant.

† It is customary to represent scalars by ordinary letters, and tensors of the first, second and third orders, and so on, by letters followed by indices equal in number to

When we come to examine the concrete significance of the g_{ik}'s, we find them to be expressive of the angles between our co-ordinate meshes, and of the general shape of our mesh-system from place to place. Since, however, the shape of our mesh-system, though arbitrary in large measure, is nevertheless restricted to certain types compatible with the structure of the space in which we trace them, these g_{ik}'s are also representative of the structure or geometry of our space. For this reason they are called **structural or fundamental tensors.**

Let us recall once more that we are not yet dealing with Einstein's discoveries or with space-time; we are discussing three-dimensional space, and all we have said refers to pre-Einsteinian science.

Other illustrations of tensors are afforded by an angular velocity and by a magnetic force. The tensor nature of these two magnitudes is less obvious because it so happens that these second-order tensors are of a special type known as *antisymmetrical*. In tensors of this type, certain components are always zero in all systems of co-ordinates, and the remaining ones are of the same magnitude but of opposite sign. The result is that in the case of three-dimensional space antisymmetrical tensors of the second order possess only three independent components just like vectors, and for this reason are often represented by arrows and confused with vectors. As a matter of fact, this confusion need not arise; for it is obvious that a rotation, for example, is better represented by a corkscrew curve than by an arrow, and the same applies to a magnetic force when we realise that it is generated by a whirling of electrons. At all events, had it not been for the accidental fact that space had three dimensions, the confusion could never have occurred at all; for, in a four-dimensional space, antisymmetrical tensors of the second order have a number of components greater than four (which is the number of components of a vector in a four-dimensional space). Still another illustration of an antisymmetrical tensor of the second order is furnished by an element of a surface.

In short, we see that the magnitudes in which classical science was chiefly interested were either invariants, vectors or tensors. At this we need not be surprised, seeing that the magnitudes to which physical science pays greatest attention are always those to which we may concede an objective existence transcending our choice of a mesh-system. Thus, an

the order of the tensor. Thus, g_{ik} is a tensor of the second order, a_{ikl} is one of the third order, and so on.

When we substitute for these indices all possible arrangements of the numbers from 1 to n, where n represents the n dimensions of our continuum, we obtain thereby the various components of our tensors.

Thus, in a space of two dimensions, the various components of g_{ik} are g_{11}, g_{12}, g_{21}, g_{22}, which reduce to g_{11}, g_{12}, g_{22}, owing to the identity of g_{12} and g_{21}, the g_{ik} tensor being symmetrical.

In order to differentiate at a glance contravariant tensors from covariant ones, the indices are placed above the letter. For instance, G^{ik} is the contravariant form of G_{ik}, and G^i_k is the mixed form. We see, then, that $ds^2 = \Sigma g_{ik} dx_i dx_k$ should really be written $\Sigma g_{ik} dx^i dx^k$, since dx^i refers to a contravariant vector.

object itself appears to be of greater significance than the shape of its shadow; for the shadow will change as we alter the orientation and shape of the surface on which it falls, whereas the existence of the object will transcend any such changes.

If now we consider the laws of nature, we can anticipate, for two different reasons, that they must be vector or tensor equations. First, because all those physical magnitudes to which we attribute an objective existence and which enter into the expression of the laws of nature are invariants, vectors and tensors; and, secondly, because a law of nature must obviously remain unaffected by a mere change of orientation or shape of our frame of reference.

It is, indeed, evident that a mesh-system is a mere artifice of calculation. It can have no intrinsic significance, and, as a matter of fact, we might dispense with it altogether. Such an absolute condition of the world as the equality of two vectors at a point or the vanishing of a vector or of a tensor would endure even had mesh-systems never been invented by Descartes and Gauss. Suppose, then, that a law appeared to hold when we selected some particular mesh-system, but no longer when our meshes were squeezed together or rotated to some new position. Obviously our law would possess no generality; it could refer to nothing objective and would be no law at all.

We may summarise these discoveries by saying that all the laws of classical science were vector or tensor laws. They remained covariant or unchanged in form when we changed the orientation or shape of our mesh-system *while maintaining it fixed in space*. This last point is important; for in classical science, time being absolute and separate from space, it was solely from the point of view of a spatial change in the shape and orientation of our mesh-system that the laws remained covariant.

We may also mention that in addition to the covariance of all natural laws to a spatial change of mesh-system, there existed a second requirement which all natural laws had to obey. We refer to the test of dimensionality. A brief discussion will make this point clear.

The numerical values of all physical quantities depend on our choice of standard units for the measurement of distance, duration and mass. If, therefore, we change our units, if we adopt the minute instead of the second, or the foot instead of the metre, or the pound instead of the kilogramme, the numerical values of both sides of the equation representing the law will be affected accordingly. For the equation to endure in spite of the change in value of its two terms, it is necessary that a change in our unit of time or of length or of mass should affect the numerical values of both sides of the equation in exactly the same way. This condition is expressed by saying that both sides of the equation must possess *the same dimensionality*. It is perfectly obvious that a law of nature must constitute an equation of this type. It would be ridiculous to suppose that a change in our units from the avoirdupois to the metric system, for example, could ever upset the validity of a law of nature.

This test of dimensionality was often useful in determining the second side of the equation representing some unknown law. We might ignore the exact form of the second side, but we could always tell in advance that its dimensionality would have to be the same as that of the first side. In this way a certain amount of *a priori* information was often obtained.

CHAPTER XXVI

We have endeavoured to show that the great change in our views of the world brought about by the theory of relativity arises from the discovery that four-dimensional space-time and not the three-dimensional space and the independent time of classical science correspond to reality. It might seem but a natural generalisation to extend to space-time the conclusions we arrived at in the last chapter when discussing the covariance of the natural laws to a change of mesh-system in space. We should then expect the laws of nature to express relationships between vectors and tensors in space-time, hence to remain covariant to a change in our space-time mesh-system.

Although this generalisation would turn out to be perfectly legitimate, its validity is by no means self-evident *a priori*. We can accept it only after we have convinced ourselves that the laws of physics, such as experiment has revealed them to be, can indeed be regarded as expressing relationships between entities (vectors and tensors) in space-time. Now, from a purely formal standpoint, it is conceivable that such might not be the case. Suppose, for instance, that the four-dimensional spatio-temporal background were characterised by some finite invariant velocity C greater than Maxwell's constant c, which is given, as we know, by the velocity of light *in vacuo*. In an extended sense we could still regard our universe as a four-dimensional space-time, since, the invariant velocity being finite, and not infinite, the fusion of space and time would hold, just as it did in Minkowski's space-time.

But, on the other hand, owing to the discrepancy between C and c, the laws of electromagnetics would cease to be covariant to a change of Galilean frame, hence to a change of space-time mesh-system. And so we could not view these laws as expressing relationships between the absolute entities of the fundamental space-time continuum.

The situation changes when we find, as a result of experiment, that the laws of electromagnetics remain covariant to a change of Galilean frame. This discovery results, of course, from the negative experiments in electromagnetics and from others of a similar nature. The first consequence of this discovery is to prove that C, the invariant velocity of space-time, and c, Maxwell's constant, must be one and the same, since only on this condition will the laws of electromagnetics remain covariant.

Einstein then extended this property of covariance to all physical laws. We thus obtain the special principle of relativity, according to which the laws of nature remain covariant when we refer them to one Galilean frame or another. In short, the special theory deals solely with Galilean frames.

From the standpoint of mesh-systems, let us recall that Galilean frames are given by Cartesian mesh-systems in space-time, the differences in their orientations representing the relative velocities existing between the frames. When we pass from measurements performed in one of these Cartesian mesh-systems to those executed in some other one of these frames, we must apply the transformations corresponding to a rotation in space-time; and these are the Lorentz-Einstein transformations. We may say, therefore, that in the special theory the principle of covariance entails the covariance of all natural laws when submitted to the Lorentz-Einstein transformations.

Now the fact that the laws of nature maintain the same form regardless of the orientation of our Cartesian space-time mesh-system, or frame of reference, suggests that they must be the expressions of relationships between such entities as vectors and tensors in space-time. If this be the case, the laws should remain covariant not merely to a change of Cartesian mesh-system, but also to a change to any species of mesh-system, whether Cartesian or curvilinear. In other words, they should be expressible as vector or tensor equations in flat space-time.

This discovery permits us to advance a step farther. For let us suppose that this erstwhile flat continuum manifests curvature in certain regions. Einstein assumed that in this case its curvature would curve along with it the entities embedded in its substance. Under the circumstances, the covariance of the laws of nature should continue to hold regardless of the flatness or curvature of space-time, hence also of the species of mesh-system we might appeal to. It is this generalisation of Einstein's which constitutes the *principle of general covariance*, or the *general principle of relativity*. We see that it represents an extension of the special principle of covariance, or of relativity, extending as it does to all regions of space-time, whether flat or curved, and to all mesh-systems, whether Cartesian or curvilinear, the covariance established originally only in a special case. Henceforth it becomes permissible to say that all the laws of nature express relationships between entities of space-time (vectors and tensors); and as a result we are able to extend to space-time the conclusions we arrived at when discussing space in the previous chapter. After this preliminary discussion, we may investigate the objectivity of our physical magnitudes.

In classical science these magnitudes were required to be invariants, vectors or tensors in space. In the light of the new discoveries, we must now substitute space-time for space. The result is that both gravitational force and potential energy turn out to be deprived of all objective significance, for by merely changing our mesh-system in space-time we can annihilate or bring into existence the components of these magnitudes at a point. Potential energy entering as it does into the expression of the law of conservation, it became apparent that this law of classical science was obviously of an artificial nature. We shall discuss the principles of conservation in greater detail in a later chapter.

Let us now examine the subject of the general laws of nature. In classical science these laws had to remain covariant to a change in the shape and orientation of our mesh-system in three-dimensional space. The new requirement is that these laws shall remain covariant to a change of shape and orientation of our mesh-system in four-dimensional space-time. This new condition will obviously exert an additional restrictive influence on the formulation of our laws, for it comprises as a special case all the classical restrictions as to covariance in space, supplementing them with the added restriction of covariance in space-time. When, therefore, we consider the laws of classical science, we shall be able to accept them without modification only provided they happen to satisfy the restrictive condition imposed by the space-time theory.

It must be kept in mind, however, that the space-time tensor test cannot aspire to tell us with certainty what the laws of the real world will be. Its action is merely one of elimination; it tells us what these laws can never be. When it comes to deciding which one of the many possible laws is to be accepted as the true law, experiment must be our guide. We can also foresee that the classical laws, though they may appear impossible in the light of the new test, must nevertheless be approximately correct since classical science had considered them to be borne out by experiment. We must not be surprised to find, therefore, that from a practical point of view very slight differences exist between the new laws and the old ones when crude experiments at least are considered. But from a theoretical point of view the difference is of momentous importance. The new laws pass the test; some of the classical ones do not.

It was soon noted that the classical equations of electrodynamics satisfied the new theoretical requirements, for they could be expressed as tensor laws in space-time. Inasmuch as they were in perfect harmony with experiment, they were left undisturbed. On the other hand, the classical equations of mechanics and, with them, Newton's law of gravitation were certainly incorrect, for they did not remain covariant to a change in our space-time mesh-system.

Accordingly, Einstein had no difficulty in writing out the correct laws for mechanics by modifying ever so slightly the classical ones so as to make them conform to the restrictive space-time condition. We have stated elsewhere that in this way mass was proved to be a relative, no longer an invariant, its value varying according to our space-time mesh-system. We have also mentioned that the most exacting experimental tests have proved the correctness of the new laws and the fallacy of the classical ones.

In the preceding chapter we stated that classical science never paid much heed to the tensor notation, because it so happened that three-dimensional space was assumed to be permanently Euclidean and flat, and hence a standard mesh-system of the Cartesian variety was applicable everywhere throughout space. In the special theory, space-time is also considered flat, so here again the tensor symbolism can be dispensed with. We have seen, indeed, that the special theory was concerned only with Galilean observers in space far from matter, and we know that observers of this type always split up space-time into space and time through the medium of Cartesian mesh-systems. These mesh-systems differ from one another merely by their various orientations in space-time; and we pass from one mesh-system to another by means of the Lorentz-Einstein transformations.

On the other hand, when we come to consider accelerated observers, we are dealing with observers who split up space-time with curved mesh-systems. Again, when masses of attracting matter are present, space-time assumes an intrinsic curvature which renders the utilisation of Cartesian mesh-systems impossible. Certain particular observers, such as those who are falling freely in a gravitational field, adopt mesh-systems which are approximately Cartesian in the immediate neighbourhood of the observer; but were we to prolong these mesh-systems sufficiently, we should soon discover that they became curvilinear, hence varied in shape from place to place, owing to the intrinsic curvature of space-time. In much the same way, a network of parallels and meridians on the surface of a sphere can be assimilated to a Cartesian one of equal squares only in the neighbourhood of the equator, since the meshes taper together as we near the poles. Thus, in a general way, we see why it is that when we leave the special theory, where space-time is flat and where the observer, being Galilean, abides by Cartesian mesh-systems, and when we investigate the general theory, where the observer is accelerated or space-time is curved, the Lorentz-Einstein transformations must give place to a more general type. Under these circumstances the tensor calculus becomes a mathematical instrument of great power.

While it is true that in the early days of the special theory Einstein never made use of this form of calculus, with which he was probably unacquainted, most modern treatises on relativity appeal to it from the very start, since it brings out the beautiful unity of the entire theory and shows that the special theory is merely a particular limiting case of the general one. This allows us to treat both forms of the theory simultaneously. We lay stress on this point, because some non-mathematical critics are of the mistaken opinion that the tensor calculus applies only to the general theory. From this belief they derive the equally erroneous opinion that the two theories constitute separate creations, having little in common. We have seen that such is not the case.

In this presentation of Einstein's theory, we have emphasised the absolute characteristics of space-time, showing how the covariance of the natural laws followed as a necessary consequence as soon as we regarded them as expressing relationships between space-time entities (tensors, vectors) transcending the mesh-system of the observer. Viewed in this way, the theory has sometimes been referred to as "the quest of the absolute." But we must remember that it is only as a result of the principle of relativity, necessitated by the negative experiments, among others, that space-time and the laws of nature can be viewed in this light. Were it not for the principle, hence for the identity of C, invariant velocity, and c,

Maxwell's constant, though there might still be an absolute four-dimensional world, the physical laws such as we know them could no longer be conceived of as expressing relations between its tensors and vectors. We might conceive of space-time entities, in this new type of space-time, as entering into laws that would remain covariant to any change of mesh-system, but these entities and laws would not be represented by the physical entities and laws that experiment has revealed.   From the standpoint of the physicist, they would be mere fictions, foreign to the world of our experience.   It would be as though the entities and laws revealed to us by experience differed from the laws that would normally be expected had the world a unity of structure and of content.   In this event the great beauty of Einstein's space-time theory would be lost, and with it the wonderful unity and harmony it has shown to exist in nature.   Thus, although the absolute viewpoint is the one we have stressed in this chapter, we should bear in mind that it was the disclosures of relativity that rendered this absolute viewpoint possible.

Furthermore we may very easily confer a relativistic aspect on the theory if we recall that by selecting curvilinear or Gaussian mesh-systems in space-time, we are merely placing ourselves in the position of an accelerated observer, who refers his measurements to some frame of reference, squirming like an octopus and who times events with clocks running wild.   So far as can be gathered, Einstein was guided originally by a desire to prove that all motion was relative; it was thus the relativistic view that preceded the absolute viewpoint in the elaboration of the theory.   There is no cause to be surprised at this precedence, seeing that we are referring to days when space-time was as yet unknown.   But when space-time was discovered by Minkowski, the theory received an added impetus (though we can realise full well that even before this discovery a theory which aspired to prove that certain permanencies endured regardless of the observer's motion would obviously lead us to suspect the underlying presence of some absolute world in which these permanencies would be situated). This absolute world could not be one of separate space and time; for in the absolute space of classical science the laws changed in form when our motion was modified.   Obviously something more was needed, and space-time was the answer.

It is imperative, however, not to misinterpret the significance of the covariance of natural laws.   The mere fact that all observers, whatever their motion, will discover the same laws, does not mean that all observers are alike in every respect.   Even in Einstein's theory, a rotating observer will realise full well, owing to centrifugal forces, that he does not stand in a Galilean frame.   The complete relativity of all motion from every point of view is not by any means embodied in the covariance of the laws; and we shall see that this complete relativity of motion, leading to Mach's mechanics, is far from being accepted by all scientists.   The discussion of this subject will have to be left aside for the present, as it is intimately connected with the difficult problem of the shape of the universe, concerning which we shall have something to say in a later chapter.

CHAPTER XXVII

THE DISCOVERY OF THE EINSTEINIAN LAW OF GRAVITATION

In the two preceding chapters we have discussed in a very general way the subject of tensors and of the laws of nature. It now remains to be seen how Einstein, guided by the requirements that accompanied all general laws, succeeded in formulating with mathematical precision the true law of gravitation.

In order to understand what is to follow we must revert to Newton's law. This law defines the intensity and distribution of the field of gravitational force in the empty space which surrounds matter. Also, it enables us in turn to ascertain how the field of force would be distributed in the interior of matter, in the interior of the earth, for instance. Thus, in its most general aspect, Newton's law of gravitation defines the intensity and distribution of the gravitational field both inside and outside of matter.

Now we have seen that the presence of a field of force connotes the existence of an uneven distribution of potentials in the space where the field exists, since any perfectly even and constant distribution of potentials would correspond to the complete vanishing of the field. Hence, Newton's law of gravitation can also be written as a law of potential distribution inside and outside of matter. Expressed in this form, it is known as **Poisson's equation** and is written

$$\triangle \Phi = 4\pi k \mu.$$

In this celebrated equation, Φ represents the potential at each point,[*] $\triangle \Phi$ is a complicated mathematical expression linking the potential at a point and its value at all neighbouring points; μ expresses the density at each point of the matter producing the field, and k is a constant of proportionality known as the constant of gravitation. And since the equation $\triangle \Phi = 4\pi k \mu$ gives us the law of potential distribution at each point of space where the density of matter is μ, in regions outside matter we obviously have

$$\triangle \Phi = 0,$$

since outside matter μ is of course zero. Expressed in this latter form ($\triangle \Phi = 0$), in which it gives us the potential distribution outside matter, the Newtonian law of gravitation is known as **Laplace's equation.**

[*] We are here assuming that the potential function for a point-mass m is defined by $\phi = -\dfrac{km}{r}$ where r is the distance from the point mass, and k is the gravitational constant.

Thus, we may say that Newton's law of gravitation inside and outside of matter is given by $\triangle \Phi = 4\pi k\mu$, in which μ is zero or has some definite non-vanishing value according to whether we are considering a region outside matter or in its interior. Now, this potential distribution in Newton's law refers to the distribution through space of one same potential, an invariant in space, known as the **Newtonian potential**. But we know that if Einstein's ideas are correct, if gravitation is to be attributed to space-time curvature, Newton's single potential Φ (an invariant) will have to be replaced at every point by the ten separate g_{ik}'s of space-time. These, as we know, are no longer invariants, they are second-order tensors;* and Newton's invariant potential Φ turns out to be none other than one of the g_{ik}'s, namely, the time one, g_{44}.†

According to Einstein's views, therefore, the expression of the gravitational potential distribution in the interior of matter will be given by an equation between some complicated expression connecting the ten g_{ik}'s at every point and at neighbouring points, and some quantity representing the characteristics of matter at every point. This equation will then constitute Einstein's law of gravitation in the interior of matter, and will represent the Einsteinian substitute for Poisson's classical equation.

Now Poisson's equation was an equation between two invariants *in space*. Einstein's law, on the other hand, in accordance with the requirements that must be satisfied by all general laws in the space-time theory, will have to be an equation between invariants or tensors in *space-time*. By analogy with Poisson's equation we see that the left-hand side of Einstein's law will be a space-time invariant or tensor which will represent the potential distribution, or g_{ik}-distribution, from point to point, while the second side will be given by an invariant or a tensor representing matter.

A tensor representative of matter and describing it fully had already been discovered by Einstein. It was a second-order symmetrical tensor generally written T_{ik} and its ten separate components, as referred to any definite space-time mesh-system, defined such physical magnitudes as the density of matter at a point, its internal stresses, *vis viva*, and momentum in that mesh-system. Matter, as a source of gravitation, was obviously described much more completely by this tensor than was the case in classical science, where the mere mention of its density of mass was considered sufficient to determine its gravitational field. Thus, Einstein was in possession of the right-hand side of the equation of gravitation in the interior matter, that corresponding to $4\pi k\mu$ in Poisson's equation (neglecting constants).

---

* In four-dimensional space-time there are sixteen of these g_{ik}'s at every point, but, the tensor being symmetrical, six turn out to be mere repetitions, so we need only speak of ten separate g_{ik}'s at every point.

† More precisely, ϕ is not g_{44}, but is connected with g_{44} by the relation $g_{44} = 1 + \dfrac{2\phi}{c^2}$.

It remained to discover the left-hand side of the equation. Owing to the condition of covariance, this left-hand side had to be represented by a tensor of the same order and nature as the right-hand side, hence, in the present case, by a symmetrical tensor of the second order. Further, it would have to be built up of the ten g_{ik}'s or potentials, expressing a relationship between these ten g_{ik}'s at every point and at the neighbouring points. It is here that Riemann's discoveries fit in with Einstein's requirements; for Riemann and his successors had discovered tensors built up from the fundamental structural tensors of a continuum, namely, from the g_{ik}'s. All these tensors served to represent various possible types of structure, geometry or curvature.

In a previous chapter we mentioned these tensors. First, there was the Riemann-Christoffel tensor B_{ikst}. It was a tensor of the fourth order, as is indicated by the four indices underlying the letter B. Then there was a second-order symmetrical tensor G_{ik}. Lastly, there was an invariant of curvature G, no longer a tensor in the ordinary sense of the word. By differentiating and combining these magnitudes, we could of course obtain others, but the ones we have mentioned are the simplest.

Einstein had still one further clue to guide him. Classical science had assumed that at an infinite distance from attracting matter the law of inertia would hold rigorously; that is to say, a body moving freely would follow a rectilinear course with constant speed when referred to a Galilean frame. Expressed in the language of space-time, this would imply that at an infinite distance from matter, space-time would become rigorously flat, as we had assumed it to be in the special theory. This condition necessitated the vanishing of the Riemann-Christoffel tensor B_{ikst}, at infinity. If this condition is maintained, the curvature of space-time around matter will have to be of a type which is capable of dying down to perfect flatness at infinity, degenerating to $B_{ikst} = 0$.

Now the law of curvature $G_{ik} = 0$ was precisely of such a type; hence Einstein, in his original paper, found it natural to select the tensor G_{ik} for the left-side term of his gravitational equations. Accordingly he wrote:

$$G_{ik} = T_{ik} \text{ (neglecting constant factors)}.$$

Outside matter, this law would of course reduce to $G_{ik} = 0$, since, outside matter, T_{ik}, representative of matter, would vanish. It is easy to see that these gravitational equations satisfy all the requirements we have mentioned. First, we are in the presence of covariant equations, since G_{ik} and T_{ik} are tensors of the same nature and order. Secondly, G_{ik} is built up from the g_{ik}'s or potentials, as it should be. Thirdly, outside matter, our equations reduce to $G_{ik} = 0$, and the curvature connoted by this relation is compatible with the existence of perfect flatness at infinity, i.e., with $B_{ikst} = 0$ at infinity.

Nevertheless, Einstein soon recognised that his equations could not correspond to reality. The fact was that energy and momentum appeared to be conserved in nature. By this we mean that energy and momentum disappear from here only to reappear there. It followed that the tensor of curvature which would be identified with the presence of matter in the law of gravitation should likewise possess the attributes of conservation. Now the tensor G_{ik} possessed no such attributes; but a very similar tensor, also symmetrical and of the second order, namely,

$$G_{ik} - \frac{1}{2} g_{ik} G,$$

fulfilled all the previously mentioned requirements. Henceforth, the law of potential distribution, or of gravitation inside matter, would have to be (neglecting constant factors)

$$G_{ik} - \frac{1}{2} g_{ik} G = T_{ik}$$

(analogous to Poisson's equation); and outside matter, as T_{ik} disappeared, there was left

$$G_{ik} - \frac{1}{2} g_{ik} G = 0$$

which reduces to

$$G_{ik} = 0$$

(analogous to Laplace's equation). From this we see that Einstein's initial error affected the law of space-time curvature only in the interior of matter. Outside matter, we have, as before, $G_{ik} = 0$.*

Let us recall that in all examples where tensor equations are discussed, the indices i and k (in the present case) are to be replaced by all manner of arrangements of the integers from 1 to 4 (since space-time has four dimensions). An equation such as

$$G_{ik} - \frac{1}{2} g_{ik} G = T_{ik}$$

or as

$$G_{ik} = 0$$

represents therefore ten separate equations. As a matter of fact, sixteen equations would be obtained were we to substitute all possible arrangements; but as the tensors are symmetrical, only ten out of sixteen are independent, hence need be retained; the other six are mere repetitions. We see, then, that in Einstein's theory ten separate equations are necessary in order to define the distribution of the gravitational field in the interior and around matter, whereas in classical science one equation, i.e., Poisson's equation ($\triangle \Phi = 4\pi k \mu$), was sufficient.

Here we may mention that if we abandon the restriction that the law of inertia holds rigorously at infinity, hence if we abandon the con-

---

* In point of fact, it was when Einstein applied the principle of Action (to be discussed presently) that he first recognised the error in his original law. Also we may note that the law of curvature $G_{ik} = 0$, when it does not reduce to $B_{ikst} = 0$, represents a non-homogeneous type of curvature.

dition of perfect flatness at infinity, we may build up other tensors satisfying all the remaining requirements. This can be done by adding a further tensor λg_{ik} (where λ represents an arbitrary constant) to the left-hand side of our equation. The presence of this additional term will imply that even for very distant regions, space-time would still manifest a trace of residual spherical curvature so that the universe would close round on itself and be finite. In the first development of his theory, Einstein rejected this additional term, but the fact remains that so far as the mathematical requirements of the theory of relativity are concerned there is no reason to exclude it *ab initio*. At all events, we can assert that if the λg_{ik} term exists at all, λ must be exceedingly minute; for even though finite, the universe is certainly of gigantic proportions, and its curvature therefore exceedingly slight. For the present we shall ignore the λ term since in any case its influence would be too insignificant to affect perceptibly such phenomena as those of planetary motions. Only when we come to consider the shape of the universe as a whole will the presence of λg_{ik} assume importance.

Now the methods whereby Einstein was led to the discovery of his law of gravitation might appear rather piecemeal, much like the fitting together of a jig-saw puzzle in which the structural tensors constituted the pieces. But this, in itself, can scarcely be regarded as a drawback to the theory. It is rather the reverse, since it proves that just as the pieces of a puzzle can only be fitted together in certain ways, so now the universe can scarcely be otherwise than it is. In this respect, there is a marked contrast between the present situation and that which endured in classical science, where practically any law of gravitation might have been expected.

Nevertheless, from the standpoint of mathematical elegance, something more was needed. When, therefore, Einstein had cleared the ground and his ideas were more fully understood, the greatest of the pure mathematicians, such as Klein, Hilbert, Weyl and subsequently Einstein himself, undertook to discover the law of gravitation by more general methods. They found that if we accepted the existence of space-time and if we agreed that the gravitational field around matter was due to a curvature of space-time, the Principle of Stationary Action showed that a large number of laws of gravitation could be entertained. If, however, it was specified that by analogy with Newton's law, the law of curvature should contain no derivatives of the g_{ik}'s or potentials to an order higher than the second, Einstein's law of gravitation with or without the λg_{ik} term was the only one possible. Furthermore, all these possible laws of gravitation were found to exhibit the attributes of conservation.

CHAPTER XXVIII

In classical science, when we wished to describe a body's motion, it was of course insufficient to describe its spatial course in our system of reference. It was necessary, also, to define its motion along this course. As soon, however, as we reason in terms of space-time instead of in terms of separate space and time, all the particularities of motion are fixed by the single space-time world track of the body. This opens up the possibility of describing the laws of motion of free bodies in a purely geometrical manner in four-dimensional space-time, so that kinematics becomes a four-dimensional statics. Let us now pass to a more precise determination of the problem.

According to Newtonian mechanics, free bodies in interstellar space far from matter, moving under no constraint, always described rectilinear courses with constant speeds. This law was known as Newton's law of inertia. We remember, of course, that a law of motion of this type remained completely indeterminate until we had defined the frame of reference to which this motion would be referred, and we saw that any Galilean frame answered our requirements. Now it was assumed that Newton's law of inertia was universally valid in the case of free bodies moving under no constraint; when, therefore, the law was found to be at fault in the neighbourhood of large masses of matter, the discrepancies were ascribed to a real force of gravitation emanating from matter and compelling the erstwhile free bodies to move under constraint. In other words, in a field of gravitation, bodies were no longer free; so the validity of the law of inertia, applying to perfectly free bodies, was not overthrown thereby. The motions actually followed by bodies in a gravitational field would thus result from a compromise between the courses the bodies would have followed had no gravitational field been present, and the perturbating influence due to the field. These laws of motion were expressed by Kepler's three laws of planetary motion, according to which planets described ellipses round the sun (Kepler's second law), with variable velocities determined by the law of areas (Kepler's first law) and by the dimensions of the ellipses described (Kepler's third law).

As in the case of the law of inertia, Kepler's laws of planetary motion held only for a Galilean or inertial frame and, in particular, for the Galilean frame attached to the sun's centre. Thus there existed a complete duality between the two types of law and motion, those for free bodies and those for bodies moving under gravitational constraint.

In Einstein's space-time theory this duality is removed. There is now but one law of motion, namely: *All free bodies pursue geodesics, or the straightest of paths, through space-time, regardless of whether we are considering regions in the neighbourhood of matter or remote regions in interstellar space.* This law is in full accord with the *Principle of Least Action*, and may be regarded as a direct consequence of this fundamental principle.*

True, owing to the curvature of space-time around matter and its flatness in distant regions, the actual courses and motions will vary according to the more or less great proximity of matter; and we might be tempted to say that we had succeeded in removing a duality in the laws of motion only by introducing a new duality referring to the structure of space-time—curved around matter and flat for distant regions. But this line of argument would be incorrect. There is but one law of space-time curvature or gravitation; for this law, which ascribes a definite curvature to space-time in the presence of matter, tells us also how this curvature dies away to perfect flatness at infinity. The existence of flat space-time far from matter is thus comprised in the one statement of the law of space-time curvature or gravitation. Thus, a precise determination of the motions of bodies, whether near matter or far from matter, is given by the one law of space-time curvature together with the law of least action, which states that the space-time courses or world lines of the bodies will be geodesics.

Let us now examine another aspect of the question. We have seen that according to Einstein's ideas it was no longer necessary to differentiate between free bodies (those moving in interstellar space) and bodies moving under constraint (those moving in the vicinity of attracting matter). Henceforth, a planet moving round the sun is just as much a free body as one moving in interstellar space. If it moves in a different way, in these two cases, it is simply because the structure of space-time is curved around matter, instead of being flat as in interstellar space. There is, therefore, no justification for invoking a physical pull or force of gravitation in the proximity of matter; this so-called physical pull is inexistent so long as the planet moves without any additional interference. In all cases the planets and bodies in interstellar space are guided through space-time by the intrinsic structure of the space-time wherein they are situated.

Now structure implies geometry, and space-time, like any other mathe-

---

* We shall see Appendix   that the geodesics of space-time are of two major varieties: the so-called time-like and the so-called space-like geodesics. The transition between the two is given by the *null-lines* or minimal geodesics; these correspond to the paths and motions of light rays. The space-like geodesics would correspond to the paths and motions of bodies moving with a speed greater than that of light. As such motions cannot exist, according to the theory of relativity, we see that free bodies can follow only the time-like geodesics. In future, therefore, when referring to the geodesics of space-time, we shall always have in mind the time-like geodesics. Also we may note that whereas the time-like geodesic defines the longest space-time distance between two points, the null-line or minimal geodesic has always a zero space-time length.

matical extension, is amorphous *per se;* hence this structure of space-time must be attributed to a foreign cause, to the presence of the four-dimensional metrical field of space-time similar to a magnetic field. It is this four-dimensional field which acts as a *guiding influence* on bodies, constraining them along the geodesics of space-time. When, therefore, the observer partitions space-time into his space and time mesh-system, the effect of the guiding field is to direct bodies along certain paths in space and to determine their motions along these paths. Were there no guiding field, or metrical field, were space-time completely amorphous, there would be no space and time partitions, motion would be meaningless and inconceivable; even were it conceivable, the bodies would not know how to move.

It would appear, then, to be more correct to state that there can exist no such thing as freely moving bodies, since all bodies are subjected to the influence of the guiding field. We can, however, retain the usual terminology by remembering that by a freely moving body we mean one submitted to the action of the guiding field and to nothing else. The important point to understand is that as regards freedom of motion, all bodies in empty space, whether situated in a gravitational field or far from matter, are on the same footing. In all cases their actions are controlled by the guiding field, and by the guiding field alone. The sole difference is that in the neighbourhood of matter, the lay of the guiding field is curved slightly. It is incorrect, therefore, to invoke a real physical pull acting on the body when it moves near matter, and to deny the existence of such a pull when the body is moving in interstellar space.

And yet it might be said: "We experience a real physical pull when we stand on the earth, whereas we should experience no such pull were we floating in interstellar space." Let us see why it is that this real physical pull appears to arise in one case and not in the other.

A freely moving body follows, as we have explained, a geodesic of space-time, regardless of whether space-time be curved by the proximity of matter or whether it be flat. So long as the body is not interfered with, it will continue to follow its geodesic, allowing itself to be guided by the space-time structure or field. But suppose now that the motion is interfered with, so that the body is torn away from the geodesic which it would normally have followed. The guiding field will immediately react and oppose this foreign interference. It is as though the geodesic were a groove from which the body could not easily be diverted.

Consider then a body moving uniformly in interstellar space, according to the law of inertia. If we interfere with this body's motion, accelerate it along its straight line or compel it to describe a curve, we shall, in effect, be tearing it away from its space-time geodesic. The guiding field will react and we shall experience a physical pull which we call a force of inertia or a centrifugal pull. In exactly the same way we may explain the appearance of the physical pull which we ascribe to gravitation when we stand on the earth's surface.

Thus, a body which is falling freely towards the earth is following a geodesic in the curved space-time that surrounds the earth. But if, for some reason or other, the free fall of the body is interfered with, if, for instance, the rigidity of the earth's crust prevents the body from falling farther along the geodesic, the geodesic will react and will strive to force the body through the earth's crust. It was this force which we formerly ascribed to the pull of gravitation. But we see once again that this force has exactly the same origin as the forces of inertia we discussed previously, so that there exists no essential difference between a force of inertia and one of gravitation. They are both real physical forces in that they will produce physical effects; and they will both disappear in exactly the same way when we cease to interfere with the natural motion of the body. In short, they are both of them but manifestations of the four-dimensional metrical field or guiding field of space-time.

CHAPTER XXIX

EINSTEIN's law, as we have expressed it, is one of space-time curvature around matter and in the interior of matter. It is also the law of distribution of the ten potentials or g_{ik}'s throughout space-time, or, again, the law of distribution of the four-dimensional metrical field or guiding field of space-time. But in order to study the implications of the law we may discuss it from yet another angle.

Being a law of space-time curvature or structure, we are able to derive from it the precise lay of the geodesics of space-time around matter. These geodesics, or paths of least resistance, are, we remember, the paths which planets, light rays and all free bodies will pursue. A knowledge of the lay of the geodesics around the sun will therefore allow us to determine the precise courses in space and the precise motions along these courses of all free bodies moving in a gravitational field. When these geodesics are submitted to mathematical analysis, it is found that the courses and motions of the planets should be very nearly identical with those required by Newton's law.

There would be slight discrepancies, however, between the requirements of the two laws. In particular, Einstein's law of gravitation would predict that the courses of the planets should be very nearly elliptical, but not exactly so; there should be a slight precession of the perihelion of the planet, increasing in magnitude with the velocity of the planet on its orbit. Now the most rapidly moving planet of the solar system is Mercury; and Leverrier had observed nearly a century ago that Mercury's perihelion displayed a precessional motion which Newton's law was unable to explain. Einstein's law accounts for this precessional motion, not only qualitatively, but quantitatively as well. Hence, a first verification of Einstein's law was obtained.

Here it must be noticed that, in the preceding example, Mercury's curious motion was already known to astronomers. But we have said enough of Einstein's methodology in his discovery of the law of space-time curvature to realise that he did not adjust his law of space-time so as to account for Mercury's hitherto unexplained motion. We saw, indeed, that the law of gravitation, as formulated by him, was practically inevitable if no derivatives higher than the second order were to be involved. However, in the following discussion, we shall see that Einstein's law was able to foretell the existence of a phenomenon totally unsuspected by classical science. We refer to the deviation of a ray of light in a gravitational field.

Classical science had always assumed that a gravitational field, such as that of the sun, would attract material bodies, but that it would be without effect on light rays, which were considered to be immune from gravitational attractions. Had classical science been thoroughly convinced of the identity of the two types of mass, it would have realised that light waves must be possessed of gravitational mass, since they were known to manifest the characteristics of inertial mass. This inertial property of light rays had been foreseen theoretically by Maxwell and proved experimentally by Lebedew, who detected the pressure exerted by rays of light on a body on which they impinge. It is a pressure of this type which is supposed to be responsible for the apparent repulsion of comets' tails away from the sun. It would have been, therefore, only a step to assume that light rays must manifest gravitational mass, and hence must be attracted by the sun.

Classical science, however, had never felt justified in taking this step. As a result, it was thought that light rays would proceed in straight lines through a gravitational field just as they would through free space far from matter. When, therefore, the first eclipse observations prompted by Einstein's predictions revealed an undeniable bend in the course of a ray of light emitted by a star and grazing the sun's limb, classical scientists did not consider it necessary to abandon Newton's law on this account. They merely recognised that light was attracted by a gravitational field in common with all other forms of matter, and calculation showed that assuming Newton's law to be correct, the angle of this deflection would be o".87 for a ray grazing the sun's limb. The first eclipse observations were conducted under very unfavourable conditions, and all that could be ascertained was that a certain amount of bending was present. The precise degree could not be determined in any reliable way, so that Newton's law still had its defenders.

Let us now pass to Einstein's attack. Of course, in Einstein's theory, even prior to his discovery of the new law of gravitation, a ray of light would certainly have been bent in a gravitational field in virtue of the postulate of equivalence. It is easy to see how this would arise. Consider a chest, floating vertically in interstellar space, and a ray of light traversing it horizontally. If now the chest be accelerated upwards along the vertical, the floor of the chest will move with acceleration across the ray and, as a result, the light ray will follow a curved path through the chest. But in this case the postulate of equivalence allows us to regard the chest as unaccelerated but permeated by a gravitational field; we may infer, therefore, that the ray of light would bend downwards under the action of this field. Hence, the mere qualitative bending of a ray of light in the sun's gravitational field would not vindicate Einstein's law of gravitation and dethrone Newton's; it would merely prove the correctness of the postulate of equivalence and the identity of the two species of masses. It is the precise quantitative bending of the ray, as determined by Einstein's law of gravitation, that constitutes the crucial test of the correctness of Einstein's law. This bending, as computed from the lay of the geodesic which a light ray grazing the limb of the sun would follow, turns out to be $1''.75$; that is, just twice as great as the Newtonian deflection $0''.87$. When this precise angle predicted by Einstein was observed during the subsequent solar eclipse of 1919, interest in relativity became widespread; for it was recognised at last that the theory was not a mere mathematical dream.

But even this verification of Einstein's law of gravitation would have been insufficient to establish the theory. As we shall see in the next chapter, other very curious phenomena were suggested by the new law of gravitation; and if these new phenomena should not be verified in turn, the theory, in spite of its wonderful achievements, would have to be abandoned.

CHAPTER XXX

WE must remember that the one fundamental continuum of the world is space-time; space and time considered separately, varying as they do with the observer's motion, have no absolute significance. The law of curvature, or of gravitation, which we have discussed is one of space-time curvature. It is the law giving the distribution of the four-dimensional metrical field of space-time in the interior of matter and around matter to infinity. It is true that we may be able to effect some sort of separation of space-time into the space and time of some particular observer; and this may enable us to represent the law of curvature of space-time as two separate laws of curvature, one affecting space and the other affecting time. But for the reasons we have just mentioned, no absolute significance could be attached to the curvatures we should obtain. Thus, in interstellar space, an observer situated in a Galilean frame would contend that neither space nor time manifested any trace of curvature, whereas an observer on the rim of a rotating disk would recognise that both space and time were curved. On the other hand, regardless of his frame of reference, either observer would always recognise that space-time possessed no trace of curvature.

It is because space-time transcends the observer that its curvature or flatness, as the case may be, can be considered absolute; whereas a curvature of space or of time alone can never represent anything but a mere relationship between the intrinsic condition of space-time and the observer's motion or frame of reference.

Of course, if we wish to split up space-time into the particular space and time directions of some particular observer we may always attempt to do so. But in the majority of cases this procedure will be impossible.

If these considerations are properly understood, there is no harm done in trying to separate the curved space-time which endures around the sun into the space and time of an observer at rest with respect to the sun and viewing the solar system as a whole. This separation, however, is possible only in the event that the masses of the moving planets can be ignored, so that the only field we have to consider is that of the sun, remaining unchanged as time passes. It is owing to this permanent condition of the gravitational field that the separation into space and time is possible. Fields of this type are called **stationary.** For non-stationary fields, such as those where the distribution of the field changes with time, the separation into space and time becomes impracticable.

We will assume, therefore, that the gravitational field of the sun is stationary; that is, that the masses and motions of the planets do not modify the field to any appreciable extent. We may then proceed to split up curved space-time into the space and time of an observer at rest with respect to the sun and non-rotating with respect to the stars. By so doing we shall be splitting up into two separate fields the metrical field of space-time, that four-dimensional field which pertains neither to space nor to time. In this way we shall obtain, first, a purely spatial metrical field, which will be responsible for the measurements of our rods when these are maintained at rest in our frame; and, secondly, a purely temporal metrical field, which will determine the behaviour of our clocks or vibrating atoms while these are at rest in various regions of the space around the sun.

When Levi-Civita and Schwarzschild had performed this separation, it was found that the curvature of the observer's space along a plane section through the centre of the sun would be identical with that of the surface of a paraboloid of revolution disposed around the sun; and that the observer's time direction would also be curved in a definite way. Both curvatures would die down gradually for regions of space farther and farther removed, to infinity. We must not forget, however, that this separation and the curvatures we have mentioned hold solely for the observer at rest with respect to the sun when surveying the solar system as a whole.

The advantage of this separation of space-time into space and time is that it allows us to express the purely formal results deduced from Einstein's space-time gravitational equations in terms of our familiar notions of space and time. It thus becomes possible to describe the world lines of the planets under Einstein's law by representing them as spatial orbits and motions along these orbits; hence we are enabled to check up Einstein's previsions with astronomical observation. Furthermore, this splitting up of space-time enables us to study separately those effects which must be attributed to the respective curvatures of space and of time, considered by themselves and independently of one another.

We will first consider the curvature of time alone. Let us assume that for some reason or other the curvature of space were to disappear, so that only the time-curvature would subsist. In that case, calculation shows that the planets would move in accordance with Newton's law, except for the fast-moving planet Mercury, which would manifest a precessional advance of its perihelion, equal, however, to only one-third of the observed value. This slower advance of the perihelion would be accounted for on the basis of Newton's law if we took into consideration the increase of mass with velocity, known to us from the special theory of relativity. It can also be shown that the curvature of time alone would entail the deflection of a ray of light grazing the sun's limb, but this deflection would have only one-half the astronomically observed value, viz: 0.87″ instead of twice that amount. This effect also could be accounted for on the basis of Newton's law if we ascribed mass to light.

The preceding considerations show that the curvature of time alone would entail consequences which would be in agreement with Newton's law, supplemented by the findings of the special theory of relativity; but it would yield too slow a precession for Mercury's perihelion and too small a deflection for a ray of light grazing the sun's limb. Inasmuch as Einstein's law of space-time curvature, or gravitation, furnishes the correct precession and the correct deflection, we see that the distinguishing feature of Einstein's law of gravitation is embodied in the additional curvature which it ascribes to the space around the sun (or around any material body).

Here it might be asked: "Since a peculiar type of space-time curvature (curved time and flat space) would lead us to Newton's law, why should Einstein have rejected the classical law *a priori* and urged his own more complicated type of space-time curvature in its place? It could not have been for the purpose of accounting for the double bending of a ray of light, since this effect was unknown at the time. Was it for the sole purpose of explaining the motion of the planet Mercury?"

The answer must be decidedly in the negative. When Einstein was led to his law of space-time curvature $G_{ik} = 0$, he never could have recognised in advance that this law would account for the motion of Mercury. He may have suspected the fact, convinced as he was that he was on the right track; but in order to establish the nature of the planetary motions when governed by this law, it was necessary to submit the law to a mathematical integration. It was only after the integration had been performed that the motion of Mercury, as required by the law, was found to be in agreement with previous astronomical observations. Hence we must discard all ideas of a law patched up for the sole purpose of harmonising with the planetary motions.

In order to understand Einstein's procedure more fully, we must recall that the general principle of covariance requires that all natural laws be expressed in the form of space-time tensor equations. This criterion alone, however, does not permit us to decide *a priori* on any particular law of space-time curvature. But Einstein, as we remember, introduces a second restriction, according to which, in the empty space surrounding matter, none but the structural tensors of space-time, the g_{ik}'s, should enter into the law of curvature.* This second restriction is a natural one to set, for were foreign tensors to be introduced at each point of space-time, the space could no longer be empty.

When these two restrictions are taken into consideration, we find that the law $G_{ik} = 0$ is practically the only one that can be constructed; at all events, it is the simplest. Theoretically, $B_{ikst} = 0$ could also be countenanced; but this law, indicating perfect flatness, would connote the absence of gravitation. Then again, we might suggest $G_{ik} = \lambda g_{ik}$, and this would lead to de Sitter's finite universe, a possibility which has not yet been discarded.

* Also constants such as λ may enter into the law of curvature in the empty space around matter; but never foreign tensors.

But in any case, with the restrictions imposed, Newton's law is barred, so that unless we wish to introduce foreign tensors, acting in the capacity of hypotheses *ad hoc*, it is hard to see how Einstein's law can be escaped.* If we convince ourselves on this score, we can see that it is quite impossible to think that we may adjust the curvatures of space and of time to suit ourselves. These curvatures are determined in advance by the law of space-time curvature, itself imposed by considerations of a rational order.

Let us revert to our separation of curved space-time into curved space and curved time. We have seen that if the curvature of space were non-existent, the curvature of time alone remaining, the motions of planets and free bodies would be in accord with Newton's law. We may understand the reason for this by following a different line of thought.

If space is flat, the g_{1k}'s which refer to space remain constant (in a Cartesian mesh-system at least).† The only g_{1k}, or potential, that can vary is therefore the time one g_{44}. This last potential is therefore the only one that can be responsible for a field of force. It follows that by considering space to be flat, we are preserving only one of the ten potentials of space-time; and we find ourselves in exactly the same position as under Newton's law, where only one potential was present, *i.e.*, the Newtonian potential. This allows us to identify the time-potential g_{44} with the Newtonain potential Φ, and the proof of the statement we made in a previous chapter is given thereby.‡

It now remains to be seen how this curvature of time may be detected directly. Here we must recall that the curvature of time is expressive of the repartition of the temporal metrical field obtained by splitting up the metrical field of space-time into one of space and one of time. The repartition of this temporal metrical field in the space and time separation we are discussing shows that its curvature will increase as we consider regions of space nearer and nearer the sun, indicating thereby a slowing

* It is well to remember, however, that the laws we have considered are all in the image of Newton's in that they contain no derivatives of the potentials to an order higher than the second. If this restriction is omitted, a number of alternative laws become possible. Inasmuch as their study presents tremendous mathematical difficulties they have not been investigated; and it is hard to say what might be the nature of their solutions.

† There are also the other time-potentials, *i.e.*, g_{14}, g_{24}, g_{34}. But as, in our mesh-system, the direction of time is perpendicular to those of space, these potentials vanish in the present case, and we are left with g_{44}.

‡ In reality the identification is a little more complicated; it is given by

$$g_{44} = 1 + \frac{2\phi}{c^2}.$$

Furthermore, it can be seen that this potential g_{44}, of varying value from place to place, is connected with the variable speed of light from place to place through the gravitational field. More precisely, the speed of light *in vacuo* is given by $c\sqrt{g_{44}}$, which is approximately $c + \dfrac{\phi}{c}$ or $c\left(1 - \dfrac{km}{rc^2}\right)$, where m is the mass of the body exciting the gravitational field, k is the gravitational constant. and r the distance from the mass.

down of time as we approach the sun.   Now it is this temporal metrical field which controls the rate of evolution of any isolated periodic phenomenon from place to place, hence which defines practical time-congruence as given by perfect clocks.   In the present case, therefore, the action of this field would be to slow down the temporal evolution of all phenomena when these have their spatial locus nearer and nearer the sun. This effect was anticipated by Einstein and is known as the **Einstein shift-effect.**

Of course, if *we*, as conscious beings, were to be situated at any particular point of the sun's gravitational field, we should never appreciate this slowing down of time and of the evolution of all things in our immediate neighbourhood; for our consciousness would participate in the same slowing down that was active around us.   It would only be when we viewed distant phenomena occurring in other parts of the field of gravitation that discrepancies in their speeds of evolution would become apparent.   Thus, although we should not be directly conscious of aging at various rates in the different parts of a gravitational field, we should discover eventually, either by direct vision or by meeting again, that our friend who had lived in the proximity of the sun had aged less than ourselves, who had lived farther away from its mass.   Gravitation would thus preserve youth, and, more generally, two men living on two different stars of unequal mass would not age at the same rate.

It is to be noted that it is not the force of gravitation itself which is responsible for this slowing down of time as we near the sun.   It is rather the decrease in the value of the potential; and in the present case we have seen that the only potential which enters into account is g_{44}, or the Newtonian potential.

All these phenomena predicted by Einstein were entirely new to science, and it was most important to verify their correctness before accepting Einstein's theory.   The method suggested was that of exploring a field of gravitation with vibrating atoms, these being assumed to behave like perfect clocks marking correct time.   For instance, if Einstein's law of space-time curvature was correct, atoms of sodium placed in different parts of the solar field would appear to be vibrating faster or slower than the atom in our immediate neighbourhood, according to whether they were farther removed from or nearer the sun than *we* happened to be situated. These variations in the rates of vibrations of the atoms would be betrayed physically by variations in the colour of the light they emitted or absorbed.   It would follow that while the atom of sodium by our side would always emit a yellow light in our estimation, those atoms nearer the sun would appear redder, and those farther away from the sun would appear greener.   In particular, an atom of sodium situated in the sun's atmosphere would beat more slowly and appear to us redder than an identical atom examined in our laboratory (the earth's gravitational field being insignificant compared with that of the sun).*

* The Einstein effect is due to a veritable decrease in the frequency of vibration

For a number of years it was questionable whether this effect really existed, because the difficulty in detecting so minute a displacement of the spectral 'lines was very great. Quite recently, however, the phenomenon has been observed on a star of enormous density, the companion of Sirius.* This is perhaps the most beautiful of all the verifications of Einstein's theory, since the effect contemplated was totally unsuspected by classical science—even more so than the bending of light in a gravitational field.

Now that we have discussed the effects due to the curvature of time and have seen that this type of curvature accounts for all the major effects of gravitation, let us consider the additional effects due to the curvature of space. The presence of this curvature of space, so far as the motions of bodies are concerned, will modify the effects due to the curvature of time. It will therefore be responsible for the small discrepancies that differentiate Einstein's law from Newton's. However, so long as the motions of bodies in the gravitational field are small compared with that of light, the curvature of space will not affect their motions to any appreciable extent.

Among the planets, Mercury alone is animated with a speed (with respect to the inertial frame of the sun) sufficient to render observable the influence due to the curvature of space; and it is this influence which is mainly responsible for the precessional advance of Mercury's perihelion. In the case of light waves, which move with a speed many times greater than Mercury, the influence of the curvature of space will be marked to a still greater degree. It will cause an increased bending of the ray of light equal to that already produced by the curvature of time alone,

---

of the atom situated nearer the sun, and this retardation is caused by the increasing departure from unity of the potential g_{44}, tending as it does towards zero as we approach the sun. It would be totally incorrect to ascribe it to a slowing down in the motion of a ray of light travelling away from the sun in a radial direction, owing to the retarding effect of the sun's gravitational pull. The gravitational pull has nothing to do with the Einstein effect; and as a matter of fact, calculation shows that a ray of light travelling away from the sun would gradually *increase* in speed till it attained its invariant speed c at infinity, as though it were repelled, *not* attracted by the sun. But over and above these results of calculation, it can be seen immediately that a modification in the speed of light would be incapable of explaining the existence of the Einstein effect. In all cases we are bound to receive the successive vibrations with the same frequency as they are emitted by the atom; for otherwise there would be a gradual accumulation or depletion of light waves travelling along the fixed distance separating us from the atom. Hence any verification of the Einstein effect could be ascribed only to a real modification in the frequency of the atom's vibrations.

* The Einstein shift in the spectral lines as seen by a definite observer will increase in importance as the atom nears the star or as the star increases in mass. For a star of given mass, the effect will therefore increase as the volume of the star decreases, and hence as its density increases. We understand, therefore, why it is that for two stars of the same mass, the best conditions of observation will be afforded by the star which has the greater density; while for two stars having the same density, the best conditions will be afforded by the star having the greater mass.

which, as we know, would correspond to the bending required by Newton's law. In this way the double bending is accounted for.

In order to exhaust this discussion, we must mention that there exists still another case, in which the curvature of space would exert a perceptible influence, even in the case of slowly moving bodies. This would be in the event of a very intense field of gravitation, many times greater than that of the sun. In such a field, even slowly moving bodies would deviate perceptibly from the course required by Newton's law of gravitation.

It may appear strange that this curvature of space should seem to be of such minor importance as contrasted with that of time, and produce observable effects only for bodies moving with great velocities with respect to the sun. The solution of this puzzle will be found, however, when we consider some of the peculiarities of the space-time continuum of relativity. In an ordinary continuum like space, where all the dimensions are of exactly the same nature, we have no difficulty in specifying some unit of length which will apply in exactly the same way to measurements computed along the dimensions of height, length and breadth. In space-time, however, the situation is somewhat modified, since the fourth dimension (time) represents something different from the other three. Thus, whereas our measurements along the spatial dimensions will be conducted with material rods, our time measurements will require time-rods, or clocks. The question is, then, how are we to co-ordinate our measurements with clocks and our measurements with rods; or, in other words, how shall we be able to establish any sort of comparison of magnitude between these measurements so different in nature?

The answer to this question is that the variable which represents measurements along the time direction is not t but ct. This means that one second in time corresponds in our formulæ to the distance which light covers during one second, namely, 186,000 miles. It follows, therefore, that if we measure time in seconds we must, in order to co-ordinate results, measure space in terms of unit rods 186,000 miles long.

Obviously there exists a vast discrepancy in actual practice between the magnitudes of our temporal and spatial units; for the great majority of bodies with which we come in contact cover far less than a unit of space (186,000 miles) in a unit of time (one second). Only in the case of bodies approaching the velocity of light in our frame are the two progressions, that in space and that in time, in any way comparable. Thus, consider our own earth. It moves in relation to the sun with a speed of some eighteen miles a second, which is equivalent to approximately one ten-thousandth part of a unit of space per unit of time. This emphasises the fact that to all intents and purposes the motions of bodies with which we are concerned reduce to a motion in time, the displacement in space being insignificant in comparison.

Now when we come to investigate the respective effects of the curvatures of time and space, taken separately, in the gravitational field of the sun, we must remember that for a curvature to manifest itself we must survey relatively large areas on the curved surface (large as compared with the intensity of the curvature). Thus, on the curved surface of the earth, for example, if we content ourselves with surveying restricted areas, no effects of its curvature will be apparent. It is only when we conduct measurements over large areas, say over a few hundred miles, that the direction of the vertical will change perceptibly and that geodesical surveys will render the curvature manifest. It is much the same with the curvatures of space and of time in a gravitational field. Only when we consider planets moving through the sun's inertial frame with a speed approaching that of light (hence covering large distances in unit time) will the space-curvature produce effects comparable to those produced by the curvature of time. We understand, therefore, how it comes that the curvature of time alone is able to produce the major effect of Newtonian attraction, such as weight, whereas the effects due to the curvature of space remain imperceptible until we consider the motion of Mercury, which is moving fairly fast with respect to the inertial frame of the sun. When we consider the propagation of light rays, the effects due to the curvature of space reassert themselves fully, and as a result the double bending of a ray of light can be attributed in equal proportion to the curvature of time and to the curvature of space.

In short, we now understand why it is that Einstein's law of gravitation deviates appreciably in its effects from Newton's only when we consider bodies moving with velocities approximating to that of light. Once again we realise that if this invariant velocity of the universe had happened to be much smaller, the deviations between Einstein's law of gravitation and that of Newton would have been far easier to detect; and, conversely, had the invariant velocity happened to be much greater than 186,000 miles a second, had it been infinite, for example, there would have been no discrepancies left between the two laws. Thus, once again we reach the same conclusions that we mentioned when discussing space-time. The fundamental difference between Einstein's theory and classical science arises from the discovery that the invariant velocity of the universe is finite and not infinite.

We may summarise all these results by stating that the major effects of gravitation are produced by the curvature of time, *i.e.*, of the time component of the metrical field of space-time; the curvature of space exerts perceptible effects only when the velocities of the moving bodies are very great or when the gravitational field is of very great intensity. Hence, for weak fields such as that of the sun, and for slow motions, Einstein's law of space-time curvature leads us to results practically identical with those necessitated by Newton's law of gravitation.

It now remains to be seen how the curvature of space could be detected by direct measurement, without our having recourse to moving bodies. It can, of course, be detected indirectly, since we have seen that it is responsible for the advance of Mercury's perihelion and for the double bending of a ray of light passing near the sun. But we now propose to detect it by direct spatial measurements.

We have said that our separation of space-time into space and time was equivalent to splitting up the metrical field of space-time into a metrical field of time and a metrical field of space. It is this metrical field of space with which we are now concerned. To this metrical field of space must be ascribed the structure of our space (curved in the present case). Were the metrical field non-existent, space would, of course, have no structure, no geometry whatsoever. As the geometry of a continuum is revealed by measurement, we see once again that it is the metrical field of space which will control the behaviour of our measuring rods at rest in our frame, hence which will determine practical congruence.

The obvious method of verifying the anticipated lay of this metrical field of space, or, what comes to the same thing, this curvature of space would be to explore space with measuring rods; we should, of course, obtain non-Euclidean results. Calculation shows that the results should be identical with those which would be obtained were we to apply rigid Euclidean rods on an appropriate paraboloid of revolution. The type of curvature, or non-Euclideanism, implied thereby is of the Riemann variety, since a paraboloid of revolution is a surface of positive curvature. Hence, we should expect the ratio of the circumference to its diameter to be smaller when some more massive body such as the sun was situated at the centre of the circumference.

But here again we are met with the same difficulty. In order to detect a curvature of this sort, we should have to survey vast areas of space, much greater than that of the solar system; and it is scarcely probable that even in future years our measurements will be sufficiently refined to permit a direct test.

Incidentally, we are now in a position to understand another erroneous assertion that has often been upheld. As soon as Riemann's discoveries were known, the possibility of interpreting gravitation in terms of a curvature of space had been suggested (chiefly by Clifford). But the great obstacle to this attitude was that the most refined spatial measurements obstinately refused to reveal the slightest trace of non-Euclideanism. Assuming, for the sake of argument, that it had been possible to interpret the curved orbits of the planets and of projectiles in terms of guided motions along curved grooves threading their way through a curved space, even then, in view of the tremendous degree of non-Euclideanism that it would have been necessary to invoke, the scheme would have had to be abandoned; the chief reason being, of course, the incompatibility of this hypothesis with the Euclideanism given by measurements with rigid rods. It was only after Minkowski had discovered space-time that a modified form of Cifford's premonitions could be vindicated; only then was it possible to reconcile the Euclideanism of our spatial measurements with that non-Euclideanism of space-time which was the source of all gravitational phenomena. In this respect it cannot be urged too strongly that the general theory and the special theory are but two chapters of one same theory. In so far as the general theory cannot work without space-time, any experiment which would conflict with the special theory would thereby conflict with the general theory.

Let us now examine another point. The velocity of light in a gravitational field is no longer an invariant, as it was in the special theory. However, if we were to measure the velocity of light from point to point, displacing ourselves from one point to the next, we should find that wherever we might be placed, this velocity would always appear the same everywhere, owing to the progressive modification in the behaviour of our rods and clocks. It would only be when viewing distant points that we should recognise variations in the velocity of light.

Thus we see that the principle of the invariant velocity of light is accurately true only in free space far from matter, and, even then, only when computed with reference to our Galilean frame. In an accelerated frame, as in the neighbourhood of matter, the cornerstone of the special theory of relativity no longer applies.

It follows that in the general theory the law of the invariant velocity of light must be replaced by some more general principle, comprising the invariant-velocity principle as a particular case. This more general principle can be expressed by saying that a ray of light moves along a world-line which is a very particular kind of geodesic called a *null-line* or *minimal geodesic*. This holds regardless of whether space-time be flat or curved.

Thus far we have studied the curvatures of space and of time outside matter in those particularly simple stationary cases where it is feasible to split up space-time into a separate space and time. Under similar conditions we may investigate these curvatures in the interior of matter. Consider a homogeneous incompressible fluid at rest in our Galilean frame. Conditions being stationary, that is, not varying with time, we can again split up space-time into space and time and study the two types of curvatures separately. We find that the time direction is curved, and this curvature once more is connected with the gravitational forces which exist in the interior of the fluid. We next find that the three-dimensional space inside the fluid possesses a curvature at every point which is proportional to the density of the fluid at every point. Assuming this density to be the same throughout, we see that the curvature of space is constant, and we obtain a spherical space of three dimensions. This curvature of space inside the sphere has nothing to do with the dimensions of our fluid sphere: the curvature is governed solely by the density of the fluid.

Now, in our ordinary three-dimensional Euclidean space, when the curvature or the radius of a sphere is given, its total area is thereby fixed. In a similar way, the density of the fluid predetermines the curvature of space and hence fixes the total volume of the spherical space. Obviously, it would be impossible for us to increase the volume of our fluid indefinitely, for there would come a time when there would be no more room for it to occupy in the spherical space which it had itself created. In the case of a fluid of the density of water, calculation shows that when the sphere of water had attained a radius of about 600 million kilometres, it would be impossible to increase its size further by adding more water to it. When this critical volume was reached, the fluid would fill the spherical space entirely. Reverting to our two-dimensional analogy, it would be as though the entire surface of the two-dimensional sphere were occupied by the water.

As a matter of fact, this critical volume could never be reached; for even before it could be attained the pressure at the centre of the sphere would have become infinite, and then negative, and calculation shows that this condition would be accompanied by an arrest in the flow of time at the centre of the fluid. In fine, we discover that the greatest possible volume our fluid sphere could ever occupy would be about $\dfrac{7}{9}$ of the volume mentioned previously.

It follows that the fluid could never fill completely the spherical volume its density had created in space. Using a two-dimensional analogy, it would be as though on the surface of an ordinary sphere the entire surface could not be covered; an uncovered portion like a cap would always remain. On this account it is sometimes said that space would not close round on itself, since part of the spherical space created by the fluid could never be filled.

Just as a fluid of given density can never increase in volume beyond a certain point, so, conversely, a fluid of given mass could never be compressed into a volume which would cause its density to increase beyond a certain limit. The argument would be the same. The pressure would become infinite, time would stop, and nothing further could happen. A sphere of fluid possessing the same total mass as the earth could never be compressed into a volume less than that of a thimble.

The numerical values mentioned show that there is only a very slim chance of putting these anticipations of the theory to an experimental test. Even in the case of a giant star such as Betelgeuse, its density is so small that the critical volume is well beyond the actual size of the star. Nevertheless, from a theoretical point of view, it is interesting to note that there must exist a definite limit to the size of a star of given density. If these critical limits should ever be reached, time would be arrested, and it is difficult to foresee what the consequences might be. Einstein has called these critical conditions the "catastrophes." When pressed to give his opinion as to what would happen were they ever to be reached, he is said to have replied that very possibly we should witness the disruption of matter into radiant or electromagnetic energy. However, he preferred not to venture an opinion, as anything he might say would be little better than a guess.

CHAPTER XXXI

THE PRINCIPLES OF CONSERVATION

CLASSICAL science had assumed that the mass of a body was an invariant, which would not change in value when the relative motion of the body was changed. It is true that highly refined experiments on electrons moving at enormous speeds (Bucherer's experiment) had shown that the mass of an electron increased with its velocity;·but it was assumed that this increase in mass was due to other causes, either to a modification in the electromagnetic field of the electron, or possibly to the mass of the ether which the electron in its rapid motion would drag along with it.

One of the first triumphs of the special theory of relativity was to prove that mass, just like time and space, must be a relative; and that a body in motion with respect to the observer would suffer an increase in mass in all ways identical with that disclosed by Bucherer's experiment. This increase could no longer be attributed to the drag of the ether, since in the special theory of relativity exactly the same increase would have to be expected, regardless of whether it was the observer or the electron that was in motion through the ether. As in all previous examples, the essential factor was the relative motion between observer and observed. The ether played no part.

Calculation then showed that the mass of a body would become infinite when the velocity of the body reached that of light; for this reason no material body could ever move as fast as light.

Now, when we discussed mass in Chapter XIV, we mentioned that mass and energy are one and the same, so that wherever there is energy there is mass, and vice versa. A distinction, however, must be made between material mass (or energy) and radiation mass (or energy). The special theory of relativity is in agreement with the conservation of energy, and so of mass, but our interpretation of results will depend on the nature of the problem investigated.

First consider the collision of billiard balls. We recall that the material mass of a ball in motion is the sum of its rest-mass and of the mass due to its kinetic energy. Now the two balls will not retain the same velocities after, as before, a collision, and therefore, the masses of the two individual balls will vary. Nevertheless the sum total of the material masses involved will remain constant, so that material mass is conserved in this example.

Next consider the example of four hydrogen nuclei combining to form a helium nucleus.* We have seen that the mass of the helium nucleus is less than the sum of the masses of the four hydrogen nuclei. A part of the material mass has therefore vanished, and so is not conserved. On the other hand the total mass, or energy, of all kinds is conserved, for the material energy, missing from the helium nucleus, will disclose itself as loose energy in the radiation emitted.

The fact that material energy (*i.e.* matter) is not conserved, has compelled physicists to reject the classical idea of the conservation or indestructibility of matter. Matter may no longer be viewed as a substance; and so although we may still speak of the conservation of mass, we must understand that the mass involved is not material mass, but a wider concept of mass.

In addition to being in agreement with the general principles of conservation for mass and energy, the special theory likewise entails the principle of the conservation of momentum. Indeed the two principles are seen to be but partial aspects of one grand principle of conservation in the world of space-time; energy, or mass, representing the time component, and momentum representing the spatial components.

Let us now study the problems of conservation in the general theory when we make use not merely of a Galilean frame of reference, but of any frame whatever, hence when our space is permeated with a field of inertial forces or, again, with a field of gravitation. We shall discover once more that conservation is required by the general theory; though, as we shall see, it will not be the type of conservation to which we were accustomed in classical science.

In order to make these points clearer we must revert to Einstein's law of gravitation. We remember that assuming the correctness of Einstein's views of gravitation as being due to space-time curvature, the principle of stationary action led us to the law of gravitation:

$$G_{ik} - \tfrac{1}{2} g_{ik} G = T_{ik}.$$

Now the tensor on the left possesses the remarkable mathematical property of expressing a condition of permanence, or of conservation. It follows that in virtue of the equations of gravitation written above, the same conservation must hold for T_{ik}, hence for mass, energy and momentum.

The mathematical identity that expresses this property, let it be noted, does not depend on Einstein's theory; it is a purely mathematical property between the G_{ik}'s of a continuum, which had been discovered by the mathematician Ricci before the theory of relativity was born. But in view of the physical relationships which Einstein had established between the curvatures of the space-time continuum and the presence of matter, it now followed as a necessary consequence of the theory

* See Chapter XIV.

that the right-hand side of the gravitational equations must present the same characters of conservation as the left-hand side. If we interpret this result physically, we see that it implies the conservation of mass, momentum and energy.

We thus reach the remarkable conclusion that the principles of conservation, which in classical science were of an empirical nature and were totally unconnected with the law of gravitation, now appear as immediate and necessary consequences of Einstein's law of gravitation. In other words, granting the correctness of the law of gravitation, conservation cannot help but exist in this universe. To many thinkers this is one of the most beautiful aspects of Einstein's theory, showing as it does the wonderful unity and relatedness of nature.

Nevertheless we must be careful to understand the deeper significance of conservation in relativity. We shall see that the type of conservation we are now discussing is somewhat different from the classical conception of conservation, under which matter, for instance, disappeared from here only to reappear there. In the present case conservation is more ethereal and abstract. We may explain these points better by examining conservation in classical science. Here, when fields of force are present, the principles of the conservation of energy and momentum would appear at first sight to be contradicted by experiment. Thus, a body falling freely towards the earth can certainly not be said to possess a constant *vis viva*, or energy, since, originally starting from rest, it acquires an increasing speed as it falls towards the ground. Rather than recognise the breakdown of the principle of the conservation of energy, classical science preferred to assume that the body possessed two types of energy: a kinetic energy given by $\frac{1}{2}mv^2$, and a potential energy depending on the position of the body in the field of gravitation. It was the sum total of these two types of energy that remained constant and was therefore conserved during the fall of the body.

It must be noted that the necessity for introducing potential energy, as an additional requirement in order to ensure conservation, exists only when we are considering a phenomenon occurring in a gravitational field or in an inertial field of force, such as would be present were we to take our stand in an accelerated frame. In the absence of such a field, potential energy would become completely meaningless, and conservation of energy would endure without it.

In the general theory of relativity, the equivalent of the classical potential energy represents the energy of the gravitational field. But this energy is not expressed by a tensor, for we can make it vanish at will by changing our co-ordinate system. Potential energy therefore represents nothing intrinsic in the world of space-time, and so the classical understanding of conservation of energy and momentum appears to be untenable. We may, however, in a purely formal sense, retain our belief in conservation by widening the concept. Or else we may obtain theorems of conservation on the classical model by intergrating over an isolated system, but in that case

we cannot localize the energy.   The entire subject is highly technical, and
we shall not pursue it further.

CHAPTER XXXII

WE remember that in classical science the Newtonian law of gravitation was expressed by Poisson's equation,

$$\triangle \Phi = 4\pi k\mu,$$

where μ represented the density of the matter exciting the field at each point of space. It followed that it was solely the mass of a body which was active in creating a gravitational field. Likewise, it was the mass of a body which was the sole cause of its being attracted in a gravitational field, so that the weight of a body at a given point near the earth was fully determined when we knew the mass of the body and the mass of the earth.

Now, in Einstein's theory, the gravitational equations become

$$G_{ik} - \tfrac{1}{2}g_{ik}G = T_{ik},$$

where G_{ik} and G refer to the curvatures of space-time, or, if we prefer, to the potential or g_{ik}-distribution, while T_{ik} represents the characteristics of the matter exciting the field. There are ten of these gravitational equations corresponding to the various values from 1 to 4 which we must assign to the indices i and k (space-time being four-dimensional). And here we must note that the tensor T_{ik} represents matter much more fully than does the μ of classical science, for the various components of T_{ik} refer to such additional characteristics as the momentum, the *vis viva* and the conditions of stress existing in the interior of matter. We may infer, therefore, that whereas in classical science the mass of the matter was solely responsible for the creation of a gravitational field, in the present case all the additional characteristics mentioned above contribute their share.

Similar considerations would hold for the gravitational field outside matter, and as a result, were the sun to be rotating with greater speed on its axis, the gravitational field it would produce would be slightly different. In the case of the planet Jupiter and its moons, it would appear possible to submit to an observational test this additional disturbance caused by the rotation of the central mass on the satellites moving around it. As yet, however, no checking up of the theory has been possible on this point.

Thus far we have considered the gravitational field, or space-time curvature, produced by matter, but it should be apparent that owing to the identification of mass and energy, space-time curvature is a necessary consequence of the presence of energy. The electromagnetic field should thus be susceptible of giving rise to a gravitational field. Briefly, we may present the problem as follows:

Minkowski and Einstein had succeeded in establishing that electric and magnetic forces were the components of the same tensor F_{1k}. From this tensor it was possible to build up another tensor E_{1k}, representing the energy and characteristics of the electromagnetic field, just as T_{1k} represented the energy and characteristics of matter. These two types of energy, which are not identical, are designated as *loose energy* and *bound energy*, respectively. If, now, we wish to discover the precise type of space-time curvature, or gravitational field, which the presence of an electromagnetic field would produce, all we have to do is to replace T_{1k} by E_{1k} in the gravitational equations. We thus obtain

$$G_{1k} - \tfrac{1}{2} g_{1k} G = E_{1k}.$$

We can see at a glance that the type of curvature generated by an electromagnetic field must differ from one due to matter, since E_{1k} and T_{1k} are different magnitudes. A more conspicuous difference is illustrated when we manipulate the gravitational equations according to the permissible mathematical rules. We then find that whereas the gravitational equations for matter yield

$$G = -T,$$

where T represents the density of mass of the matter, the gravitational equations for electromagnetic fields yield

$$G = -E = 0,$$

for E happens to be a quantity which vanishes in all electromagnetic fields. Hence, we realise that in regions of space where electromagnetic fields are present, but all matter is absent, the Gaussian curvature G of space-time always vanishes; whereas in regions occupied by matter it is equal to the density of the matter, and does not vanish.

Before proceeding to compare these various types of curvature, let us recall that there exist three types of tensors of curvature of especial importance: the Riemann-Christoffel tensor B_{1kst}, the contracted tensor G_{1k} and the invariant G. If we write these three magnitudes in line: G, G_{1k}, B_{1kst}, we find that the vanishing of B_{1kst} over a region signifies the complete flatness of the continuum over this region. As a result, both G_{1k} and G will certainly vanish everywhere within the region. If it is G_{1k} that vanishes, G will automatically vanish but B_{1kst} will not necessarily vanish. Hence $G_{1k} = 0$, while comprising perfect flatness as an extreme case, permits a certain degree of curvature. Finally, if it is G that vanishes, G_{1k} need not vanish; and if G_{1k} does not vanish, neither will B_{lkst}; whereas if G_{lk} vanishes, we are thrown back on the previous case, *i.e.*, B_{lkst} may or may not vanish. Hence, $G = o$, while compatible with perfect flatness as a special case, connotes a law of curvature still less stringent than $G_{lk} = o$.

We may now contrast the various types of space-time curvature produced by matter and electromagnetic energy. By writing the gravitational equations in a slightly different though equivalent way, we obtain, in the interior of matter,

$$G_{lk} = T_{lk} - \tfrac{1}{2}g_{lk}T,$$

whence

$$G = -T.$$

Thus, in the interior of matter, none of the curvatures need vanish, and hence no limitation is imposed on the curvature of space-time.

In regions where there are electromagnetic fields but no matter, we have

$$G_{lk} = E_{lk} - \tfrac{1}{2}g_{lk}E = E_{lk} \quad \text{(since } E \text{ always vanishes)},$$

whence

$$G = -E = o \text{ (since } E \text{ always vanishes)}.$$

The curvature is thus less thorough, since G vanishes.

In regions outside matter, where no electromagnetic fields are present, the curvature is still less pronounced, for there

$$G_{lk} = o.$$

B_{lkst}, however, vanishes only for points infinitely distant from matter (in the hypothesis of the infinite universe).

In absolutely free space, remote from matter and where there is no electromagnetic energy, all the tensors and invariants vanish.

We thus find that there exists a progressive curvature of space-time as we pass from empty space-time far removed from all matter and energy to empty space-time in the neighbourhood of matter, then to regions where there are electromagnetic fields but no matter, and finally to regions filled with matter or electrons. If we regard the successive gradations of curvature of space-time as corresponding to successive grades of materialisation, starting from emptiness, we must place electromagnetic fields, or loose energy, at a stage of materialisation preceding that of matter proper, or bound energy.

Still another discovery which Einstein was able to deduce from his gravitational equations refers to the propagation of a gravitational force. Classical science had held to the view that gravitation must be propagated with a speed much greater than light, perhaps even with infinite speed. But Einstein, by integrating the equations of gravitation, proved that in a weak field gravitational waves would be propagated with the speed of light. In a strong field, of course, the presence of the field itself would influence the speed of propagation of the waves, and the problem would assume much greater complexity.

CHAPTER XXXIII

THE FINITENESS OF THE UNIVERSE

EINSTEIN's original speculations on the form of the universe were published in 1917; they led him to the so-called *Cylindrical Universe*. Soon after, de Sitter suggested a rival theory: The *Spherical Universe*. Today, as a result of the explorations of the remoter parts of the heavens by the giant telescopes, both universes of Einstein and de Sitter have been discarded in favour of a third type of universe: the *Expanding Universe* of Friedmann and of the Abbé Lemaitre. At the end of the chapter we shall review the more important features of Lemaitre's universe, but inasmuch as in the course of its expansion Lemaitre's universe passes through the Einstein form to the de Sitter form, we shall give a prior account of the two discarded universes.

To appreciate better the sequence of discoveries, we must first turn to the older conceptions of the universe. Some fifty years ago, the exploration of the heavens was limited in the main to a relatively insignificant region of space. In the regions explored, the stars appeared to be more or less uniformly distributed, and it was believed that however far we might wander into space, the same quasi-uniform distribution would manifest itself.

Gradually, it was realized, however, that the stellar universe which we perceive with the naked eye, and even with the smaller telescopes, does not extend indefinitely, but forms an island of matter. The name *galaxy* was given to this agglomeration of matter to which all the visible stars belong. As seen on a clear summer night our galaxy appears to be bordered by the Milky Way (which is nothing but an assemblage of stars, too far distant to be resolved by the naked eye); and the name "galaxy", derived from the Greek work "gala" (*i.e.* milk) expresses this feature. In addition to containing stars, our galaxy contains vast clouds of cosmic matter, which blot out parts of the heavens from our vision. On other occasions these clouds of dust are illuminated by the light from neighbouring stars and thus reveal their contour. Still other faint luminosities were observed, some of them visible to the naked eye. Viewed through the telescope they appeared to have a spiral form, but little was known about them.

With the advent of the larger telescopes the faint patches of luminosity just referred to, were found to lie in the extra-galactic regions of space, far far beyond the limits of our own galaxy. Futhermore, some of these faint patches were resolved by the telescope into individual stars, and were seen to constitute agglomerations of matter comparable in size to our own galaxy. The name spiral nebulae was given to these island universes, because many of them exhibit the spiral form indicative of rotation, though they do not all have the spiral form. Our own galaxy is one of the spiral nebulae, and it has the spiral form. In the course of time astronomers discovered that there were millions of these spiral nebulae strewn more or less uniformly through space as far as the telescope could explore. And so our own galaxy, though constituting an island of matter, was seen to be but one of the islands of an archipelago of islands more or less uniformly distributed. Thus although our conception of the heavens has been modified considerably as a result of the discoveries mentioned, the earlier belief in an approximately uniform distribution of matter throughout the spatial universe is still retained today. The only change in our outlook is that we now view the individual spiral nebulae instead of the indivdual stars as representing the units in the distribution.

With this preliminary information disposed of, let us revert to the state of affairs around the close of the last century, at a time when very little was known of the spiral nebulae, and when the stars were believed to be more or less uniformly distributed through infinite Euclidean space. Now a difficulty which beset classical science was the incompatibility of Newton's law with any such uniform stellar distribution. According to Newton's law of universal gravitation, the mutual attractions of material bodies should cause the stars to become grouped together into a configuration of uneven density. At the centre of the configuration the stars should be more closely packed, whereas they would thin out as we moved away from the centre. This configuration would thus exhibit a kind of nucleus at its centre, and would therefore be in conflict with a uniform distribution. We need not insist further on the exact form which the configuration of stars would assume, since all that we are concerned with here is the impossibility of a uniform distribution. Let us then examine this latter point more carefully.

Consider a quasi-uniform distribution of stars. On a grand macroscopic scale we may view the distribution as one of dust particles, packed together densely enough to simulate a continuous cloud of matter.* Newton's law, when applied to the interior of a continuous distribution of matter, is expressed by Poisson's equation.

(1) $$\Delta\Phi = 4\pi k\mu,$$

where Φ is the Newtonian potential, k the gravitational constant, and μ the density of matter. For our dust cloud to retain a uniform distribution to infinity, no forces, on an average, must be acting on the dust particles: and this requires that the Newtonian potential Φ should retain a constant value throughout infinite space. If, however, we set $\Phi =$ constant in Poisson's equation (1), we obtain $0 = 4\pi k\mu$, and so $\mu = 0$. We conclude therefore that the only possible uniform distribution of matter is one in

---

* In a similar way, on our commonplace level of experience the atmosphere appears as a continuous fluid, although from a microscopic point of view, it is composed of discrete molecules separated by relatively large distances.

which there is no matter; and this is but another way of saying that a uniform distribution is impossible.

In view of the incompatibility of Newton's law with the observed uniform distribution of matter, Neumann and Seeliger, in 1896, suggested a slight modification of the law of gravitation, which would have for effect to render a uniform distribution possible. The equivalent of Poisson's equation when the Neumann-Seeliger law is used, is

$$(2) \qquad \Delta\Phi - \lambda\Phi = 4\pi k\mu,$$

where λ is a positive constant, which may be as small as we choose. Assuming the density μ to be constant, this modified law, contrary to Poisson's, is compatible with the constancy of the potential, and so is compatible with a uniform distribution of homogeneous matter.

Let us now turn to Einstein's law of gravitation. We recall that in the general theory, as developed to this point, space-time as a whole was flat, the flatness being departed from only here and there, inside matter and in the proximity of matter, as per the Einsteinian law of space-time curvature, or gravitation. Space-time extended to infinity, and at infinity it was perfectly flat. Owing to these peculiarities, the universe was referred to as the infinite *quasi-Euclidean* universe; the qualification of *quasi* was introduced on account of the local curvatures around and inside each agglomeration of matter.

Einstein was averse to the idea that space-time could retain a definite, flat structure at spatial infinity. He stated: "it is certainly unsatisfactory to postulate such a far-reaching limitation without any physical basis for it." With a view of remedying the situation, he explored several possibilities. None of them satisfied him, and for that reason among others, he finally eliminated the idea of spatial infinity altogether by assuming that space curled round on itself, yielding a finite Riemann universe. For the present we shall confine our attention to another clue which prompted Einstein to the theory of a spatially closed universe.

Einstein observed that not only was the matter of the universe uniformly distributed (on a cosmic scale) but in addition the relative velocities of the stars appeared to be small, so that it was possible to select a frame of reference relative to which the stars would be approximately at rest. The quasi-Euclidean universe of the general theory was incompatible with this state of affairs, for the quasi-Euclidean universe, just like classical infinite space and Newton's law of gravitation, required the formation of a nucleus of stars. The problem was therefore to conceive of a space-time universe which would be consistent with a uniform distribution of matter at rest. Now, if we are to account for the absence of relative motion between the material particles of the universe, we must assume that the world-lines of the particles (which are geodesics along the time-direction of space-time) can neither taper together nor diverge. These world-lines must therefore be straight parallel lines; and this is possible only if the structure of space-time (on a cosmic scale) admits the same time-direction at every point (for an observer at rest in the privileged frame). We thus obtain a straight time-direction for the universe as a whole.

Consider next the evenness in the distribution of matter. We may account for this feature if we assume that the space of the universe closes round on itself in the fashion of a three-dimensional Riemann space. We may intuitively understand why a spatially finite universe solves the problem of the absence of a nucleus. The two-dimensional analogue of a star distribution spread out uniformly throughout the entire volume of a three-dimensional Riemann space is given by a homogeneous distribution of stars *over the two-dimensional surface* of an ordinary sphere situated in three-dimensional Euclidean space. The *surface* of the two-dimensional sphere, and *not the volume* enclosed within the sphere, is the analogue of the finite space of the universe. Reasons of symmetry suggest that a uniform distribution of matter over the spherical surface should be in equilibrium, since there is no reason for a nucleus to form *here* rather than *there*.

Thus the conception of the universe to which Einstein's speculations led him was that of a spherical, finite universe of three-dimensional space, from which rises a straight time-direction. If, for simplicity, we retain only one space direction and the time-direction, the two-dimensional simile is that of a straight circular cylinder, the axis of the cylinder representing the time-direction, and any circular section representing space. By reason of the characteristics just described, Einstein's universe has been called the **Cylindrical Universe**

Now a cylindrical universe, filled with matter uniformly distributed, at rest, is incompatible with the earlier gravitational equations*, *viz* with the equations

$$(3) \qquad G_{ik} - \tfrac{1}{2} g_{ik}\, G = T_{ik}$$

Einstein, however, overcame the difficulty in a manner reminiscent of Seeliger's modification of Newton's law, *i.e.* by adding to the equations (3) a corrective term involving a new universal constant, λ. But whereas Seeliger's adjunction of a corrective term to Newton's law was entirely arbitrary, Einstein's procedure was in full agreement with the requirements of the theory of relativity. Let us justify this last statement.

When we discussed Einstein's law of gravitation, we noted that the inclusion of an additional tensor λg_{ik} (λ being any constant) would not destroy the tensor character of the law, and so would not interfere with the general requirements of covariance that all natural laws must satisfy. Einstein therefore, added the term λg_{ik} to his earlier equations, and obtained in place of his original law (3) for the space-time curvature inside matter the amended law

$$(4) \qquad G_{ik} - \tfrac{1}{2} g_{ik}\, G + \lambda g_{ik} = T_{ik}$$

The added term, λg_{ik}, which is responsible for the difference between the

---

* At least, if we assume the matter of the universe to be represented by a cloud of dust, and introduce no universal pressure.

new law of gravitation and the earlier one, is called the "cosmological term"; and the constant λ itself, of unknown magnitude, is referred to as the *cosmological constant.*

The amended gravitational equations (4) admit among their solutions a cylindrical universe, so that they are consistent with our requirements. The equations show that the spatial radius of the cylindrical universe is equal to $\sqrt{\dfrac{1}{\lambda}}$. Furthermore the equations require that the mean density μ of the uniform distribution of matter in this cylindrical universe be equal to the cosmological constant λ (when appropriate units are used), and from this we find that the total amount of matter in the universe must be proportional to the radius of the spherical universe.

Now it would seem most unlikely that the mean density of matter should just happen to be equal to λ. Einstein, however, dismissed the thought of any miraculous coincidence. In his opinion (in the earlier days of the theory at any rate), the density of matter, or equivalently the total amount of matter, determines the radius and volume of the spatial universe. If there were more matter, the universe of space would be larger, just as the size of a rubber balloon is increased when we force more air into it. If there were no matter in the universe, space and space-time would have no structure, no shape or size, and space would become meaningless. To this extent we may say that matter generates space. The equality between the mean density of matter and the cosmological constant λ thus expresses a mere relation of cause and effect, and no miraculous coincidence is involved. We shall revert later to the consequences of Einstein's attitude. For the present we shall consider other peculiarites of the cylindrical universe.

It is only on a cosmic scale that the universe of space can be perfectly spherical, and the time-direction perfectly straight, for it is only on a cosmic scale that the distribution of matter can be viewed as uniform and at rest. If we examine the situation from a less macroscopic standpoint, the distribution of matter will no longer be perfectly even, and the curvature of space-time will vary from place to place. The local irregularities that develop around each lump of matter will be responsible for the mutual attractions of bodies, as was the case in the general theory. Reverting to our two-dimensional simile, we must liken our universe of space not to the surface of a perfect sphere but rather to the irregular surface of an orange. Einstein assumed that notwithstanding these local irregularities his cylindrical universe would remain stable. Apparently, this was a mistaken belief, as will be explained later when we consider the *Expanding Universe.*

In Einstein's universe, a ray of light, following a geodesic, will circle round the universe and return to its starting point. This feature has prompted the speculation that the rays of light emitted from a star would circumscribe the universe of space in all directions, then meet again at the point where the star was originally situated. Since a few billion years would be required for this circumnavigation to take place, the star would have moved away from its original position, and the converging rays of light would form a real image at the point whence the star had moved. This real image might be mistaken for a true star, whereas in reality, it would be a mere ghost. Accordingly some of the supposedly true stars seen in the heavens might turn out to be ghosts. In view, however, of the local irregularities of structure, it is extremely unlikely that the rays of light could converge with sufficient accuracy for this phenomenon to occur.

Having surveyed some of the notable features of the cylindrical universe, we are led to inquire: Can the correctness of Einstein's speculations be submitted to a test? It is certain that the intrusion of the constant λ in the gravitational equations, modifies these equations, and it might seem that more refined measurements of the planetary motions would disclose the influence of λ. But we must remember that λ being equal to the mean density of matter in the universe, is an extremely small magnitude, far too small to be disclosed by measurements over such relatively restricted areas as are available in our solar system. In much the same way it would be impossible to reveal the curvature of the Earth by measurements over the area of a city block. Only measurements over much larger areas can detect the Earth's curvature; and the same situation, but in a much aggravated form, confronts us when we wish to detect the presence of λ. Explorations of space over distances accessible only to the larger telescopes would be required.

Today, Einstein's universe, in its original form, has been abandoned, chiefly on account of the recently discovered large recessional speeds of the distant spiral nebulae; these high speeds conflict with the belief in a universe of matter at relative rest.

Soon after Einstein had suggested his cylindrical universe, an alternative universe was proposed by de Sitter. De Sitter criticised Einstein universe on the grounds that it introduced a definite time-direction for the universe (the axis of the cylinder), and thereby reestablished the discarded notion of absolute time.* Other criticisms were also made by de Sitter and were countered by Einstein. At all events de Sitter conceived of a universe in which space and time would be on the same footing. He retained the amended gravitational equations (4) on which Einstein had based his speculations, but so as to avoid the special status which matter had conferred on the time direction in Einstein's universe, de Sitter envisaged a space-time universe empty of matter. De Sitter thereby did not follow Einstein in assuming that matter was the generator of space.

Now if we revert to the equations (4) and if we postulate the absence of matter, we must set $T_{ik} = 0$ in these equations, for we recall that T_{ik} represents matter. When this is done, the equations can be shown to reduce to

(5) $$G_{ik} = \lambda g_{ik}.$$

These then are the equations which de Sitter's empty universe must

---

* This universal time-direction holds only for the universe as a whole. For restricted regions we have the multiplicity of time directions as required by the Lorentz-Einstein transformations.

satisfy. The most symmetric solution consistent with these equations is represented by a four-dimensional spherical universe of space-time, of radius $\sqrt{\dfrac{3}{\lambda}}$. De Sitter adopted this solution, and his universe has accordingly been called the *Spherical Universe*.*

The time-direction in the de Sitter's universe is no longer unique or straight, as it was in Einstein's universe, and so the appearance of an absolute time on a cosmic scale is avoided. As explained in the note, de Sitter's universe, in spite of the misleading name "spherical" that has been given to it, is open at both ends in the time-direction; and so there is no danger of time curling round on itself, like a serpent swallowing its tail, and causing the past to become the future.

Although the spherical universe was obtained on the assumption that the world was empty, this did not convey to de Sitter the idea that the absence of matter was essential to his theory.† De Sitter was of course perfectly well aware that our universe does contain matter, but he assumed that the amount of matter present was too trifling to affect the spherical structure otherwise than locally. The local curvatures generated by material bodies in de Sitter's universe would give rise to local gravitational influences, as was the case in Einstein's universe. The principal differences between the cylindrical and the spherical universes, as it appeared at the time, was thus that Einstein's universe was generated by matter, and the value of the cosmological constant λ was determined by the density of matter, whereas de Sitter's universe could exist without matter, and the cosmological constant was a characteristic of space-time itself.

An important feature of de Sitter's universe, differentiating it from Einstein's, is that its time-directions are not parallel and straight, but diverge from one another. As a result, particles of matter, initially at relative rest, will tend to scatter, as though they exerted mutual repulsions. Of course if the particles are at sufficiently small distances apart, their mutual gravitational attractions will overcome the repulsions, but for large distances the repulsion will predominate. This repulsion, which arises from the structure of de Sitter's space-time universe, is called the *cosmic repulsion;* the repulsive force which it embodies is proportional to the cosmological constant λ and to the distance. We shall have more to say of the cosmic repulsion when we consider the Expanding Universe.

A further consequence of de Sitter's universe refers to the slowing down of time in the remoter regions of space. When we consider points further

---

* De Sitter's universe is truly spherical only when we argue in terms of imaginary time it. In this case, for any given observer, both time and space close round on themselves. When, however, we use real time t, as indeed we should, we find that de Sitter's universe yields a three-dimensional spherical Riemann extension, for the space of a given observer, but that real time no longer curls round on itself. If we use a two-dimensional analogy by retaining the time dimension but only one of the space dimensions, we find that when real time is introduced, the universe can be represented on the surface of a one-sheeted hyperboloid, open at both ends in the time direction.

† As we shall see, de Sitter was mistaken on this point.

and further removed from where we stand, the curvature of the time-direction causes the time-g_{1k}, *i.e.* g_{44} to decrease and finally to vanish. The decrease and the eventual vanishing of g_{44} indicate, respectively, the gradual slowing down and the eventual complete arrest of time in the distant regions of the universe. The slowing down of time is reciprocal, however, for if we transported ourselves to a distant point, we would find that things went on as usual, whereas it would now be where we formerly stood that the slowing down of time would appear to manifest itself. Thus, for every observer there would exist a locus of distant points where nothing would appear to change or move. A ray of light could never circle round the universe, but would gradually slow down and ultimately be arrested when this passive horizon of the observer was reached. We also see that the nearer an incandescent atom is situated to this passive horizon, the slower would its vibrations appear; hence, in the case of a sodium atom (emitting a yellow light under normal conditions), the redder would its light appear to us, till finally the light would cease completely when the atom was situated at our passive horizon.

Soon after de Sitter had suggested his universe as an alternative to Einstein's astronomers turned their attention to the spiral nebulae, which appeared as faint objects in the sky. They found that the spiral nebulae were far beyond the limits of our own galaxy. They also found that the majority of the nebulae studied emitted radiations that were shifted towards the red end of the spectrum. This phenomenon was to be expected according to de Sitter's theory for two reasons: first on account of the slowing down of time in the very remote regions of space, but more especially on account of the scattering tendency manifested by particles of matter, a scattering increasing with the distance. As a result of this scattering, the more distant nebulae would be receding with large speeds from our galaxy, and their recessional motions would cause their spectral lines to be deflected towards the red owing to the well-known Doppler effect .* For these reasons some physicists were inclined to accept de Sitter's universe in preference to Einstein's. Judgment was reserved, however, until a systematic study of the still more distant nebulae had been performed.

Today, both de Sitter's and Einstein's universes have been discarded, but before examining the type of universe that has replaced them, we shall mention further peculiarities of Einstein's universe on account of their philosophic interest.

According to the general theory of relativity, prior to Einstein's investigations on the form of the universe, matter was known to influence space-time, since it imposed a curvature on it. On the other hand, space-time in the absence of matter still retained a flat structure, so that matter did not create a structure but merely modified a preexisting one. But this situation gave rise to a dualistic conception of space, part existing in its own

---

* The Doppler effect is illustrated in acoustics by the lowering in the pitch of a whistle emitted by a receding locomotive, or by the increase in the pitch when the locomotive is approaching us.

right, and part modified by its content. To many this dualism appeared objectionable.

A second dualism affected the problem of motion. The theory of relativity had proved that velocity through space was relative, but it had failed to extend this relativity to accelerated motion. After having made a promising start towards the relativisation of all motion, it had abandoned the attempt halfway before the final goal was reached.

Then again, a similar duality existed in the case of mass and weight. Here we have *weight* depending on the distribution of surrounding matter, so that weight would be inexistent in a world containing but one body. On the other hand, here was the *inertial mass* of the body which would subsist even in an otherwise empty world. Thus, were we to be transported far from all large masses, we could hold a ton of bricks above our heads without the slightest effort, since the bricks would have no weight. And yet, when we tried to push these same bricks aside, we should encounter the same inertial resistance as on earth. What rendered this duality particularly displeasing was the fact that the general theory had proved that even mass was influenced in a minute degree by the proximity of matter, so that the mass of a body appeared to be due in part to the distribution of surrounding matter and in part to the body itself. In other words, there existed a partial relativity of inertia, just as there existed a partial relativity of motion. These dualities are not equally objectionable to all thinkers; but it must be admitted that for those who crave unity in nature, every effort to remove them will be welcomed. With the cylindrical universe, however, these difficulties were overcome, and, for this reason among others, the cylindrical universe satisfies a natural philosophical craving. Inasmuch as the foregoing considerations present an intimate connection with Mach's mechanics, we will discuss the significance of Mach's ideas at some length.

As is well known, on account of the existence of centrifugal force, of the inertial frame and of the rectilinear paths followed by free bodies according to the law of inertia, Newton felt himself compelled to postulate the existence of absolute space. Einstein summarises Newton's attitude in the following lines:

"In order to be able to look upon the rotation of the system, at least formally, as something real, Newton objectivises space. . . . What is essential is merely that besides observable objects, another thing, which is not perceptible, must be looked upon as real, to enable acceleration or rotation to be looked upon as something real."

The opposite stand was taken by Mach, who endeavoured to account for the dynamical manifestations of acceleration and rotation without appealing to that suprasensible entity, "absolute space." Instead of holding, then, that centrifugal force was generated by a rotation in absolute space, he urged that we should attribute it to a rotation with respect to the material universe as a whole. Inasmuch as the vastest agglomeration of matter is given by the totality of the stars and nebulae, it would follow that the dynamical effects would be generated by the rotation of an object relatively to the stars, or, what amounts to the same thing, by a rotation of the stars relatively to the object. In fact, these two different alternatives would represent but a difference in phraseology, since with absolute space banished, apart from stars and matter generally, there would exist no other terms of comparison.

To this rotation relatively to the star-masses would be due the protuberance of the equator, the splashing of water in Newton's bucket, the bursting of a rapidly revolving flywheel, etc. In short, all those dynamical effects which in Newton's opinion betrayed rotation in absolute space, are held by Mach to betray a rotation relative to the star-masses of the universe. Now, it should be apparent that Mach's relativity of rotation entails more than a mere kinematical representation. As he is careful to specify, this active rôle of the stars is due *not* to the accidental circumstance that they happen to be visible, but to universal gravitational actions which would be generated between matter and matter when a state of relative acceleration or rotation was present. In this way, forces of inertia and centrifugal force are no longer to be attributed to the intrinsic structure of absolute space; they now become akin to gravitational forces. Were the stars invisible, no change in the dynamical manifestations would be observed, but if their masses were to be annihilated, if the stars ceased to exist (not merely ceased to shine), all the telltale dynamical effects would disappear; a flywheel would not burst, water would not splash, and so on. In much the same way, the elliptical orbits of the planets are due to the gravitational attraction of the sun's mass; *not*, of course, to the accidental circumstance that the sun happens to be luminous rather than dark.

From these conclusions further important consequences follow; in particular, we must assume that the inertial mass of a body must be generated solely by the mutual actions existing between this body and the other bodies of the universe. We can understand the reason for this statement when we realise that were all the stars to be annihilated, all inertial forces would disappear; but it is precisely the existence of inertial force which is responsible for the effort required to set a body in motion or to arrest it when started. For this reason the disappearance of inertial forces would automatically entail that of inertia or mass. We may note the contrast with Newton's views, for in classical science the inertial forces arose from the structure of absolute space itself, and mass was regarded as an intrinsic property of matter. In the present case, mass joins weight in being a relative; for just as a body has no weight in the absence of other bodies, so now we see that, according to Mach, it would also present no inertial mass if situated all alone in an otherwise empty world.

When we consider Mach's mechanics, we see that its net result has been to identify Newton's absolute space with the space generated by the stars; absolute space has become materialised, so to speak. We may obtain a more concrete representation of the new ideas by imagining a crisscross of threads extending between each individual star and all the others. Rotation, which develops dynamical effects, is, then, the rotation of an object with respect to this network of threads; and it is obviously a matter of indifference to suppose that the object is rotating among the threads and stars, or that threads and stars are rotating around the object. Thus the inertial frame is any frame in which the threads appear to be non-rotating.

Mach's mechanics might seem to draw too much on our credulity. We can grant that the sun exerts a gravitational effect on the earth, but to assume that the bursting of a flywheel is due to the presence of the stellar universe would appear to be stretching matters too far. Nevertheless, it should be remembered that Newton's introduction of that supra-sensible entity, absolute space, was also extremely distasteful to a number of thinkers, so that it is a question of choosing the lesser of two evils.

These ideas were upheld by Mach for a number of years; and although they can lay claim to no special originality, they are always referred to under his name. Mach's stand has been so consistently misrepresented by philosophers that we feel justified in warning the reader by mentioning some of the erroneous assertions that have been made with respect to it.

Two centuries ago, Berkeley had defended views which present a certain superficial similarity to those of Mach. He argued that if there were no stars, no permanent points of comparison in the heavens, we could not even imagine the earth's rotation, hence this rotation would be deprived of all meaning. To this argument the Newtonian would reply that it was, after all, not impossible to imagine absolute rotation as a rotation in a rigid jelly-like substance called space. Furthermore, he would add that it was the dynamical facts, the appearance of unsymmetrically distributed forces, that seemed to confirm the absolute nature of rotation. Absolute rotation was, then, totally irrelevant to our perception of luminous reference points in the heavens, or even to our ability to imagine their existence. So far as the Newtonian was concerned, even were the stars to be annihilated entirely, it would still be quite easy to determine whether or not the earth was in rotation. A number of experiments, such as that of Foucault's pendulum, the gyroscope, and, more generally, the presence of centrifugal forces on the earth's surface, would demonstrate the existence of this absolute motion. In view of these dynamical effects, which appear to have been unknown to Berkeley, his views received no consideration from men of science. Only when, with Mach, we maintain that were the stars to be annihilated, the dynamical effects would also disappear, only when we conceive of the stars not merely in a visual capacity or as a support for the imagination, but as producing causal gravitational influences, can an argument be presented against Newton's absolute space. In short, according to Mach, it would be the gravitational action of the totality of the material universe, including the stars, whether visible or dark, the nebulae, our sun, the planets and moons, that were responsible for centrifugal force.

Before we proceed farther, a number of objections must be discussed. It might be claimed that whereas the earth could be conceived of as rotating among the stars, it would be quite impossible to consider the stars as rotating round the earth, since in this case centrifugal force would cause the stars to fly apart to infinity. A criticism of this sort arises from a confusion between the Newtonian attitude and the new one. With Newton's ideas the argument would hold, for the stars would indeed be describing gigantic circles in absolute space. But in the present instance the space of the stars themselves has usurped the position of absolute space, since it is in this star-conditioned space that circular motion generates centrifugal forces. And in their own space the stars are obviously at rest whether or not we regard them as rotating, together with their space, around the earth. Hence, centrifugal force does not arise on the stars, though it will be felt on the earth's surface. An equivalent way of explaining the same point would be to say: The stars as rotating round the earth are not subjected to centrifugal force, for they are rotating with respect to nothing; the mass of the earth being insignificant in comparison with the totality of the star-masses.

Now, although a number of thinkers were in sympathy with Mach's attitude, owing to their belief that all motion in a perfectly empty universe must be meaningless, yet the physical difficulties confronting his idea were so great that classical science had never taken it seriously. Euler had discarded the possibility of any such stellar action, and even quite recently Planck had criticised Mach for his stand. It was not that in theory a possible stellar action should be excluded on general principles; for physics had made us acquainted with foreign actions of a similar nature, the best-known of which would be illustrated by the gravitational action of the sun on the planets. Then again, when we wish to cut through an apparently empty space with a knife, we may experience considerable difficulty if the space happens to be permeated by a powerful magnetic field. We cannot, therefore, exclude the possibility that the stars might in some way generate a field endowing space with a structure which we had mistakenly taken to represent the absolute intrinsic structure of space. Science, however, cannot content herself with vague qualitative analogies; and before a mechanics such as Mach's could be entertained, it would be necessary for it to succeed in accounting for the existence of centrifugal and Coriolis forces on a body in rotation with respect to the stars. This accounting would have to be made, *not merely in some vague qualitative way, but quantitatively also, to a high order of precision.* Such, indeed, are the requirements that all hypotheses in physics must satisfy. Now, all that Mach had done had been to postulate a possible stellar influence; he had not even attempted to justify his ideas rigorously. Indeed, even had he investigated the problem more deeply, he would have failed; for in his day space-time was unknown.

Furthermore, Mach's ideas presented insuperable difficulties. Newton's law of gravitation stated that the attractions exerted between bodies depended solely on their masses and mutual distances; the conditions of relative motion or rest of the bodies being considered irrelevant. But if Mach's ideas are accepted, we must assume that in addition to the ordinary Newtonian attraction, a supplementary field of attraction, hitherto unsuspected, should arise if the bodies were in a state of relative acceleration, as in the case of the stars rotating round the earth, or the planets moving round the sun.* Mach's mechanics would entail, therefore, the downfall of the Newtonian law of gravitation. Now, when Mach suggested his mechanics, Newton's law appeared to be justified indirectly by the accuracy with which it had permitted mathematicians to foresee numerous astronomical occurrences (with the possible exception of Mercury's motion and certain lunar disturbances). In addition, Newton's inverse-square law of the attraction of matter on matter had been verified by direct measurements in the laboratory, in what is known as Cavendish's experiment. So far as experiment could detect, there was no trace of an increase in the mass of a body when other bodies were placed in its proximity; hence, on this point it appeared impossible to detect the variations in mass demanded by Mach's mechanics. As for attempting Cavendish's experiment with masses mutually accelerated, technical difficulties rendered its success impossible. For all these reasons the majority of scientists viewed Mach's relativity of rotation with the utmost suspicion.

But when we come to consider the theory of relativity, the situation changes. In the first place, Newton's law of gravitation is seen to be only approximate, and the relative motion existing between bodies is found to modify their apparent mutual attractions. This in itself is, of course, insufficient to substantiate Mach's mechanics, but when we examine the implications necessitated by Einstein's cylindrical universe, we find that Mach's views may turn out to be correct after all. First, let us recall certain features of the general theory of relativity.

Einstein's gravitational equations without the λ term, *i.e.*, the equations applying to the infinite universe, namely $G_{1k} = o$, had received brilliant support in having enabled him to account for the motion of the planet Mercury, and to anticipate such unsuspected phenomena as the double bending of a ray of light and the Einstein shift-effect. It was only natural, therefore, to endeavour to extract from these equations all

---

* The situation would be somewhat similar to that which exists in the case of electric and magnetic actions. Here, also, we know that an electron at rest develops a purely electric pull according to the Newtonian law, whereas, when it is in relative motion, a magnetic pull is superadded at right angles to the line of motion and to the electric pull.

the remaining treasures they contained. But here integrations of an extremely difficult nature awaited mathematicians.

When the equations were treated in a very approximate way, they yielded Newton's law of gravitation and the Einstein shift. When the approximation was increased, they yielded the double bending of a ray of light. When the approximation was increased still further, we obtained the precessional advance of Mercury's perihelion. When treated in a more general way, they enabled Einstein to predict that gravitation would be propagated with the speed of light. Thirring then attacked the equations, carrying his calculations to a still higher degree of approximation, and at last, the effects necessitated by Mach's mechanics appeared. Thirring found that the rotation of a central body would modify the nature of its attraction on the planets or satellites circumscribing it. (These anticipations would appear to be susceptible of astronomical verification in the case of Jupiter and one of its satellites.) But, most important of all, he discovered that Mach's belief in the gravitational influence of the stars rotating round a fixed body was finally justified. In particular, the equations proved that even in the hypothesis of an infinite universe, a rotation of matter would produce around a test-body a field of force disposed in exactly the same peculiar way as are the centrifugal and Coriolis forces in the space attached to a body in rotation. Furthermore, the mass of a test-body would be affected by the proximity of surrounding matter. This last consequence, however, can be obtained without resorting to difficult calculations.*

Now the gravitational equations which have yielded these interesting results have most certainly not been doctored up for the sole purpose of justifying Mach's ideas. As explained in previous chapters, they were discovered by Einstein as a result of a totally different train of thought, long before the relativity of rotation was even considered. Hence, it cannot be denied that by vindicating, as they did, the correctness of Mach's ideas, a fact of the utmost importance had been established. Mach's mechanics received at last that rational justification which until then had been lacking; and the first inkling of a possible physical connection between the origin of inertia and gravitation was obtained. However, it is important to note that although from a qualitative standpoint

---

* The variation of mass in a gravitational field can be anticipated most easily as follows: Consider a disk rotating in a Galilean frame. As referred to this frame, the points of the rim will be moving with a certain velocity; hence a mass fixed to the rim will increase when its value is computed in the Galilean frame. But the postulate of equivalence allows us to assert that conditions would be exactly the same were the disk to be at rest in an appropriate gravitational field. Inasmuch as in this case the gravitational force would be pulling outwards from centre to rim, just as though a massive body had been placed outside the rim, we may infer that the mass of a body increases as it approaches gravitational masses. For instance, the inertial mass of a billiard ball would be increased were the ball to be placed nearer the sun.

the agreement was perfect, yet when we view the problem quantitatively, calculation shows that the increase of mass and the centrifugal forces generated by the rotation of the stars would produce but an infinitesimal fraction of the magnitudes observed in practice. In order to obtain magnitudes corresponding with observation, it would be necessary to assume that the stars were many trillion times more numerous than we believed them to be, and this supposition would lead us into other difficulties.

The reason for this discrepancy is not hard to discover. Centrifugal force, as we know, is produced when a body is torn away from the space-time geodesic which the metrical field would compel it to follow. To say, therefore, that a rotation of the firmament round a test-body would affect the field of force surrounding the body is equivalent to stating that a rotation of the stars would be capable of affecting the distribution of the metrical field round the body and of dragging the geodesics along. There is no reason to be surprised at some such effect taking place, seeing that the phenomenon of gravitation connotes that matter, whether at rest or in motion, must affect the curvature of space-time, hence the lay of the metrical field. But, on the other hand, until Einstein had discovered the cylindrical universe, the action of matter could only modify the lay of a pre-existing metrical field, and could not create such a field out of nothing. Space-time of itself was assumed to be flat, and matter could only endow it with a certain degree of curvature. The result was that the major part of centrifugal force was due to the intrinsic texture of flat space-time, and only an infinitesimal portion of this force could be attributed to the additional action of the stars. Centrifugal force thus still appeared, in the main at least, to betray absolute rotation in empty, absolute space-time.

None the less, an important point had been established. Mach's ideas, though still refuted quantitatively, yet seemed to be acceptable qualitatively, whereas in classical science they had appeared utterly impossible from every standpoint. The thin end of the wedge had been driven in. From then on, it became permissible to assume that with an increase in our understanding of the universe the complete relativity of all motion would be established, not only qualitatively, but also quantitatively. In other words, not merely an infinitesimal portion of centrifugal force and of mass, but their totality, might be accounted for by appealing to the interaction of matter and matter. It is easy to understand how this achievement would be realised.

All we should have to do would be to assume that the totality of the metrical field, and not merely a portion of it, was created by the distribution of matter throughout the universe. As we have seen, this condition is precisely the one realised in the cylindrical universe; so that with Einstein's universe, Mach's ideas may be vindicated and the bugaboo of absolute rotation dispelled.

Both de Sitter's universe and the quasi-Euclidean one would be incompatible with Mach's mechanics, since in either universe the metrical field and the geodesics of space-time would exist in the complete absence of matter. This means that in the empty world of de Sitter, and far beyond the nucleus of stars in the quasi-Euclidean universe, bodies would be submitted to forces of inertia when torn away from their natural geo-desical world lines. In either case the space-time void would appear to be endowed with dynamical properties even in the absence of all matter.

The following passage quoted from Einstein's writings expresses his views on the subject:

"The idea that Mach expressed, that inertia depends upon the mutual action of bodies, is contained to a first approximation in the equations of the theory of relativity. It follows from these equations that inertia depends, at least in part, upon mutual actions between masses. As it is an unsatisfactory assumption to make that inertia depends in part upon mutual actions and in part upon an independent property of space, Mach's idea gains in probability. But this idea of Mach's corresponds only to a finite universe bounded in space, and not to a quasi-Euclidean infinite universe. From the standpoint of epistemology it is more satisfying to have the mechanical properties of space completely determined by matter, and this is the case only in a space-bounded universe."

THE EXPANDING UNIVERSE

WE recall that Einstein's speculations on the form of the universe were prompted by his desire to account for the quasi-uniform distribution of matter, which he believed to exist throughout space. The original gravitational equations of the general theory seemed to be incompatible with a uniform distribution of matter, unless they were supplemented by an additional term. Einstein accordingly introduced the additional term, *i.e.* the cosmological term, involving the cosmological constant λ. He thus was led to the cylindrical universe. Einstein, however, was dissatisfied with his solution, because he viewed the introduction of the constant λ as a hypothesis *ad hoc*. In a subsequent paper, he reverted to his original gravitational equations and rendered them consistent with a uniform distribution of matter, by postulating in place of the constant λ, a universal pressure of a peculiar kind. By this means he obtained once again the cylindrical universe. However, the universal pressure, which he had introduced, was extremely hypothetical and was eventually discarded by him.

In 1922, Friedmann showed that it was possible to account for a uniform distribution of matter, while retaining the original gravitational equations, (and so dispensing with the cosmological constant λ) and without postulating the existence of the universal pressure. But to secure this result Friedmann discarded the idea of a static universe of invariable size (such as was Einstein's universe) and replaced it by a universe of varying radius.

Einstein has indorsed Friedmann's solution of the problem, and has given an account of it in the second edition of his book, "The Meaning of Relativity." Friedmann's universe may contract or expand according to conditions. It may have a positive or negative curvature, or none at all, the various possibilities depending on the mean density of matter in the universe, whose value is not as yet known with sufficient accuracy to permit of a decision.

From the preceding explanations, we see that Einstein's main desire was to rid his theory of the cosmological constant λ, which he himself had introduced in the earlier days. Other physicists, however, did not share Einstein's aversion to the constant λ, and saw no objection to complicating the original gravitational equations by the adjunction of that constant. In 1927 the Abbé Lemaitre formulated a theory of an expanding universe in which the constant λ plays a leading part. We shall be concerned with Lemaitre's theory in the remaining pages of this chapter.

So as to avoid all misconceptions from the start we must stress that Lemaitre's expanding universe is a possible consequence of the theory of relativity, just as were the universes of Einstein and de Sitter; but the problem of deciding whether the expanding universe corresponds to reality is one which only astronomical observation can settle. Like the universes of Einstein and de Sitter, Lemaitre's universe, insofar as its space is concerned, is a bounded and finite Riemann space, so that we must not fall into the error of viewing it as a material universe expanding through an infinite space.

We have seen that the universes of Einstein and de Sitter are solutions of the equations (4) page 301. These equations in common with all differential equations admit an infinity of solutions, the static universes of Einstein and de Sitter being but two among the possible solutions. Lemaitre submitted Einstein's and de Sitter's static universes to a critical analysis and found that both universes lacked stability. Consider de Sitter's universe first. De Sitter's spherical universe was obtained on the assumption that the world was empty of matter. Since, however, the presence of matter in our actual universe is a fact that cannot be denied, any theory of the universe which is acceptable must necessarily be compatible with the existence of at least some matter. De Sitter never doubted that his spherical universe was compatible with the existence of a certain amount of matter, the presence of this matter merely affecting the space-time·curvatures locally. We shall see, however, that de Sitter was mistaken on this point: his static universe, such as he had pictured it, can exist only if it is completely empty of matter.

In this connection, we recall that in de Sitter's universe there is a cosmic repulsion, due to the universal structure of space-time itself. So long as de Sitter's universe contains no matter, it may remain static. But suppose that matter is introduced into a de Sitter universe. The cosmic repulsion will now act on the particles and will tend to scatter them in all directions. On the other hand, the mutual gravitational attractions between particles will oppose the scattering. A scattering will take place, therefore, if the matter is so rarefied that the gravitational attraction is too small to overcome the cosmic repulsion.

What this scattering implies is best illustrated by adopting the two-dimensional simile of particles of matter distributed like specks of dust over the surface of a fixed sphere. Since the distances between particles must increase constantly all over the surface of the sphere, as appears to be required by the cosmic repulsion, we are faced with an impossibility, for the surface of the sphere being constant in area, mutual distances between particles cannot increase in one region of the surface without decreasing in others. The only way to avoid a contradiction is to assume that the spherical surface itself is expanding, and so is becoming larger and larger. Translating these results into the language of three-dimensional space, we see that if a small density of matter is present in de Sitter's universe, the cosmic repulsion will cause a progressive expansion of space—a static de Sitter universe containing some matter is thus impossible.

Next consider Einstein's cylindrical universe, filled with a uniform distribution of matter, and in which the mean density of matter, μ, is equal to the cosmological constant λ. Of course, it is only on a grand cosmic scale that matter may be viewed as evenly distributed, for we know that matter is concentrated here and there into galaxies and stars. Einstein assumed, however, that such local condensations would merely affect the structure locally without disturbing the general stability of the universe. It is this belief which Lemaitre challenges. Lemaitre points out that the cosmic repulsion, which we have seen to be a feature of de Sitter's universe, is also present in Einstein's universe, but in Einstein's universe so long as matter is uniformly distributed, the cosmic repulsion is exactly balanced by the gravitational attraction, with the result that no cosmic repulsion will manifest itself. If, however, condensations of matter arise, the neat balance will be destroyed, and the cosmic repulsion will predominate. The condensations of matter will therefore move away from one another, and the radius of the spatial universe will expand—at least in those parts of space where the condensations have occurred.

Einstein's universe is thus in equilibrium provided matter be uniformly distributed with density $\mu = \lambda$, but this equilibrium is *unstable* (much as is the equilibrium of a cone balanced on its vertex) and the slightest condensation of matter will start an expansion. The expansion in turn will increase the mutual distances between bodies, and so will weaken still further the gravitational attractions, thereby upsetting still further the initial equilibrium. The expansion, once started, will thus proceed automatically with increasing rapidity, and will continue indefinitely under the action of the cosmic repulsion.

Suppose then that at some period of the dim past, our universe was an Einstein universe. Inasmuch as our present-day universe obviously exhibits condensations of matter, we may assert that it has expanded since the early days. Lemaitre assumes that the total mass of the universe remains unchanged; consequently the progessive expansion entails a constant decrease in the mean density of the mass of the universe, and matter becomes more and more rarefied. The Einstein form is thus departed from, and the universe evolves towards other forms. Now note that however great the rarefaction, the universe still contains the same total amount of matter and so never becomes empty. The form towards which the universe evolves in thus a form in which the mean density of matter is extremely small, and tends to become insignificant as in a de Sitter universe. We conclude that the universe is evolving from the Einstein stage in the direction of the de Sitter form.

Lemaitre's expanding universe, like its static predecessors, is spatially finite, and involves the cosmological term in λ with which we are already acquainted. Although the Lemaitre universe is expanding, the cosmological magnitude λ is still a constant as before, but it is no longer directly connected with the size of the spatial universe as it was in Einstein's universe.

Lemaitre, in his mathematical approach, reverts to the same equations (4) which formed the starting point of Einstein's and of de Sitter's investigations, but instead of seeking a solution representing a static isotropic space, he seeks a solution corresponding to a varying universe. Proceeding in this way, he was led to an equation previously obtained by Friedmann. This equation connects the time-rate of expansion of the universe with the constant λ, the total mass of the universe, and the radius of the universe at the instant contemplated.

Let us next examine how this progressive expansion of our finite space will manifest itself to us. Suppose our Earth were made of rubber and were to expand. The distances between cities would progressively increase, and the people in each city would believe that all the other cities were moving away from them. Furthermore, the greater the distance between two cities, the greater would be the recessional velocity. A city twice as distant as another would be receding twice as fast. Interpreting this result in terms of our expanding three-dimensional space, we see that the expansion of the spatial universe will reveal itself as a recession of all other cosmic masses from our galaxy, the recessional speed increasing with the remoteness of the celestial objects considered.

And this brings us back to the red shift which astonomers had observed in the spectra of the extra-galactic nebulae. The shift was first noticed around 1920, and the natural explanation was to ascribe it to a recession of the nebulae involved. Since de Sitter's universe had predicted the shift, the phenomenon was generally interpreted as lending support to de Sitter's universe. Today, for the reasons previously listed, de Sitter's invariable universe has been discarded, and the shift is now ascribed to the expansion of space. Since 1920, with the installation of still more powerful telescopes and following lengthy research, a number of spiral nebulae have been examined and Hubble has found that the recession is a general phenomenon and that its speed appears to increase proportionately to the distance, in agreement with Lemaitre's theory. The ratio of the velocity of recession of a nebulae to its distance is called the Hubble ratio; its present value is about 432 km per sec. per megaparsec.*

Lemaitre's theory interprets this recession as due to an expansion of space

* 1 megaparsec is 3.26 million light-years.

provoked by the cosmic repulsion but partly resisted by the mutual gravitational attractions of the nebulae. As the expansion proceeds, the distance between nebulae increases and the gravitational restraining force force decreases. As a result the rate of expansion, and with it the Hubble ratio, will be constantly increasing until such time that the nebulae are so far apart that their mutual gravitational actions are negligible. From then on, the value of the Hubble ratio will remain fixed, and Lemaitre shows that this limiting value is equal to $c\sqrt{\dfrac{\lambda}{3}}$ where c is the velocity of light.

In the present state of the universe, the Hubble ratio must be close to its limiting value in view of the large distances between the nebulae. Consequently, the limiting value cannot exceed the present-day value by any large amount. The limiting value is believed to be approximately 530 km per sec. per megaparsec, and since this value is equal to $c\sqrt{\dfrac{\lambda}{3}}$ according to Leimatre's theory, the value of the unknown constant λ may be determined. We find $\lambda = 9.8 . 10^{-55}$ cm.$^{-2}$ approximately.

From our knowledge of λ the mathematical theory enables us to compute the radius of the universe in the Einstein stage, and thence the constant total amount of matter in the universe. All this information, he it noted, is obtained in Lemaitre's theory from the observed recessional speeds of the nebulae.

In his subsequent interpretation of his theory, Lemaitre has abandoned the idea that the Einstein form marks the birth of the universe. Basing his speculations on the principle of entropy and on the quantum theory, he assumes that originally the radius of the universe was no larger than some eight times that of our present-day sun.* The matter of the universe was thus compressed into a space of relativity insignificant dimensions. Matter in those days was more or less undifferentiated; it had an enormous density, and conditions were similar to those reigning inside the nuclei of the heavier atoms. Lemaitre refers to this original universe as the "atome primitif," best translated into the *primordial atom*. We must remember, however, that compared to our modern atoms, the primordial atom had an enormous volume.

According to a recent lecture given by Lemaitre, the subsequent history of the primordial atom was as follows:

The primordial atom was highly unstable and existed only for an instant. It immediately exploded into radioactive fragments, which started to scatter and to swell the volume of the spatial universe. These fragments exploded into others, hurling out electrons, protons, and γ-particles, and emitting radiation in a super-radioactive display. Gradually the disintegrating process broke matter down into smaller and smaller fragments. Most of the radioactive substances existing at that time have long since disappeared; some few, however, such as radium and uranium had a

* Estimates on the initial dimensions of the universe vary.

slower rate of disintegration, and their unexploded remnants subsist to the present day. The constantly increasing space was filled with radiation and with particles moving at tremendous speeds. According to Lemaitre, the primitive radiation and the fast-moving particles of that catastrophic era are still with us today and manifest themselves as cosmic rays.

After a few billion years of expansion most of the commonplace atoms that had been formed lost their enormous velocities and settled down into a state of statistical equilibrium, forming gases. The Einstein form of the universe was then reached, and with it a condition of temporary equilibrium, as a result of which the expansion ceased. Space was then uniformly filled with matter and radiation. The Einstein stage lasted two billion years or more, but eventually the gases began to condense locally into stars and nebulae, and under the action of the cosmic repulsion, space resumed its expansion. Some regions, however, were subjected to no condensations and did not participate in the general expansion. Lemaitre suggests that such regions may be identified today with those occupied by the ill-defined clusters of nebulae and clouds of cosmic dust, which he regards as remnants of the equilibrium distribution existing in the Einstein stage. The present density of matter in these exceptional regions would thus be the density that existed throughout the universe in the Einstein stage, and from this density we may obtain an alternative computation of the cosmological consant λ. Lemaitre finds that the value of λ, thus obtained, coincides with the value derived from the recessional speeds of the nebulae, and he views this concordance as providing strong support for his theory.

The Einstein stage came to an end some three billion years ago, and the universe has expanded ever since and is still expanding, evolving towards an expanding de Sitter universe. Lemaitre estimates that the present radius of the universe is about ten times the Einstein radius.

It is scarcely necessary to mention that the entire subject of the finite universe is extremely speculative, and the primordial atom still more so.

CHAPTER XXXIV

THE IMPORTANCE OF SPACE-TIME, AND THE PRINCIPLE OF ACTION

In the course of this book we have seen how a number of difficulties were overcome as soon as we recognised the existence of four-dimensional space-time in lieu of the world of separate space and time of classical science. There are still other aspects of the problem, however, and we shall now mention them briefly.

In the first place, all vectors such as velocities and forces, and all tensors such as Maxwell's stresses of the electromagnetic field, erstwhile expressed in three-dimensional space, must now be extended and supplemented with additional components so as to yield vectors and tensors in four-dimensional space-time. These new components do not define new physical magnitudes unknown to classical science; in the general case they give us well-known magnitudes, such as work, power, and so on. As a case in point, let us consider electric and magnetic intensities. In classical science these were entirely distinct magnitudes. But when we represent these intensities in terms of space-time, we find that they are given by the various components of *one same* space-time tensor F_{1k}; the time components give the electric force, while the space components yield the magnetic force. The advantage of the new point of view is chiefly to show us that what we once considered different entities now reveal themselves as the different components of the same space-time existent. The reason for the observed connections between these different entities then becomes apparent.

In particular, let us examine the case of the equations of electrodynamics. In the study of electricity and magnetism we may consider phenomena in which conditions do not vary as time passes by; the electric charges and the magnets remain at rest, and currents flowing in fixed wires do not vary in intensity. Conditions are then termed stationary; it is as though time played no part. The laws which govern this type of phenomena were discovered empirically over a century ago, and were expressed mathematically in terms of spatial vectors. The problem of ascertaining how electric and magnetic phenomena would behave when conditions ceased to be stationary was one that could not be predicted; further experimental research work was necessary before the general laws could be obtained. Even so, the difficulties were considerable, and it needed Maxwell's genius to establish the laws from the incomplete array of experimental evidence then at hand. All this work extended over nearly a century; it was slow and laborious.

Yet, had men realised that our world was one of four-dimensional Minkowskian space-time, and not one of separate space and time, things would have been very different. By extending the well-known stationary laws to four-dimensional space-time, through the mere addition of time components to the various trios of space ones, we should have written out inadvertently the laws governing varying fields, or, in other words, we should have constructed Maxwell's celebrated equations. Electromagnetic induction, discovered experimentally by Faraday, the additional electrical term introduced tentatively by Maxwell, radio waves, everything in the electromagnetics of the field, could have been foreseen at one stroke of the pen. A century of painstaking effort would have been saved.*

It is also interesting to note that the four separate relations which constitute Maxwell's equations of electromagnetics are now merged into two; and the cumbersome aspect of the equations gives place to forms of great simplicity.

Still another simplification which the discovery of space-time has conferred upon us is to be seen when we study the principles of the conservation of energy and of momentum. We find that these two distinct principles constitute in reality but one. Conservation of energy is given by the time component, while conservation of momentum is given by the three space components, of one same space-time tensor law of conservation.

In a general way, all problems of dynamics or kinematics in three-dimensional space can be treated as problems of statics in four-dimensional space-time. All these advantages of space-time appeal very strongly to the theoretical investigator who deals with mathematical equations; but they all go to show how much simpler it is to understand the workings of the universe when we realise that the fundamental continuum of the world is space-time, and not separate space and time.

We must now mention another most important conception of classical science, namely, **Action**; and we shall see that this important mathematical entity finds its place in a perfectly natural way in the world of space-time. In order to understand the significance of Action, let us consider any mechanical system passing from an initial configuration P to a final configuration Q. Classical science defined the action $\mathcal{A}$ of this system as the difference between its total kinetic energy T and its total potential energy V, taken at every instant and then summed over the entire period of time during which the system passed from the initial state P to the final state Q.

Now the total kinetic and potential energies of the system at any instant are given by

$$\iiint T\,dx\,dy\,dz \quad \text{and} \quad \iiint V\,dx\,dy\,dz,$$

---

* We are assuming that a four-dimensional vector calculus would have been in existence; but this is a purely mathematical question.

where T and V represent the densities of the kinetic and potential energies at every point throughout the space occupied by the system. Accordingly, the expression of the action will be given by

$$\mathcal{A} = \int\int\int\int (T - V) \, dx \, dy \, dx \, dt \text{ or } \int\int\int\int L \, dx \, dy \, dz \, dt.$$

In the second formula we have merely replaced $(T - V)$ by a single letter L. Henceforth $(T - V)$ or L will be referred to as the *function of action*.*

Roughly speaking, action was thus in the nature of the product of a duration by an energy contained in a volume of space. On no account may this action be confused with the action dealt with in Newton's law of action and reaction, also expressible as the principle of the conservation of momentum. Still less may it be confused with the term "action" which appears in philosophical writings. The importance of the conception of action arises from the fact that the laws of mechanics can be expressed in a highly condensed form when the concept of action is introduced. Various forms may be given to the principle of Action; here we consider only the form which embodies the concept of Action as defined above; it is called **Hamilton's Principle of Stationary Action.** If we restrict our attention to the very simplest case, we may state Hamilton's principle as follows:

If we consider all the varied paths along which a conservative system may be guided, so that it will pass in a given time from a definite initial configuration P to a definite configuration Q, we shall find that the course the system actually follows, of its own accord, is always such that along it the action is a minimum (or a maximum).

Now the principle of action issues, as we have said, from the laws of classical mechanics. Were these laws different, the principle of action would cease to apply, at least in its classical form. *A priori*, we have no means of deciding whether the laws governing physical phenomena of a non-mechanical nature—those of electromagnetics, for example— would also issue in the same principle of action.

But when Maxwell had proved that his equations of electromagnetics could be thrown into a form compatible with the principle of action, and when he had succeeded in amalgamating electricity, magnetism and optics into one science, the universal validity of the principle was accepted. Inasmuch as this principle includes that of the conservation of energy, we can understand why the principle of action was often referred to as the supreme principle of physical science. Incidentally, we may mention that when the principle of action is satisfied by a phenomenon, an indefinite number of different mechanical interpretations of the phenomenon are theoretically possible. In the case of electrodynamic phenomena,

---

* Also called Lagrangian Function.

however, in view of the complicated hypotheses which he was compelled to postulate, Maxwell abandoned all attempts to discover the precise mechanical interpretation which would correspond to reality.

Thus far, we have limited ourselves to stating the general validity of the principle of action, as applying to all physical processes. It now remains to be seen how the principle will be of use to us in furthering our knowledge of the laws of nature. This we may understand as follows: The principle, as we have seen, imposes the condition that the natural evolution of any system must be such as to render the action a maximum or a minimum. Could we but express this condition in terms of the usual physical magnitudes, we should be enabled to map out in advance the series of intermediary states through which the phenomenon would pass. From this knowledge we should derive the expression of the laws which governed the evolution of the phenomenon. Here, of course, a twofold problem presents itself. First, we must succeed in finding the correct mathematical expression for the action; and, secondly, we must be in a position to solve the purely mathematical problem of determining under what conditions the action will be a maximum or a minimum.

Now all problems of maxima and minima are solved by means of the *calculus of variations,* a form of calculus we owe chiefly to Lagrange. According to the methods of this calculus, we establish under what conditions a magnitude is a maximum or a minimum by discovering under what conditions it will be *stationary.*

Let us explain what is meant by the word "stationary." When a stone is thrown into the air, it ascends with decreasing speed, then seems to hesitate for a brief period of time as it hovers near the point of maximum height before it starts to fall back again towards the earth. During this brief period of hesitation at the apex of its trajectory, the stone is said to remain "stationary." We can recognise a stationary state by observing that when it is reached no perceptible changes take place over a short period of time. In this way, we may understand the connection which exists between the stationary condition and the presence of a maximum or a minimum. In mathematics small variations are represented by the letter δ; hence the stationary condition of the action, or again, the principle of action, is expressed by

$$\delta \mathcal{A} = 0, \text{ i.e., } \delta \int \int \int \int L \, dx \, dy \, dz \, dt = 0.$$

The calculus of variations enables us to solve such purely mathematical problems. Under the circumstances the problem of determining the laws which govern any particular phenomenon reduces to discovering the expression of the function of action L pertaining to the phenomenon in question. Larmor applied this method to the phenomena of electricity and magnetism and showed how Maxwell's laws of electrodynamics could be deduced from a suitable mathematical expression L defining the electromagnetic function of action.

When the theory of relativity supplanted classical science, it was recognised that the classical equations of mechanics were only approximate, and it became necessary to reformulate the principle of action so as to render it compatible with the mechanics of relativity, hence also with space-time. This work was carried out by the pure mathematicians—by Klein and Hilbert in particular. It was then found that a principle of action differing but slightly from the classical one could be obtained. But inasmuch as the difference involved presents only a theoretical interest, we shall discuss the bearing of the principle of action on the theory of relativity without taking the slight aforementioned modification into consideration.

In classical science, it was strange to find that action, though so important in its physical significance, should yet present the artificial aspect of an energy in space multiplied by a duration. As soon, however, as we realise that the fundamental continuum of the universe is one of space-time and not one of separate space and time, the reason for the importance of the seemingly artificial combination of space with time in the expression for the action receives a very simple explanation. Henceforth, action is no longer energy in a volume of space multiplied by a duration; it is simply energy in a volume of the world, that is to say, in a volume of four-dimensional space-time. Designating a volume of space-time by $d\omega$, we have

$$d\omega = dx\,dy\,dz\,dt,$$

so that our principle of action, $\delta \mathcal{A} = 0$, becomes

$$\delta \int L\,d\omega = 0.$$

There is now perfect symmetry between the rôles of space and time.

Now there was still another aspect of the principle of action which was extremely displeasing in classical science: By making the present course of a phenomenon contingent on a condition which could be determined only at some future date, it acquired a teleological character. But, as Planck points out, with the action as given by an energy in a volume of space-time, we may, by placing space and time on the same footing, regard duration as static; and the unsatisfactory teleological aspect of the principle may be obviated.

The space-time theory offers yet further advantages. It suggests that the function of action L must be a *space-time* invariant. If, then, we know the elementary physical magnitudes which must enter into this invariant, purely formal mathematical considerations enable us to construct all possible invariants, and one of these will represent the function of action. Assuming the world to be built up of electricity and metrical field, the possibility of obtaining the "function of action" L of the universe would seem to be within our grasp. Could this result be accomplished, a momentous discovery would have been made, since all the laws of the physical world would be comprised in the expression

$$\delta \int L d\omega = 0.$$

The application of the principle of action, expressed in the preceding formula, would enable us to derive all the laws of the physical universe. We should understand many things that are still mysterious; the elusive reason for the existence of two types of electricity in nature would be revealed, and the strange tendency of matter and energy to congregate into atoms and quanta would be explained. Unfortunately we are far from being able to determine the "function of action" of the universe, for there may be physical elements entering into its constitution whose nature we as yet ignore. (Furthermore, we might mention that the very validity of the principle of action is called into question by the quantum phenomena.) The following passage quoted from Weyl gives an idea of what is at stake:

"Whereas, in mechanics, however, a definite function L of action corresponds to every given mechanical system and has to be deduced from the constitution of the system, we are here concerned with a single system, the world. This is where the real problem of matter takes its beginning: we have to determine the 'function of action,' the world-function L, belonging to the world. For the present it leaves us in perplexity. If we choose an arbitrary L, we get a 'possible' world governed by this function of action, which will be perfectly intelligible to us—more so than the actual world—provided that our mathematical analysis does not fail us. We are, of course, then concerned in discovering the only existing world, the **real** world for us. Judging from what we know of physical laws, we may expect the L which belongs to it to be distinguished by having simple mathematical properties. Physics, this time as a physics of fields, is again pursuing the object of reducing the totality of natural phenomena to **a single physical law**: it was believed that this goal was almost within reach once before when Newton's *Principia*, founded on the physics of mechanical point-masses, was celebrating its triumphs. But the treasures of knowledge are not like ripe fruits that may be plucked from a tree." *

Whatever may be the ultimate fate of such speculations in revealing to us the mystery of matter, it cannot be denied that they possess a grandeur which is compelling. At all events, in more restricted domains, the principle of action has led to important discoveries.

Hilbert, Klein, Weyl and later Einstein himself, by applying the principle of action to the space-time metrical field, were led to the law of

---

* This deification of the principle of action which is traceable to the influence of Hilbert and Weyl is resisted by Eddington and Silberstein, who point out that the principle has none but a formal significance.

space-time curvature in the interior of matter and outside matter; that is to say, to Einstein's law of gravitation.

The problem resolves itself into the selection of an appropriate expression for the action; but in the case of the law of gravitation a peculiar difficulty is encountered. We shall best understand the nature of this difficulty by considering the simpler case of the electromagnetic action. Electricity and magnetism present themselves in a substantial form as electrons, magnets and electrons in motion, or currents; and in a more ethereal form as fields, illustrated by the electromagnetic field. Inasmuch as both these manifestations of electricity and magnetism should enter into the laws of electrodynamics, the total electromagnetic action will have to comprise the two corresponding types of action. These can easily be determined, and an application of Hamilton's principle of stationary action yields the laws of electromagnetics without further ado. But when we consider matter, classical science knew of but one form, the substantial form; the field form was lacking. To be sure, the gravitational forces surrounding matter appeared to present a certain analogy with the electromagnetic fields surrounding electrons and magnets. But, mathematically, at least, the analogy was superficial, for in contradistinction to the electromagnetic field, the gravitational field was not expressed by differential equations indicating continuous action through a medium. Only in name was it a field; and it appeared impossible to determine its action. With Einstein's discoveries this difficulty was overcome. The gravitational field now appeared as a field analogous to the electromagnetic field. It was the metrical field of space-time, defined, as we know, by the g_{1k} potential distribution analogous to the potential distribution of the electromagnetic field. It can also be regarded as representative of the structure, or geometry, of space-time. Henceforth, the total material or gravitational action will be given by the action of matter proper plus that of the metrical field; the missing action has been found.

Thus the solution of the problem of gravitation reduces to establishing the mathematical expression of these two separate actions. Now the action of matter in a space-time region is the energy of matter contained in a volume of space-time. According to Einstein's theory, therefore, it is given by the density of the mass of the matter multiplied by a space-time volume. As for the action of the gravitational field, inasmuch as it is now seen to be the action of the metrical field, it can be discovered without much difficulty. In this connection, we must recall that the relativity theory suggests that the action in a volume of space-time must be an invariant; its numerical value, whatever it may be, must be independent of our choice of mesh-system. In other words, the action of the metrical field must be an invariant built up from the fundamental tensors g_{1k}, whose distribution defines the metrical field.

We mentioned, when discussing tensors, that several such invariants exist, but that the simplest of all was represented by the Gaussian curvature G. To this G we may add any scalar or number λ without modifying its invariance, since numbers are, of course, invariants. Hence, the simplest expression for the function of action of the metrical field is obtained by writing it: $G + \lambda$. This was, indeed, the expression selected by Einstein.

From now on, it is a mere question of mathematics (calculus of variations) to express the stationary condition of the total action. As a result, we obtain Einstein's law of space-time curvature, inside and outside of matter; that is to say, Einstein's law of gravitation. According to whether we assume that λ is zero or has a non-vanishing value, we obtain the infinite universe, or a finite universe.

Now there remains one other important point which we must mention before abandoning the discussion of the law of gravitation. We saw that the principle of action alone did not enable us to decide whether the constant λ, which was adjoined to the action G of the metrical field, was zero or had some non-vanishing value. In other words, it did not enable us to decide whether the universe was infinite or finite. But this is not the only source of indeterminateness. We mentioned that Einstein chose G for the function of action of the metrical field because it was the simplest possible invariant made up of g_{ik}'s; but we have also seen that other more complicated types of invariants could be constructed. Had these been substituted in the expression of the action, we should have obtained alternative laws of gravitation.

Finally, we may note that by applying the principle of action to the total action formed by the gravitational action and the field-action of electricity, we should obtain the precise type of space-time curvature which would exist in an electromagnetic field free of electrons and magnets; for instance, in an empty region flooded with light. It is then found (as was mentioned in Chapter XXXII) that the electromagnetic field also produces a curvature of space-time, less pronounced than that reigning in the interior of matter, but more pronounced than that existing in the empty space around matter (that is to say, in a gravitational field due to ordinary matter).

CHAPTER XXXV

THE MYSTERY OF MATTER

ONE of the greatest merits of the theory of relativity has been to allow us to represent gravitation as a direct consequence of the curvature of space-time. It would certainly be the crowning achievement of this superb theory if it could enable us to interpret all the manifestations of the physical universe in terms of the various types of curvature of one sole fundamental continuum, space-time. But at the present stage two separate reasons render this solution highly improbable.

We remember that in the course of our wanderings we came across two foreign tensors: T_{ik}, the matter-tensor, and E_{ik}, the electromagnetic tensor. It seems impossible to express these tensors in terms of the basic structural tensors of space-time, namely, the g_{ik}'s. However, in the case of the matter-tensor T_{ik}, the situation is not yet clear; and we shall see that opinion is divided on the subject. In order to understand the point at issue, let us turn once more to the gravitational equations in the interior of matter:

$$G_{ik} - \tfrac{1}{2} g_{ik} G = T_{ik}.$$

There are ten separate equations corresponding to the ten separate components of the second-order symmetrical tensors G_{ik} and T_{ik}. If we assume that a certain definite mesh-system has been chosen, or, again, that the observer is specified, each one of the ten components of T_{ik} will represent some physical magnitude relevant to matter at each point. As an example, the component T_{44} will represent the density of mass of the matter at a point; T_{14}, T_{24} and T_{34}, the three components of momentum of the matter at a point, measured in the frame of reference specified; T_{11}, T_{22} and T_{33} will represent the three components of *vis viva*, or energy of motion, coupled with the stresses in the matter at the point considered, etc.

The equations of gravitation thus signify that whenever we recognise the existence of one of these physical magnitudes it is always accompanied by corresponding curvatures of space-time. It is usual to assume that the curvatures are produced by those concrete somethings which we call mass, momentum, energy, pressure. In this way, we must concede a duality to nature; there would exist both matter and space-time, or, better still, matter and the metrical field of space-time. Einstein, when he elaborated his hypothesis of the cylindrical universe, attempted to remove this duality by proving that it was possible to attribute the entire existence of the metrical field, hence of space-time, to the presence of matter. This attitude led to a matter-moulding conception of the universe, elevating matter over the metrical field of space-time. And, as we recall, only when this attitude was adhered to could Mach's. belief in the relativity of all motion be accepted.

Eddington's attitude is just the reverse. He prefers to assume that the equations of gravitation are not equations in the ordinary sense of something being equal to something else. In his opinion they are identities. They merely tell us how our senses will recognise the existence of certain curvatures of space-time by interpreting them as matter, motion, and so on. In other words, there is no matter; there is nothing but a variable curvature of space-time. Matter, momentum, *vis viva*, are the names we give to these curvatures on account of the varying ways they affect our senses.

Up to this point the motive that has guided us has been the desire to discover unity in nature by reducing all things to some fundamental form of existence. There is, however, another way to approach the problem; and this time our guiding principle will be no longer solely mathematical harmony, but also experimental facts.

We must realise that the tensor T_{1k}, whose components represent those physical magnitudes which our senses perceive and our instruments measure, can at best be considered to portray effects observed only in a macroscopic way. Phenomena such as radioactive explosions and all we have learnt indirectly from the exploration of the atom go to show that matter in the final analysis is highly complex. It follows that a microscopic investigation of matter would yield a very different picture from that which we construct from our crude sense perceptions, and that could we but view the atom in an ultra-miscroscopic way it would appear to us as a region subjected to electromagnetic and other actions. It would seem, therefore, that our macroscopic tensor T_{ik} must comprise the electromagnetic one E_{ik} in its constitution.

At any rate, all attempts to reduce matter to electromagnetic fields have thus far presented insuperable difficulties owing to the impossibility of accounting for those ultimate aspects of matter, the electron and proton, in terms of E_{ik} alone. The difficulty consists in understanding how it is that the electron does not explode under the mutual repulsion of its negatively charged parts. Some kind of counterbalancing pressure, the Poincaré pressure, seems to be necessary, and it appears to be impossible to account for this mysterious pressure in terms of the electromagnetic Maxwellian stresses E_{ik} alone. Following Poincaré, Mie attacked the problem. His procedure was to introduce additional electromagnetic-field magnitudes which would be effective in the interior of the electron, but whose action in outside space would be negligible. By this means, he succeeded in building up matter out of purely electromagnetic quantities. Einstein comments on Mie's theory in the following words:

"In spite of the beauty of the formal structure of this theory, as erected by Mie, Hilbert and Weyl, its physical results have hitherto been unsatisfactory. On the one hand, the multiplicity of possibilities is discouraging, and, on the other hand, those additional terms have not as yet allowed themselves to be framed in such a simple form that the solution could be satisfactory."

Following this attempt, Einstein showed that it was possible to obviate the introduction of Mie's supplementary electromagnetic terms provided the space of the universe were assumed to possess a slight residual curvature, such as would exist in the cylindrical universe. It would then be this residual universal space-curvature which would act as a negative pressure preventing the electron from exploding. If we note that the universal curvature concerns the metrical field, hence the field of gravitation, we see that the energy of the electron must be composed of an electromagnetic and of a gravitational part. In Einstein's own words: "Of the energy constituting matter, three-quarters is to be ascribed to the electromagnetic field, and one-quarter to the gravitational field."

Later, Einstein postulated a universal pressure, which would fulfill the same role as the cosmological constant λ. We mention these investigations on account of their historic interest. In his subsequent papers, however, Einstein has abandoned his earlier attempts in favour of other avenues of exploration.

Let us now pass to a study of the electromagnetic tensor itself. We can see that even if the preceding attempts had proved successful, there would still remain a duality between gravitation and matter or electricity, or, we might say, between the metrical field of space-time and the tensor E_{ik}. This latter tensor would still appear as an irreducible foreign entity present in space-time, marring the unity of nature. It was the aim of Weyl's theory to rid science of this duality.

In order to understand the nature of the problem, we must state that just as the fundamental structural tensors of the metrical field B_{ikst}, G_{ik} and G at every point could be reduced to highly complicated expressions between the ten fundamental g_{ik}-potentials at every point, so now the electromagnetic tensor E_{ik} can be reduced to a complicated expression between the four electromagnetic potentials, the four ϕ_i's (namely ϕ_1, ϕ_2, ϕ_3 and ϕ_4). The unification demanded by science was therefore to express the g_{ik}'s in terms of the ϕ_i's, or vice versa. Unfortunately, not the slightest similarity appeared to exist betwen the ten g_{ik} potentials and the four ϕ_i's. The ϕ_i's appeared as foreign entities embedded in the metrical field of space-time.

All attempts to overcome this duality were unsuccessful so long as we confined our speculations to the space-time of Einstein's theory. But there always existed the possibility that space-time was of a less simple category than had been assumed by Einstein. If, then, we could conceive of a generalised type of space-time possessing a more complicated metrical field, it might be possible to identify the fundamental ϕ_i's of electromagnetics with certain particularities of structure of this generalised space-time and of its metrical field. Weyl's theory is an attempt in this direction.

CHAPTER XXXVI

THE THEORIES OF WEYL AND EDDINGTON

THE general theory of relativity reduces to an application of Riemann's purely mathematical discoveries, to the real world of space-time in which we live. A further extension of the theory became possible when Weyl had succeeded in carrying the geometrical study of manifolds beyond the point where Riemann and his successors had left it. Weyl describes the situation in the following words:

"Inspired by the weighty inferences of Einstein's theory to examine the mathematical foundations anew, the present writer made the discovery that Riemann's geometry goes only halfway towards attaining the ideal of a pure infinitesimal geometry. It still remains to eradicate the last element of geometry 'at a distance,' a remnant of its Euclidean past." *

In order to understand Weyl's theory, which may be regarded as a generalisation of Einstein's, it will be necessary to enquire into the nature of Weyl's geometrical contributions. Let us recall that when discussing the geometry of space, we said that the geometry we attributed to conceptual space was defined by the numerical results of measurement which were obtained when we had defined the type of conceptual measuring rods which we intended to employ. The behaviour of these measuring rods was arbitrary to a great extent, and according to the behaviour we attributed to them, the space was Euclidean, Riemannian or Lobatchewskian.

However, certain limitations were imposed by Riemann. Riemann assumed that two unit rods which coincided at a point A would continue to coincide when displaced to another point B, regardless of whether the two rods had followed the same or different paths through space during their displacement from A to B. Now it is possible to remove this restriction imposed by Riemann without falling into any logical inconsistency; and we may perfectly well assume that two unit rods which coincide at a point A may cease to coincide when brought together at another point B, provided they have followed different routes while moving from A to B. From a purely mathematical angle, such an assumption is by no means absurd; and it is this generalisation of the possible behaviour of our conceptual measuring rods which was studied by Weyl. It must be borne in mind, however, that for the present we are not discussing the problem of real space; we are not arguing about the actual behaviour of our rigid measuring rods as defined by practical con-

---

* "Space, Time and Matter."

gruence. We are merely speculating, as mathematicians, on a logical extension of geometry.

Let us therefore review our premises. We assume that two unit rods which coincide at a point A cease to coincide when brought together at another point B, after having followed different paths of transfer from A to B. Obviously, if this be the case, the statement that two lengths situated at different points A and B of space are equal in magnitude is necessarily indeterminate. The two lengths might measure out as equal if we displaced our unit rod from A to B along a certain path, and as unequal if we displaced our unit rod along some other path. In other words, length is non-transferable. In view of the indeterminateness that surrounds the comparison of lengths in different places, we must confine ourselves to the comparison of lengths at any one place or at points separated by infinitesimal intervals. We therefore agree to consider that at every point in space there is fixed a rod which is to serve as unit of length when we measure lengths situated by its side. The totality of these unit rods constitutes what is known as a *gauge-system*. These gauges may be selected arbitrarily; and it is not meant to imply that they are necessarily of equal length with respect to one another, since we have seen that comparison of lengths in different places is ambiguous. Their sole utility is to serve as standards of measurement for lengths situated by their sides, and thus to enable us to compare lengths situated at the same point.

A gauge-system is the natural extension of a mesh-system, with which we are already acquainted. Just as we may vary the shape of the mesh-system, so now we may vary the nature of the gauge-system.

Let us endeavour to ascertain the significance of this non-transference of length as relating to the nature of space itself. In particular, let us suppose that two rods coincide at a point A; then, while one rod remains fixed at A, the other is made to follow a closed curve, returning finally to its starting point. Owing to the difference in routes followed by the two rods, one having remained fixed and the other been made to describe a curve, the two rods will no longer coincide when we place them side by side after the completion of the circuit.

This non-transference of length is very similar to what is known as the non-transference of direction which holds for the non-Euclidean spaces that we have studied so far. Thus, consider a curved surface in three dimensional space. Draw a very small closed curve on the surface, and let A be a point of the curve. If at this point A two short rods applied to the surface coincide in direction and if, while one rod remains fixed, the other is displaced from A along the curve in such a way that for each successive infinitesimal displacement the rod maintains the same direction, it would be erroneous to suppose that when the second rod had been returned to its starting point its direction would still coincide with that of the first rod. A distinct discrepancy would be found to exist between the directions of the two rods, and the size of this discrepancy would depend on the area and location of the closed curve which the second rod had been made to follow over the surface. This discrepancy has nothing to do with our choice of a mesh-system; it transcends this choice completely, and so represents some absolute characteristic of the surface in the neighborhood of the small region considered. Calculation shows that for a given small closed curve, the discrepancy is proportional to the curvature of the surface at the center of the loop. If we had operated on a plane sheet of paper instead of on a curved surface, the discrepancy would not have arisen, since a plane surface has no curvature.. Thus the magnitude of the discrepancy, when we operate with one small loop or another, here or there on the surface, measures the intensity of the curvature, or non-Euclideanism, of the surface from point to point.

Just as non-transference of direction is caused by the curvature, or non-Euclideanism, of space (which we might also call *direction-curvature*), and disappears with the vanishing of this curvature, so in an analogous manner the non-transference of length is caused by a new type of curvature (called *Weylian curvature* or *distance-curvature*), and disappears completely when this type of curvature is non-existent. In a general way we may call a space a *Weyl space* when it possesses a non-vanishing distance-curvature; the appellation *classical spaces* will be reserved for all the categories of space which we have studied thus far (Euclidean, Riemannian, Lobatchewskian, semi-Euclidean), since in all these spaces a Weylian or distance curvature is non-existent.

We may note that just as ordinary non-Euclideanism, or direction-curvature, transcended our choice of mesh-system, so now, in similar fashion, the Weylian characteristics—the non-transference of length, and, therefore, the presence of distance-curvature—transcend our choice of a gauge-system.

Now we have seen that the Euclideanism or non-Euclideanism of a manifold was ascribed to the metrical field which it contained. In a similar way we must assume that the Weylian characteristics must also be ascribed to the metrical field. However, the metrical field due to the g_{ik}'s which has been discussed so far, is incapable of accounting for Weylian peculiarities. It is necessary to complete the description of the field in order to render it accountable for these new characteristics. When this is done, it is found that in a four-dimensional continuum, in addition to the ten g_{ik}'s which suffice to describe the metrical field of a four-dimensional classical continuum such as space-time, it is necessary to add four new quantities which we may call the four k_i's. The generalised metrical field then contains fourteen separate magnitudes in place of ten as in Einstein's space-time. Just as the direction-curvature of space characterising non-Euclideanism at a point could be expressed by a tensor (the Riemann-Christoffel tensor), built up from the changes in values between the g_{ik}'s at a point and at the neighbouring points in any mesh-system, so now the distance-curvature at a point can be given by a tensor built up from the changes in values between the k_i's at a point and at the neighbouring points in any gauge-system.

In particular, the Weylian curvature f_{1k}, at a point, is given by the tensor equation

$$f_{1k} = \frac{\partial k_1}{\partial x_k} - \frac{\partial k_k}{\partial x_1}$$

whence we may derive

$$\frac{\partial f_{k1}}{\partial x_1} + \frac{\partial f_{11}}{\partial x_k} + \frac{\partial f_{1k}}{\partial x_1} = 0.$$

If the space manifests no trace of distance-curvature, hence no Weylian characteristics, f_{1k} vanishes completely and the preceding equations are superfluous; we are then in the presence of a classical space, whether Euclidean, Riemannian or Lobatchewskian.

Now it is a remarkable fact that these purely geometrical relations or equations which Weyl had discovered, expressing as they do the value of the Weylian curvature, *are none other than Maxwell's equations of the electromagnetic field*, provided we identify the Weylian curvature f_{1k} with the tensor F_{1k} whose components represent electric and magnetic force, and provided we identify the elements of structure, the k_1's, with the electromagnetic potentials, the ϕ_1's. To be sure, this unexpected parallelism between the equations of electromagnetics and the curvatures of a Weyl space might have been attributed to mere chance. But, as must be admitted, the temptation to view matters otherwise was great. Weyl assumed, therefore, that space-time was of the Weylian type and not of the more restricted classical species, as Einstein had supposed. The electric and magnetic field tensor F_{1k}, which in Einstein's space-time appeared to be a foreign entity intruding on the continuum, was then in reality none other than the Weylian curvature of this generalised space-time. These views led Weyl to assume that when electromagnetic fields were present, hence when F_{1k} did not vanish, the Weylian characteristics of space-time manifested themselves, whereas, in regions free of electromagnetic fields, space-time resumed its classical characteristics and became once again the space-time of Einstein's theory. For similar reasons, the four electromagnetic potentials, *i.e.*, the four ϕ_1's were identified with the four k_1's of the Weylian structure. In this way, electromagnetic phenomena, to the same extent as gravitational phenomena, appeared to be necessary manifestations of the metrics of the fundamental space-time continuum.

Let us consider the further modifications which these new views will impose. First, we remember that in the classical spaces, hence in Einstein's space-time, it was possible to build up fundamental or structural tensors (B_{1kst}, G_{1k} and G) from the ten g_{1k}'s. These magnitudes, being tensors, transcended, of course, our choice of mesh-system. But now that we are considering a Weyl space, we must superimpose the additional requirement that our magnitudes should transcend a choice both of mesh-system and of *gauge-system*. We thus obtain the magnitudes $*B_{1kst}$, $*G_{1k}$, $*G^2\sqrt{g}$, F_{1k}, which are built up from the ten g_{1k}'s and the four k_1's (now identified with the four ϕ_1's) and which in addition satisfy all our previously mentioned requirements. In order to distinguish these magnitudes from ordinary tensors, which transcend solely our choice of mesh-system, Eddington calls them *in-tensors* and *in-invariants*, and indicates them by asterisks. Henceforth it will be to these in-tensors and in-invariants that we shall have to appeal in order to express those magnitudes which (depending neither on mesh-system nor on gauge) are intrinsic to the world itself.*

Now, assuming space-time to be a Weylian manifold, a change of gauge-system will modify the expression ·of the fundamental tensors of Einstein's theory (except F_{1k}), hence will modify the aspect of the mechanical laws and of that of gravitation. A natural extension compels us to assume that all the Einsteinian laws should be generalised and expressed in terms of in-tensors and in-invariants, so as to yield in-tensor equations instead of tensor equations. Only thus can we render them indifferent to a change of gauge-system, just as their original tensor form had already rendered them indifferent to a change of mesh-system. This modification in the form of the Einsteinian laws will affect only the mechanical and gravitational laws, since the electromagnetic ones defining the Weylian curvature are already in-tensor equations.

However, we may assume that the Einsteinian tensor laws, in the form in which they were given by Einstein, are in harmony with some particular gauge-system; and this Weyl calls the *natural gauge*. We shall have to conceive of this natural-gauge system as one imposed upon us by nature and defined by certain intrinsic properties of the universe. But the only determinate magnitude which is offered us by nature at every point appears to be the radius of curvature of the universe at every point. Hence Weyl defined the natural gauge at every point as given by a unit of length equal to the radius of curvature of the universe at that point or to some definite fraction thereof. If these ideas are correct, the universe cannot help but possess a curvature at every point, and hence must be finite, since were it infinite there would be no natural gauge whereby to measure distances.

Weyl's theory, as outlined so far, represents the theory in its original form; but we shall see that Weyl was led to modify his original views in certain respects. The fact is that empirical results would appear to conflict with his theory in its initial form. The theory compels us to assume

---

*We cannot insist on numerous niceties such as the distinction between tensors and tensor-densities, etc. We may note, however, that whereas the in-magnitudes $*B_{1kst}$, $*G_{1k}$, $*G^2\sqrt{g}$ differ from the tensors B_{1kst}, G_{1k}, and from the invariant density $G^2\sqrt{g}$, yet F_{1k}, the tensor of the electric and magnetic forces, is the same as in Einstein's theory. This is because F_{1k} in Einstein's theory already happened to be an in-tensor.

that in regions where electromagnetic fields are present, in regions, there-fore, where space-time is Weylian, transference of length is an indeter-minate process. Of course, when we speak of a length in space-time, we are primarily considering a space-time distance, that is, an Einsteinian interval. But an Einsteinian interval, we remember, can be split up into a spatial distance, or into a duration, or into both, as the case may be. This result is achieved when the observer refers the Einsteinian inter-val to his space and time mesh-system. It would follow, therefore, that in a Weylian space-time not only would the transference of length of an Einsteinian interval yield indeterminate results, but that the same would apply automatically to the transference of spatial, as also of temporal, magnitudes. Thus, two identical electrons situated, first, at any one point then separated, would in general differ in size and in charge when brought together again. Likewise, the same indeterminateness would apply to duration. Hence, it would be quite in order to suppose that two identical atoms situated at the same point and beating with the same frequency would present different frequencies of vibration after they had been separated and brought together again. It would be as though the size of an atom and its frequency of vibration depended on its previous life history. Incidentally, these occurrences would deprive the Einstein shift-effect of all value, since any slowing down in the rate of vibration of the solar atom could be attributed to a difference in the previous life histories of the solar and terrestrial atoms, and not necessarily to the influence of the sun's field of gravitation.

Now, all these anticipations of Weyl's theory in its original form are in conflict with precise observation. So far as can be detected by experi-ment, the charges and sizes of all electrons are exactly alike regardless of their past history; and this also applies to the frequencies of vibration of atoms of the same chemical element.* It would appear, therefore, that if space-time is a Weyl space, the Weylian characteristics are, for some reason or other, completely obliterated; so that, in fine, space-time, though Weylian, manifests only the properties of the more restricted Riemann type of space.

Weyl was thus compelled to draw a distinction between *adjustment* and *persistence,* and to assume that bodies adjusted themselves automatically to the natural gauge, that is to say, to the radius of the universe. If bodies followed the tendency of persistence, the Weylian discrepancies would be observed; but with bodies adjusting themselves afresh from place to place and from instant to instant to the form of the universe, and with the frequencies of vibration of atoms acting likewise, these dis-crepancies would be banished automatically.

These views also enable us to attain a better understanding of the sig-

---

* Eddington has shown that the discrepancies which might be expected to arise would in all probability be too small to be observed.

nificance of the natural gauge. Since objects in equilibrium, such as rigid objects, electrons and atoms, all adjust themselves to the radius of curvature of the universe, it is only natural that results of greatest simplicity should be obtained when we select this radius or some definite fraction thereof as a gauge at every point.

Now, with the restrictions we have imposed, it may appear that there is nothing left of Weyl's theory, since we have deprived the Weylian space-time of all its characteristics by invoking the tendency of adjustment in place of that of persistence. This view would, however, be too extreme. It is true that Weyl's identification of the electromagnetic potentials, the ϕ_i's, with the k_i's or particularities of structure of the continuum, now appears more in the light of a graphical representation; but as such it remains useful. Indeed, we must remember that the original urge which rendered a theory such as Weyl's desirable was the necessity of reaching a more unified understanding of nature by including at least the representation of electrical manifestations in the structure of the fundamental continuum of space-time, instead of having to regard them as foreign occurrences. From this standpoint, Weyl's theory appears secure.

But it is when we come to consider the problem of the *Action* of the universe, that Weyl's theory opens up the most interesting possibilities. We remember that the action would constitute, so to speak, the nucleus whence all the laws of nature would issue. In fact, were the action known, all we should have to do would be to express its stationary condition, and the laws of matter and electricity would be revealed. But the trouble is, we do not know what the action may be; we can only guess at it. In this respect Weyl's theory proves of help. For whereas Einstein's theory restricts the action to being a space-time invariant, Weyl's theory allows us to narrow down the choice by imposing the more stringent conditions of in-invariance. As a result, the Gaussian curvature G or $G\sqrt{g}$, which Einstein had assumed to constitute the function of action of the gravitational field, must be discarded and replaced by $*G^2\sqrt{g}$, an in-invariant.* We are thus led to the law of gravitation with the λ term, that is to say, to the finite universe.

Furthermore, the ideal of unity towards which science tends incessantly, demands that the action be built up from the structural elements of the fundamental continuum. In Einstein's theory this was impossible, since the electromagnetic potentials (*i.e.*, the ϕ_i's) were foreign to the space-time structure. It followed that we were faced with two separate actions, the gravitational action and the electromagnetic one. Weyl's theory permits us to obtain a unification of these actions, since the electromagnetic potentials, the ϕ_i's, are now represented by the k_i's in the generalised space-time structure. Accordingly, Weyl constructed an in-invariant which he suggested tentatively as the function of action of the world. In addition to the gravitational and electromagnetic equations

* See note, page 326

which were derived therefrom, the stationary condition of the action (see Chapter XXXIV entailed certain electromagnetic equations bearing on the constitution of matter. In the light of these equations, and taking into consideration the fact that matter is always of positive density, Eddington concludes: "It would seem to follow that the electron cannot be built up of elementary electrostatic charges, but resolves itself into something more akin to magnetic charges."

At this stage, a further generalisation of great elegance was given to Weyl's geometry by Eddington. Here let us recall that whereas in Einstein's theory the fundamental elements of space-time structure were ten in number (the ten g_{ik}'s), and whereas in Weyl's theory they were fourteen (the ten g_{ik}'s and the four k_i's), in Eddington's still more general manifold we are confronted with 40 quantities (the forty Γ^i_{kt}'s). It is not that Eddington has introduced new elements of structure into the manifold. Far from it: these elements were present in all the types of manifolds we have discussed. But they were crystallized by restrictions which removed them from the mathematician's control; and what Eddington has done has been to loosen them. For instance, starting from Eddington's generalised manifold, if we impose certain restrictions on the forty Γ^i_{kt}'s, we are led to Weyl's manifold; by imposing still others, we are led to the classical manifolds, of which Einstein's space-time was a particular instance. As a matter of fact, even in Eddington's manifold, certain restrictions of symmetry defining what is known as the *affine* condition are still imposed upon the Γ^i_{kt}'s; were it not for these restrictions, there would have been 64 of these Γ^i_{kt}'s. So far as is known to the writer, the highly generalised manifolds which would result have not been studied; but it is interesting to note that they can be conceived of, and no one can foresee whether their introduction may not prove necessary eventually. Since Eddington's geometry is a generalisation of Weyl's, we may anticipate certain novel features in it. Thus, in Weyl's geometry, we have seen that the length of a rod would vary were it to be displaced from one point to another. With Eddington's generalisation, we must assume that a change of length would also arise if the rod were rotated round one of its points, hence if its orientation were modified, even though the rod were not displaced as a whole.

And here a few words may be said with respect to the purpose of these successive generalisations. The ultimate goal is one of unification. We wish to succeed in representing the entire physical universe in terms of the relationships of structure of the fundamental space-time continuum. Einstein's space-time enabled us to account for gravitation in this way, but the electromagnetic field appeared as a foreign invasion. None of the known elements of structure seemed capable of accounting for it. It seemed as though the world of space-time, gravitating masses and energy might just as well have existed in the absence of this electrical intrusion. Now, we have every reason to believe that such cannot be the case, for since matter is built up of protons and neutrons, the annihilation of electricity would entail that of matter, gravitation and energy, and possibly also of space and time. In order to overcome this difficulty, it appeared necessary to generalise our conception of space-time, obtaining thereby a continuum presenting additional elements of structure which might be identified with electric and magnetic forces. And so we were led to Weyl's generalisation, which permits the electromagnetic field to enter into the general synthesis, no longer as a foreign adjunct, but as a constituent element of space-time structure. But, besides the field, there is the electron, and the field alone does not seem to afford us the possibility of building up the electron; we cannot account for atomicity or for the quanta of action. Therein resides the problem of matter, and it remains to-day an outstanding challenge to science. Neither can we understand the significance of the two different kinds of electricity, the positive and negative. So long as all these mysteries remain unsolved, it is only natural that we should seek new elements of structure by proceeding with further generalisations. Eddington's theory was conceived of with this object in view.

There exist certain interesting features about Eddington's theory which we may mention briefly. Inasmuch as the intrinsic magnitudes of the universe must be in-tensors and in-invariants, instead of tensors and in-variants, the g_{ik}'s lose their position of pre-eminence; for the g_{ik}'s, while transcending our mesh-system, fail to transcend our choice of a gauge-system. Accordingly, in the expression of a distance, we must replace g_{ik} by some in-tensor of a similar nature. The only fundamental structural in-tensor that presents itself is $*G_{ik}$, or some definite fraction of it such as $\dfrac{*G_{ik}}{\lambda}$ where λ is an arbitrary constant. Hence, we must put

$$ds^2 = \frac{1}{\lambda} \sum *G_{ik}\,dx_i\,dx_k$$

in place of

$$ds^2 = \sum g_{ik}\,dx_i\,dx_k.$$

The identification of the new expression with that employed by Einstein leads to the tensor equation $*G_{ik} = \lambda g_{ik}$, which, when no electromagnetic fields are present, reduces to $G_{ik} = \lambda g_{ik}$. This is the equation of a finite universe, so we see that the finiteness of the universe appears to constitute a necessary consequence of the theory.

A further point is also worthy of note. Gravitation, which involves the g_{ik}'s, is given by a symmetrical tensor; electricity, on the other hand, is given by an anti-symmetrical tensor F_{ik}. Now, in the theories of both Weyl and Eddington, the in-tensor $*G_{ik}$ takes the place of Einstein's tensor G_{ik}. But whereas G_{ik} is symmetrical, $*G_{ik}$ splits up into a symmetrical and an anti-symmetrical part; these two parts are given by g_{ik} and F_{ik}. In other words, the world tensor splits up of its own accord into gravitation and electricity, showing thereby the deep union between the two.

As late as 1922, Einstein did not show much enthusiasm for these new generalisations, but subsequently he appears to have modified his views, going so far as to investigate the problem of the world-action in the light of Eddington's geometry.

He first studied a possible mathematical expression for the action, one which had already been suggested by Eddington. It was a strikingly simple expression representing certain of the generalised space-time curvatures; no longer a mere stringing together of the various actions. When the principle of stationary action was applied to it, it fell apart, yielding the law of gravitation on one side and the laws of electricity on the other.

The mathematical results obtained by Einstein in this respect point to a general diffusion of electricity throughout the world; and as in the case of Weyl's treatment, magnetic and not electric charges appear to lie at the basis of matter. Subsequently, Einstein proved that other expressions for the action would yield the same results.

Yet, in spite of all these remarkable successes, the outlook is none too promising. The problem of matter is no nearer a solution than before. Indeed, it is highly probable that some totally new departure will have to be considered. Here it must be remarked that the attempts we have been outlining follow the methods of field physics; that is to say, the entire physical universe is assumed to be built up of the g_{ik}'s and ϕ_1's or other magnitudes of the same kind. But a number of facts and phenomena render the success of this attitude doubtful.

Field physics, together with the principle of action, encounters further difficulties when we consider the quantum phenomena and the problem of the atom. In a number of cases, therefore, there appears to be something deeper behind the field itself. If so, we should have to limit the rôle of the field to a purely passive part, to that of transmitting effects, not of engendering them. If, however, we abandon all attempts at interpreting matter in terms of the field quantities, this does not mean that we must relapse into the discarded idea of substance. In the present state of our knowledge all these problems are so confused that it seems useless to state anything definite about them. All we can say is that we are faced with a something which we do not understand. Weyl expresses these ideas in the following words:

"If Mie's views are correct, we could recognise the field as objective reality, and physics would no longer be far from the goal of giving so complete a grasp of the nature of the physical world, of matter and of natural forces, that logical necessity would extract from this insight the unique laws that underlie the occurrence of physical events. For the present, however, we must reject these bold hopes. The laws of the metrical field deal less with reality itself than with the shadowlike extended medium that serves as a link between material things, and with the formal constitution of this medium that gives it the power of transmitting effects."

And elsewhere: "We must here state in unmistakable language that physics at its present stage can in no wise be regarded as lending support to the belief that there is a causality of physical nature which is founded on rigorously exact laws."

All in all, we must realise that the farther we go, the more inextricable our difficulties become. The existence of matter still remains a mystery.

PART IV

THE METHODOLOGY OF SCIENCE

CHAPTER XXXVII

THE METHODOLOGY OF SCIENCE

AN opinion commonly expressed by philosophers is that the function of physicists should be to weigh, measure, tabulate, discover new chemical elements, new facts; and the aim of mathematicians, to solve equations. That so long as the scientist restricts himself to investigations of this sort, he is proceeding along strictly scientific lines and his efforts should be encouraged, but that when he proceeds to discuss the implications of his discoveries and to draw conclusions on the subject of space, time, energy, laws of nature, he is getting beyond his depth, and encroaching on the field of the metaphysician.

Bergson expresses this attitude very clearly in his book, "Duration and Simultaneity." He informs us that the reason mathematicians have been so deceived by the significance of the theory of relativity, to the point of wishing to substitute space-time for separate space and time, lies in their lack of philosophical insight. They err by wishing to attribute a metaphysical significance to a concept which (according to Bergson) represents a mere mathematical fiction.

To these and kindred accusations, scientists (when they retort at all) will answer, with Kelvin: "Mathematics is the only true metaphysics." Clifford, in his popular essays, voices the scientific attitude with increased emphasis when he writes: "The name philosopher, which meant originally 'lover of wisdom,' has come in some strange way to mean a man who thinks it his business to explain everything in a certain number of large books. It will be found, I think, that in proportion to his colossal ignorance is the perfection and symmetry of the system which he sets up; because it is so much easier to put an empty room tidy than a full one." These opinions prove that there exists a definite misunderstanding between scientists and philosophers; a misunderstanding which might easily have been avoided had philosophers possessed a proper realisation of their inevitable limitations when discussing scientific matters. The simplest way to approach the source of the trouble appears to be to analyse the methods of scientists and ascertain in what respect they differ from those of philosophers.

The development of all the branches of physical science has proceeded roughly along the same lines. First we witness an accumulation of experimental and observational facts, furnished by crude observation and by the discoveries of the laboratory workers. Then, thanks to the efforts of the theoretical investigators, this raw material is co-ordinated into a consistent whole. In this way, out of a disconnected series of facts, a coherent doctrine or science is born. History proves that with very few exceptions (illustrated by such men as Newton, Archimedes and also Hertz), the most brilliant experimenters had but little theorising ability; and that, vice versa, the ablest theoretical scientists made poor laboratory men. We are thus led to differentiate between two distinct types of scientists: the practical and the theoretical workers. For example, in physics, Maxwell, Planck and Einstein must be placed under the heading of theoretical investigators, whereas Hertz and Michelson are splendid illustrations of able experimenters.

In modern science, at least in physical science, these works of co-ordination can be attempted only through the medium of advanced mathematical analysis; hence the theoretical physicists must necessarily possess a profound knowledge of mathematics. Yet they are not, properly speaking, great mathematicians. Mathematics, for the theoretical physicists, constitutes but a tool, a means of arriving at a co-ordination of the physical facts. These men have never entered into the study of mathematics as an art in itself; they have never forged new mathematical instruments, never made mathematical discoveries. Thus, they cannot be classed with mathematicians of the calibre of Lagrange, Gauss, Riemann or Poincaré, to whom modern mathematics owes its existence. Just as we were compelled to make a distinction between experimenters and theoretical physicists, so once again we must make a distinction between theoretical physicists and pure mathematicians. Again, with very few exceptions, as exemplified by Fourier, Poisson, Poincaré, Minkowski and Weyl, pure mathematicians have rarely contributed directly to the advancement of theoretical physics, although indirectly, of course, their discoveries have been made use of by the physicists. Since they are neither great mathematicians nor able experimenters, what are we to call such men as Maxwell, Lorentz and Einstein?

If we concede that the name philosopher should apply to those who are concerned more especially with a harmonisation of the whole than with the seeking of individual facts, or, again, with a general view of things rather than with a restricted view, we must agree that the theoretical physicists must be called philosophers. They are, then, the philosophers of the inorganic world, just as the pure mathematicians might be called the philosophers of abstract relations.

Now, as we have said, the facts which these scientific philosophers are seeking to co-ordinate are of a restricted species; they are mathematical, physical and chemical in nature; hence it is clear that there is room for a more general type of philosopher—a super-philosopher, as it were—whose facts would comprise all the spheres of human knowledge, including consciousness, emotions and the relationships between mind and matter. The traditional philosophers—or shall we say the lay philosophers, since we are discussing scientific matters?—aspire to be placed in this category of thinkers.

It would appear, then, that theoretical scientists, and lay philosophers have much in common; they differ only in the scope of the facts they are seeking to co-ordinate. But here is where the first breach arises. The theoretical scientist proceeds with the utmost caution and considers himself at liberty to theorise only after a sufficient number of facts have been established by experiment and observation; till then he remains silent. This cautious attitude is evidenced even in the most revolutionary theories, such as those of relativity and of the quanta. It was not one, nor two, nor even three of the negative experiments in electromagnetics that drove Einstein towards his revolutionary theory; it was the whole body of electrodynamics. Even so, he formulated his solution only after a number of more classical attempts to solve the same difficulties had failed.

But when we examine the procedure of the lay philosopher who discusses scientific matters, we see that his procedure is entirely different. Taking examples at random, we find a contemporary philosopher telling us that the spatio-temporal system has to be regarded as limited, since "all existent things, as distinguished from pure forms or orders, are finite." * But the average scientist would retort that with the same degree of plausibility we might say that the spatio-temporal system must be infinite, since the concept of infinity exists. At all events, it is instructive to contrast these loose arguments with Einstein's painstaking labours which led him to the finite universe.

Another example is afforded by Bergson's comments on the invariant velocity of light. He explains that inasmuch as the corpuscular theory of light has been rejected, light is a propagation and not a transference of matter. It is, then, only natural (according to Bergson) that its speed should be invariant. In his own words: "Why should it be affected by a certain too human way of perceiving and conceiving of things?" Bergson apparently forgets that it is only light *in vacuo* that moves with an invariant speed, and, even so, only when far from matter. A ray of light passing through water is also a propagation, and yet its speed is not invariant, as Fizeau's experiment proved seventy years ago. Also, we might with equal plausibility apply Bergson's arguments to sound waves. They also constitute propagations, and yet, once again, their speed is not invariant. In short, it would appear that the philosopher, even when he interprets the facts correctly, appears to be ignorant of so large a number of other facts that his philosophical conclusions rest on no solid foundation.

So much for the facts; but the facts are not all, since the theoretical scientist's chief aim is to co-ordinate these facts. But, here again, the philosopher's criticisms would suggest that, being unable to follow the series of mathematical deductions which lead the theoretical scientist to his conclusions, he fails to understand the necessity for these conclusions, and accordingly considers them to be wild guesses over which he will invariably suggest his own as an improvement.

Possibly a few definite illustrations will bring out with greater clarity some of the points we are endeavouring to stress. Let us revert, for example, to space-time, which proves itself so offensive to Bergson. His position, as already stated, is that space-time is a pure mathematical

* J. S. Mackenzie.

fiction in no wise demanded by the theory of relativity, which he accepts as sound physically, provided its interpretation is abandoned to lay philosophers. His arguments proceed somewhat as follows: You physicists have performed certain electromagnetic experiments; you mathematicians have deduced therefrom the Lorentz-Einstein transformations on purely formal grounds. Then you have plastered these transformations together and discovered a certain mathematical invariant. So far, well and good; now you must stop. It is for us philosophers to interpret these results. We will begin by ruling out your faulty inferences as to the existence of a four-dimensional continuum, space-time, of which space and time are mere abstractions. Follow our arguments carefully and we shall convince you. Possibly you mathematicians have not noticed that every observer is perfectly aware of a separate space and time wherever he goes, just as he is aware of a separate space and temperature. Were space-time one sole continuum, as you maintain, we could never be aware of space and of time, but only of space-time. This is contrary to the facts. Hence, space-time is a hoax, and your mathematics has led you astray.—Q. E. D.

But it is obvious that arguments of this sort can never have any merit, for the reason that they are based on a total misconception of the significance of dimensionality and manifolds. These are highly technical concepts which it requires more than a crude elementary knowledge of mathematics to grasp. However, even without going into details, we can easily convince ourselves of the fallacy of Bergson's argument by applying it to the three-dimensional space of classical science. Thus, we know that wherever we are situated in three-dimensional space, we can always split it up into height, length and breadth; yet this does not preclude the fact that length, breadth and height are but abstractions from one single entity, namely, three-dimensional space. And, in the same way, as has been demonstrated by Minkowski, length, breadth, height and duration are but abstractions from four-dimensional space-time.

Another philosopher, Professor Broad, in his book, "Scientific Thought," informs us: "(a) No matter what frame we choose, we shall need *four* independent pieces of information to place and date any instantaneous point-event. This fact is expressed by saying that Nature is a four-dimensional manifold; and nothing further is expressed thereby. (b) In whatever frame we choose we shall find that our four pieces of information divide into two groups; three of them are spatial and one is temporal. Thus we must be careful not to talk, or listen to, nonsense about 'Time being a fourth dimension of Space.' "

Now it is agreed that whoever speaks of time as a fourth dimension of space is expressing himself very loosely. What we should say is: "Time is a fourth dimension of the space-time world." But Broad's argument does not suggest that it is this looseness of phraseology that offends him. His words would imply that it is the reference to time *as a fourth dimension* that must be branded as "nonsense." His error in this respect is thus exactly the same as Bergson's. As for his contention that the four-dimensional structure which relativity ascribes to nature means no more than that events need four co-ordinates to be placed and dated, the statement is scarcely correct. Were this 'trivial piece of information all that Minkowski was conveying when he referred to the world as four-dimensional, his discovery would have excited neither admiration nor criticism. It would have passed unnoticed by scientists, as expressing a mere platitude that could have been no news even to a child. The point that philosophers so persistently fail to understand is the difference between an *amorphous continuum,* such as a manifold of sounds or colours, and a *metrical continuum,* i.e., one with which a definite geometry is associated. What Minkowski did was to prove that, contrary to the belief of classical science, the world was a four-dimensional "metrical" continuum, *i.e.,* one with which a four-dimensional space and time geometry was associated. It was this novel aspect of the world that implied the revolutionary conception of the fusion of space and time. And it was this aspect that entitled Minkowski to speak of time as a fourth dimension in a profound sense as well as in the trivial sense which, though never disputed by classical science, was never stressed on account of its artificial nature. Indeed, had it not been for this interpretation placed upon his words by Minkowski himself and by all scientists, the four-dimensional world would have appeared to be as artificial as a four-dimensional space-temperature continuum. It is granted that the various meanings attributed to the word "dimension" by mathematicians may cause some trouble to beginners, but, on the other hand, unless the four-dimensional aspect of nature in the relativistic sense is understood, it is quite useless to philosophise on the more advanced aspects of the theory. Before we try to run, we should at least learn to stand on our feet.

To take another illustration at random, it is the same when Professor Broad discusses the law of gravitation. Notwithstanding his condemnation of the conception of space-time, as one sole four-dimensional continuum, he sees no inconsistency in discussing the geometry of this four-dimensional continuum. At any rate, after telling the reader that a space may be flat or curved like an egg or like a sphere, he decides (p. 224) that the law $G_{1k} = 0$ denotes a *spherical* curvature. Then, on a later page (p. 485), we are informed that the curvature is *homaloidal.* This taxes the imagination of the reader, for as "homaloidal" means flat, a space cannot be both homaloidal and spherical at one and the same time. However, it is to be presumed that Broad meant "homogeneous," not "homaloidal," and we shall interchange the two words accordingly. But even when amended in this way, his premises are totally incorrect, since the law $G_{1k} = 0$, which is to account for gravitation, is neither spherical nor homogeneous nor homaloidal. This law represents a heterogeneous species of curvature; indeed, were the curvature spherical, it could never account for gravitation.

Starting from his mistaken premises, our philosopher soon falls into further confusion. After telling us (p. 224) that the supposedly spherical law of curvature, $G_{1k} = 0$, is able to account for gravitation, and for all the gravitational effects predicted by Einstein, he next informs us (p. 485) that this law cannot be the law of gravitation in the real world because the space round the sun is never completely empty, but is filled with tenuous cosmic matter, electromagnetic fields, and light waves. This casual presence of matter and electromagnetic fields would (always according to Broad) modify the law of spherical curvature, $G_{1k} = 0$, causing it to become heterogeneous. But all these statements are utterly erroneous, since, as we have said, even in an ideally empty space-time, the law of curvature, $G_{1k} = 0$, round the sun would still be heterogeneous, quite apart from the minute superadded effects that might be occasioned by a casual matter distribution. In other words, the heterogeneity is essential, not accidental.

We now come to the Professor's philosophical conclusions. He notes that this passage from a space-time which is spherical and homogeneous in a gravitational field to one which is heterogeneous is brought about by the casual presence of matter round the sun. Einstein never stresses this point which Broad has brought to light, whence Broad concludes that this important transition from homogeneity to heterogeneity has been "slurred over." As a philosopher he deprecates this loose treatment, pointing out that a totally new conception of space is involved, one that should be submitted to careful philosophical scrutiny.

Of course, this charge is totally unjustified and issues solely from our philosopher's faulty understanding of the facts. The heterogeneity was never introduced in the casual way Broad believes. It was introduced by the front door the moment Einstein wrote out his gravitational equations $G_{1k} = 0$, as any one familiar with the theory of spaces would have recognised immediately. Far from having been slurred over, this heterogeneity of space-time constitutes the very essence of Einstein's general theory. Whereas in classical science the sun developed a field of force around itself in a perfectly flat space, in the general theory of relativity the sun develops a *heterogeneous* space-time curvature. The field of force is then but a manifestation of this curvature. As for the new conception that this heterogeneity forces upon our understanding of the world, it has been subjected to critical enquiry by Weyl and Einstein. Furthermore, it is not so new a conception as Broad appears to believe, for it is one we owe to Riemann following his epochal discoveries in non-Euclidean geometry seventy years ago. As Weyl tells us:

"Riemann rejects the opinion that had prevailed up to his own time, namely, that the metrical structure of space is fixed and inherently independent of the physical phenomena for which it serves as a background, and that the real content takes possession of it as of residential flats. *He asserts, on the contrary, that space in itself is nothing more than a three-dimensional manifold devoid of all form; it acquires a definite form only through the advent of the material content filling it and determining its metric relations.*"

And again, elsewhere:

"It is upon this idea, which it was quite impossible for Riemann in his day to carry through, that Einstein in our own time, independently of Riemann, has raised the imposing edifice of his general theory of relativity." (We may mention that the reason Einstein was able to carry through Riemann's ideas is because he applied them to space-time instead of to space alone.)

Of course, it is granted that Riemann's discoveries and non-Euclidean geometry are not of easy access; yet, on the other hand, the man who ignores at least the implications of non-Euclidean geometry, is in no position to discuss Einstein's theory or the problem of space even from a purely philosophical point of view. Indeed, it may be said that the philosophical importance of non-Euclidean geometry is even greater than its scientific importance. Such has always been the contention of Lobatchewski, Riemann and other great mathematicians.

So far as the writer is aware, the only philosopher who made any reference to non-Euclidean geometry, prior to Einstein's discoveries, was Lotze; and Lotze expressed the hope that philosophy would never allow itself to be imposed upon by it. But a perusal of Lotze's writings on the subject proves that he had a very superficial understanding of what was meant by "curvature." His opinions seem to have been based on Helmholtz's more or less successful attempts to popularise the new doctrine by easy illustrations. Unfortunately, the subject is too deep to be explained in any loose way. At any rate, the effect of this negative attitude on the part of philosophers has reacted to their disadvantage in that it has deprived them of a very powerful insight into the problem of space.

Our sole purpose in mentioning these few examples (which we might have multiplied *ad infinitum*) has been to show how wary we should be of criticising the conclusions of scientists before proceeding to acquire more than a superficial schoolboy knowledge of the physical and mathematical facts which they are endeavouring to co-ordinate. If, as a result of misinterpretation or ignorance on our part, we are acquainted with only a small number of these facts, the conclusions of scientists may well appear strange and unwarranted; but it should be remembered that had not all these additional facts, which we ignore, been known to the scientist, it is quite certain that he never would have been driven to the conclusions that offend our natural views so deeply.

Unfortunately, in many cases, these facts are not of an elementary character; they cannot be explained in an hour or so. More often than not, an appreciation of what they represent would require years of preliminary study, for they are often in the nature of conclusions derived from other facts through the medium of laborious mathematical analysis. These statements, of course, must not be construed as applying solely to theories of mathematical physics. They apply with equal force to the discussion of numerous concepts such as those of infinity, continuity and atomism, dimensionality, number, measurement, rigidity, etc. Discoveries in higher mathematics and in physics have thrown a new light on all these subjects. Students of advanced mathematics know only too well how crude was their understanding of continuity and infinity before they were apprised of the mathematical discoveries of Riemann, Weierstrass, Cantor, Dedekind and du Bois Reymond. Furthermore, no one can have studied advanced mathematics without realising its intimate connection with problems of psychology, for mathematics has brought to light mind-forms of which we were only dimly conscious.

The thorough remodelling of our ideas of the universe which the discoveries of Planck and Einstein appear to be rendering inevitable, makes an understanding of these fundamental points imperative. For example, let us consider the quantum theory. As we shall see, its necessity arises only when we take into consideration a number of empirical discoveries. pertaining to the various realms of physical science. Here, for instance, is a heated enclosure. We make a pinhole aperture in its wall and examine the colour and intensity of the light rays streaming out. The experimenter notes that as the temperature increases, the colour of the light rays passes gradually from red to white. He then studies the density of energy of the radiation emitted, and finds it proportional to the fourth power of the absolute temperature of the enclosure, regardless of the material of the enclosure. This empirical discovery is known as *Stefan's law*. Then, by splitting up the light through a prism, he finds that the radiation which exhibits the maximum intensity is of a frequency which is proportional to the absolute temperature. This discovery is comprised in *Wien's displacement law*. These are the facts of the case; the physicist has accomplished his task and may now retire. What is the significance of these facts? Were science to limit itself to the bare discovery and cataloguing of facts, there would be nothing more to do. But regardless of what philosophers may say to the contrary, science should not and does not limit itself to any such humble rôle. The discoveries of the experimenters are handed over to the theoretical investigators, and it is these who are called upon to interpret their significance.

In the present case, the outstanding names connected with the problem of radiation are those of Boltzmann, Wien, Rayleigh, Jeans and Planck, all of them theoretical physicists. Their object was to co-ordinate the facts of radiation with what little was known of the phenomenon of light emission. This task involved mathematical calculations which we should be unjustified in criticising unless we possessed an extensive knowledge of mathematics; hence we may assume that this part of the work is not disputed by the critic.

As a result of these calculations, it was found to be quite impossible to reconcile the existence of the facts disclosed by experiment with any continuous emission theory of light. Whenever it was assumed that the atoms in the heated enclosure emitted their radiations continuously, it always followed that, for any given temperature, the maximum intensity of the radiations should be found in the infinitely high frequencies, and not in the visible spectrum; whence it followed that a heated enclosure could never emit *visible* light. This was, of course, contrary to common experience. The literature on the subject extends over a number of years. Various explanations of the discrepancy between theory and experience were suggested, but for one reason or another none was found satisfactory.

Then Planck noticed that if, contrary to all previous opinion, we assumed that the atoms in the heated enclosure emitted their radiations by discrete *quanta*, it was possible to obtain a mathematical formula of radiation in perfect agreement with experiment. According to Planck's calculations, it was necessary to assume that light was emitted in bundles,' or quanta, possessing an energy $h\nu$. In his expression, h represents a universal constant called Planck's constant, and ν represents the frequency of the light. Obviously, the greater the frequency of the light, the greater the value of its quantum of energy. These discontinuities in the light-energy emission necessitated the introduction of probability and entropy considerations into the theoretical treatment. Inasmuch as entropy is an abstruse concept drawn from thermodynamics, we see that the facts entering indirectly into the problem of radiation were considerably increased in number. Obviously, it would be quite impossible either to justify or to criticise Planck's hypothesis by any general line of talk, for without the aid of the mathematical instrument, and a knowledge of thermodynamics and mechanics, no conceivable connection could be seen to exist between a discontinuous emission and the facts disclosed by experiment; hence the necessity of the hypothesis could not be gauged.

Thus far, however, nothing very revolutionary appeared to be involved. All we had to do was to assume that the emission of light by a heated atom was due to the breaking up of some intra-atomic structure. Nevertheless, on deeper investigation, even assuming Planck's emission theory to be correct, a number of theoretical difficulties were noted. These are of so technical a nature that we shall not dwell on them. Suffice it to say that they relate in a general way to the exchange of energy between the various atoms and to the conditions of equilibrium. The names of Poincaré, Lorentz and Einstein are encountered at this stage, and the final result was that somewhere, somehow, discontinuities had to be introduced.

The next great advance was due to Einstein. He noticed that the well-known anomalies in the specific heats of bodies at very low temperatures could be explained, provided we assumed the discontinuity of energy. As applied to solids, this would connote that the vibrating systems could take up or emit energy only in definite quanta; and, as applied to gases, we should have to assume that the molecules could rotate only with definite frequencies. All these problems dealing with specific heats bring us into contact with the kinetic theory of gases, as also with other realms of physics, such as the problems of anomalous dispersion and selective reflection in optics. Again basing his deductions on the existence of quantum phenomena, Einstein deduced a formula for the photo-electric effect, and this formula was found to be in harmony with experiment. Finally, we may mention that Bohr's atom, which is also based on the same idea of discontinuity, allowed him to account for certain spectral series.

We now get to the important point. From the expression for the atom of energy or quantum $h\nu$, where h is a constant and ν is the frequency of the radiation, it is obvious that there exist as many different types of quanta of energy as there exist different frequencies of radiation. There is no one unique type of quantum of energy in nature. That which is universal is not the quantum of energy $h\nu$, but the constant h. It can be shown that Planck's constant h is not a mere number; it represents some definite abstract mathematical entity, and that entity is *action*.

We must assume, therefore, that there exist atoms of action in nature, just as there exist atoms of matter. But what is the deeper significance of this atom of action? First of all, has it any deep significance, or is its discovery on a par with that of some new variety of flower, or some new mineral? But to decide on a question of this sort we must possess a fairly thorough understanding of what is meant by *action*, as also of the part this important entity plays in science. Now, action is a highly abstruse concept taken from analytical dynamics. It would be absurd, therefore, to criticise the conclusions of scientists, whatever these might be, unless we were in a position to discuss the numerous facts which were involved in their arguments.

The verdict of scientists is that the atomicity of action entails a gigantic revolution in our understanding of nature; and it would appear that our insight into the laws of nature, even in spite of the great advance due to relativity, was still far too crude. We would have observed merely macroscopic average effects; the deeper laws, the underlying microscopic ones, would have escaped us completely. Conclusions of this sort had, of course, already been arrived at in other ways, as, for example, in the kinetic theory of gases; but the atomicity of action extends these views still farther. It suggests that change is always discontinuous; that a system passes from one state to another, not in a continuous way, but by a series of jerks or jumps. When we wish to decide how the jumps will follow one another, no exact laws can be formulated; and we are compelled to appeal to statistical considerations and probabilities. No rigid deterministic scheme is apparent in nature, or, in Weyl's words, "no causality of physical nature which is founded on rigorously exact laws." In the realm of the microscopic, we appear to be confronted with total chaos and anarchy. The past does not entail the present, as it would in a purely deterministic scheme. Free will appears to be rampant; and our sole means of prevision is to establish averages, just as a life-insurance company does when it fixes its premiums. Statistics and probability, blind chance and uncertainty, replace rigid determinism. As Weyl expresses it: "Above all, the ominous clouds of those phenomena that we are with varying success seeking to explain by means of the quantum of action, are throwing their shadows over the sphere of physical knowledge, threatening no one knows what new revolution."

Confining ourselves to the positive accomplishments of the quantum theory, we find that it has established the high stability of certain energy states, as opposed to the instability of others. De Broglie and Schrödinger initiated a new mathematical method for investigating quantum phenomena and were thus led to predict wave properties for matter. Heisenberg and Dirac, by following a different method, arrived at similar conclusions. Either method can be used to obtain Heisenberg's famous "Uncertainty Relations," which symbolize the breakdown of strict causality and the absence of rigid natural laws. The ubiquitous atom of action has been found to inject itself in the kinetic theory of gases, compelling thereby a thorough rehauling of Boltzmann's statistical theory. The quantum theory has led to the discovery of the electron spin, and this spin in turn has shed a new light on the significance of chemical valency; it has also furnished an interpretation of the binding together of neutral atoms into molecules (e.g. $H + H = H_2$) a phenomenon which classical theory was unable to explain. The quantum theory has also accounted for the peculiarities of crystal structure, as evidenced in the stability of the spacings and arrangements of the atoms in crystals; and there is every reason to suspect that the atomicity of matter is itself a quantum manifestation, though what the exact connection may be remains a mystery.

The quantum theory suggests that space and time are approximate concepts, which may have to be abandoned when the infinitely small is contemplated. In much the same way, the concept of temperature as due to an agitation of molecules loses its meaning when the molecules themselves are considered.

What the future may hold in store is any one's guess at the present time. But one thing is certain: we are faced with a gigantic revolution; and the new ideas will undoubtedly conflict with the common-sense instinct which the rationalist often erroneously attributes to "reason." But what if they do? Did not the existence of men walking upside down at the antipodes conflict with the crude common sense of our ancestors? Or, again, consider the less trivial illustration of the wave theory of light. When Fresnel defended it, Poisson pointed out that it would imply the existence of a bright shadow behind a body of a certain size situated at a certain distance from a wall. On the strength of this argument Poisson dissented from Fresnel's views. Yet when the experiment was actually performed, the bright shadow was seen to be there. And why did Poisson regard his argument as valid? Because it was based on common sense; but we see that this common-sense instinct that was misleading him was by no means the product of reason; it was founded merely on partial and crude experience.* It is probable that the more we study of nature, the more the common sense of our day will be submitted to disagreeable jolts; and this is only natural, since the more refined our investigations, the farther we shall be wandering from the familiar world of common experience.

* In Poisson's case the argument had some weight, for there was reason to suppose that were Fresnel's views justified, we should at some time or other have observed bright shadows in the course of our daily experience.

To revert to the new ideas that suggest themselves as a result of the quantum theory, we see that starting from facts which by common consent it is the function of the physicist to establish, we are gradually led through a series of theoretical deductions to considerations which border on metaphysics. But whereas the metaphysician will claim *a priori* knowledge or else, whether he admits it or not, will be deducing his knowledge from the crudest facts of daily experience, the theoretical scientist will base his deductions on extremely accurate observations embracing all the phenomena known to science.

For this reason, these deductions will constitute knowledge; for knowledge springs from an *accurate* investigation of a *large number* of facts, not excluding those revealed by ultra-refined experiment. More precisely, a knowledge of nature consists in the co-ordination of all known facts, and this co-ordination must be performed so as to account for phenomena not merely in some vague qualitative way, but with the utmost accuracy. Only after this preliminary synthesis has been accomplished can general philosophical conclusions be forthcoming. Until then all we can do is guess, and past experience proves that ninety-nine times out of a hundred our guesses will be wrong.

In short, it is in the accuracy with which the facts are studied and co-ordinated, and it is in the number of facts considered, that the scientific method differs from all other methods. Therein resides its superiority. Had the great scientists confined themselves to a general line of talk based on a crude survey of a number of conspicuous facts, we might still to-day be defending Plato's idea that earth, air, fire and water constituted the four elements, or Aristotle's contention that the heavenly bodies described circles because circular motion was the most noble of motions, or, again, Kant's belief that the axioms of geometry were in the nature of *a priori* synthetic judgments, and so forth.

Thus it may be realised that a discussion of the philosophical significance of the discoveries of physical and mathematical science must be left to the theoretical physicists and to the mathematicians. They alone, in view of their wide knowledge of facts and their mastery of the rigorous mathematical mode of thinking, are in a position to co-ordinate the apparently disconnected results furnished by experience and by reason. If, then, a super-philosophy is to be attained, it would appear that the most successful results would ensue from a work of collaboration between the scientists of the various branches of knowledge. Such collaborations are continually in progress. We need only mention the contributions of physicists like Einstein and Nernst to problems of physical chemistry; and physical chemistry is closely allied to the chemistry of colloids, hence to that of the living organisms. All these spheres of science, from mathematics to psychology, dovetail into one another, as terms like biochemistry, psychophysics and physical chemistry indicate.

It might be urged that the results embodied in the special sciences do not embody all of nature; that there is mind and consciousness, that there are emotions, values in quality, religious instincts, and so on. But here it should be remembered that whatever transcends the sphere of the special sciences transcends it precisely because it is vague and only dimly apprehended. And where facts are so vague and poorly established as to be refractory to the scientific method, we generally find that there are so many different ways of co-ordinating them loosely that almost any opinion can be expressed with the same degree of plausibility. As a result, the opinion of the wise man is on a par with that of the ignoramus.

If we wish to advance at all, it would seem that the best method would be to start by considering the facts we know something positive about, rather than reverse the procedure. The results embodied in the special sciences are so numerous and (relatively speaking) so well established, that we cannot afford to consider seriously any system of philosophy which would enter into conflict with them, or ignore them.

At any rate, granted that, in the present state of our knowledge, no co-ordination of facts can be anything but extremely fragmentary, it may be presumed that the theoretical scientists who have shown such marked ability in their co-ordinations of natural facts are the men best fitted to construct a philosophy of nature. We need not be surprised, therefore, to find that their study of nature has yielded them a philosophy which, with very few exceptions, they all share in common.

In the present chapter, we shall discuss the methodology together with the philosophic views that are accepted by the vast majority of theoretical physicists. In particular, the philosophy expressed will be exemplified by the attitude of the two leading scientists of our generation: Poincaré as a mathematician, and Einstein as a physicist.

The problem of knowledge, though generally considered independently of science proper, is yet so involved with scientific considerations (some of which are by no means elementary) that it appears necessary to approach it by the usual methods of scientific investigation. By knowledge we do not wish to imply our bare awareness of sensations, such as the worm and oyster probably share in common with human beings. Sensations constitute one of the original sources of knowledge, but knowledge entails considerably more; judgment and inference enter into knowledge, and knowledge is a construct or co-ordination in which our sensations enter merely as fundamental elements. For instance, when as a result of our visual and tactual sensations we recognise that a table is there before us, we are claiming knowledge; but the visual and tactual impressions by themselves do not constitute knowledge.

The natural uncritical view would be to assume that knowledge was given directly by the disclosures of perception. Thus, when we see the table, we know it to exist "there-now" in the spatio-temporal background. But if we analyse the situation, we cannot subscribe to this view, for our recognition of the table's existence and position in space and time appears to have been brought about in a far more complicated manner than a cursory examination might lead us to suppose. What we really experience is an aggregate of colours with a well-defined contour.

Only when we wish to ascribe a cause to these sensations does the idea of a table enter our consciousness. But here again, in this search for a cause, it is only in view of past experience, in view of an association of ideas, that we feel justified in postulating the existence of the table situated there-now in the spatio-temporal background. Our judgment may be correct, but it might also be erroneous, as it would be had we been gazing at a skilful painting or at the image of a table in the mirror. Quite independently of such possible errors of judgment, the main point is that there is a decided difference between our awareness of the visual sensations and our belief in the presence of a table located there-now in space. Whereas the visual sensations do not entail as an inevitable corollary the existence of the table, yet our belief in its existence does presuppose our awareness of certain sense impressions or our memory thereof, since otherwise there would be no reason for us to postulate the table's existence. For this reason we must credit to our awareness of sensations primacy over the so-called immediate disclosures of perception, just as we must concede to the recognition of experimental facts priority over the scientific theories constructed with a view to bringing about their co-ordination. Accordingly we shall refer to our awareness of sensations as facts.

A further illustration may make this point clearer. For example, we feel hot, and exclaim: "What a warm day!" Our awareness of a sensation of hotness constitutes a fundamental fact, but our claim that the day is warm is but an inference. We may test the warmth of the atmosphere with a thermometer, hence we may test the validity of our inference, but we cannot test our awareness of warmth. The feeling of warmth may be due to the temperature of the air, to fever or to sheer imagination, but, regardless of its cause, we know that we feel warm, and our assurance of this fact can neither be disputed nor analysed.*

Then again, we experience very similar sensations when our skin comes into contact with a substance at high temperature or with liquid air. If from this similarity in our sensations we should infer the identity of the temperatures involved, we should be led into error; but it would be our inferences that were at fault, and not the sensations, which subsist regardless of whether or not we choose to draw inferences therefrom.

It is scarcely necessary to mention that those philosophers who deny the legitimacy of the views just expressed, claiming our knowledge of

---

* It is scarcely necessary to add that our awareness of sensations does not presuppose any knowledge of space or of our human body as an object situated in space. When, for instance, an infant who is beginning to emerge into consciousness feels a pain in its leg, then one in its arm, it is not supposed that it succeeds in localising the two sensations in the two respective limbs. Only much later will it succeed in localising its sensations. For the present we are assuming that the infant knows nothing of space or of its body; it is merely registering sensations, and the pain in its leg will appear to it to differ in some obscure qualitative way from the pain in its arm.

external reality to be a matter of direct apprehension, find themselves in a very embarrassing position when such commonplace phenomena as mirror-images are considered. It is all very well for them to say that the results of visual perception may have to be corrected subsequently by varying the conditions of observation and taking into consideration all other perceptions. But the fact is that if this visual perception is to be regarded as revealing reality directly, *i.e.*, as fundamental, its disclosures should be beyond dispute and it would be meaningless to correct them. For we can correct a datum only in virtue of other criteria, which are then automatically regarded as more fundamental. In the present case the philosopher's argument would indicate that the result of a co-ordination of perceptions in general, is to be taken as more fundamental than bare visual perception or bare auditory perception, etc. But inasmuch as the disclosures of no one of the individual perceptions that enter into the co-ordination can be accepted without a possible danger of correction, we must go back farther, behind the individual perceptions, to the sensations whose recognition is untainted by any trace of inference. We then arrive at our original conclusion, namely, that knowledge springs from a co-ordination not of perceptions, but of sensations.

Let us consider another example. The neo-realist would probably state that when we say, "I hear a bell," no element of inference need enter into this knowledge. But a contention of this sort would of course be untenable, for the only reason the auditory sensation yields us any knowledge of the bell is because, in view of an association of ideas issuing from past experience, we have come to connect a certain class of auditory sensations with the presence of a bell. A man who had seen bells, but had never heard them clang would not attribute the sound to the bell any more than a man who had never encountered rattlesnakes or been told of their existence would ever say: "I hear a rattlesnake." On the other hand, a man who had previously come into contact with these reptiles would be just as apt to say, "I hear a rattlesnake," as the philosopher to say: "I hear a bell." Obviously, past experience, and not direct recognition, is at the root of this knowledge. In the absence of past experience, all we could say would be, "I sense a sound," and then try to describe the sound by reproducing it, as nearly as possible, with our tongue and lips.

With visual perceptions our conclusions will be exactly the same, but at first sight it might appear that the problem was slightly different, for even had we never seen a snake before, we should still be able to say, "I see a creature with a long, thin body," and then proceed to coin some name for it. But it should be remembered that although we might never have seen a snake before, yet a snake is a particular instance of an object, and unless we had always been blind we should have seen objects ever since infancy. If, then, we wish to obtain a parallel to the case of the rattlesnake (as we applied it to auditory perceptions), we must start with a blind man who has felt objects but never seen any.

Then, assuming he were suddenly endowed with eyesight, the question is whether he would identify immediately, by a mere act of intuition, the colours and shapes he would now see, with the table he had previously explored with his hands. This is precisely what we have every reason to doubt. We may now proceed to a more detailed discussion of the knowledge-problem.

Here it should be clearly understood that we are merely seeking to establish a hierarchy of knowledge following the psychological order according to which human knowledge appears to have arisen. We may therefore consider the faculty of unreasoned sense-awareness as furnishing the initial data which it is necessary to accept before we are in a position to speculate any further. Together with our awareness of sensations we have to consider our awareness of thoughts or ideas. Whether, in the absence of sensations, thoughts would still have originated, is a question which does not concern us. At all events, an answer to it seems impossible, for we cannot study the thoughts of a living corpse.

Next we have to consider our awareness of the passage of time. This awareness appears to be closely allied with the faculty of memory. Indeed, it is only thanks to the memory of sensations or ideas that we can differentiate the present from the past. It is probable, therefore, that to a being devoid of all trace of memory, to a being sensing but the present, past and future would convey no meaning and the passage of time would be unthinkable. Again, whether, in the absence of our awareness of sensations or ideas, memory would have anything left to remember, is a question which need not detain us. The reason for this omission is, of course, that in the present survey we are considering normal human beings; and with these an awareness of sensations, ideas, and of the flowing of time is known to exist.

Our recognition of the simultaneity of two sensations is fundamental, in that it cannot be analysed further. The same fundamental nature must be credited to our recognition of a succession of sensations or ideas. For this reason a sense of simultaneity and succession (when referring to our *awareness* of sensations or ideas, and *not,* of course, to a simultaneity of spatially separated events) must serve as a basis for subsequent knowledge.

We have also to mention our judgments of constancy, invariancy or sameness. Thus, we recognise two sensations or two groups of sensations as identical, or, again, we judge them as subsisting unmodified through time. Although all these seemingly *a priori* basic sources of information are of a widely different nature, we may consider them as a whole in our future discussions.

Now, had men confined themselves to registering sense impressions and to sensing the flow of time, there could never have arisen such a thing as scientific knowledge or even crude commonplace knowledge. Somehow or other the Initial facts had to be co-ordinated and interconnected. Only thus could a coherent form of knowledge leading to prevision become possible. It is to this synthesis of our sense impressions, supplemented by the fundamental forms of recognition we have mentioned, that we owe our belief in an external universe of space, matter, force and colour. In this universe, time was assumed to be flowing, events were regarded as exhibiting causal relationships; and our sense impressions were attributed to common causes existing in a public objective universe.

It is probable that a belief in causal connections arose from our ability to produce and to arrest certain sense impressions at pleasure through the conscious action of our will. Thus, we "will" a certain effort (which turns out to correspond to the closing of our eyes), and the luminous impression ceases. It was then but a short step to extend our belief in causal connections to sequences of events in the outside world, even though in this case the human will was known to play no part. The same with force. Though suggested originally by our awareness of effort, it was soon exteriorised and credited with an existence in the outside world; and a similar process of exteriorisation would appear to have been responsible for our belief in physical time (no longer the "I" stream of our consciousness), in a time regarded as enduring in the outside universe regardless of our presence.

It does not seem to be of great interest to question whether these exteriorisations of the dehumanised concepts of causal connection, force and time were justified or not. In any case it cannot be denied that the exteriorisation of causal connections, for instance, has proved of inestimable service in allowing us to account for the routine of our experience. Therein resides its justification, since causality appears to be an indispensable condition for the sequence of phenomena to be intelligible, hence for prevision to be possible. But it would seem scarcely correct to state that the understanding imposes causality on an indifferent world; for, as we know, there are regions of science where, although causal connections may be suspected, none have yet been established.

Finally, we are led to enquire into the nature of this synthesis or coordination of facts which seems to be a prerequisite condition for knowledge to be possible. It is here that an analysis of the more sophisticated syntheses of science will be helpful in enabling us to understand how the mind proceeds. The reason for this is because scientific knowledge can be traced back to crude commonplace knowledge by a series of insensible gradations, the same incentives appearing to be active in all cases.

Now, when we consider the procedure of the scientist, we find that it consists in co-ordinating and linking together in a rational manner a number of experimental facts, *with the maximum of simplicity*. By "a rational way" we mean primarily "according to the rules of logic." Irrespective of whether these rules are assumed to have been derived from experience or to reduce to *a priori* judgments, all normal men appeal to them, and all scientific theories, even the most revolutionary ones, are based on their acceptance. As for the criterion of simplicity, which enables us to select one co-ordination rather than another, it appears to be linked with our valuing of the expenditure of effort. Thus, even a dog finds it simpler to enter a house by the front door rather than clamber in through the back window. At any rate, inasmuch as, with small variations, all human beings agree unanimously on those co-ordinations which are to be regarded as the simplest, we may assume the urge towards simplicity to be fundamental. There are, of course, a number of other factors which enter into the construction of scientific syntheses, but for our present purpose the ones mentioned will suffice.

It will be noticed that in our list of fundamental facts, though our awareness of the passage of time was considered to be unanalysable, no mention was made of our awareness of space and of objects, including our own human bodies, situated in space. The concept of space was assumed to have been generated in an indirect way when we sought to co-ordinate our sensory impressions with maximum simplicity. It would thus be in the nature of a mental construct arrived at *a posteriori*. By this we do not wish to imply that the concept of continuity and of extension may not have pre-existed in the mind in a latent form; all we wish to assert is that the necessity for appealing to the concept of space seems to have arisen from the totality of our sensory experience.

A co-ordination of our sensations will yield us at the same time empty space and what we will interpret as filled space representing material objects, including our own human bodies, which are then recognised as being situated in space. We cannot attempt to justify these views here, but those who are interested in the subject will find the ideas expounded in Poincaré's masterly discussion of the problem in his book, "The Value of Science," and more especially in "Last Thoughts."

When it comes to differentiating and locating the portions of filled space occupied by bodies as against the empty portions, we again proceed in the same way, by co-ordinating sense impressions. Our belief in the existence of a table, for instance, is attributed to the fact that a simple co-ordination of the complex of our tactual, visual and muscular sensory impressions is possible only when we concede the existence of the table as a concrete reality, situated in the space before us. It is true that in common practice we may see a table and recognise it as such without touching it. But it is to be presumed that at earlier stages of our life we have exercised all our sensory faculties and have already recognised by a synthetic process the existence of space and of material objects. The table is then immediately recognised as existing, by a mere association of past impressions, without our having to explore it tactually.

Up to this point we have restricted our attention to space as mere extension. But space, as understood in common practice, implies considerably more: it represents a three-dimensional Euclidean continuum. When thus particularised, Kant's arguments as to its *a priori* character are no longer tenable in the light of modern discovery; and we must assume that this special form we credit to space arises entirely from our co-ordination of sense impressions conducted in the simplest way possible. On no account may we consider three-dimensional Euclidean space to be imposed *a priori* either by sensibility or by the understanding.

These discussions on the empirical origin of space are not mere philosophic fancies having no bearing on science. They are in many respects vital; and it is generally conceded by scientists that the *a priori* doctrine of three-dimensional Euclidean space is one of the most pernicious teachings that philosophy has ever attempted to impose upon science. Similar arguments hold for "time," when by "time" we are referring, not to our awareness of the time-stream in our consciousness, not to the "I" time, but to physical time or duration throughout the universe, which our consciousness has exteriorised and projected into space. As Einstein remarks in a passage previously quoted:

"I am convinced that the philosophers have had a harmful effect upon the progress of scientific thinking in removing certain fundamental concepts from the domain of empiricism, where they are under our control, to the intangible heights of the *a priori*. This is particularly true of our concepts of time and space, which physicists have been obliged by the facts to bring down from the Olympus of the *a priori* in order to adjust them and put them in a serviceable condition."

We may mention that these views on space as professed by the greatest scientists are in large measure to be attributed to the discoveries of non-Euclidean geometry supplemented by the investigations of the psychophysicists. Still, in view of the difficulty of imagining hyperspaces and non-Euclidean spaces, the views presented might appear difficult to accept, and it might be held that three-dimensional Euclidean space imposes itself *a priori* regardless of experiment. But it should be noted that by the time men are of an age to philosophise, they have been subjected for so many years to beliefs based on inferences from experience, that the beliefs have remained, whereas the inferences, owing to the monotony of their repetition, have become second nature and appear intuitive.

And yet a moment of reflection should suffice to convince us that were three-dimensional Euclidean space an *a priori* condition of the understanding, it would have been quite impossible for mathematicians to wend their way through the non-Euclidean hyperspaces of relativity. Neither can three-dimensional space be considered to be imposed by sensibility, since, as Poincaré tells us, after a certain amount of perseverance, he was aided to a considerable degree by sensibility when investigating the problems of Analysis Situs of four dimensions.

This empirical origin of the spatial concept is stressed by Einstein in the following lines:

"We now come to our concepts and judgments of space. It is essential here, also, to pay strict attention to the relation of experience to our concepts. It seems to me that Poincaré clearly recognised the truth in the account he gave in his book, 'La Science et l'Hypothèse.' Among all the changes which we can perceive in a rigid body, those are marked by their simplicity which can be made reversibly by an arbitrary motion of the body; Poincaré calls these, changes in position. By means of simple changes in position we can bring two bodies into contact. The theorems of congruence, fundamental in geometry, have to do with the laws that govern such changes in position."

When we realise that it is precisely these laws governing changes in position which govern our choice of a space among all those which the mathematician has to offer, we see how utterly dependent we are on experience when the problem of space is considered.

We may present these problems in a more vivid form. Suppose all we had ever seen of the world were given by its image in a reflecting spherical surface, such as a large door knob. The world of our visual perception would be very different from the one in which we normally live; the shapes of objects would squirm in a variety of ways as we displaced them before the curved mirror. And yet, however different our world might appear from the one of common observation, we should eventually succeed in co-ordinating our perceptions. We should still conceive of an outside space, but this space would no longer be Euclidean. If, then, all of a sudden the mirror were to be removed, and we to behold the world as other men perceive it, we should be completely at sea, accustomed as we were to the laws of our non-Euclidean world. In fact, the situation would be very similar to that of the man who tries to ride a bicycle through a crowded thoroughfare while crossing his arms over the handle-bars. He probably would come to grief; and yet had he always ridden his bicycle in this peculiar way, he would find it just as hard to alter his habits and ride it in the normal way.

Summarising, we may say that a belief in an outside universe of space, matter and change is arrived at as a result of a synthesis of sense impressions. These conclusions, which apply to commonplace knowledge, will be substantiated further when we consider illustrations taken from the more advanced fields of knowledge of the scientist. As mentioned previously, it is impossible to draw a line and say: "Here scientific knowledge begins and commonplace knowledge ends." And since the methods of the scientist are easier to dissect, a study of the scientist's procedure cannot help but shed light on the more obscure problem of the genesis of commonplace knowledge.

In scientific syntheses we do not restrict ourselves to co-ordinating mere sense impressions; we must also co-ordinate scientific facts. But scientific facts are themselves the results of previous co-ordinations of other scientific facts, and these in turn are traceable to a co-ordination of sense impressions. A few examples taken at random from science will make these points clearer. Why, for instance, does the astronomer maintain that the sun is spherical?

It is, as we know, in order to account for the continued circular aspect of the solar disk, for the passage of sunspots, for their flattened appearance when nearing the sun's edges, suggesting that they are seen in perspective, for the protuberance of its equator, for the Doppler effect exhibited on its equatorial rim, for the brilliancy of the planets when illuminated edgewise. It is also in order to render compatible the sun's shape with its fluidic nature imposed by its high temperature. In other words, the aim of the scientist is to frame one single hypothesis which will permit him to co-ordinate this wide variety of facts. We may note that all the facts that the astronomer is seeking to co-ordinate presuppose a knowledge of space and of material objects situated in space. But of course we are assuming that by the time men began to worry about the shape of the sun, they had advanced beyond the primitive stage of recognising the existence of objects in space. Now, when we decide that the sun is spherical, our first argument is based on its circular aspect. In order to account for this, we appeal to probability, arguing that it is very improbable that the sun should always turn the same face towards us. Of course the argument in itself does not carry much weight, since it is refuted in the case of the moon. Still, it serves as a suggestion, if nothing more.

Next consider the case of the planets. Were the sun a flat disk, it would appear strange that their brilliancy should remain appreciably the same regardless of their positions relatively to the sun. But this argument, be it noted, is highly sophisticated, for the natural view would be to assume that all bright points in the heavens shine of their own accord; and there would be no reason to differentiate between planets which reflected the solar light, and the stars which were in no wise dependent on the sun's presence. It was only at a later stage that a differentiation between stars and planets became necessary. In short, the facts the astronomer is seeking to co-ordinate are of a highly sophisticated nature; it is only when we dissect them further and further, analysing the previous syntheses of science, that we are finally thrown back on our immediate awareness of sense impressions. We see, then, that science appears as an unending series of syntheses of other syntheses, but that in every case the synthetic method is the same.

Let us now pass to a less simple example, namely: "We know that molecules exist." In the case of the existence of the table or the chair, all we needed was to co-ordinate certain immediate sense impressions. In the case of the shape of the sun, the procedure was more complicated, since we could not explore its surface with our hands and it was only by inference that we were led to believe we were viewing a spherical object from various positions in the course of a day or a year. But with molecules it is far worse, for no one has even seen or felt them. The inferences which we are led to make are based on others, these others again on others. The possibility of our co-ordination being proved incompatible with future discovery is therefore increased, and for this reason again our knowledge of molecules loses much of its certainty. Apart from questions of degree, however, this knowledge was arrived at in precisely the same way, by conceiving the simplest rational synthesis capable of co-ordinating a wide variety of facts of observation and experience.

It is probable that at a very remote stage in human history men noticed the difference in texture which existed between sand, which was grainy, and water, which appeared continuous and smooth. It would have been natural for them to wonder whether water would not turn out to be grainy if viewed microscopically. Some guessed one way, others another. There existed, however, a number of elementary facts of observation which had to be taken into consideration. For instance, a phenomenon on which the Greek thinkers laid due stress was the ability of wine and water to intermingle. The simplest manner of accounting for this was to assume that water and wine were formed of discrete particles which would exchange positions, much as two powders, one black and one white, would yield to a uniform grey mixture, when shaken together.

But Democritus went farther. Democritus appears to have been a man of exceptional scientific ability (as science went in those days); the geometrical solution of the volume of the pyramid and cone are attributed to him, and Pliny mentions that he spent his life among experiments. At any rate, he appears to have been the first thinker of antiquity (indeed, one of the very few) to display the scientific spirit, that of seeking unity in the various manifestations of nature by reducing quality to quantity. In this respect he initiated what was to become the necessary method of scientific investigation. Accordingly he suggested that all the elementary particles of matter were of the same substance. The qualitative differences which bodies reveal would then be due to differences in the shapes and sizes of their constituent elements or atoms. In this way unity was conceivable; but for this unity to endure, it was imperative that these elementary atoms should themselves constitute imperishable units. They could not be microcosms whose internal parts might suffer changes of position; hence they would have to be indivisible plena. As for cohesion, it was attributed to the atoms hooking on to one another.

As a scientific aspiration, Democritus' scheme was perfect, but the trouble was that the facts known to him were too few in number. And so his theory was a crude guess at best; and it was only natural that a wider survey of facts should render it untenable. The co-ordination of facts known to modern science has proved, indeed, that atomism, as understood by Democritus, *is* untenable: for whereas beyond the atom of the Greeks there was no mystery, nothing further to look for, the atoms of matter are now known to be divisible. They differ qualitatively from each other, contain heterogeneities, are subject to change and decay. In other words, the atoms of modern science are new microcosms of baffling complexity, so that the appellation "atom" (meaning *indivisible* in Greek), retained by custom, is no longer appropriate. When we get beyond the atoms to the electrons and protons and other particles more recently discovered,* we are in no position to assert that these constitute

---

* Neutrons, positrons, mesons, and possibly neutrinos.

imperishable units. Indeed, today we know that when an electron and a positron combine, they cancel each other and give rise to radiation: matter in the ordinary sense is thus annihilated. Even if we adhere to the view that with these electrons, protons, and other particles the ultimate atoms have been reached at last, we know so little about them that we cannot even be certain that they possess a definite size or shape. And so we see that atomism, as upheld by Democritus, is far from having been established by modern science. In the present state of our knowledge all we could do would be to guess, just as Democritus did in his day; but in view of the paucity of facts there are to guide us, no interest could be attached to our guesses.

Nevertheless, if by atomism we mean merely the tendency of matter and electricity to congregate into entities of great stability, we are on safe ground and we may consider the doctrine proved. When understood in this more restricted sense, a wide variety of phenomena drive us to the atomic theory. In addition to the mixing of liquids, mentioned by the Greeks, we have to consider the diffusion of gases and of solutions, the compressibility of gases and the phenomenon of osmosis. All these phenomena appear to demand the existence of molecules or atoms. As an illustration let us consider the phenomenon of the compressibility of gases, studied by Boyle in the seventeenth century. We know that when a gas is compressed its volume is decreased. Yet its mass or weight remains unchanged. We cannot assume, therefore, that matter has vanished through compression; hence the simplest alternative is to suppose that the gas is made up of atoms floating or moving in the void. Compression will then result in crowding these atoms into smaller spaces while leaving their total number unchanged. Similar arguments may be advanced in the case of liquids. Thus, when we mix one pint of alcohol and one pint of water, we do not obtain two pints of the mixture, but appreciably less. The simplest way to account for this partial disappearance of volume is once again to assume that our liquids possess a grainy constitution and that vacant spaces exist between the grains or molecules. We may then assume that when water and alcohol are mixed, some of the molecules of water squeeze into the vacant spaces between the molecules of alcohol, or vice versa. Again, we find atomism cropping up in chemistry when Dalton sought to account for the empirical law of constant proportions. Nevertheless, although the corpuscular nature of matter seemed to impose itself if we wished to co-ordinate a large number of phenomena, something more was needed for this hypothesis to be accepted without reserve.

The major reason for the present-day belief in atomicity arises from the following considerations: A celebrated hypothesis due to Avogadro, the legitimacy of which we need not discuss here, suggested that if molecules existed, equal volumes of all gases maintained under the same conditions of pressure and temperature should always contain the same number. This number, called *Avogadro's number N*, corresponds to the case where the volume selected is 22,400 litres, the temperature zero centigrade, and the pressure equal to 760 millimetres of mercury (*i.e.* atmospheric pressure). Then it was shown as the result of complicated mathematical syntheses that, if molecules existed, a wide variety of phenomena should be influenced by the precise value of Avogadro's number. The phenomena referred to deal with Brownian movements in fluids (Einstein), the viscosity of gases (Maxwell, Boltzmann, Einstein), the blue colour of the sky (Rayleigh, Keesom), equilibrium radiation (Planck), the specific heat of solids (Einstein), the phenomenon of critical opalescence (Smoluchowski), and many other phenomena which it is not necessary to mention. Accurate observations and experiments conducted on these phenomena should therefore permit us to deduce the value of Avogadro's number. The results of all these experiments were in striking agreement, always yielding the same value. As this number runs up into sextillions ($6.06 \cdot 10^{23}$), it was scarcely feasible to attribute the marvellous agreement to mere chance; hence we were forced to conclude that our suppositions were correct and that the existence of molecules had been established. It was then an easy matter to determine their masses and to obtain at least some information as to their sizes and other characteristics.

Although we may be repeating ourselves unnecessarily, we must again draw attention to the fact that this knowledge of molecules, which is obtained through a co-ordination of experimental results, is in all respects (other than in degree of certainty) of the same type as our knowledge of the spherical shape of the sun or of our knowledge of space and of the table. On the other hand, it is essentially different from our awareness of feeling hot or tired.

Finally, let us consider a last example, namely, our belief that light is atomic. In this case our knowledge is still more uncertain. The justification for a belief in quanta was arrived at by Planck (as we have explained elsewhere), and arose from the peculiar phenomenon of blackbody radiation. The mathematical treatment of problems of this sort is based on the calculus of probabilities. When calculations were conducted on the assumption of the continuous emission of light, we obtained *Rayleigh's law of radiation,* which was refuted by facts. Planck noticed that by substituting discontinuous probabilities for continuous ones, the results of observation would be anticipated with great precision. When, a few years later, Nernst and then Einstein showed that the extension of these same discontinuities to the energy of molecular motions would account for the curious anomalies and variations of the specific heats of gases and solids, and when in addition Einstein succeeded in accounting with high precision for the photo-electric effect, Planck's ideas appeared to gain in probability. The belief was enhanced still further when Bohr presented science with his model of the quantum-emitting atom.

On the other hand, there were a number of phenomena which the quantum theory was unable to explain. Most of them were connected with the interference effects of light, which seemed to suggest continuity rather than discontinuity of emission. What is required is a more general synthesis, combining these two conflicting forms.*

The purpose of these different illustrations has been to show how slight and gradual is the transition from our commonplace knowledge of space and of the table to the loftiest forms of scientific knowledge. The methodology is ever the same, consisting in the formulation of a mental construct capable of co-ordinating in a rational and simple manner the sum total of our sense impressions. With the child and his knowledge of space and of the table, this co-ordination is so simple that he obtains it without perceptible mental effort; whereas, in the more sophisticated cases of scientific knowledge, the synthesis is the result of abstruse mathematical speculations. Apart from this difference in degree, however, there exists no essential change.

And now let us examine the significance of these co-ordinations and investigate what they have revealed. We have seen that what we call reality reduces to the simplest co-ordination of the facts of observation, and that these facts, in the last analysis, represent co-ordinations of sense impressions. Hence we may say that it is these mind co-ordinations of sense impressions which yield us our knowledge of the so-called external universe. This scientific interpretation of reality may offend those who believe in the existence of some more absolute type of reality, the reality of the world in itself. To be sure, it appears natural enough for us human beings to assume that there must exist some sort of real world representing a reality of a more concrete category than the mental construct we have been discussing; a world which subsists in the absence of all human observers and in which the causes of our sense impressions originate. But what the nature of this real world of things in themselves may be, whether it could ever be described in human language, whether it is identical with the idea we have formed of it, are questions which are probably meaningless, and in any case are insoluble. As soon as we start arguing on these problems, it is a case of every man for himself; no one seems to be able to convince any one else. Fortunately, discussions of this kind are of no interest to science. In this chapter, the word reality will therefore be held to connote scientific reality, which means the simplest co-ordination of scientific facts, hence, ultimately, of sense impressions. That so restricted an interpretation of reality is ample for the needs of science issues from the following considerations:

The mere fact that men appear to understand one another when discussing the external world is sufficient proof that they have reached some common conception as to its nature. Hence, even if we assume that this so-called external universe, which each one of us believes he has discovered, represents but an individual dream which exists only in

* Recently considerable progress has been made by Schrödinger in the interpretation of quantum phenomena within the atom, by means of a wave theory of matter, known as wave mechanics.

our respective minds, the fact remains that we must conceive of this dream as one shared in common by all men. As such, it manifests itself to all intents and purposes as an objective reality which we may regard as pre-existing to the observer who discovers it bit by bit.

It follows that the distinction between idealism and realism is purely academic in science, for our rule of action will be the same whichever of the two opposing philosophies we may prefer.* Thus the physicist who studies the properties of matter will proceed in precisely the same way, co-ordinating his results with the maximum of simplicity, regardless of whether he believes in matter as a metaphysical reality or as a mere mind construct. We have thus arrived at an understanding of what is meant by the objective world of science.

Now it is obvious that had it been impossible to discover or create a common objective world the same for all men, of which the various observers would obtain private perspectives, science would have been quite impossible; for science deals with the general, not with the particular or the individual. Were it not for the fact that the objective world of John is also that of Peter, John and Peter would never agree on the most elementary subjects. An exchange of commonplace knowledge, and, to a still greater degree, of scientific knowledge, would be quite impossible. We must recognise, therefore, that the very existence of science proves that the co-ordinations of the sensations of John possess the same structure as those of Peter. This is all we can say. We have no means of discovering whether what one man sees as red the other might not see as blue were our observers to exchange eyes and brains while retaining their memory of past sensations. On the other hand, we *can* assert from experience that, for normal human beings, two objects which appear to be of the same colour to one observer will appear to be the same colour to the other.

When the importance to science of one same common objective world is realised, the necessity of ridding it of all purely individualistic appearances becomes obvious; and we are thereby led to the problem of illusions or hallucinations. By an illusory object we mean one that has only a private existence, yet whose appearance can be accounted for scientifically, in terms of other phenomena, to which we may concede a common objective existence. Thus, in the case of a mirage, if we insist upon saying that water lies before us in the desert, our objective world will differ from that of our friend who, standing where we see the water, will claim that no water is present. The unfortunate consequences of our error might be very great, as in the case of the dog who dropped the bone for its image.

Hence, either we must abandon a belief in a common objective universe and relinquish all attempts at framing a science, or else we must

---

* By realism we mean "common-sense realism," and not that monstrous distortion known as "neo-realism."

succeed in interpreting the differences in our opinions by appealing to some phenomenon, such as refraction, which all men will accept. In the case of the mirage, for instance, the introduction of refraction saves the objective universe, hence saves science. Finally, we may say that the bright band in the desert is a reality from the point of view of one particular individual, but that it is an illusion from the point of view of the generality; whereas refraction is a reality from the point of view of all.

In the case of hallucinations, we are dealing with delusions of another kind. Consider the fever-stricken man who sees snakes around him. Of course, the snakes, no more than the water in the desert, can have a place in the common objective world; so that in this respect the hallucination and the illusion appear analogous. Nevertheless, there exists an important difference between the two. Thus, in the case of the mirage, all we had to do was to stand at a given point to behold the bright band in the desert as easily as any one else. A certain community in the illusion would still subsist. The hallucination, on the other hand, is essentially private. We cannot step into the patient's place and suffer from his hallucination. Something more deep-seated is at stake than a mere change in our post of observation in a common objective world.

While it is true that the physiologist and brain specialist may succeed in accounting for a hallucination in terms of objective changes in our organic condition (blood pressure, etc.), yet even so we should be dealing with problems far more complex than those which the modern physicist is required to investigate. For all these reasons, while it is the aim of the physicist to account for illusions in terms of the common objective world and the laws of physics, the study of hallucinations belongs to another field of scientific research. In many cases the distinction we have established between a hallucination and an illusion lacks definiteness. For instance, the stars that we see when punched on the nose would not be considered the result of a hallucination, and yet, according to the division established above, the phenomenon would be more in the nature of a hallucination than of an illusion. Similar considerations would apply to dreams.

There exists a more popular interpretation of illusions. It results from a confusion between ignorance and illusion. Suppose, for instance, that we were to receive visual sensations which led us to assert that a table was present, but that on approaching the table we were to find that we could walk through it. If *we*, and *we* alone among men, could see the table, we should be justified in assuming, in view of past experience, that we were suffering from some hallucination, since any other alternative would deny us the possibility of conceiving of an objective world the same for all men. But if all who came into the room perceived the same table and failed, as we had done, to derive any tactual impressions from its presence, it would be taking too much for granted to assert that we had all of us been the victims of error. The mystery might be cleared up without difficulty were we to find it possible to account for this appearance in terms of phenomena already known; but if this procedure failed, the wiser course would be to agree that some new mysterious manifestation had been discovered. At all events, it would be wrong to maintain that the appearance must necessarily be the result of a collective illusion on the ground that its reality would conflict with the laws of matter. Our knowledge of natural laws is obtained only by generalising from experience; and where experience is incomplete, as it must always be, the laws can lay claim to no measure of certainty. For example, a man knowing nothing of atmospheric pressure might well assume that a balloon could never rise in the air without coming into conflict with the law of gravitation, or that a firefly could not emit light without getting burnt. The reason why the theoretical scientist is compelled to approach problems of a seemingly mysterious nature with great open-mindedness is because on so many previous occasions he has been confronted with discoveries which yield nothing in strangeness to the one we have mentioned; and strangeness is not necessarily the product of illusion. More often than not, it is the result of ignorance and prejudice. In this respect, we have only to conceive of the surprise we should experience, had we never heard of an electromagnet, on discovering that distant objects flew towards it without being pulled by strings.

And now let us return to our objective world. We will assume that illusions have been barred therefrom; and the world we have thus obtained can be conceived of as existing in precisely the same way for all observers. If we assume that there is a real world of things in themselves, constituting the underlying cause of our sense experience, the point we wish to ascertain is whether scientific procedure can throw any light on its nature.

In the first place, when discussing the outside world as an existent reality, we must differentiate between its substance and its structure, or form. If, with science, we consider that our knowledge of the external universe can be arrived at only by a rational synthesis of facts of experience, we must recognise that substance escapes us completely; all we can hope to approach is structure, or relationships. Of course, a view of this sort presupposes that we are justified in asserting that knowledge can be acquired only by means of rational co-ordinations. If we dispute the legitimacy of this point the entire argument collapses. Hence the scientific argument can constitute no refutation of the views of the mystic. A man who "knows" of things through intuition or faith, or a man who tells us that he knows that in his previous incarnation he was Julius Cæsar, is a type of opponent with whom the scientist cannot even argue; for whatever can be neither proved nor disproved by observation or reason constitutes a form of knowledge which is meaningless, or at least completely irrelevant to science. If we did not place some kind of limitation on what we were to regard as knowledge, there would be no reason to prefer the opinions of a Newton or an Einstein to the ravings of an ignoramus or a lunatic; and human knowledge would become so conflicting as to lose all significance. It is not denied that intuition may often lead to discovery. Even in mathematics examples are numerous; but the fact remains that intuition alone has so often led men astray that unless its disclosures can be submitted to some kind of test, no reliance can be placed in them.

Furthermore, it is not asserted that intelligence is everything. We know that with animals and, still more, with insects, instinct plays a major rôle. But it is to be noted, first of all, that instinct is scarcely apparent in human beings. And in the second place, with bees, for example, which have selected the hexagonal shape for their combs, a choice which mathematical calculation proves to have been the best they could have selected, instinct has yielded exactly the same results that intelligence would have done; so that aside from the lack of consciousness which accompanies instinct, no difference is detected in the results to which it leads. Hence, it can scarcely be deemed to yield a new form of knowledge. If these points be granted, we may return to our subject and enquire why it is that substance escapes us completely when by "knowledge" we refer to a rational co-ordination of facts and sense impressions.

Rational co-ordinations have their most perfect prototype in mathematical co-ordinations—in those of mathematical physics, for example. Not all co-ordinations can be constructed mathematically; chemistry, and of course, to a still greater degree, biology, afford us illustrations where mathematics is comparatively useless. Nevertheless, as the methods of co-ordination are essentially the same in all cases, the mathematical ordinations, in view of their greater clarity, can be studied to advantage as typical instances in this respect.

Now mathematical equations are nothing but relations, and from initial relations all we can deduce are other relations. In other words, our equations can never yield us more than we originally put into them. It follows that were all relationships in nature to be preserved and the substances changed, no observable difference could be detected; and we should never be able to differentiate between a whole class of worlds identical in structure but differing in substance. If, then, we discard the procedure of the mystic, or of the metaphysician who claims a knowledge that cannot be submitted to the control of experiment, we must recognise that substance escapes us completely and that our knowledge of the real world can at best reduce to a skeleton or structure.

An illustration given by Eddington presents the same problem in a more concrete manner. He assumes that in some future age the game of chess may be dead and forgotten; and he compares our position in respect to the real world with that of archæologists who will have unearthed curious records of the game written in the usual obscure symbolism. Eventually they would succeed in reconstructing the moves of the different pieces, the two-dimensional ordering relation of the partitions, and the rules of the game. To this extent they could claim to have understood the game of chess. Yet certain aspects would escape them completely. The shape of the partitions, whether square, round or oblong, the aspect of the pieces, the nature of the board, whether of wood or of stone, would remain unknown. But it is to be noted that these elements of knowledge that escaped them would be irrelevant. The board might be of wood or stone, the partitions of diamonds or squares, and yet the game could be played as before. If, on the other hand, the ordering relation of the partitions were changed, if the permissible moves of the pieces were modified, the written records would make no sense. It is the same with the real world. The substance may change in nature, yet no difference will be perceived; but if the relationships are modified, a new world will divulge itself to our observation. To this extent, therefore, we are justified in saying that the most our observations can reveal reduces to the structure of the real world. As the French mathematician Bertrand once said, the age of the captain, the number of the crew, the height of the mast can yield us no information about the position of the ship.

Structure, as outlined above, resolves itself into those mathematical relationships or equations which appear to account for the sequence of phenomena. But structure in the more usual sense would apply to the hidden mechanisms themselves, to the deep-seated, underlying causes. When, however, we wish to attain this more concrete knowledge of structure, we encounter a number of difficulties. As Poincaré points out, every time conservation of energy and the principle of least action are found to be satisfied (as is the case in electromagnetics, for example), an indefinite number of different mechanical models are possible; so that our knowledge of the structure of the world is vitiated by the fact that the substructure remains indefinite. Clifford expresses the same idea when he says:

"Whatever can be explained by the motion of a fluid can be equally well explained either by the attraction of particles or by the strains of a solid substance; the very same mathematical calculations result from the three distinct hypotheses; and science, though completely independent of all three, may yet choose one of them as serving to link together different trains of physical enquiry."

In a certain measure, difficulties of this sort may prove temporary. It may happen that further experiment will enable us to determine which of the various hypotheses corresponds to concrete reality. Such, indeed, was the hope of the scientists of the eighteenth and early nineteenth centuries. And in our days the practical isolation of electrons and molecules is a proof that these hopes were not always unfounded. Yet, even so, we should have advanced but a step, and should again be faced with the problem of determining the structure of these electrons, and so on *ad infinitum.* It would be of no avail to say that our aim had been realised when the ultimate constituent elements were isolated. For either these supposedly ultimate elements would possess no structure, in which case we could never know anything further about them, or else they would present a structure; but then they would not constitute ultimate elements, since this structure would imply relationships between their various parts, and we should not be at the end of our journey.

With the discoveries of modern science, still other difficulties bar the route to success. It would now seem that as we wended our way into the microscopic, exact laws might vanish. If this be the case, the microscopic structures must ever elude us. Yet, even if we waive these latter difficulties, it is apparent that the scientific method can yield no knowledge of substance; the structure of reality is all we can approach.

But this reality, once again, is nothing but the expression of the simplest co-ordination of scientific facts; no genuine metaphysical reality is implied. If we discard the criterion of simplicity, we may conceive of our world in as many different ways as we please. Consider, for example, Einstein's theory. One of the principal discoveries of the theory has been to disclose unsuspected relationships between space and time, leading to the four-dimensional space-time structure of the universe. And yet, if we wished, we might account for all of Einstein's previsions while retaining our traditional belief in a three-dimensional structure of space and in an unrelated time. Thus, Michelson's experiment would be accounted for by Ritz's or Stokes' hypothesis; Bucherer's experiment, by the dragging along of a certain mass of ether by the moving electron just as a ship drags along part of the water surrounding it; Fizeau's experiment, by a modification of the ether in transparent bodies (and this was, in fact, Fresnel's original explanation). The double bending of a ray of light, and Mercury's motion, would be accounted for by a modification of the Newtonian law of gravitation; the Einstein shift-effect, by a modification in the intra-atomic structure of the solar atom, and so on. It is a pure waste of time to consider the various hypotheses that might be suggested, for we can invent as many different ones as we choose. Needless to say, all the simple beauty and unity of nature revealed by Einstein's theory would be lacking; and prevision would be impossible, seeing that new postulates *ad hoc* would be needed at all further stages. But when all is said and done, what have simplicity, beauty, unity and prevision to do with the reality of the metaphysician?

Thus the reality of structure which the metaphysician would defend (quite aside from the reality of substance) presupposes his *a priori* belief in the inevitable simplicity of nature. But simplicity is, after all, but an expression of human appreciation. Furthermore, even if this view were contested, there would seem to be no reason to suppose that nature should prefer simplicity to complexity in the first place. As a matter of fact, what little we know of nature proves that the reverse is true when we cease to view phenomena in an approximate macroscopic way and adopt a more microscopic standpoint. To illustrate: There is no more simple law in physics than that of perfect gases; and yet we know that this apparent simplicity is due to our macroscopic observations and that it conceals the most bewildering chaos and uncertainty. It is the same when we consider numerous other phenomena, such as intra-atomic changes.

From this it follows that it is at least a questionable procedure to identify reality with simplicity of co-ordination. And if we discard this tentative identification, reality (even reality of structure) escapes us completely. Of course, the type of reality which we have been branding as elusive is the absolute reality, the "true being" of the metaphysician. The scientist also appeals to the word "reality," but he employs it in a different sense. For him, reality is identified with simplicity of co-ordination, and he states his views explicitly, realising full well that a reality of this type is far from being absolute and that it is essentially pragmatic. It is for reasons such as these that the vast majority of scientists are agnostics at heart, not on account of any *a priori* predilection, but because a proper understanding of the limitations of scientific knowledge leaves them no other alternative, refusing as they do to accept the knowledge of the mystic or the metaphysician as of any significance whatsoever. But scientific agnosticism must not be confused with that extreme form of idealism which denies the existence of any world apart from consciousness. It merely contents itself with stating that the objective world of science (that of space, time, matter, motion, in classical science, and of space-time, intervals and tensors, in relativity) is nothing but the embodiment of the simplest co-ordination of sense impressions, for which some unknowable supra-intelligible world is assumed to be responsible.

If, therefore, our minds worked differently, there would be no reason to assert that we should form the same conception of the world. We do not know whether, in the event of a Martian stepping down on earth, his mind would form the same mental construct of the universe as ours has done. For a Martian, there might be no such things as space, time and matter. Yet he might form a conception of the universe which would be perfectly intelligible to him. Those who believe in the metaphysical reality of space and time enduring in the world of things in themselves, under the plea that we can only conceive of existence in space and in time, appear to be guilty of the same error as those who, never having heard of fishes, would assert that no such creatures could exist, since they would be unable to breathe. Anthropomorphic arguments of this sort have no place in science.

In short, it is not denied that there may exist a world of things in themselves; all that is asserted is that a world of this sort defies all description. A few quotations from the writings of the leading men of science will make this attitude clearer. Thus, in Larmor's book, "Æther and Matter," we find: "Laws of matter are, after all, but laws of mind." Again, in Poincaré we read: "Are the laws of nature, when considered as existing outside of the mind that creates or observes them, intrinsically invariable? Not only is the question insoluble; it is also meaningless. What is the use of wondering whether laws can vary with time in the world of things in themselves, when in a world of that type time may have no meaning? Of this world of things in themselves we can say nothing; we cannot even think of it. All we can discuss is how it would appear to minds similar to our own." And, again, elsewhere: "The Bergsonian world has no laws. The world that may have laws is merely the more or less deformed image that scientists have conceived of it." There is no need to give further quotations, for with slight variations the same philosophy is expressed by all scientists.

However, it cannot be emphasised too strongly that from a practical standpoint these questions are purely of academic interest. The physicist and the mathematical physicist are compelled to operate and reason as though they believed in the real existence of a real absolute objective universe, one of space and time, according to classical science; one of space-time, according to the theory of relativity. In fact, as we have said, were it impossible to conceive of a common objective world, one existing independently of the observer who discovers it bit by bit, physical science would be impossible.

Incidentally, we may recall that such problems as the relativity or the absoluteness of space and motion have nothing in common with idealism and realism. Space may be subjective, yet absolute, or real and yet relative. It is important not to confuse the meaning of the words, "absolute," and "relative" as used by physicists, with their meaning as understood by philosophers. Much of the difficulty that philosophers appear to have experienced in understanding the attitude of science seems to have arisen from confusions of this sort.

Now, if we revert to that rational synthesis of our sense impressions which yields us the objective world of science, there remains to discuss the methodology whereby this synthesis is achieved. In the more elementary cases, the procedure is one of commonplace, logical and inductive reasoning; but in the more advanced cases, the process is exclusively mathematical. There is a close resemblance between the procedure of development pursued by pure mathematics and that of theoretical physics. In either case the tendency has been a search for unity through progressive generalisation.

Mathematics is, of course, so vast a subject that it is impossible to expose even its characteristic aspects in a few pages; but for the purpose we have in view it will be its generalising aspect which we shall mention. Mathematics deals with generalities. It seeks to obtain general laws and general theories as opposed to particular results. For instance, the sum of the first two odd numbers is found to equal the square of

2; the sum of the first three odd numbers is found to equal the square of 3. This, however, does not tell us what the sum of the first 10,000 odd numbers would be. We might proceed to add up these numbers and discover the result; but the aim of mathematics is to obviate a laborious undertaking of this sort. It succeeds in this attempt by demonstrating the existence of a mathematical law according to which the sum of the first n odd numbers is equal to n^2 however great n may be. When a general law of this sort is established, we can apply its formula to any particular case and thus obtain a solution without further ado.

The procedure whereby this law is obtained is that of *mathematical induction*. We verify the fact that the sum of the first two odd numbers is 2^2; then we prove that *if* this is the case, the sum of the first three odd numbers must necessarily be 3^2. Finally, we prove that *if* the sum of the first $(n-1)$ odd numbers is $(n-1)^2$, then the sum of the first n odd numbers must necessarily be n^2. We assume that this law will hold however great n may be, or, in other words, must always hold; and our mathematical law is established. In short, mathematics aspires to give us general laws in place of particular facts, or, again, to proceed from the particular to the general. A search for generality and unity constitutes the *leit-motif* of mathematics.* A few specific examples may make this point clearer.

Suppose we start from the natural sequence of integers and labour under the impression that these integers alone constitute true numbers. We should notice that if we divided one of our numbers, say the number 10, by the number 5, we should obtain one of our integers, namely, 2. On the other hand, if we divided 10 by 4, the result would not yield one of our integers. Accordingly, we should be compelled to state that the division of one number by another might or might not yield a true number. In order to re-establish generality, mathematicians were com-

---

* Of recent years certain philosophers known as logisticians, Bertrand Russell in England, Couturat in France, among others, have stressed the logical aspect of mathematics. The question is whether they have not overstressed it. That mathematical reasoning implies clear thinking and complies with the rules of logic has never been denied; nevertheless the assertion that all mathematical reasonings are of a purely deductive nature, and are reducible to the rules of logic, is an opinion which is by no means unanimous. Some of the greatest among the modern mathematicians, notably Poincaré and Borel, have protested vigorously against this view and have pointed out numerous cases of circularity in the arguments and lack of rigour in the definitions presented by the logisticians. Over and above this aspect of the matter, they have maintained that the rules of a game are not everything in its make-up. To say that mathematics and logic are one and the same would be equivalent to maintaining that poetry was nothing but grammar, syntax and rules of versification, or that music was nothing but counterpoint and harmony. It is conceivable that we might acquire as thorough a knowledge of counterpoint and harmony as Beethoven may have possessed and yet be unable to compose a work rivalling any of his symphonies. We should have no hesitancy in granting that a Beethoven must obvi-

pelled to assume the existence of fractional numbers, just as actual or true as the integers. Thanks to this introduction it became possible to assert that the division of one number by another would *always* yield a true number. Now suppose we attempted to extract square roots. We should find that whereas the square roots of some of our true numbers were themselves true numbers(*i.e.*, $\sqrt{4}$ = 2), yet in the majority of cases we should obtain no such result (*i.e.*, $\sqrt{2}$). Hence, we should have to say that in certain cases the square root of a true number would be a true number, whereas in other cases it would correspond to nonsense. Once again, in order to obtain generality, we should have to incorporate these seemingly nonsensical magnitudes along with the true numbers, obtaining thereby the so-called irrational numbers. In a similar way negative numbers would have to be introduced in order to confer sense *in all cases* on the subtraction of one true number from another. Imaginary numbers would follow when we wished to generalise the significance of the square root of a number, whether positive or negative; then complex numbers, when we wished to confer generality on the significance of the addition of numbers, whether real or imaginary. A further generalisation would yield quaternions and hyper-complex numbers generally. Following still another line of generalisation, Cantor introduced transfinite numbers into mathematics.

ously have been gifted with some mysterious faculty which had been denied us; and this faculty, whatever its essence, would relate to music, would be a part of music, since were all men lacking in it, there would be no great music. Under the circumstances, in spite of what might be called our logical knowledge of music, could we truthfully claim to have as thorough a knowledge of it as a Beethoven?

And it is exactly the same with pure mathematics. We know from experience that many persons, though possessing highly logical minds, are yet refractory to advanced mathematics. Were mathematics nothing but logic, this situation would seem extraordinary. Logistics, from a failure to see in mathematics anything but a series of rules and regulations with no creative faculty behind it, has been christened "thoughtless thinking" by its adversaries. But without wishing to take sides in a controversy for which the majority of persons evince but little interest, there is a point which the unprejudiced onlooker must perceive. With the sole exception of Hilbert, who, though opposing Russell's views, defends opinions of a somewhat similar nature, none of the logisticians have contributed to the constructive side of mathematics. This again appears somewhat strange when we recall that one of the earliest boasts of this school of thinkers was that logistics would give them wings. One cannot help but suspect that the logisticians are lacking in some creative faculty of which they may not be conscious, and that as a result they perhaps occupy in mathematics a position analogous to that of the professor of counterpoint and harmony in music. Under the circumstances, it is questionable whether they possess a sufficient understanding of this difficult science to contribute any information of value.

Of course a charge of this sort cannot apply to Hilbert, whose great work in the creative regions of mathematics has proved him to be gifted with the creative faculty in addition to the purely formal dissecting faculty which all mathematicians, regardless of their tendencies, must necessarily possess. Owing to Hilbert's attitude, the problem is generally regarded as controversial.

A further illustration would be afforded by geometry. Thus, if in a plane we trace a straight line and a second-order curve (circle, ellipse, parabola or hyperbola), it may happen that the line intersects the curve in two points, but it may also happen that no intersection takes place. Generality can be re-established, however, provided we introduce imaginary points and consider them as true points. With this extended concept of a point we may say that the straight line *always* intersects the second-order curve in two true points, whether real or imaginary. Further generalisations require the introduction of *ideal points, points at infinity, imaginary lines* and *angles*.

Also, when we consider a second-order curve, it is necessary to state the positions of five of its points for the curve to be determined without ambiguity. But a circle is also a second-order curve; yet, in the case of a circle, when we state the positions of three of its points the circle is completely determined. Hence, our rule lacks generality and cannot be claimed to hold for all second-order curves, whether ellipses, parabolas, hyperbolas or circles. Once again, however, when imaginary points are introduced, this duality disappears and our rule becomes general. It is found that all circles, without exception, pass through the two same imaginary points called the *circular points at infinity*. When we take these two imaginary points into consideration we are able to state that *all* second-order curves, whether circles or ellipses, etc., are determined when five of their points are given. It is also interesting to note, as Cayley and Klein have shown, that when imaginary points, lines and angles are introduced, the various geometries (non-Euclidean and Euclidean) can be treated in terms of projective geometry. We see that in all cases mathematicians attempt to replace the word *sometimes* by the word *always*, the particular by the general, thereby revealing unity where diversity once held sway.

Let us mention yet another geometrical example. It is always possible to draw a triangle circumscribed to a circle, that is to say, one whose three sides are tangent to the circle. Also it is always possible to inscribe a triangle in a circle; which means that the three summits lie on the circle. Now, Euler noticed that if two circles were drawn at random, it was impossible in the majority of cases to draw a triangle which would be inscribed in one circle while circumscribed to the other. When, however, one such triangle could be found, an indefinite number of others could also be drawn.

Poncelet then gave a celebrated generalisation of this theorem, and Jacobi showed its intimate connection with elliptical functions. We may understand the nature of this generalisation as follows:

A circle is a particular instance of a conic, and a triangle a particular instance of a polygon; whence Poncelet extended Euler's theorem, which deals with circles and triangles, to one dealing with conics and polygons.

In this illustration, again, we witness the same tendency, progress towards generalisation.

Possibly it is in the theory of functions that the most beautiful examples will be found. Mathematical functions are magnitudes whose values vary with that attributed to the variable they contain. We can construct as many different types of functions as we please by annexing additional terms, and the functions thus constructed differ in their behaviour from one another.

For instance, if we call y the value of the function and represent the variable by x, then

$$y = a + bx$$

is a mathematical function (a and b being assigned constants). For every value of x a definite value corresponds for y. A more complicated function would be given by

$$y = a + bx + cx^2.$$

It was soon discovered that, if instead of annexing merely a large number of successive terms to the expression of our function we added an infinite number of terms selected in an appropriate manner, we often obtained functions presenting exceedingly simple properties; so simple, indeed, as to warrant our coining separate names for them. These various functions may be likened to different beings, possessing various peculiarities and tendencies. Certain mathematicians (Borel, for instance) have gone so far as to differentiate between wholesome functions and contaminated ones; and the appellation "pathology of functions" has been introduced into mathematics. As an example of the very simplest type of functions, we may mention those given by the unending series

$$x - \frac{x^3}{1!} + \frac{x^5}{5!} - \frac{x^7}{7!} + \text{etc.}$$

$$1 - \frac{x^2}{2!} + \frac{x^4}{4!} - \frac{x^6}{6!} + \text{etc.}$$

$$1 + \frac{x}{1!} + \frac{x^2}{2!} + \frac{x^3}{3!} + \frac{x^4}{4!} + \text{etc.}$$

where $3!$ stands for $1 \times 2 \times 3$, and $n!$ for $1 \times 2 \times 3 \times \ldots \times n$. These functions were known, respectively as, "sine x" (or sin x), "cosine x" (or cos x) and the *exponential function* e^x.

It was then found that the functions sin x and cos x presented very rigid relationships as though they belonged to the same family. The name *circular functions* was given to the family; the relation we have in view being given by

$$\sin x = \sqrt{1 - \cos^2 x}.$$

On the other hand, no such relationships appeared to exist as between these circular functions and the exponential function e^x. But the entire aspect of the problem changed when Euler appealed to imaginary magnitudes, such as ix; where i stands for $\sqrt{-1}$ and x stands for any real number, as before. Of course an imaginary number, as its name indicates, was conceived of by its inventors as an unreal magnitude, a mere mathematical fiction having nothing but a symbolic significance. And yet, as we shall see, without imaginary numbers some of the most practical industrial problems could never have been solved.

As we have said, Euler substituted imaginary numbers for the real ones in the functions we have discussed, and discovered that when this was done a surprising connection was found to exist between these erstwhile unrelated functions. The relation was

$$e^{ix} = \cos x + i \sin x.$$

In this formula we have only to make $x = 2\pi$, and we obtain

$$e^{2\pi i} = 1.$$

Thus, not only has the introduction of imaginary numbers revealed an unsuspected relationship between what were widely different types of functions, but the two numbers π and e, taken from totally different regions of mathematics, from geometry and from logarithms, are suddenly found to manifest extraordinary relationships.

But this introduction of imaginary numbers into mathematical analysis was soon to lead to still more wonderful discoveries. Cauchy undertook a general study of functions in which the variable would be a complex number, that is to say, a real number plus an imaginary one. He uncovered thereby a new universe of functions, called functions of a complex variable, possessing the strangest properties; and it was seen that our older functions were but particular instances of these more general types. But for these new functions to possess the essential attributes of our older ones, a certain mathematical restriction had to be placed upon them. The restrictions turned out to be expressed by the same mathematical relationship which defines the potential distribution in Newton's law of gravitation, namely, Laplace's equation.* This equation crops up again

* In this case, however, Laplace's equation is of the two-dimensional variety and not of the usual three-dimensional type which defines the distribution of the Newtonian potential in the empty space around a gravitational mass of finite dimensions. However, the two-dimensional Laplace equation also gives the distribution

in the theory of conformal representation, in the theory of heat, in the theory of elasticity, etc. Unsuspected relationships thus appear to be springing up on all sides. Incidentally we may note that Riemann's great discoveries in the theory of functions take their start from this analogy existing between the functions of a complex variable and the laws of the potential distribution around matter.

The theory of the functions of a complex variable constitutes one of the most extensive domains of pure mathematics, with which the names of some of the greatest mathematicians are associated, such as Cauchy, Riemann, Weierstrass, Jacobi, Abel, Hermite and Poincaré. Numerous problems pertaining to the ordinary functions receive a very rapid solution when we consider these functions as restricted cases of the more general functions of a complex variable. As time went on, a number of these functions manifesting strange yet simple properties were brought to light, always by the same process of successive generalisation. First came the *elliptical functions* discovered by Abel and Jacobi, then the *modular functions* of Hermite, and finally the *automorphous functions* of Poincaré. Once again the discovery of these new functions revealed unexpected relationships between domains of mathematics which had appeared to be totally estranged; and Poincaré established surprising relationships between the automorphous functions, certain problems of the theory of numbers, and non-Euclidean geometry.

Incidentally, the popular belief that mathematicians, by appealing to imaginary magnitudes and the like, are abandoning the world of reality for one of shadows, is belied by the fact that one of the most elementary mechanical problems, namely, the oscillatory motion of an ordinary pendulum, can be solved mathematically only by an appeal to elliptical functions, in spite of the fact that imaginary quantities are part and parcel of their make-up.* It is the same with a number of electrical problems of widespread industrial importance; likewise, with Einstein's law of gravitation, the orbits of the planets cannot be calculated unless we appeal to elliptical functions. Similar considerations would apply to Poincaré's automorphous functions. However, this point being granted, it is of course obvious that the mathematician is not primarily interested in knowing whether his abstract speculations have any counterpart in the real objective world, any more than the poet wishes to know whether his dreams will come true. The sole object of his investigations is to explore all rational possibilities and to co-ordinate into one consistent

of the potential around matter in special instances. Such is the case when we consider an attracting cylinder of uniform density, finite section and infinite length.

* In the particular case of the oscillating pendulum, we may restrict our attention to the real (*i.e.*, non-imaginary) realm of the elliptical function considered. But the very existence of elliptical functions is dependent on the introduction of imaginary quantities.

whole various mathematical edifices which at first blush might appear to be totally estranged from one another.

We have insisted particularly on this generalising tendency of mathematics, as it is the one which will play a prominent part when we consider the methodology of mathematical physics as applied to the world of reality. Here we must recall that however mysterious it may seem, nature appears to be amenable to mathematical investigation and to be governed by rigid mathematical laws, at least to a first approximation. So far as scientists are concerned, this belief is not the outcome of religious or philosophical presuppositions. Rather is it a belief which is forced upon our minds by the triumphs of theoretical physics, the first grand example of which was afforded by Newton's celestial mechanics. As soon as this susceptibility to law was recognised in nature, the avowed aim •of science was to discover the unknown laws and in this manner allow us to foresee and to foretell, hence also to forestall, and to cease living in a world of unexpected miracles. In those realms where laws were finally established, science displaced superstition; and wherever, for one reason or another, laws could not be found, superstition continued to reign supreme. A case in point is afforded by meteorology, which in spite of recent progress remains an extremely backward science. As a result an astronomer, who would never think of praying for a solar eclipse, might still pray for a cloudless day favourable for his observations.

Now, in order to render nature amenable to mathematical treatment, it is necessary that we should succeed in reducing the various natural phenomena to common terms. This is done by seeking differences of quantity beneath differences of quality. When and only when this quantitative reduction has been accomplished can science proceed with its investigations, its deductions and inductions. In the case of the objective universe of physics, this process of reducing quality to quantity leads us to conceive of the objective universe as one of electromagnetic vibrations and of molecules, atoms and electrons acting on one another according to fixed laws, rushing hither and thither in space or vibrating round fixed points. The objective universe of science is thus noiseless, lightless, odourless. All the qualities and values are conceived of as arising from the interactions existing between our sense organs and the motions of the outside world, just as the impact of a body would reveal itself as sound to a gong gifted with consciousness.

This quantitative reduction is, however, but a first step; and even after it has been accomplished, nature might still defy mathematical investigation. The fact is that mathematics consists in compounding the similar with the similar; wide heterogeneities even among quantities would render an appeal to human mathematics useless. Thus, the game of chess, where the various pieces can move in widely different ways, is not subject to mathematical treatment. The fact that theoretical science is possible proves that similarity and unity can be found in nature.

Then again, nature must be simple, or at least simple to a first approximation. Theoretically, simplicity cannot exist in nature, since the whole influences the part and the part influences the whole. But in certain cases it has been found permissible to neglect a number of influences owing to their minuteness, and science has thus been able to progress. It is because in meteorology this restriction of active influences to a minimum appears impossible that long-range weather prediction is any man's guess.

But there is yet another condition which must be found in nature, and that is continuity. Mathematics is capable of attacking problems where discontinuities are present, but the technical difficulties are very great; and this is the chief reason why the theory of numbers presents such obstacles. On the other hand, where we are dealing with problems of continuity the mathematician feels more·at ease; and that superb mathematical instrument known as the differential equation becomes applicable. Incidentally we can understand why it was that Planck's discovery of quantum phenomena, or discontinuous jumps in the processes of nature, was such unwelcome news to theoretical physicists; the differential equation had lost its power.

Even now we are not at the end of our difficulties. Assuming that nature manifests unity, simplicity and continuity, we can only investigate her secrets provided we are faced with mathematical problems that we can solve. Even such relatively simple problems as that of the motion of four bodies attracting one another under Newton's law, have thus far defied all attempts. Again, our differential equations are of a very simple species. They suffice only for the simplest type of problems, when, for instance, the future position of a body moving under the action of a given force depends solely on its circumstances of motion in its initial position. But what if the future position of the body were to depend on its entire past history? Phenomena of this type, subjected as they are to hereditary influences, are well known in physics, being illustrated by the hysteresis, or fatigue, of metals; and of course in biology they are the rule. Such problems can be attacked to-day, at any rate in the most elementary cases, following Volterra's discoveries in integro-differential equations. This opens up a new domain of science, that known as "hereditary mathematics."

When we consider all these technical difficulties, and realise that solutions of these problems have been obtained thanks only to the most abstract speculations of pure mathematicians, to the invention or discovery of mathematical entities which at first sight would appear to be totally estranged from the world of reality, we must not belittle the practical importance of these seemingly unreal entities.

As a matter of fact, it is little short of a miracle, in view of the insignificance of our mathematical ability, that theoretical physics should have been possible at all. Had our solar system contained two suns such as are present in the double-star systems, or had one of the planets been comparable in size to the sun, it is safe to say that Newton, in spite of his genius, would never have discovered his law of gravitation; for he would have been thrown back on the tremendous problem of the three bodies.

Leaving aside these technical difficulties, the fact remains that in a number of cases mathematics has been found applicable to nature at least in an approximate way, and as a result mathematical laws have been discovered. We shall now proceed to show how these theories of mathematical physics have evolved. To begin with, a large number of empirical facts are first discovered by the experimenter. These facts are then co-ordinated into a consistent whole, general relations or laws are established, and in this way we obtain various groupings of phenomena, such as electricity, magnetism, optics, mechanics, chemistry, biology, etc. In certain cases it may be impossible to get beyond the initial stage; though experimental data are accumulated, the co-ordination of facts and the discovery of general relations and laws may defy analysis. In such cases we can scarcely regard our knowledge as constituting a science.

The next task is to co-ordinate and establish relations between these various realms of science. Confining ourselves to physics, we find that, first of all, relationships were discovered between electric and magnetic phenomena; then between light propagation and the vibrations of elastic solids (Fresnel, MacCullagh); then between electromagnetic phenomena and optical ones (Maxwell); then between heat and molecular motion (Maxwell, Boltzmann). But in spite of these lofty syntheses, one mysterious influence appeared to remain estranged from all the others; this was gravitation. It has been one of the triumphs of the relativity theory to succeed in establishing the connections between gravitation and optics or electricity.

Finally, an amalgamation of all these various realms is sought for in the expression of a gigantic universal law, the principle of action, governing one unique mathematical world-function, the function of action of the universe, which the theories of Weyl and Eddington appear to suggest.

In short, we see that the development of theoretical physics has followed that of pure mathematics in its generalising characteristics. In either case the breaking down of barriers, the discovery of unity in diversity, has been the guiding motive. And yet, in spite of this mathematisation of physics exemplified in the works of the theoretical physicists, physics is not mathematics, and truth in physics is not the same as truth in mathematics.

In the first place, physics progresses by successive approximations and does not attain its goal at one stroke, as is often the case with mathematics. Thus, in the days of Galileo it would have been correct to say: The centre of gravity of a projectile moving *in vacuo* describes a parabola. Half a century later, Newton recognised that this statement was only approximate. The correct statement then became: Under ideal conditions of isolation, the centre of gravity of the projectile will describe an ellipse, the earth's centre being situated at one of its foci. But, according to relativity, Newton's statement in turn is approximate. We must now say that the trajectory lies along an ellipse whose axis is slowly rotating. There is every reason to believe that even Einstein's mechanics is but an approximation to truth; and Schrödinger's wave mechanics is already supplanting it in the infinitely small. Indeed, there is every reason to suppose that however far we go, we shall always be dealing with approximations. Here, then, is an essential difference between physics and mathematics. Thus far, it would appear that absolute truth might exist but that our present means of investigation had not allowed us to attain it. But we shall see that more is at stake than a temporary admission of failure.

Consider, for instance, the constants of physics, such as R, the gas constant, or h, Planck's constant; and contrast them with the constants of mathematics, such as π or e. Whereas the constants of mathematics can be calculated to any degree of approximation we choose, the values we assign to the constants of physics can never be considered absolutely rigorous. It is not merely because physical observations are necessarily inaccurate or because conditions of observation may not be perfectly ideal owing to the presence of contingent influences; other reasons of a deeper order are involved. The fact is that we are by no means certain that nature is amenable to rigorous mathematical laws and that the so-called constants of physics represent other than average values. The triumphs of theoretical physicists suggest that a mathematisation of physics is permissible as a first approximation; but a closer microscopic survey might prove that this appearance of mathematical purity and simplicity in nature was due to our crude macroscopic survey of phenomena. For instance, it is a well-known fact that if conditions are sufficiently chaotic, the chaos will generate simplicity when we view things from a macroscopic standpoint; it is only when we wish to view things microscopically that the chaos appears and mathematical methods become impossible.

As we mentioned earlier in the chapter, the kinetic theory of gases furnishes us with an apt illustration. The molecules are rushing hither and thither, bounding off one another in the most capricious way.* It would be quite impossible to foresee the history of a given molecule; and yet, thanks to this very chaos, the macroscopic laws of perfect gases are exceedingly simple. We may infer, therefore, that the constants of physics which enter into our macroscopic laws represent mere average effects and differ essentially from those of pure mathematics. More generally,

---

* This is of course merely a figure of speech. It is not assumed that the molecules come into actual contact. Furthermore, it would be difficult to specify exactly how contact should be defined for molecules.

even under ideal conditions of observation the world of physics can never be assimilated to the world of pure mathematics.

The simplification which ensues from a macroscopic survey of nature allows us to understand the reason for those periodic swings which take place in theoretical science between the physics of the general principles and the atomistic viewpoint. If we view nature atomistically, it will be our desire to interpret phenomena in terms of attractions and repulsions between molecules, atoms or electrons. Following the success of Newton's treatment of planetary motions, the scientists of the eighteenth and early nineteenth century endeavoured to pursue this method by reducing the entire world of physical science to attractions and repulsions between discrete particles. All sorts of laws were appealed to: inverse-cube laws and laws involving still higher powers. But nature was not so simple as scientists had hoped, and paid very little heed to the difficulties of mathematical analysis. So there arose an opposing school of scientists, who substituted general principles embodying wider concepts such as energy, entropy, action, for the atomistic theories of their predecessors.

The argument of this new school of thought ran somewhat along the following lines: "By concentrating more and more upon your world of atoms and molecules you are losing sight of those general principles which experiment has revealed. Would it not be better, therefore, to abandon your hopelessly complex speculations, and accept as a starting point these fundamental principles, which can be deduced from macroscopic investigation? We grant that by so doing you will be substituting a macroscopic for a microscopic view of nature; but by reason of the mathematical difficulties into which you have been led and the numerous hypotheses you have had to make in order to compensate for your ignorance of the microscopic mechanisms, it might be the safer course to content yourselves, for the time being, at any rate, with the macroscopic viewpoint."

We may illustrate the new viewpoint somewhat as follows: A life-insurance company does not know and has no means of foretelling when one of its clients will die; yet, when instead of considering the case of an individual client (which would correspond to the microscopic or atomistic viewpoint) we consider men as a whole (the macroscopic point of view), certain general death statistics can be established. By attaching too much importance to one individual, we should lose sight of this general principle, and premiums could never be fixed. The analogy is far from perfect; but it is helpful in that it shows us that in certain cases, at least, the macroscopic point of view can yield us the information we require. A better illustration would be given in the following very free translation of a passage from Poincaré. As he tells us:

"Suppose that we are in the presence of some machine. The primary and terminal gears are alone observable, while the intermediary systems of transmission are concealed. We have no means of knowing whether the motion is transmitted by means of gears or straps, by means of pistons or by other means. Does this signify that we can never understand anything about the engine unless we take it to pieces? We know that such is not the case, for the principle of the conservation of energy enables us to determine the most important point. We observe, for example, that the last wheel turns ten times more slowly than the first, both these wheels being visible. We may infer therefrom that if a couple is applied to the first wheel it will balance a ten times greater couple applied to the last one. This information is obtained without our having to consider the mechanism whereby this state of equilibrium is realised and without our having to know how the various forces will balance one another inside the machine. When we consider the universe, similar arguments will apply. Most of its workings are beyond the sphere of observation; but by observing those motions which we can perceive, we can, thanks to the principle of the conservation of energy, derive conclusions which will remain true regardless of the structural details of those parts which are invisible."

The tendency of science was therefore to lose interest in the microscopic mechanisms and to concentrate on the general principles. It was not long, however, before a return to the atomistic attitude was again forthcoming; but this time our knowledge was more advanced, and greater success was the result. We witness this return to the atomistic procedure in Lorentz's theory of the electronic structure of matter and electricity. Marvellous anticipations resulted from this epochal theory, and it seemed as though a great advance had been made. But, once more, difficulties arose; and here we are referring to those mysterious phenomena which were the original starting point of the theory of relativity.

Einstein again reverts to general principles. Instead of endeavouring to account for the mysterious negative experiments by ascribing all kinds of curious properties to the electrons and to matter, he accepts these negative experiments as significant of some general principle, namely, the relativity of Galilean motion through the ether, entailing the invariance of the velocity of light.

Recent theories on the nature of the atom furnish instances of a similar sort. Thus, Bohr's theory of the atom has atomistic tendencies in that it ascribes definite paths and motions to the electrons moving round the nucleus. In this way it was able to attribute to helium certain spectral lines which before then had been ascribed to hydrogen. Accurate experiments have since confirmed the correctness of Bohr's anticipations. But insuperable difficulties were soon forthcoming when attempts were made to extend the theory to the heavier atoms. Half-quantum numbers and spinning electrons were introduced with the result of complicating the theory considerably.

So Born and Heisenberg abandoned the atomistic outlook, stressing the necessity of following a strictly phenomenological procedure. The motions and orbits of the electrons were disregarded entirely, since these had never been observed, and nothing but the repartition, polarisation and intensities of the spectral lines was taken into account. This new departure, known as the *matrix method,* led to remarkable results, removing many of the difficulties which had beset Bohr's atom.

Still more recently we have witnessed a return to the hidden-mechanism viewpoint with Schrödinger's wave mechanics. We can scarcely refer to it as an atomistic method, since waves take the place of discrete particles. Nevertheless, from the standpoint of methodology, the general idea involved is the same—that of basing our deductions on things that cannot be observed.

This dual tendency in science, that of atomism versus the general principles, that of the microscopic versus the macroscopic, must not be thought to arise from any peculiarities in the philosophies of the various scientists. It is due to the circumstances under which they happen to find themselves. Thus, we see a theoretical scientist such as Einstein adopting a phenomenological attitude when investigating the difficulties attendant upon the negative experiments; and we see the same scientist, Einstein, adopting an atomistic attitude when studying the problems of Brownian movements, the problems of radiation, or the quantum theory of specific heats. In the same way, when we climb a ladder we raise our left leg, then our right one. It is not because we conclude we were wrong in raising our left leg; it is because we cannot progress unless we allow the right leg to catch up with the left one. So it is in science; and we may be quite assured that in the years to come, as in the past, we shall continue to witness these periodic swings from the macroscopic to the microscopic view, from the general principles to atomism. We must realise, however, that as each swing takes place we are advancing higher and higher towards that unattainable ideal, perfect knowledge.

And now let us revert to the common objective world of classical science, to the world of separate space and time in which molecules, atoms and electrons move and vibrate, and in which other types of realities, such as electromagnetic fields, are present. Let us recall once again that this real objective universe is the world as the scientist must assume it to be if he wishes to co-ordinate the complex of his experiences (reducing in the final analysis to sense impressions) in a simple and consistent manner. Whether or not this objective universe can be identified with the real world of the metaphysician is a subject we need not discuss further, since it is of no interest to science. The point we wish to investigate is somewhat different. It is to be noticed that the so-called secondary qualities, such as colour, sound and smell, have been banished as self-supporting entities. They are now ascribed to the reactions of our brain. to the neural disturbances which occur when our sense organs are submitted to the action of the realities of the common objective world.

The criticism is often directed against the scientific attitude that by reducing quality to quantity we are eliminating values, substituting, for example, electromagnetic vibrations for the colour red. Of course, it is scarcely necessary to state that, aside from all philosophical considerations, the reduction of quality to quantity is a prerequisite condition if

science is to exist. To illustrate: In Newton's day the vibratory nature
of light was unknown. Red light differed from green light, but this
qualitative difference manifested itself as an irreducible fact for which
it was impossible to account. Under the circumstances, if the observer
were to rush towards a red light or to move away from it, it was quite
impossible for science to anticipate what effects would arise. As soon,
however, as Fresnel discovered the vibratory nature of light, red light
was found to differ from green light owing to its slower rate of vibration; prevision then became attainable. It was possible to anticipate
that were we to approach a red lamp with sufficient speed it would
appear green, that with greater speed it would appear violet, and that
with still greater speed it would become invisible. Likewise, were we to
recede from the light with sufficient speed it would also cease to appear
visible. This was the celebrated Doppler-Fizeau effect, which astronomical
observations soon suceeded in detecting; it is thanks to this effect that we
are able to determine the radial speed of approach or of recession of the
stars.

We may presume, therefore, that the practical utility of the scientist's
constant endeavour to reduce differences of quality to differences of quantity is not called into question by the critic, and that his objections are
directed solely against the supposedly unjustified philosophical outlook
which this reduction of quality to quantity has exercised on the scientific
mind. But, in order to investigate the problem, we must proceed to
a more detailed discussion of the reasons that compelled scientists to
differentiate between the so-called primary and secondary qualities.

We shall first restrict our attention to the primary qualities of classical
science, and shall furthermore consider the matter solely from the
standpoint of visual perceptions. For instance, let us assume that we
are viewing what would commonly be called a cone. When we view
this supposedly conical object, we perceive it with various shapes according to the relative position we may occupy: we may see it as a
circle, or as a triangle with a base which is either straight or elliptical;
and so, by calling the object a cone, we are giving it a name which
corresponds to none of its many perceived shapes. The conical shape
is thus never perceived directly, but is the resultant of a synthesis of
all the private views we may obtain when viewing the object from various
directions. Now the question arises: Which of the various shapes
should we call the *real shape* of the object? One of the many shapes
it manifests when viewed from successive positions? Or that shape at
which we have arrived in an indirect manner as a result of a co-ordination
of private views? Of course the answer to this question will depend
on the meaning we wish the word "reality" to convey. If by "reality"
we wish to refer to what we apprehend directly, it would be absurd to
refer to the conical shape as the real shape, since this shape is never apprehended directly, but merely inferred. The trouble, however, with this tentative interpretation of reality would be that we could never attribute any

definite shape to an object, for there would be no reason to favour the shape as seen from *here* rather than the shape as seen from *there*. On the other hand, if we ascribe to the word "reality" the alternative meaning, *i.e.*, that pertaining to the entity issuing from a synthesis of private views, it becomes possible (so classical science believed) to attribute a definite shape to the object, a shape transcending the relative situation and motion of the percipient. And so, according to classical science, shape was no longer indefinite, no longer a matter of point of view: it became impersonal, and we were enabled to conceive of the object as having a definite shape even in a world devoid of all observers. This was the objective space-and-time world of classical science. It would appear, then, that if we wish to retain the word "reality" in the two different capacities mentioned, we should at least underline the difference in meaning by referring to a "private reality" in the first case and to a "common or objective reality" in the second.

We now come to the so-called secondary qualities—colour and sound. Here, for instance, is what would commonly be called a red light. We apprehend it as red wherever we may be situated, so that, in contrast to the case of apparent shape, it might appear as though we were justified in claiming that the light was really red, *i.e.*, red in an objective world devoid of all observers. But if now, instead of occupying a succession of various positions at rest with respect to the luminous source, we move towards or recede from the light with sufficient speed, it will change colour. Colour, when considered from an impersonal objective standpoint, is thus just as indefinite as apparent shape. Can we at least combine these various colours, as perceived by the various observers, into one common colour, of which our private perceptions would constitute but different perspectives? Needless to say, the task is quite impossible. In other words, there exists in the case of colour no parallel to the objective cone of classical science, no possibility of speaking of objective colour: colour remains private. Here, then, is a first reason for differentiating colour from real objective three-dimensional shape, or, again, secondary from primary qualities.

In the preceding discussion of the primary qualities of classical science, we have considered solely objective shape. But, in a more extended sense, we must class with the primary qualities all those which may be deemed to manifest an impersonal existence. With this more general understanding, we find that the most important of the primary qualities of classical science were given by such concepts as shape, size, duration, mass, force and electric charge.

When we consider the new views that are forced upon us by the theory of relativity, we find that certain refinements are necessary. The special theory of relativity has shown that there was no reason to dispute the status of those qualities thought to be primary by classical science so long as we restricted our investigations to the points of view of observers at rest with respect to the object; but that when observers in relative motion were considered, it became impossible to effect a synthesis of all the apparent shapes so as to obtain one sole impersonal or real spatial shape. Real shape was thus removed from the status of a primary quality, and there was no sense in speaking of the shape of an object in a world devoid of all observers. Similar conclusions applied to size, duration and all the other primary qualities mentioned, with the sole exception of electric charge, which retained its impersonal status. Of course, had the theory of relativity deprived us of the possibility of conceiving of a sufficient number of primary qualities, hence of an objective world, it would have led us to a stone wall. But we know that such is not the case, and that the theory has merely modified the list of the primary qualities of classical science, assigning the same fundamental rôle to others in their stead. These are given by the space-time intervals and, more generally, by space-time configurations and tensors. In other words, the objective universe disclosed by relativity is no longer one of shape and duration, or space and time, but one of space-time.

Yet, regardless of the actual list of qualities that are to be classed as primary, the significance of these qualities remains the same in either science in that they are deemed to represent entities to which an impersonal existence may be conceded, if for none other than methodological reasons.

Sufficient information has been given in this book to enable the reader to understand the similarity which exists between the comparative status of the objective and the private view, on the one hand, and between tensors and tensor components, on the other. Thus, to take an illustration from electromagnetics, we know that what our instruments detect at a given point in a given frame of reference are the electric- and magnetic-field intensities. When the frame is changed, the measured values of these quantities are modified. Hence here we have the analogue of the various apparent shapes of the cone. And just as the apparent shapes of the cone could be attributed to an objective or real three-dimensional cone (in classical science) or to a space-time configuration (in relativity), so now, in the theory of relativity, the electric and magnetic intensities can be regarded as the six variable components of a real space-time tensor, the electromagnetic-field tensor F_{ik} subsisting in the objective space-time world. When we change frames, the tensor remains unaffected in the objective space-time world, but its six components in the space-and-time frame we happen to occupy are subjected to definite variations when our frame is changed; and for this reason fields that appear purely electrical when we are moving with a certain relative velocity will appear electromagnetic when our relative velocity is modified.

The general classification of objective, absolute or public realities, on the one hand, and of relative or private ones, on the other, might yet appear to present none but a methodological interest and to have no bearing on the problem of the genuinely real. But, assuming for argument's sake, that the concept of the genuinely real has a meaning, it can scarcely be denied that if we ever hope to discover this genuine reality, it is assuredly more reasonable to seek it in those realities which we have cause to regard as public rather than in those which remain essentially private.   For, as we have seen, a private reality, such as colour or apparent shape (or three-dimensional spatial shape in relativity), is indefinite until such time as the conditions of observation are specified; hence it is not self-supporting, but entails the mention of a relationship to something else.   On the other hand, the generalised primary qualities transcend this specification of relationships, that are always of a contingent nature.   Hence, quite aside from considerations of a methodological order based on convenience, there appears to be every reason to assume that the generalised primary qualities approximate more nearly than the secondary ones to the unattainable ideal of the genuinely real, or the metaphysical reality.

We have now to consider a further type of difficulty which would confront the metaphysician who claims colour to be a reality having an existence *per se* in a world devoid of observers.   In the preceding pages, when discussing the relativity of colour to the percipient, we have dwelt on the relativity of colour to motion, embodied in the Doppler effect. But it is well known that a distinction between primary and secondary qualities had imposed itself long before the Doppler effect was discovered. One of the reasons responsible for this differentiation is to be found in the fact that colour, in many cases, is also relative to position.   Berkeley gave an illustration of this type of relativity when he wrote of a cloud which, though it looks crimson from *here,* is a dark mist when we penetrate its interior.   But his illustration was a poor one, for when we enter the cloud we are no longer viewing its surface.   It would not be difficult to improve upon Berkeley's example, but for our present purpose it will not be necessary to dwell further on this type of relativity, so we shall examine another species of relativity, namely, relativity to gauge.

Suppose, then, that a green object is placed on the table.   The metaphysician who states that colours are realities which would exist in a world devoid of percipients would presumably state that the object's surface was green in its very essence.   But it might well happen that, on viewing the surface with a powerful microscope, we should find it to be made of a patchwork of blue and yellow spots and that all appearance of green had vanished.   Unless the metaphysician proceeds to ignore the microscopic view entirely, he will be placed in the dilemma of wondering whether the green quality or the yellow-and-blue one is to be ascribed to the object in the real world.   Even so, his troubles would not be at an end, for we might conceive of an ultra-microscopic vision compared with which our erstwhile microscopic vision would be macroscopic, and so on indefinitely.   Now the change in colour that accompanies a passage from the macroscopic to the microscopic suggests that changes of equal importance might ensue from each successive passage to the following microscopic view. Indeed, it might be that for organisms whose size was of the order of wave-lengths no colour would manifest itself at all. Inasmuch as it would be highly arbitrary to assert that the ultra-microscopic vision is in any wise less worthy of consideration than the more usual macroscopic one, it appears to be impossible to ascribe definite colours to existents in a world devoid of all observers.

It is true that in the present discussion we started from an example where the microscope enabled us to detect a difference in colouring; and cases might arise where even under the microscope the object would still appear green. But, as we said before, the microscopic vision is itself ultra-macroscopic when contrasted with the infinite regress of ultra-microscopic visions we can conceive of; hence, even in this case, the difficulty would be only postponed, not obviated. After all, our ability to pass from microscopic to macroscopic views is contained within comparatively narrow limits. The microscope brings us to the limits of microscopic vision; yet it does not even enable us to see some of the tinier micro-organisms. As for the ultra-microscope, it yields but a blur. Turning to the ultra-macroscopic vision, it appears to be best exemplified when we view the Milky Way; but this vision in turn would be microscopic as contrasted with others of which we might conceive. The upshot of this discussion is that when the metaphysician speaks of red as inhering in the flower of a geranium, he is making a statement that cannot withstand scientific criticism.

Contrast these ambiguities and difficulties with the primary quality of shape in classical science (not in relativity). For in classical science, if a coin be recognised as round as a result of a synthesis of certain macroscopic points of view, it will remain round when percipients situated in any position, moving with any velocity, and observing it with any macroscopic vision co-ordinate their impressions. *Owing to this comparative irrelevance of the percipient,* there appears to be a certain justification (in classical science) for speaking of a round coin existing in a world devoid of percipients, whereas there is none for speaking in the same sense of the brown colour of the coin.

The next stage we have to consider deals with the reduction of secondary to primary qualities. Whereas, up to the present, our arguments based on the relativity of perceptions applied solely to such secondary qualities as colour and sound, we shall now be in a position to advance arguments holding for the remaining secondary qualities, such as heat and smell.

We said, when discussing the red light, that it was impossible to effect a synthesis of all the colours perceived by the variously moving observers and obtain therefrom one definite, impersonal colour. Nevertheless, we may succeed in obtaining objectivity if we find it possible to regard sensations of colour as due to certain specific perturbations suffered by the retina as a result of the impact of electromagnetic waves. According to the relative motion of the observers, the objectively real electromagnetic waves will impinge upon the retina with one frequency or another, and corresponding sensations of colour will ensue. Colour would thus manifest itself as the result of a relationship.

The realistic metaphysician might urge that this may be a convenient way of accounting for colour, but that it is artificial and that, in spite of all, colour should be considered a reality existing *per se*. But if this stand were adopted it should also be permissible to say that the fizziness of the Seidlitz powders already exists *per se* in the powders, and results in no wise from a relationship, *i.e.*, from the chemical interaction of the two powders. Or, to take Galileo's illustration, we should say that the tickling quality must be in the feather, since a feather passed over our skin may tickle us.

The metaphysician's stand appears untenable, for it is impossible to deny the existence of relationships giving rise to effects which were non-existent before the relationships were established, and which often differ appreciably from their generating causes. And every time we sense a colour or a sound or a smell, a relationship is established between some objective entity and that particular sense-organ which is affected. The fact is that the human organism is a highly complex piece of workmanship and that as a rule it reacts differently (though sometimes in the same way) to the various perturbating influences; while at other times, so far as we know, it does not react at all and no sensations are registered. To speak of colour and sound as existing independently of the organism is thus as nonsensical as to discuss a one-sided triangle.

To take an example at random, let us consider a problem of heat. The sun emits rays which are commonly supposed to be luminous and hot. If, then, as the metaphysician asserts, hotness is inherent in the rays, we should expect the space between the earth and the sun to be flooded with heat, hence warm. But such is not the case. The temperature of interplanetary space is in the neighbourhood of absolute zero. On the other hand, the atmosphere becomes warm under the sun's rays. Why this difference? Because the energy of the electromagnetic waves is transformed into molecular agitation when the waves impinge on matter, just as a spark is produced when two flints are struck together. The more rarefied the matter on which the waves impinge, the smaller will be the percentage of the electromagnetic energy transformed or degraded into molecular agitation. It is partly for this reason (owing to the progressive rarefaction of the atmosphere) that the temperature gradually falls as we rise to high altitudes in an airplane. Now science claims that on a hot day, when standing in the shade, we experience a sensation of heat, our sensation results from an interaction between the extremities of our sensory nerves and the impacts of the molecules constituting the atmosphere. The metaphysician will presumably balk at this statement, claiming that the heat we sense exists independently of us in the air and that we merely detect its presence. But if his contention were accepted, we should also have to assume that the heat existed in the sun's rays and had been transferred directly from them to the air. And we have shown that this view is untenable, since the light rays produce heat merely as a result of a transformation of their energy into one of another type (molecular agitation). And since it appears impossible to maintain that the heat was already in the rays prior to their encounter with matter, what justification is there for asserting that the heat was already in the air prior to its encounter with our sensory organs? In every case we are witnessing transformations of energy, and these transformations are always accompanied by degradation (except in the ideally reversible transformations). This fact is expressed by the law of entropy, according to which, though the energy is conserved in quantity, its potentiality, or quality, is lowered, *i.e.*, degraded.

Now, to revert to the problem of colour, it is possible that when a light-wave impinges on our retina some chemical change is produced (as in the case of the photographic plate, where the molecule of chloride of silver is loosened); possibly, too, a liberation of electrons takes place by a photo-electric effect. At all events, it is more than likely, on the strength of other phenomena of which we know something definite, that by the time the energy of a supposedly red electromagnetic vibration has reached our brains it has been subjected to all sorts of transformations. Even if we feel justified in saying that redness hits our eye, what right have we to maintain that, contrary to all the known laws of physics, this redness is perceived by our consciousness without any transformation having taken place? And if we admit that physics is not a complete hoax, how do we propose to assert that the redness we sense is already present as redness in the electromagnetic radiations?

The metaphysician's attitude will lead to still further difficulties when a number of other examples are considered. In the first place, science has discovered that the electromagnetic vibrations accompanied by colour constitute an insignificant minority. There exist electromagnetic vibrations of varying frequencies identical in all respects with the luminous ones, except that they do not happen to influence the human eye. If we adhere strictly to the ultra-realistic viewpoint, we shall have to assume that these invisible radiations contain no properties of colour. And yet, as we know, though invisible to the human eye, they are detected by the photographic plate and, in all probability, by a number of animals. Were the photographic plate possessed of consciousness, or, again, were our eye provided with a photographic plate in place of the human retina and optic nerve, etc., the colour red would be unknown; whereas a new colour, ultra-violet, would be sensed with ease.

It thus becomes impossible to escape the conclusion that our precise perception of colour is affected by the idiosyncrasies of the eye and organism. We need not dwell on a number of well-known illustrations, such as the effect produced by certain drugs. If, therefore, when defending the reality of colour, the metaphysician is referring to those colours which the human eye can detect, he is making the very existence of his realities dependent on the accidental circumstance that the human eye is sensitive to one particular range of vibrations and to no other. He can emancipate himself from this difficulty only by assuming that all electromagnetic waves possess colour, whether visible or invisible. This would lead him to ascribe colour to radio waves. But why stop even there? Why not attribute colour, sound and smell to everything while we are about it, and talk of the colour of sound waves and the sound or smell of electromagnetic ones? We should thus be led into complete intellectual anarchy. Then again, consider interference phenomena. Red light plus red light yields obscurity. If we consider redness as an objective reality, we must at least concede that positive redness and negative redness must exist, so that a mutual destruction becomes possible. In similar fashion, there would be positive and negative sound, and all kinds of other absurd notions.

That all these points should have appeared controversial in the days of the Greeks is only natural, since they knew very little of the transformations of energy, and nothing of the invisible radiations or of the Doppler effect or of interference bands, etc. But nowadays the ultrarealistic standpoint is quite untenable for the scientist. The reader acquainted with scientific thought may feel that this long discussion was quite unnecessary. Such would also have been our own opinion had it not been for the profusion of contemporary philosophical literature purporting to defend the reality of secondary qualities. Were it not for a few scientific words interspersed here and there in their writings, one would have no difficulty in believing that these productions were those of men who had lived more than two thousand years ago.

Another type of objection to the method of quantitative reduction is that a philosophy of nature must be based on a consideration of qualities as well as of quantities. But the critics overlook the fact that by postulating too many conditions they may be rendering a consistent philosophy quite impossible. The aim of a philosophy of nature should be to increase our understanding of nature. Vagueness has ever been an obstacle to knowledge; it prevents us from differentiating between concepts which present but a superficial similarity, and obscures the interconnection of phenomena. For this reason, science has had to reduce qualitative differences whenever possible, since it was they that interfered with the possibility of obtaining any precise co-ordination.

In a similar spirit, science has ever endeavoured to eradicate such statements as "nature abhors a vacuum" or "chemical elements possess various affinities." Statements of this sort, just like the Aristotelian qualities, are worthless, for they lead nowhere. To clothe our ignorance in the guise of a phrase may flatter our vanity, but it does not advance us one iota. In short, although the scientist may not have taught us much about colour and light, he has at least increased our powers of prevision; and that is no mean achievement. Of course prevision is not everything. The enjoyment of the present is also of interest; and it might always be claimed that the quantitative reduction had eliminated beauty and noble aspirations from the world. However, if 'this were the case, why not add that it has also eliminated filth and misery? But such considerations are quite beside the point. There is no reason why a physicist should not derive just as much pleasure from the beauties of music as a man ignorant of the laws of acoustics. Hence, we must make up our minds once and for all and decide whether in our quest for knowledge we intend to be guided purely and simply by our feelings, or whether we intend to seek our inspiration in the abstract intellectual ideal of unity and simplicity with which theoretical science dwells.

Let us now return to a more scientific subject. We have seen that a theory of mathematical physics reduces to a rational co-ordination of a large number of facts and phenomena. The interconnections between these phenomena permit us to pass by a continuous chain of reasoning from one to another, and enable us thereby to anticipate the future or the past from a knowledge of the present. We have also seen that various syntheses of facts were possible; but that, in general, one particular synthesis stood out prominently, owing to its greater simplicity and to the smaller number of auxiliary hypotheses it contained. Thanks to the absence of disconnected hypotheses *ad hoc* in this simplest synthesis, we are able to find our way about without being bewildered at each turn by gaps and hiatuses. Owing to our ability to foresee and to foretell the future activities of nature, we feel that we have succeeded in unravelling her hidden scheme and in understanding her guiding motives. To this extent we consider ourselves justified in assuming that the knowledge we have uncovered or created corresponds to reality, although, of course, the reality contemplated remains essentially pragmatic. This pragmatic conception of reality has been stressed repeatedly by Poincaré. But it is by no means peculiar to the great mathematician and philosopher. We find it in the teachings of Euler and Riemann, and of all the great scientists of the last two centuries; in short, it is the philosophy of science.*

Obviously, there is nothing in the nature of an explanation in a scientific theory. Phenomena are not explained; they are merely interconnected, or described in terms of their mutual relations. As a matter of fact, there is no cause to be surprised at this failure of science to explain phenomena, for the failure arises not from the limitations of science, but *from the limitations of the human mind itself.* All we can ever do is to interpret A in terms of B, and B in terms of C. However far we

---

* Thus, Euler, when discussing absolute space and time, writes: "What is the essence of space and time is not important; but what is important is whether they are required for the statement of the law of inertia. If this law can only be fully and clearly explained by introducing the ideas of absolute space and absolute time, then the necessity for these ideas can be taken as proved." Again, Riemann, when discussing the possible non-Euclideanism of space, maintains a similar attitude. We read: "It is conceivable that the measure relations of space in the infinitesimal are not in accordance with the assumptions of our [Euclidean] geometry, and, in fact, we should have to assume that they are not, if, by doing so, we should ever be enabled to explain phenomena in a more simple way."

go, we can never avoid an ultimate unknowable. This perfectly obvious contention is sometimes obscured because the end of the series may be so far removed as to escape our attention; but this is merely a mental subterfuge, a form of mental laziness.

Thus, the Greeks wondered why the earth did not fall. An explanation was soon forthcoming. The giant Atlas holds it on his shoulders. But what, then, prevented the giant Atlas from falling together with the earth he was carrying? The nether regions on which he stands give him support. This is as far as the explanation went, for no theory seems to have been advanced explaining why the nether regions did not fall, together with Atlas and the earth.

We might consider more scientific examples; but we should find that the case was always the same. A is described in terms of B, B in terms of C, and so on. Consider the phenomenon of gravitation. Does any one really imagine that Newton or Einstein has ever attempted to explain gravitation? To say that gravitation is a property of matter or a property of space-time in the neighbourhood of matter is just as much of an explanation as to say that sweetness is a property of sugar; for in the last analysis, what is matter—what is space-time? If we say that matter is an aggregate of molecules, atoms, electrons, protons, what of it? What are electrons? What are protons? We can only confess our complete ignorance and, while attempting to reduce the number of these unknown fundamental entities to a minimum, content ourselves with describing the properties which appear to characterise them and the relationships that appear to connect them. Clearly, those who seek explanations will find no comfort in science. They must turn to metaphysics.

And yet, as a matter of fact, these rather gloomy conclusions are gloomy only because we are expecting too much. If we content ourselves with what we can obtain, we shall find that the descriptions of science are creative and fertile, and not sterile, as descriptions usually are. As we have already explained, a scientific description is creative in that it allows us to foresee and to foretell. A description that could not confer this power on us would be of little use. Thus, we might describe the petals and stamens of a flower, describe its colour, odour and general appearance. A description of this sort would be useful in enabling us to recognise the same plant when we came across it again. But apart from this, it would teach us nothing new. It would not tell us, for instance, whether the root was bulbous or fibrous. On the other hand, if as a result of prolonged botanical study we could establish certain recurrent relationships between the characters of flowers and the other characteristics of the plants, we might be able by a mere examination of the flower to anticipate the nature of the root. In palæontology it is the same. The discovery of a fossilised bone has often been sufficient to enable scientists to reconstruct the entire skeleton of the unknown animal. It is in this respect that a scientific description is creative and differs from the ordinary type of description which we encounter in everyday life.

Now we have already attempted to explain that among all the syntheses of experimental facts known to science some stood out prominently, owing to their extreme simplicity. In these privileged syntheses simple relationships were discovered, and the expression of these permanent relationships constituted the laws of nature. If the experimental facts we are seeking to co-ordinate are few in number, and if the relationships between phenomena, as established by experiment, are more or less crude and uncertain, we may be led to a certain definite synthesis which will appear to be the simplest. But if ultra-refined experiment leads us to more precise relationships at variance with our former crude ones, or, again; if a wider array of facts has become known, it may well happen that our original synthesis will be incapable of co-ordinating all these results unless we appeal to a number of artificial corrective hypotheses. In this case we may be compelled to widen our synthesis or even to reconstruct it entirely. We must realise, therefore, that a synthesis can never be considered final, for we never know what the future may hold in store.

It is the same with our hypotheses and theories. Theories which would appear plausible when a small number of facts are appealed to, often become untenable when a greater number are taken into account. Consider, for instance, Lesage's theory of gravitation. Lesage had suggested that gravitation might be accounted for by assuming corpuscles to be rushing hither and thither with enormous speeds in all directions through space. Under this hypothesis two bodies, say the sun and earth, would screen each other mutually from the impact of some of the corpuscles. The residual action caused by the impact of the other corpuscles would then be to press the sun and earth together; in this way gravitation would be accounted for mechanically. Needless to say, Lesage's hypothesis was nothing but a wild guess, but, even so, it was felt that there might still be some truth in it.

The astronomer Darwin attacked the problem and proved that gravitation might be accounted for in this way, not merely qualitatively (for this would have been quite insufficient), but quantitatively as well, provided the corpuscles possessed no elasticity whatever, hence did not rebound after impact. Thus far, then, everything was in order, and the theory was acceptable. But now let us take into consideration a few additional facts. Gravitation, as we know, exerts its action even when bodies are interposed. Hence we must assume that only an insignificant percentage of the corpuscles would be arrested by the earth's surface. Now, when account is taken of the sizes of the molecules of matter constituting the earth, and of the vacant spaces which separate them, certain limitations are imposed on the size, mass and velocity of the corpuscles. When, furthermore, we realise that the earth moves round the sun without slowing down in any perceptible way, as though its motion encountered no friction, we are compelled to conclude that the density of the gas formed by the corpuscles cannot surpass a certain limit. When all these facts are taken into account, we find that Lesage's theory cannot be countenanced, for calculation proves that under the impact of the corpuscles, the earth would become white-hot in a very short time.

The same difficulty would arise were we to seek to save the hypothesis by substituting for the impacts of material corpuscles the pressure of invisible electromagnetic radiations. Rather than introduce further hypotheses *ad hoc.*, for the sole purpose of saving Lesage's theory, Lorentz preferred to abandon it.

We see, then, that it is owing to the incessant accumulation of new facts that scientific theories are constantly submitted to change and revision. Until the advent of Einstein's theory, however, although theories of mathematical physics had to be subjected to incessant modification, our basic concepts of space, time and matter had remained unaffected. Indeed, there was no reason to deviate from the original synthesis in this respect. Only when ultra-precise experiments were performed was it found impossible to retain the classical spatio-temporal foundation. As a result, Einstein was compelled to construct a totally new synthesis built on an entirely different foundation, that of space-time. Had it not been for our knowledge of ultra-precise experiments, there would of course have been no reason to modify the essential characteristics of the classical synthesis. From this we see that the synthesis which we adopt depends on the extent of our knowledge, or, what is more to the point, on the extent of our ignorance of nature.

There is very little similarity between the erection of a building and the creation of a scientific theory. When a building is erected, we know beforehand the type of structure we propose to erect, and we can lay our foundations accordingly. But in an empirical science such as physics, the structure is never completed. The superstructures are subject to incessant change as our experiments become more precise and as new phenomena are revealed. The result is that we can never lay our foundations once and for all with perfect security. At all times we must be prepared to abandon our theory, remodel the foundations and erect a totally new structure.

It would appear to be unnecessary to stress these points further, for they must appear obvious to those who are at all familiar with the achievements of science. Certain metaphysicians, however, hold that similar reconstructions should never endanger the traditional concepts of space and time. But an attitude of this sort is equivalent to removing the study of space and time from the control of experience; for all concepts which we approach empirically are necessarily subject to change as our experimentation increases in precision. Furthermore, if we adopt this attitude, why not go the limit and assert that the discoveries of the laboratory can be figured out *a priori?* The entire trend of modern science is hostile to the philosophy of the *a priori*. Non-Euclidean geometry undermined its foundations a century ago, and to-day the physical theory of relativity has completed the work. To assert in the name of logic or of reason, as so many philosophers have done, that space and time must be separate,* or that space cannot manifest variations of curvature from place to

* Bergson, "Durée et Simultanéité."

place, or that space-time must always be flat,* or that matter must be a substance which is conserved, is to impose narrow philosophical doctrines which are likely to cramp the entire synthesis, rendering a simple scheme of natural relations impossible. To quote Weyl:

"Matter was imagined to be a substance involved in every change, and it was thought that every piece of matter could be measured as a quantity, and that its characteristic expression as a 'substance' was the Law of Conservation of Matter which asserts that matter remains constant in amount throughout every change. This, which has hitherto represented our knowledge of space and matter, and which was in many quarters claimed by philosophers as *a priori* knowledge, absolutely general and necessary, stands to-day a tottering structure. First, the physicists, in the persons of Faraday and Maxwell, proposed the 'electromagnetic *field*' in contradistinction to *matter*, as a reality of a different category. Then, during the last century, the mathematician, following a different line of thought, secretly undermined belief in the evidence of Euclidean geometry. And now, in our time, there has been unloosed a cataclysm which has swept away space, time and matter, hitherto regarded as the firmest pillars of natural science, but only to make place for a view of things of wider scope, and entailing a deeper vision."

Summarising, we may say that in natural science it is essential to beware of fundamental premises which are claimed to be forced upon us by the requirements of logic or of crude perception.

Yet, on the other hand, it must be admitted that when we have succeeded in constructing some lofty scientific synthesis which takes in the facts of experience, our synthesis must always be such as to permit us to account for the facts of crude observation; for these, to the same extent as the results of ultra-precise experiment, constitute facts which must be co-ordinated. A theory such as Einstein's must not go counter to the apparent separateness of space and time when phenomena are investigated only in the usual crude way by commonplace observation. In other words, the scientific synthesis must reduce to the primitive synthesis when too great accuracy is not imposed, so that the primitive synthesis must appear in the light of a first approximation. Einstein's theory satisfies this general condition in every respect. As we have seen, for low velocities or for crude observation the classical separateness of space and time reasserts itself; hence no conflict exists between the relativity theory and commonplace observation. Similar conclusions apply to the classical addition of velocities. This also appears to be verified to a first approximation.

Now it might be contended that since all finality seems to be denied a scientific theory, it would be better to cease theorising and content ourselves with a bare accumulation of facts. But an attitude of this sort

* Whitehead, "The Principle of Relativity "

would be untenable, for scientific theories, though necessarily inaccurate from a long-range point of view, are nevertheless inevitable if knowledge is to exist. A bare accumulation of disconnected facts would not constitute a science, precisely because, the facts being disconnected, they would afford us no means of foretelling the future sequence of events. Under the circumstances, it is better to be guided by a faulty synthesis which we can improve upon as experience demands, rather than be guided by nothing at all. Furthermore, co-ordinations and theories which have since turned out to be patently inaccurate, have yet been successful in leading to the discovery of new facts.

Thus, Newton's law, to-day recognised as inaccurate, brought about the discovery of Neptune; it permitted astronomers to assert that it was the same comet which reappeared at definite intervals. Again, it was thanks to Newton's law that Römer, in 1675, discovered the finite velocity of light, thereby exploding the belief that the existence of an event and our perception of it were simultaneous occurrences. All these discoveries would have been arrived at by other means at some later date; yet it must be agreed that, owing to the law of gravitation, they were considerably precipitated. Indeed, it was not until the nineteenth century that the finite velocity of light was established by direct terrestrial measurement.

In other cases, a scientific theory, whether true or false, directs physicists towards crucial experiments of which they might never have thought. Eddington's astronomical expedition was prompted by Einstein's discoveries; had the theory been unknown, it is probable that the bending of light, when finally observed, would have been regarded as an error of observation or as due to some casual disturbing influence. Under the circumstances it might never have become a permanent scientific fact; and it is the same with the Einstein shift-effect. Regardless, then, of whether Einstein's theory will endure or be scrapped to-morrow, it has led, at least, to the discovery of new facts.

Furthermore, we must remember that each successive theory retains certain elements of the one it has supplanted. It rises to a higher plane, but it rises from a rung that has been set in position by its predecessor. It is not in the physical hypotheses that we shall find these elements of permanence, but in the equations themselves; for it is these that constitute the relations of reality. The physical hypotheses, as originally formulated, are but convenient scaffoldings of a more or less temporary nature. This is especially noticeable in the successive theories of electricity and magnetism. From the initial theories of Coulomb and Ampère to the present ones of Larmor and Lorentz, as modified by Einstein's discoveries, we pass through the intermediary ones of Maxwell, Helmholtz and Hertz. We may follow the equations through these transformations and see how they were supplemented by the adjunction of additional terms or by the refinement of those already present. Thus, Maxwell's equations differ from Ampère's by the adjunction of a new term. Lorentz accepts Maxwell's equation of the free ether without modification; it is only when matter is present that refinements are superadded. Again, Einstein's gravitational equations degenerate into Newton's when the field is·weak and the velocities are small.

When we realise that all these successive refinements in the theories were brought about by a desire to co-ordinate an ever-increasing number of facts, it would appear that the scientist has very little control over the course his theory will follow, since in every case it is the criterion of simplicity of co-ordination which guides him.

However, it may be of interest to examine whether, in addition to these formal considerations, there may not exist other reasons which prompt scientists towards their solutions. It might be claimed that the various thinkers start from different initial philosophical presuppositions, and that these presuppositions, rather than the number of facts co-ordinated, are responsible for the differences in the nature of the theories and conclusions. Opinions of this sort have been expressed by numerous philosophers who have devoted their attention to what they call the "metaphysics of science"; but we may hasten to say that the opinion appears to be entirely unwarranted. It is solely the criterion of simplicity of co-ordination which decides on the orientation of science. Where the personal idiosyncrasies of the thinker may come into play will be in the directing of his initial attempts of co-ordination along certain lines rather than along others; but if a simple co-ordination is found to be impossible along the lines originally chosen, other attempts in totally different directions will be undertaken in due course. The final result may issue in a co-ordination presenting a philosophical import totally different from the one preferred originally; and this change is due to the facts of the case and not to a change in the philosophic fancies. Of course, it is not easy for the layman to realise this state of affairs, for scientists do not usually write a diary, mentioning all their abandoned attempts. When a theory becomes widely discussed, it is generally presented in its final form, and the numerous hesitations and partial attempts that may have extended over a period of years are lost sight of entirely. And yet, if we revert to the original scientific papers, these hesitations and changes of attitude are apparent.

It would be tedious to give illustrations; they abound everywhere. However, we may quote a few paragraphs of Einstein's from his paper, "Cosmological Considerations on the General Theory of Relativity." We shall see that he hesitated a long time between a number of alternatives before even considering the finite universe. We read:

"In the present paragraph I shall conduct the reader over the road that I have myself travelled, rather a rough and winding road, because otherwise I cannot hope he will take much interest in the result at the end of the journey. At this stage, with the kind assistance of the mathematician J. Grommer, I investigated centrally symmetrical gravitational fields, degenerating at infinity in the way mentioned. But here it proved that for the system of the fixed stars no boundary conditions of the kind can come into question at all, as was also rightly emphasised by the astronomer de Sitter recently. . . . At any rate, our calculations have convinced me that such conditions of degeneration for the g_{1k}'s in spatial infinity may not be postulated.

"After the failure of this attempt, two possibilities next present themselves.* . . . From what has been said it will be seen that I have not succeeded in formulating boundary conditions for spatial infinity. Nevertheless, there is still a possible way out, without resigning as suggested under (b). For if it were possible to regard the universe as a continuum which is *finite (closed) with respect to its spatial dimensions,* we should have no need at all for any such boundary conditions."

And it is the same in every other case. A number of attempts are made to co-ordinate the facts, and the simplest solution is the one which is accepted by the scientific world, until such time as some newly discovered experimental result conflicts with it.

Of course, all these statements would suffer from an incurable vagueness, were any argument to develop as to what constitutes simplicity. But whatever the reason may be, we find that the agreement among scientists is striking in this respect. There have never been two highly developed theories co-ordinating the same number of facts, in which by common consent one was not recognised as vastly simpler than the other; just as a mathematical expression containing one term is simpler than one containing two.

A case in point would be afforded by a comparison of Einstein's and Newton's co-ordination of gravitation and mechanics. If we are concerned solely with the facts of planetary motion and mechanics known to Newton, then Newton's laws of mechanics and his law of gravitation are by far the simpler. But if we supplement the facts known to him with those more recently discovered, Einstein's synthesis has the advantage, for Newton's co-ordination would necessitate a number of additional hypotheses *ad hoc.* In the same way, so far as our everyday experience is concerned, a non-rotating earth is the simpler solution, whereas, when the planetary motions are taken into consideration, the reverse holds true. It may happen, of course, that the simpler synthesis will conflict with certain philosophical prejudices, but objections of this sort will always give way eventually before the higher criterion of simplicity. Indeed, we know that men finally accepted Galileo's stand.

If, then, we may assume that the criterion of simplicity presents no further difficulty, the problem reduces to determining whether or not the theoretical physicist introduces presuppositions and assumptions uncritically. We are here concerned solely with the physicist and theoretical physicist, not with the pure mathematician, whose procedure is entirely different. In order to clear up ·these points, we shall give a rapid survey of some of the basic presuppositions that are common to physicists. Although we fear that it will be impossible to avoid a certain amount of repetition, yet the importance of the subject is such that a mere matter of elegance of presentation will have to be sacrificed.

* Both attempts failed.

The purpose of science is, as we have said, to co-ordinate facts and to obtain thereby a common knowledge which will hold for all men. But our co-ordinations, deductions and inductions must be conducted according to rigid rules which all men will recognise. In the absence of such common rules, two men starting from the same premises would arrive at different conclusions; as a result, a common knowledge, and therefore a science, would be impossible: We shall assume that the rules of logical reasoning will serve our purpose in this respect.

Here, of·course, it might be argued that we are unwittingly introducing an assumption when we state that beings who neglected to conform to the same rules of reasoning would fail to agree. Of course, if we adopt so hypercritical an attitude, there is nothing left to argue about. The instant we start to talk we are assuming from past experience that sounds will be emitted and that our words will convey at least some meaning to our neighbour. It is the same with the rules of logic. Regardless of what the ontological status of these rules may be, we may assume that in view of past experience it is convenient for men to abide by them if they wish to construct a common science.

Let us consider now what scientists mean by the *uniformity of nature*. The uniformity of nature may be interpreted in various ways. It may mean that nature is regulated by universal laws. In this form, it is obviously not a presupposition adopted uncritically. The primitive man does not feel the necessity for it; he prefers to believe in miracles. If the modern physicist accepts this uniformity it is because it has been confirmed *a posteriori* in so large a number of cases that it seems simpler to accept it than to reject it, at least until further notice. Furthermore, it is an assumption which is perfectly well recognised as of limited validity, and scientists are the first to suspect that after a more microscopic survey of nature it may prove untenable. Nature may not be rational and our logic may be unable to cope with it. If this be the case, statistical knowledge may be the only type possible, though even statistical knowledge may fail us.

We can also understand the uniformity of nature in a different sense. We may wish to imply that an experiment attempted on a Monday would yield the same result if repeated on a Tuesday. In other words, the passage of time, though affecting us in a number of ways (aging, etc.), can be disregarded in certain important cases. Suppose, for the sake of argument, that we were to dispute the legitimacy of this assumption; what would be the result? Obviously science would become impossible. A geometrical proof holding in Euclid's time would no longer hold to-day; knowledge would be intransmissible. There would be no object in performing an experiment to-day, since to-morrow our conclusions would have lost all value. In fact, it would be futile to attempt any experiment at all, since every experiment is spread over a certain time. In a world of this sort, prevision would be quite impossible; a rule of action would be out of the question. Inasmuch as one of the most important functions of science is to serve us with a rule of action, we can realise that science would become unthinkable. This brings out an important difference between science and history, for example. History tells us of events past and gone which, for all we care, may never reoccur; but in science a phenomenon which could never occur again would be of little interest. In short, we must agree that the very existence of a common science proves that our assumption of the uniformity of nature, whether it be justified or not *sub specie æternitatis,* is justified at least in that limited field of experience which is ours. On no account, then, must our belief be confused with a presupposition adopted uncritically.

As a matter of fact, even were we assured that the uniformity of nature was a mistaken belief, we should continue to be governed by it just the same. We know perfectly well that the uniformity of our stream of terrestrial life does not hold; we know that though we have risen every day, yet at any instant we may die. But unless we wish to exist in a state of complete stagnation, we must act as though our life would continue, and so we plan for the future. Under the circumstances, it appears futile to seek any rational justification for the uniformity of nature, basing our arguments on probability considerations. The fact that things have existed until to-day does not prove they will exist to-morrow; but it suggests that we act as though they would.

Among the principles, special mention must be made of the "principle of sufficient reason." In physical science, the application of this principle is subject to all manner of difficulties. For instance, when we assert that there is no reason for a body to move *here* rather than *there*, we are assuming that we possess at least a general knowledge of all the possible influences which happen to be present.

Obviously, our assumption is apt to lead us to erroneous conclusions. We mentioned a case in point when discussing the problem of space. Assuming space to be perfectly empty, the principle of sufficient reason suggested that it should be homogeneous and isotropic; but as we were by no means certain that physical space was truly empty, the application of the principle was of no great value.

Difficulties of this sort are by no means restricted to problems of physics. Even in pure mathematics it is not always easy to decide in what way the principle should be applied. The mathematical problems in which we are compelled to appeal to sufficient reason are those dealing with probabilities; and the folowing elementary illustration, discussed by Bertrand, will afford us a sample of the type of difficulties which we may encounter. Suppose, for instance, we are asked: "What is the probability that on tracing a straight line at random, the line-segment contained within a given circle should be greater than the side of the equilateral triangle inscribed in the said circle?"

Now, if we conceive of all possible straight lines intersecting the circle, sufficient reason tells us that any one of these lines is as likely to be traced as any other. Hence, the probability we are seeking will be given by the ratio of the number of different lines that satisfy our stipulated condition, to the total number of different lines that can be drawn to intersect the circle. Clearly, this last number will be given by considering all the lines passing through a given point of the circumference and then conceiving this point as occupying in succession every position on the said circumference. Under these conditions the probability turns out to be ⅓. But we might, with equal justification, have proceeded in another way. We might have considered all the straight lines parallel to a given direction and then conceived of this direction as suffering in succession every possible orientation. In this case, the probability would work out as ½. This example warns us that a large measure of arbitrariness lies concealed even in the most elementary problems where sufficient reason is appealed to.

As for the regulative norms which guide scientific discovery, we have seen that they reduce to a search for unity in diversity; to a desire to generalise, to detect uniformities or laws, and thus to economise effort, as Mach has aptly said. These same regulative norms are just as apparent in mathematics as in physics and have been heeded by all scientists from Democritus to the moderns.

We might also mention the *principle of continuity* and the *principle of causality* as understood by modern scientists. The principle of continuity implies the eradication of all action at a distance such as is exemplified in Newton's law of gravitation. It is probable that a belief in the principle of continuity arises from the fact that in daily life all action seems to be transmitted by contact. Again, an object moves from "here" to "there" through a continuous series of intermediary positions, or at least when such is not the case we cease to recognise it as the same object. Newton himself felt the necessity of adhering to the principle of continuity, but in his day the development of theoretical science was not sufficiently advanced to permit a realisation of his hopes. For the same reason, physicists during the eighteenth and nineteenth centuries paid little heed to the principle. Particles were assumed to attract one another according to certain laws, but how the attraction was transmitted was never discussed.

Only when, as a result of Maxwell's discoveries, electromagnetic induction was found to be propagated by continuous action through a medium, did the popularity of the principle increase. This leads us to **field physics,** to the discovery of a new category differing from matter, namely, the electromagnetic field. Einstein succeeded in placing gravitation on the same field basis as electromagnetics, in this way obviating Newton's action at a distance. Even here, however, it was not necessarily the principle which guided him to his equations; it was the study of the facts co-ordinated in his theory which drove him to the principle. With such signal initial successes of field physics, it was only natural to seek to extend them still 'farther, and to interpret matter in terms of the field. To-day, the attempt seems to have been abandoned, but, once more, not owing to a change in the philosophical viewpoint, but owing to the tremendous technical difficulties that have been encountered. As Weyl tells us: "Meanwhile I have quite abandoned these hopes, raised by Mie's theory; I do not believe that the problem of matter is to be solved by a mere field theory."

Today, broadly speaking, the phenomenological outlook is the rule among physical scientists; and there is a very good reason for this preference. For so long as we can interpret sensible effects in terms of sensible causes, there is no advantage in invoking suprasensible ones about which anything can be postulated with impunity. True, it may be impossible in many cases to discover sensible causes, and it is not assumed that the whole world may be built up of sensible entities alone. Still, whenever suprasensible causes can be escaped, the procedure will always be to avoid them. It should be mentioned, however, that this last principle does not appeal with equal force to all physicists.

And now we come to a totally different type of assumptions, the so-called scientific hypotheses. These are never imposed on us *a priori* as inevitable; they constitute mere postsuppositions conceived of with a view to co-ordinating the known facts of experience. As such they are contingent on our knowledge of these facts, are never adhered to blindly, and may vary from time to time as this knowledge is increased or becomes more refined. Nevertheless, if the phenomena to which they apply seem more fundamental than the ones we propose to investigate ,the hypotheses will automatically enter into our further co-ordinations, and will then be present in the light of assumptions. For instance, long before the kinetic theory of gases (based on a physical hypothesis) was established as firmly as it is to-day, all new phenomena pertaining to gases were interpreted in terms of the kinetic theory.

As we have said, the primary object of these physical hypotheses is to enable us to co-ordinate facts; the molecular hypothesis, the kinetic theory of gases, the quantum hypothesis, the ether, and the absoluteness or the relativity of space, are all cases in point. In a number of instances, the formulation of these physical hypotheses was arrived at as a result of highly technical theoretical considerations bearing on the equations of mathematical physics. By this we mean that their justification can be made clear only when a mathematical representation of the phenomena under consideration has been elaborated. In the absence of this mathematical representation, they would never even have suggested themselves and their utility could never have been anticipated. Planck's hypothesis of the discontinuous nature of light-emission is of this sort.

In a number of other cases, physical hypotheses suggest themselves in a much simpler way, without our having to consider the mathematical treatment of the problem. Fresnel's ether hypothesis, for example, was formulated as an obvious method of accounting for the wave nature of light which his celebrated experiment of the two mirrors seemed to render inevitable. Newton's absolute space is another illustration of a physical hypothesis which appears to impose itself almost immediately when the dynamical facts of motion are taken into consideration.

It often happens that these physical hypotheses, conceived primarily for the purpose of co-ordinating facts, are subsequently proved to correspond to physical reality (atoms, electrons). In other cases no subsequent experiment has appeared to corroborate them. As an illustration, Ritz had imagined the existence of an atom of magnetism, the *magneton,* in order to account for the spectral series. Weiss also was led to a similar hypothesis when studying certain magnetic phenomena. Thus far, however, all attempts to isolate the magneton have failed, and it is quite possible that no such entity exists as a physical reality.* Nevertheless, although physical hypotheses often fail to be vindicated by further experiment, they may serve a useful purpose. Fresnel's ether has been discarded, yet it was of use to Fresnel and his immediate successors.

To summarise, we see that a clear-cut distinction must be drawn between such tentative assumptions as physical hypotheses and the more basic assumptions that appear to be demanded by the very requirements of scientific knowledge. This does not mean that the latter assumptions present any absolute certainty; it simply means that they must necessarily be accepted, at least tentatively, unless we agree to throw up our hands and abandon all attempts at formulating a science. If, then, we note what appear to be sweeping changes in the scientific viewpoint, we must not attribute these changes to variations in the metaphysics of science; quite the reverse is true. The changes in the scientific viewpoint are necessitated by the discovery of new facts, and it is then these changes in the scientific attitude which are the cause of the variation in the scientist's philosophical outlook upon the world. For this there should be no cause for surprise. Men might have philosophised about space, matter and infinity for ever and ever, giving logical definitions to their hearts' content; but unless they had taken into consideration the facts which mathematics and physics and the sciences generally had revealed, we of to-day should be as ignorant of all these subjects as were our forefathers.

Finally, a requirement demanded of any physical theory, is that it form a consistent whole. By this we mean that it be free from hypotheses *ad hoc,* postulated as and when required, in order to account for every new fact of experience which appears to be in conflict with our theory. It is indeed obvious that if we introduce hyotheses of that kind we can account for anything we choose but at the same time we can predict nothing, since we never know what additional hypotheses we shall not have to imagine in the sequel. An *experimentum-crucis* becomes impossible.

A properly constructed theory is thus exceedingly fragile and might be overthrown at any moment; but this is precisely what militates in its favour. In a general way, it may be said that the fragility of a theory expresses a measure of its coherence. For a coherent theory, uniting, as it does, into one doctrine a large number of facts and anticipations, must in-

---

* Quantum phenomena would now appear to account for what Ritz ascribed to that new entity, the magneton.

evitably collapse, owing to its very coherence, when any one of its anticipations is proved incorrect by experiment. A theory full of hypotheses *ad hoc* is never fragile; whenever an experiment does not bear out the theory, all we need do is vary the hypotheses *ad hoc* or introduce new ones. But a theory of this sort lacks the essential requirements of a scientific theory, since no reliance can ever be placed on its anticipations. The great merit of the relativity theory resides precisely in its freedom from hypotheses *ad hoc;* therefrom arise both its fragility and its value.

One of the reasons why Einstein's theory commands such respect among scientists is precisely because, by predicting definite occurrences and no others, it allows itself to be submitted to a test. As Einstein wrote in 1918, several years before the spectral shift was finally observed on the companion of Sirius: "If the displacement of spectral lines towards the red by the gravitational potential does not exist, then the general theory of relativity will be untenable." It is theories permitting such categorical statements that scientists demand. The theories of Newton, of Maxwell and of Einstein are of this sort.

In order to clarify the very general notions developed in this chapter, we shall now consider a specific illustration. We shall examine the manner in which our concept of space and motion has evolved from the days of Ptolemy to the present time. We shall see that the vast changes that have taken place were brought about always in the same way—by the advancing pressure of the facts of experience and by attempts to co-ordinate these facts as simply as possible.

Suppose, for instance, that we were starting science all over afresh, and that in conformity with our crude belief we assumed the earth to be fixed and the stars to be circling around it. After centuries of observation we should probably have discovered that the stars were not fixed to one another, for successive charts of the heavens would have shown that the shapes of the constellations were not immutable. Probably we should have ended by assuming that the stars were luminous bodies moving freely in infinite space. It would then have been only natural for us to follow the Greeks, and to assume that free bodies described circles, circular motion being the noblest of all motions. But, on the other hand, when we watched a stone gliding over a smooth surface of ice, it would become evident to us that the motion of free bodies on the earth's surface appeared to be rectilinear. So here would be a duality in the laws of motion which would complicate our understanding of mechanics and our synthesis of natural phenomena.

Eventually some thinker would assert that the free motion of bodies was rectilinear, but that the stars, being attached to concentric spheres of crystal, were compelled to move in circles. In this way the duality in the laws of motion of free bodies would be removed; but it is questionable whether the cure would not be worse than the evil. For with our spheres of crystal an artificial hypothesis having no palpable connection with any other of the known phenomena would have been introduced *ad hoc*.

Experimenting still further, we might notice that on the surfaces of bodies rotating with respect to the earth, strange forces would make their appearance. We might eventually plot out the lay of these forces, and divide them into centrifugal and Coriolis forces. Later, a number of curious phenomena occurring on the earth's surface would be discovered. We should notice that the trade winds and sea currents invariably followed slanting courses, that cyclones always whirled anti-clockwise in the northern hemisphere and clockwise in the southern hemisphere, that rivers flowing northward in the northern hemisphere had a tendency to eat into their east banks. The protuberance of the equator, and the decrease of weight, as we moved towards the tropics, would also be detected. Later still, curious experiments such as Foucault's pendulum and the gyroscopic compass would be studied; and all these separate discoveries would remain disconnected unless we accumulated one hypothesis after another.

Finally, some genius would come along and say: "Why, all these phenomena will be in order if we assume the earth to be in rotation, the magnitude of this rotation being computed with reference to a system of axes fixed with respect to the stars. This rotation will entail the existence of a field of centrifugal and Coriolis forces on the earth's surface as on all rotating bodies; and, as a result, all the strange, disconnected phenomena noted will fit quite naturally into a perfectly simple and consistent synthesis."

Thus far we have seen that by assuming the earth to be in rotation and by referring motion to an extraterrestrial frame, a considerable simplification was introduced into our synthesis when purely mechanical experiments conducted on the earth's surface were considered. But this simplification will be still more apparent when account is taken of the planetary motions. So long as in our description of these motions we refer our measurements to the earth frame, the wanderings of the various planets will appear very erratic. However, after long periods of observation certain uniformities would be disclosed, and we might succeed in constructing some highly complicated clockwork mechanism which would portray the motions of the sun, moon and planets. Such was indeed Ptolemy's accomplishment. We might then, by accelerating the motion of our mechanism, foretell the positions of the sun, moon and planets at some future date. In this way the advent of eclipses might be forecast in a more or less accurate way. But the point is that this model could only yield us what we had put into it. We might predict the future positions of the planets because we had detected uniformity in their motions. But if perchance some stray comet were to wander into the solar system, it would be utterly impossible for us to anticipate its motion. Our powers of prevision would thus be extremely limited.

Then we have to consider another type of phenomenon first noticed by

Hipparchus. We refer to the precession of the equinoxes, according to which the Pole Star appears to wander away gradually from the direction of true north, finally returning after having described a circle in the course of 28,000 years. To complete our model it would be necessary to take this phenomenon into consideration by assuming that the entire firmament was wobbling like a spinning top. All such phenomena and many others would appear as so many separate empirical discoveries, and no connection between them could be invoked unless one artificial hypothesis were piled up on another, until nature was deprived of all unity.* In short, we should merely have succeeded in describing phenomena just as we might describe the patterns on a butterfly's wings, but our description would in the main be sterile. In a general way, prevision would be impossible; and any such discovery as that of Neptune by Leverrier, deduced from the peculiar motion of Uranus, would be completely out of the question, since no mathematical connection would have been found to exist between the motions of the various planets.

But just as all the difficulties we encountered in mechanics vanished when we substituted a description of motion in terms of the inertial frame (a frame fixed to the stars), so now, once again, by taking the same frame, we are enabled to overcome all the astronomical difficulties we have mentioned, and re-establish harmony and unity in nature. Kepler, as we know, referred the successive positions of the planets (as obtained from Tycho Brahe's observations) to an inertial frame. He found that, as referred to this frame, each planet described an ellipse round the sun, the sun being situated at one of the foci of the ellipse. Kepler's laws of planetary motions were then as follows:

1. The planetary motions obey the law of areas, by which is meant that the radius vector joining the sun to a planet sweeps over equal areas in equal times.

2. The planets describe ellipses. the sun being situated at one of the foci.

3. The square of the time required for a planet to describe its orbit is proportional to the cube of the major axis of the ellipse described.

These are Kepler's laws. They constitute a distinct advance over Ptolemy's description, since in addition to the numerous advantages entailed by the choice of the inertial frame in terrestrial mechanics, these laws appear extremely simple and are capable of being expressed in concise mathematical form. But the great importance of Kepler's laws is that they rendered Newton's discoveries possible.

Newton argued that, since in an inertial frame free bodies appeared to follow rectilinear courses with constant speeds, it was probable that the planets were being acted on by forces which prevented them from following their natural courses. Now, Kepler's first law, the law of areas, is compatible only with the existence of a force at all times directed

* We might also mention the annual variation in the angle of aberration of a star, and its relationship with the star's parallax. This relationship would appear to be utterly mysterious.

along the line joining the sun and planet. Inasmuch as the planets described closed curves round the sun, this force was obviously one of attraction towards the sun. Kepler's second and third laws, considered jointly, enabled Newton to determine the precise numerical law of solar attraction, the law of the inverse square. Further, the elliptical motion of the moon round the earth, as it appeared when referred to the inertial frame, connoted the existence of a similar law of attraction existing between moon and earth. It appeared probable, therefore, that we were in the presence of a general action of matter on matter, and Newton found that the phenomenon of weight could be accounted for, *not only qualitatively, but also quantitatively*, by assuming that matter always attracted matter. Hence, he was led to the formulation of his law of universal attraction.

It is easy to see that Newton's description is more creative by far than Kepler's, and still more so than Ptolemy's. Thus, Kepler's laws merely enable us to foretell the future positions of the planets, and nothing more. If we abided by these laws, we should probably assume that elliptical orbits were the only ones that could exist. Newton's law, on the other hand, enables us to predict that other types of orbits (parabolas and hyperbolas) are equally capable of existence, and in this way the motions of certain comets are brought into the general synthesis. All we have to do is to note the instantaneous position and velocity of a body with respect to the sun, and we can foresee immediately the precise orbit it will follow. Hence, our powers of prevision, as deduced from Newton's law, are much greater than they were in Kepler's day.

Again, consider the phenomenon of the precession of the equinoxes. Prior to Newton's discoveries this phenomenon was always a mystery, unconnected with all other phenomena. Now its cause is clear. It is due to the uneven pull exerted by the sun and moon on our planet, brought about by its radially unsymmetrical shape (illustrated by the protuberance of its equator). The result will be a periodic to-and-fro oscillation of the earth's axis with respect to the sun. Owing to the earth's rotation this swinging motion will generate a gyroscopic couple which will cause our planet to wobble like a top. The net result is that even had the phenomenon of precession been unknown to astronomers, mathematicians would have anticipated it as an inevitable consequence of the earth's unsymmetrical shape and of the law of gravitation. And it is the same in innumerable instances. Newton's description has thus widened the scope of the general synthesis of nature. It has shown the intimate connection which exists between such apparently independent phenomena as the motions of cyclones, the phenomenon of the tides, the feeling of weight, planetary motions, the precession of the equinoxes, the annual variations of the angles of aberration, etc. It has allowed us to anticipate the existence of one phenomenon from that of another. All these discoveries were rendered possible by our choice of the inertial frame for the reference of motion; so we realise that a description in terms of the inertial frame is creative, whereas one in terms of the non-rotating earth leads us nowhere.

Of course, now that we have learned so much, thanks to the introduction of the inertial frame, there is nothing to prevent us from interpreting all our results and all our relationships backwards, as it were, in terms of a non-rotating earth.   But by transplanting, in this way, all the results obtained from the first description into the new description, no really new description would have been obtained; all we should have accomplished would have been to obtain a parody on the first one.   In the same way, a record run backwards on the pianola does not constitute a new tune, but merely a parody on the original one.   Obviously, this is not the method we must follow when we seek to construct a new synthesis.   We must not merely transpose the results of the old synthesis in terms of the new one; we must operate directly.

But then, in the problem we are here discussing, had the description in terms of the inertial frame been unknown, the motion of the sun with respect to the fixed earth would in itself have been so complex as to defy mathematical investigation.   As judged from the earth, the sun would appear to circle round it every 24 hours (neglecting niceties); and as at different seasons of the year its distance from the earth would vary, its course would be a spiral expanding, and then contracting, every six months.

Under the circumstances, the law of areas would not be satisfied: that is to say, if we considered the line joining sun to earth, it would not sweep over equal areas in equal times.   Now, a general theorem of mechanics proves that whenever a body is moving under the attraction of a central body, the law of areas is always verified; whence we must conclude that the sun could never be considered as rotating round the earth under the sole action of the earth's attraction.   Supplementary forces would have had to be brought into play, but their introduction would have been so arbitrary, and the mathematical difficulties so great, that even Newton might have been nonplussed; and we should have remained as ignorant of the harmony of the heavens as were the Greeks.

In the examples discussed, we have limited ourselves to problems of extreme simplicity; but even in these simple problems we have seen that different descriptions, though equivalent in theory, were far from equivalent in practice.   When we come to consider problems of greater complexity, such as those with which modern science deals, the uniqueness of one particular type of description stands out with increased definiteness, and it is this particular simple continuous description where everything is harmonious and connected that is deemed to constitute the real description.   Only when we follow this unique description do we find it possible to forestall nature by forecasting her actions in advance, and as a result claim that we have discovered reality.

The conclusions to be drawn from the preceding discussions are then as follows: From a purely kinematical point of view, it is simpler to take an inertial frame as frame of reference when the planetary motions are considèred; this is the essence of Copernicus' discovery. Also, when dynamical phenomena‘ are considered, it is simpler to select an inertial frame with respect to which rotation will be measured. In short, the inertial frame imposes itself as conducive to the simplest interpretation of all phenomena whether mechanical or astronomical, so that we may say that there exists a privileged frame of reference in space. Here we are merely stating facts which are in no wise subject to controversy.

We might, of course, content ourselves with this discovery and neglect to consider the "why" and "wherefore." But if we wish to pass beyond, if we wish to seek a palpable cause for the existence of a privileged frame in space, we are driven to one of two conclusions. Either we must say that the existence of a privileged frame arises from the nature of space itself, in which case we are driven to a space which possesses dynamical properties *per se;* and this is absolute space in which an inertial frame is privileged because it is non-rotating. Or else we must assume that space itself has nothing to do with the matter, and that the existence of the privileged frame arises from extraneous conditions, from the presence of the stars, the will of the Supreme Being, the shape of the earth, the number of existing planets, or anything else we may care to imagine.

There is not the slightest doubt that so long as we restrict our attention to the limited number of facts known to Newton, the simplest solution is the first, and its adoption constitutes the essential characteristic of the Newtonian position. There is, however, no necessity to conceive of this absolute space in the light of a metaphysical reality. All we need say is: Everything occurs as though there existed an absolute background of space, hence as though absolute space existed. This was the attitude of classical science until the advent of Einstein's theory. We hope to make these vital points clearer in the course of this chapter.

Prior to Newton, the problem of space and motion appears to have been any one's guess. To mention only two of Newton's contemporaries, Descartes guessed that motion was relative, while More guessed that motion and space were absolute. These guesses were unsupported by any kind of scientific evidence, one way or another. It is true that More, for instance, seems to have realised the necessity of considering scientific examples in support of his claims, but his illustrations were too poorly interpreted to be of any scientific interest. Thus, in a letter to Descartes, he notes that a man walking at a rapid pace experiences fatigue, whereas his friend lying in repose experiences none; whence More concludes that motion and space must be absolute. But had he considered a man standing motionless for hours while his friend was being borne away, sitting at ease in a carriage, one may wonder what conclusion he would then have reached.

A second argument presented by More has no greater merit. A simplified form of it would be as follows: Two trains pass each other, moving in opposite directions. If motion were relative, we could maintain that, as referred to the first train, a telegraph pole was speeding north, while, as referred to the second train, it was moving south. As the pole cannot be moving north and south all at the same time, relative motion cannot exist, and the pole must therefore be at rest in absolute space; whence More concludes that space must be absolute.

It is scarcely necessary to state that Newton's arguments were of a totally different calibre, and even to-day it appears extremely difficult to refute them in their entirety. Newton was one of the earliest exponents of the scientific method. As he tells us himself:

"For the best and safest method of philosophising seems to be, first, diligently to investigate the properties of things and establish them by experiments, and then later seek hypotheses to explain them. . . . For hypotheses ought to be fitted merely to explain the properties of things and not attempt to predetermine them except so far as they can be an aid to experiments."

These statements of Newton appear perfectly clear and merely reflect the spirit of scientific procedure. It has been held, however, by numerous philosophers who have written on the "metaphysics of science" that Newton departed from his avowed empirical method when he formulated his theory of absolute space and time. It is scarcely credible that any one at all conversant with mechanics could ever maintain an opinion of this sort, but as it appears to be widespread in certain quarters, it may be of interest to examine its claims.

Let us then return to Newton's exposition. He starts out in the *Principia* by expressing his belief in absolute immovable space everywhere the same, and in absolute time. These statements are followed by a description of the experiment of the rotating bucket of water and others of a kindred nature. According to the critic, Newton evidently intended absolute space and time to be taken in the light of necessary presuppositions. But apart from the fact that in the order written the statements on absolute space and time precede any reference to experiment, there appears to be no basis for any such belief. Here we must remember that Newton was a scientist writing for the benefit of fellow scientists; and it is a method commonly followed by mathematicians to state a proposition and then show why the statement must be accepted as correct. Although the statement of the proposition precedes the proof, no one would be misled into believing that we were asked to accept the proposition in the light of a philosophical presupposition which might lead to controversy, and then regard the demonstration as an argument of secondary importance. And it is the same with absolute space and time. Just as Euclid follows up his statement of a geometrical proposition by a proper demonstration, so does Newton proceed to demonstrate the existence of absolute space by showing that absolute rotation is revealed by a number of experiments (by the experiment of the rotating bucket of water, among others). Had manifestations of absolute motion been unknown to Newton, had absolute motion eluded experiment, there never would have been any reason for him to postulate its existence in physics; and as a result its direct consequences, namely, absolute space and time, would have been reduced to meaningless hypotheses and no longer to necessary conclusions.

As a matter of fact, Newton mentions explicitly two methods of presentation which may be adhered to. Thus, we read in his *Opticks:*

"By this way of analysis we may proceed from compounds to ingredients, and from motions to the forces producing them; and in general from effects to causes, and from particular causes to more general ones, till the argument end in the most general. This is the method of analysis: and the synthesis consists in *assuming the causes discovered, and established as principles,* and by them explaining the phenomena proceeding from them, and proving the explanations." *

As an illustration of the two methods, we may consider the case of the law of gravitation. Either we may say: The planets have been found to describe conics round the sun, with certain definite motions, and in order to account for these motions we must assume the existence of a solar attraction; or else we may say: The law of gravitation is such and such, and the proof of it is that the planets describe conics round the sun. From the standpoint of neatness of presentation, the deductive method is preferable, but from the standpoint of the chronological order of discovery, the first method describes the situation.

Let us not, then, waste any more time in dwelling on these perfectly obvious points which Newton explains time and again in answer to the criticisms of his contemporaries. The important point to decide lies elsewhere. Is it true that, as Newton imagined, the dynamical facts of motion, established empirically and illustrated in the experiment of the rotating bucket render absolute motion, hence absolute space, inevitable? Here we may be permitted to point out that a question of this sort must be left for scientists, trained in the school of rigorous thinking, to decide. It is indeed obvious that critics, like Berkeley and Kant, who possessed such hazy ideas of mechanics as to confuse velocity and acceleration, momentum and force, were too poorly equipped to express any opinions of value. † Turning, then, to the verdict of subsequent scientists, men

---

* Italics ours.

---

† Kant's attitude towards Newton's absolute space is somewhat confused. At times he defends the absoluteness of space, making extensive use of the arguments of Newton and Euler. At other times he presents his own arguments in favour of the relativity of space and motion. Finally, in his last work (*Metaphysische Anfangsgründe der Naturwissenschafften*), he writes: "Absolute space is, then, necessary not as a conception of a real object, but as a mere idea which is to serve as a rule for considering all motion therein as merely relative." How motion can be relative while space is absolute is a problem that Kant fails to elucidate. At any rate, the problem of the absoluteness of space and time in classical science refers *not to the essence of space and time* (a problem which would degenerate into one of metaphysics, hence which would be meaningless to the scientists), but solely to a discussion of those conceptions which are demanded by the world of experience. Hence we may realise that a man ignorant of mechanics is in no position to pass

of the calibre of Euler, Laplace and Poincaré, we find the conclusion unanimous. Absolute space is recognised by all as the simplest physical hypothesis that will account for the observed mechanical facts. To that extent it corresponds to reality. Even Einstein, to whom the downfall of the Newtonian position is due, recognises that with the facts at Newton's disposal, his solution was the only one that could be defended.

We shall now consider these basic problems in greater detail, and endeavour to understand why it was that Newton was compelled to accept absolute motion. Inasmuch as the absoluteness or the relativity of motion cannot be settled by *a priori* reasoning, we are compelled to appeal to experiment. Suppose, then, we perform mechanical experiments in the interior of some gigantic rigid box. The facts disclosed by experiment will be as follows: So long as, with respect to this box, the stars appear to occupy fixed positions, the course of our experiments will remain unchanged regardless of whether we operate in this part or in that part of the box. In other words, our mechanical experiments offer us no means of deciding as to our location in space. We may infer therefrom that experiment suggests the *relativity of position,* which means that position can be defined only with respect to other bodies, or at least to observable points of reference. This relativity of position entails as a direct consequence the *homogeneity of space.* In precisely the same way we should find that the orientation of our box, together with that of the mechanical apparatus it contained, would also fail to manifest itself by any variation in the course of our experiment; whence we infer the *relativity of orientation,* or *the isotropy of space.*

But what about motion through space? The mere fact that position in space was relative might suggest at first blush that motion, being a mere change in position, should also be relative. But, here again, the problem we are dealing with is one of physics, not one of pure mathematics or of metaphysics. Now, if we perform our mechanical experiment first in a box which is non-rotating with respect to the stars, then in one which is in relative rotation, a considerable change can be detected in the course of the experiment. Yet the relativity of motion would imply that this change in the nature of the box's relative motion should entail no perceptible difference; hence we are compelled to conclude that in contradistinction to position and orientation, motion cannot be relative.

---

an opinion one way or the other. And Kant's knowledge of Newtonian mechanics was extremely poor, to say the least.

Thus, in his *Allgemeine Naturgeschichte und Theorie des Himmels,* we find him giving incorrect formulæ for the most elementary facts concerning falling bodies. Then again, basing his arguments on what he claims to be the laws of dynamics, he tells us of a nebula which would set itself into rotation owing to its outer parts falling towards the centre and rebounding sideways against the inner parts. But this hypothesis is in flagrant opposition to the principles of dynamics, and had Kant spoken of a man pulling himself up by his bootstraps he would have given expression to no greater absurdity. Whereas this latter statement would violate the principle of action and reaction, Kant's violates the principle of the constancy of the angular momentum of an isolated dynamical system.

Motion through space is not a mere matter of point of view; we can detect it even in the absence of any perceptible landmarks in space.

A more thorough investigation of this problem led Newton to differentiate between two grand categories of motion: one of which appeared relative, in that it was impossible to detect it in the absence of landmarks in space; the other of which appeared absolute, since it was accompanied by physical disturbances and dynamical manifestations which could be detected and measured without our having to take landmarks in space into consideration. A train running smoothly along a straight track with constant speed is an illustration of the relative type of motion, while an object rotating with respect to the stars, or a ship tossed at sea, affords us an illustration of the absolute type. In short, position, orientation and a certain type of motion manifested themselves as relative, whereas certain other types of motion manifested themselves as absolute. These were the facts of experiment, and they were summed up by the *law of inertia*, or again by the *Galilean* or *Newtonian principle of relativity*. It is this duality in the manifestations of motion that renders classical mechanics so unsatisfactory.

Two courses were open to Newton:

1. Either he might have assumed that space was absolute; that all motion was absolute, but that accelerated and rotationary motions alone could be detected by mechanical experiments;

2. Or else he might have assumed that space possessed a dual nature; absolute for rotationary motions, but relative for uniform translationary motion.

Newton preferred the first alternative; and the Newtonian principle of relativity, which stressed the impossibility of detecting absolute velocity *through mechanical experiments*, acted as a damper, *not* on the absolute nature of space and motion, but on our ability to detect absolute velocity mechanically.

There were a number of reasons that most assuredly prompted him to this choice. In the first place, a duality in the nature of space and of motion was not easy to conceive of. Furthermore, a circumstance which influenced Newton's successors was the fact that, after all, the Newtonian principle of relativity concerned solely mechanical phenomena. It was still possible that electromagnetic experiments would reveal the absolute velocity in which Newton believed but which had ever eluded science.

Nevertheless, many of Newton's successors preferred to adopt the second alternative. They assumed that experiment had revealed a duality in the nature or structure of space and had proved its relativity for uniform translationary motion. They were not perturbed over the fact that electromagnetic and optical experiments might finally succeed in detecting absolute velocity; for they argued that were such experiments successful, all they would reveal would be velocity through the ether, not absolute velocity through empty space. According to this attitude the Newtonian principle of relativity emphasised no longer the inability of mechanical experiments to detect absolute velocity through empty space, but stressed the fact that *absolute velocity was entirely meaningless*.

Now the important point is the following: Whichever of the two previous attitudes we accept, that of Newton or that of many of his successors, space must still remain absolute in either case. For, in either case, space would be absolute for acceleration and rotation, regardless of what it might turn out to be for velocity. In much the same way, if an object is faintly coloured it *is* coloured; the faintness of the colouring cannot alter this fact. Hence, when we consider space, the entire question centres round the following problem: Is rotational motion truly absolute? If not, hence if centrifugal force cannot be attributed to rotation in space, whence does this force arise?

Of course, here a problem of extreme difficulty confronts us. When by varying external conditions at will we can produce variations in a magnitude, we may maintain that this magnitude is relative to surrounding conditions; for instance, the weight of a body varies with its distance from the earth. When, on the other hand, nothing that we can do appears to produce the slightest effect, we claim that the magnitude is absolute; mass in classical science was a case in point. Obviously, however, a test of this sort has a purely negative value for the simple reason that many of the external conditions lie beyond our power to vary. We cannot, for instance, annihilate the stars or the sun or the electrons. Thus, whereas a magnitude which has been established as relative will presumably remain a relative for all time to come, the same measure of assurance can no longer be claimed for our determinations of absolute quantities. Nevertheless, unless we are to fall into a state of complete agnosticism, we are compelled to establish a difference between quantities which appear absolute in the present state of our knowledge and those which are known to be relative. If we prefer, therefore, we may refer to absolutes as "relative-absolutes," where the word "relative" implies the limitations of our present means of investigation.

It is possible to present Newton's arguments in favour of absolute space in a number of different ways. The following illustrations may serve to clear up a few additional points. Consider a circular disk of gigantic proportions around whose central axis the stars would appear to be rotating in a clockwise direction with an angular velocity ω. We shall assume the disk to be inhabited by beings living around its centre; and we shall further suppose that dense clouds conceal the stars from their gaze, so that there would be no incentive for them even to suspect that their disk might be in rotation. These men would naturally refer all motion to their disk, just as, prior to Copernicus and Galileo, men were wont to refer all motions to the earth's surface.

Suppose, now, that they were to perform mechanical experiments. They would soon discover that no object could remain motionless on the disk unless it were placed at the very centre or else fixed by artificial means. In particular, all objects originally fixed to some non-central point, then suddenly abandoned, would start moving away radially from the centre, then gradually follow a curved course, circling clockwise along an expanding spiral. In certain cases, however, a body might describe a true circle round the centre; but whenever this circular motion occurred, it would always be directed clockwise, and, furthermore, its angular velocity would invariably be given by a certain constant quantity ω. Then again, if billiard balls were shot out in all directions with equal initial speeds from some non-central point, no two of their trajectories would be alike; not the slightest symmetry would be observed, and the clockwise motions would always be present. It may well be realised that for men who claimed space to be homogeneous and everywhere the same, this dissymmetry in the motions of bodies would be hard to account for.

Let us assume that just above the first disk a second one is rotating relatively to the first in a clockwise direction with angular velocity ω. The inhabitants of this second disk would find that bodies remained at rest wherever placed on their disk, that bodies set in motion would pursue straight courses with constant speeds, and, in short, that the asymmetry characteristic of the lower disk would give place to perfect symmetry of motion, no special direction being privileged. Furthermore, whereas the inhabitants of the lower disk would have the greatest difficulty in returning to the centre if peradventure they ever wandered away from it, and whereas they would always feel the action of unsymmetrical forces when they moved about, the inhabitants of the upper disk could move as they pleased and never be subjected to the action of forces.

Such would be the facts of the case. Now, if motion were relative, we could not assume that there existed any absolute difference between the motions of the two disks; either disk would be rotating, but only with reference to the other disk taken as standard. Yet here we see that whether or not a metaphysical difference is assumed to exist, it is quite certain that a vast physical difference is present in the conditions reigning on the two disks.

And so we are naturally led to enquire: What causes this dissymmetry in a homogeneous isotropic space? It is always preferable in science to search for causes in phenomena that are, so to speak, palpable, rather than in invisible agencies, for less guesswork is involved. In the present case, however, no visible cause can be countenanced. True, if the clouds were to lift, it might strike the inhabitants of the lower disk that, with respect to their world, the stars were rotating clockwise with an angular velocity ω; whereas the inhabitants of the upper disk would note that the stars appeared to be fixed. If our observers were firmly convinced of the relativity of motion, arguing that absolute motion in space was inconceivable, they might attempt to search for the cause of all these unsymmetrical phenomena in the state of relative rotation or rest of the stars.

But unless they were able to attribute a definite causal influence to this star rotation, the solution would be no solution at all. Although it now appears that they might have guessed right, yet in Newton's time it would have been quite impossible to justify this attitude; not for

THE EVOLUTION OF SCIENTIFIC THOUGHT

philosophical reasons, but on account of the still undeveloped condition of mathematics and physics. This solution being denied them, there would be no other recourse but to appeal boldly to suprasensible absolute space and to recognise that the lower disk was rotating anti-clockwise in absolute space with an angular velocity ω, whereas the upper disk and stars were at rest. Henceforth, the causal influence would be attributed to the structure of absolute space itself; and the discrepancies noted above would be accounted for with perfect mathematical precision. (This illustration is, of course, merely an alternative way of presenting Newton's experiment of the bucket of water.)

We are again driven to the same conclusions when we consider the problem of inertia. If here, there and everywhere are one and the same thing, why do we have to expend effort to move a body from here to there?

Again, if all motion were relative, the only significant type of motion would be motion relative to a frame of comparison. But then, if, as referred to a certain frame, the path of a free body were straight, with respect to another frame it would be crooked or curved. In other words, there would be no sense in discussing the body's absolute course, since by an appropriate choice of a frame of reference, we could make it anything we pleased. Incidentally, the law of inertia would thereby be deprived of all meaning.

Consider, then, a number of free bodies taken at random, in relative motion with respect to one another. These bodies might ignore their mutual presences entirely (assuming even they were gifted with consciousness). If, therefore, a frame of reference were selected, there would be no reason to suppose that, as referred to this frame, the motions of the various bodies should be in any wise connected. And yet the law of inertia, deduced from experiment, proves that our anticipations would be incorrect.

For if we select a certain privileged frame, an inertial or Galilean frame, all the bodies will be found to describe straight lines with constant velocities as though there existed some secret understanding among them. Is it our frame that imposes this understanding on the unintelligent bodies? Do the bodies see the frame and agree to behave in the same way with respect to it? But the frame might be millions of miles away; furthermore, it only exists as a result of our free will; and the motions of the bodies with respect to one another would obviously remain the same even were some new frame to be chosen, or several different ones to be selected simultaneously. If we refuse to agree that it is absolute space itself which directs and guides the free bodies and which is thus the cause of this wonderful pre-established harmony, where else shall we look for a regulating cause? If we give up the search we fall back into the miraculous. Shall we appeal to the stars once more? But we have said that in Newton's day this was impossible. There appears to be but one solution, and that is absolute space.

If we summarise all these results we see that whether we like it or not, the dynamical facts of motion and the law of inertia render an escape from absolute motion and space impossible. To be sure, as Einstein tells us, Newton might just as well have called his absolute space "Ether," or anything else. The essential point was that something acting as an absolute background appeared to be required in order to account for the undeniable reality of centrifugal force generated by certain types of motion.

At all events, it was for the dynamical reasons outlined above, and for these alone, that Newton was driven to absolute space. So far as science is concerned, this is the only aspect of Newton's discoveries that is of interest. Newton, however, after he had established the existence of absolute space scientifically, proceeded to weld it in with his theological ideas, identifying it with the existence of the Divine Being. This was, of course, his privilege, but these further speculations were extra-scientific and were placed on a par with his writings on the horned beast of Babylon or the bottomless pit of the Apocalypse.

We may be quite certain that had absolute space been based on considerations of a purely philosophical, theological or theosophical character, it would never have survived the criticisms of subsequent scientists. Some philosophical schools would have championed it, others have attacked it, but scientists without exception would have rejected it as possessing no empirical or rational foundation. That Newton's great authority was insufficient to ensure the acceptance of his ideas is seen when we consider the fate of his corpuscular theory of light or of his belief in the impossibility of constructing an achromatic lens. But in the case of absolute space, we find that precisely because it was based on the facts of experience, it survived till quite recently, when, under the attacks of Einstein, it began to crumble. Even so, however, we must remember that it was only thanks to the very recent discovery of new facts, unknown to Newton, that Einstein was able to overthrow the absolute doctrine, and, furthermore, as we shall see, the cornerstone of Newton's mechanics, the absolute nature of rotation, is far from having been destroyed even to-day.

As Weyl puts it when discussing the dynamical facts of motion: "It is just these facts that, since the time of Newton, have forced us to attribute an absolute meaning, not to translation, but to rotation."

Again, in the following passage, Euler expresses the same opinion, amplifying it by stating that the metaphysical nature of space and time is irrelevant to Newton's stand. We read:

"We do not assert that such an absolute space exists."

And again elsewhere:

"What is the essence of space and time is not important, but what is important is whether they are required for the statement of the law of inertia. If this law can only be fully and clearly explained by introducing the ideas of absolute space and absolute time, then the necessity for these ideas may be taken as proved."

Euler's very clear statement brings out the pragmatic and anti-metaphysical attitude towards reality which is characteristic of scientific thought. From a failure to understand the significance of scientific hypotheses, the philosophers of the idealistic school felt it their duty to do away with absolute space. Had their attacks been limited to Newton's theological expressions of opinion, there would have been no reason to criticise their arguments, for all theological opinions can be either defended or attacked. But their failure to grasp the deeper scientific reasons that drove Newton to absolute space, led them to the erroneous belief that mechanical phenomena might be accounted for equally well under the hypothesis of the relativity of space and motion. As a matter of fact, they should have realised that the scientific hypothesis of absolute space is quite independent of our philosophical preference for realism or idealism. In classical science space was absolute in the scientific sense for idealists and realists alike. Indeed, many of Newton's scientific successors viewed nature in an idealistic way; yet they never saw fit to refute Newton's absolute space and motion on this score.

And now let us consider one of the typical objections that have been presented against Newton's absolute space.

Berkeley tells us that absolute motion cannot be imagined and for that reason should be banished from science. He then proceeds to point out that motion can only mean a displacement with respect to something sensible and that, space being suprasensible, motion through space is meaningless. Now it is perfectly correct to claim that men originally came to conceive of motion through visual experience by observing the motion of one body with respect to another. In other words, the *phoronomic* conception of motion is the natural one.

But when these points are granted, it still remains a fact that we cannot force our scientific co-ordinations into a mould which would satisfy too narrow a phenomenological attitude. In physics, as in pure mathematics, we are often confronted with conceptions which it may be difficult to imagine. Thus, although our understanding of continuity is derived from experience, for instance by passing our finger over the table, yet the concept of continuity has had to be extended and elaborated profoundly by mathematicians. Continuous curves with no tangents are not easy to imagine; yet we cannot deny their existence merely because it puts too‑great a strain on our imagination. Neither can we deny that there are as many points in a cube as on one of its sides, for Cantor has proved that such is indeed the case.

It is the same in physics. Experiment presents us with certain facts which we may often interpret in various ways. But we cannot ignore the facts merely because they happen to conflict with our own particular philosophy. All that we may do is to repeat the saying of a certain king of Spain, that had we been God, we should have constructed the universe with greater simplicity. Not being God, however, we must make the best of a bad job. Now, in the case of motion, any philosophy which starts to rule out absolute motion is forthwith confronted with the difficulty of accounting for centrifugal force. Berkeley, as might be expected, falls down on this point completely. His arguments reduce to a criticism of absolute space without affording us any better solution. Furthermore, his premises are scientifically incorrect.

For instance, when criticising Newton's proof of absolute space as illustrated by the experiment of the rotating bucket of water, he remarks that owing to the earth's velocity along its orbit, the particles of water cannot possibly be describing circles. And so Newton's argument purporting to have detected absolute circular motion would thus be at fault. Berkeley's argument reveals that same ever-recurring confusion between velocity and acceleration; he fails to realise that velocity through absolute space is not claimed by Newton to be detectible by mechanical experiments; the Newtonian principle of relativity states this point explicitly. Acceleration alone gives rise to forces. Now, so far as the particles of water are concerned, they possess exactly the same acceleration, whether they be whirling in circles or describing cycloids. Hence, Newton's experiment discloses absolute acceleration, and this is all it was ever intended to disclose. A further argument presented by the same critic is based on the tangential force which he claims to be acting on the revolving water particles. He hopes thereby to account for centrifugal force without introducing absolute rotation. The reasoning is so obscure that we do not pretend to have fathomed it, but one thing is quite certain —the premises are totally incorrect; for there is no tangential force acting on uniformly revolving water particles, and centrifugal force manifests itself in this case. Obviously Berkeley is confusing momentum and force.

Other critics have speculated on the possibility of our space revolving in a more embracing one, this other in still another, etc. But what if it does! A hypothesis of this sort would never endanger the Newtonian position. Suppose, for argument's sake, that our universe of space were likened to a bubble of ether rotating in a superspace; this bubble of ether would still represent absolute space. In every case the essential feature of absolute space and motion resides in the empirical fact that a certain definite set of dynamical axes appears to be present in nature; and the above-mentioned argument could never banish the existence of these absolute axes. All it could do would be to prove that nothing could be predicated of the motion of these axes in the superspace. But in any case the axes would be non-rotating in the ether bubble. This would thereby assume the position of absolute space as understood by science, and we should have to conceive of the more embracing space as rotating round it. Needless to say, the entire speculation is unscientific in the extreme, since it reduces to a hypothesis *ad hoc,* beyond the control of experiment, conceived of for the sole purpose of complicating a co-ordination of facts; whereas the only possible justification for this type of hypothesis is to permit us to introduce simplicity into our co-ordinations.

We may mention yet another type of argument because it brings out an important point of terminology. It is asserted, for instance, that acceleration must be relative since we measure it relatively to an inertial frame or to space. But here, of course, we have a mere confusion of words. The critic is assuming that absolute motion should mean *motion with respect to nothing*. But absolute motion in science does not mean this at all. It means motion which cannot be regarded as *relative to something observable*, such as matter; hence it becomes automatically motion with respect to *something suprasensible, i.e.,* to space. Were it not for this interpretation placed on absolute motion in science, the expression would become meaningless, for motion with respect to nothing is itself nothing.

We may summarise the scientific attitude towards space and motion as follows: If motion can be detected otherwise than in relationship to things observable, or, more precisely, if a co-ordination of scientific facts renders it simpler to assume that such is the case, then motion, and hence space, must be considered absolute. If, on the other hand, no trace of absolute motion, as defined above, can be detected by experiment, two courses are open to us. Either we may assume that absolute motion has no scientific significance and that motion is always motion between observable existents, in which case motion and space are held to be relative; or else we may follow Lorentz's method of dealing with the ether and claim that absolute motion would be detected were it not for a number of compensating physical effects which just happen to conceal its presence from our experiments. But if we adopt this very artificial attitude, it is imperative that we follow up our argument as Lorentz has done, and succeed in specifying the precise mode of action of these compensating effects and also their precise numerical magnitudes —and this would entail a considerable knowledge of advanced mathematics. So much for the scientific status of the problem of space and motion.

Now metaphysicians have a habit of confusing this scientific aspect of the absoluteness or relativity of motion and space with that other problem dealing with the *essence* of space. But, as Euler pointed out two centuries ago in the passage quoted previously, this metaphysical problem is of a totally different order, and has no bearing on the one that scientists are discussing. Even if we were prepared to attribute any meaning to such metaphysical inventions as Leibnitz's monads, and agreed that space might be conceived of as the result of a relationship between them—even so, in view of the dynamical facts stressed by Newton, motion and space would still be absolute. It would be of no avail to hold that motion was relative since it was relative to the monads, for these, whether fictitious or real, are at all events suprasensible. Conversely, even were we to accept the metaphysician's claim that, metaphysically speaking, space is an absolute existent, nevertheless, were it impossible to detect any trace of absolute motion, we should have to say that motion and space were scientifically relative, or else follow Lorentz's method of dealing with the ether.

When, in addition to all these facts, we remember that the meta-physician's theories reduce to mere expressions of opinion affording no possibility of proof or disproof, we can well understand that the scientific attiude towards space and motion has been governed solely by the empirical evidence.

And now the query will naturally be raised: If the dynamical facts of motion impose the belief in absolute rotation, how is it that Einstein, and even before him Mach, should have considered it possible to escape Newton's solution? Let us first consider Mach's argument. Mach's aim was to co-ordinate the dynamical facts of motion in terms of sensible factors alone and thus obviate the introduction of that suprasensible entity, absolute space. And so he was led to conceive of rotation as rotation with respect to the universe, hence with respect to the stars. In contrast to Berkeley, however, Mach realised the great difficulty of accounting for centrifugal forces under this view. He, at least, made an attempt to solve the puzzle by attributing a direct dynamical influence to the relative rotation of the star-masses. But, since his attempt was not followed up mathematically, it was nothing more than a loose, unsupported suggestion. The fact is that in physical science the only convincing theories are those we can defend with quantitative arguments; mere undeveloped guesses are of no value.

Even Newton speculated on the cause of gravitation, attributing it to some species of ether pressure, yet in spite of his illustrious name these ideas were not even mentioned by subsequent scientists. Why? Because they afforded no quantitative treatment and were so vague as to be worthless. It is the same when we consider the hypotheses that have been suggested in order to account for the origin of the solar system: we find these to be extremely numerous and varied. But not one of them carried any weight until it had been submitted to mathematical investigation, and then, according to the results obtained, it was declared possible or impossible. And it is precisely this mathematical work that is the crux of the difficulty, calling for the genius of a Laplace or a Poincaré. In much the same way, we may guess that the sequence of prime numbers will eventually be included in a concise mathematical formula, or that some day interplanetary communications will be established. The difficulty resides not in the guess, not in the desire, but in its realisation. As a matter of fact, it is questionable whether loose guesswork has ever been of any use in science. Jules Verne's idea of ships that travelled under water can scarcely be claimed to have contributed to the invention of submarines, any more than his story of a trip to the moon can be of much assistance in enabling us to set foot on our satellite. Taking a more scientific example, Democritus' guess about atoms most certainly never advanced their discovery by a single hour.*

---

* The development and gradual acceptance of the kinetic theory of gases is particularly instructive in this connection. Some two centuries ago Daniel Bernoulli had suggested that the tendency of a gas to expand might be attributed to a rushing hither and thither of its molecules. But inasmuch as the idea was not worked

And so it is with Mach's guess (we can scarcely dignify it with any other name). Indeed, in Mach's day, it would have been quite impossible for any one to justify his idea, since the necessary material, *i.e.*, space-time, was then unknown. When we realise the important rôle played by space-time in our attempts to avoid a belief in absolute rotation, we can well understand that the doctrine of the relativity of all motion would have been absurd in Newton's day. In fact, any thinker prior to, say, the year 1900 could never have anticipated the discovery of space-time, for its sole justification arose from the negative experiments in optics and electrodynamics attempted at about that time. As for Newton, not only did he know nothing of the non-mechanical negative experiments, but in addition, the equations of electrodynamics had not been discovered in his day. Furthermore, even had he conceived of space-time through some divine inspiration, he could never have utilised it for the purpose of establishing the relativity of all motion. His ignorance of non-Euclidean geometry would have rendered the task impossible. In fact, space-time, in the seventeenth century, would have been a hindrance, and the sole result of its premature introduction into science would have been to muddle everything up and render the discovery of Newton's law of gravitation well-nigh impossible.

And this brings us to another point which is often true in physical science. Premature discoveries are apt to do more harm than good. For instance, had the astronomers of the seventeenth century possessed

out quantitatively, no great attention was paid it. Not till a century or so later was it investigated in a mathematical way by Maxwell and by Boltzmann. For this reason these scientists receive the credit for the kinetic theory; and Bernoulli's name (in connection with this theory) has lost all but a historical interest. Maxwell's and Boltzmann's theoretical anticipations were borne out *quantitatively* by the experiments performed in their day, and a large number of scientists accepted the theory as sound. Even so, the doctrine still had its detractors, for subsequent experiment proved that in the matter of specific heats at low temperatures, theory and observation were in utter conflict. Further difficulties related to the problem of the equipartition of energy. The net result was that other physicists (Kelvin and Ostwald in particular) were hostile to the kinetic theory.

At this stage we must mention that as far back as 1827 an English doctor named Brown had noticed that fine particles suspended in a liquid appeared to be quivering when viewed under the microscope. Brown attributed these curious motions to the presence of living organisms; others suggested that they were due to inequalities of temperature brought about by the illumination of the microscope. On the other hand, the adherents of the kinetic theory maintained that these Brownian movements were due to the impacts of the molecules of the fluid on the suspended particles. Now the important point to understand is that this latter hypothesis could serve only as a suggestion. Before any reliance could be placed in it, it would be necessary to prove that the precise Brownian movements actually observed were in *complete quantitative agreement* with the theoretical demands of the kinetic theory; and a quantitative theory of this sort had never been formulated. Such was the state of affairs when Einstein, in one of his first papers, gave an exhaustive quantitative solution of the problem of Brownian movements (in the case both of translations and of rotations), stating what the precise movements would have to be *if* the kinetic theory of fluids corresponded to reality.

Perrin then submitted the Brownian movements to precise quantitative measurements with a view to checking up on Einstein's anticipations. The result was a dismore perfect telescopes, had they recognised that the planets (Mercury, in particular) did not obey Kepler's laws rigorously, Newton's law might never have been discovered. At all events, its correctness would have been questioned seriously and mathematicians might have lost courage and doubted their ability to discover natural laws. Leverrier, for example, might have lacked the necessary assurance to carry out his lengthy calculations leading to the discovery of Neptune. In short, physical science proceeds by successive approximations, and too rapid jumps in the accretion of knowledge are liable to be disastrous.

We may now follow up Einstein's investigations step by step and see how Newton's absolute rotation was gradually eliminated from science. The situation facing Einstein was somewhat different from the one that had confronted Newton. In Newton's time, there was no *a priori* reason why motion through empty space should be regarded as absolute rather than relative. Whichever way experiment pointed would therefore be equally acceptable. But when motion through the ether was considered there was every reason to anticipate that it could not be meaningless. The fact was that the ether seemed to present the properties of an elastic material medium, so that it was difficult to anticipate a marked difference between motion through the ether and motion through matter. More important still, the equations of electromagnetics proved that phenomena should be affected by motion through the ether. It followed that when the negative experiments in electrodynamics were being performed, there was every reason to suppose that absolute velocity through the ether would be detected. Had this been the case, the stagnant ether

appointment; for Perrin found a considerable discrepancy between those anticipations and experiment. And so the kinetic theory of Brownian movements appeared to be untenable, and, more generally, the whole kinetic theory of gases and fluids to be in peril. When informed of Perrin's results, Einstein went over his calculations afresh and discovered a numerical mistake in his computations. On rectifying this error he found that the theoretical anticipations were in perfect agreement with Perrin's measurements. About the same time, by means of mathematical calculations based on the quantum theory, Einstein succeeded in accounting for the specific-heat difficulty mentioned previously. As for the arguments directed against the kinetic theory by reason of the "equipartition-of-energy theorem," they in turn were answered, thanks to the quantum theory, itself a product of quantitative investigation. The net result was that the kinetic theory was finally established, the principle of entropy ceased to be regarded as an absolute principle, and Ostwald surrendered.

Our purpose in giving this brief historical sketch of the problem of Brownian movements has been to show that loose guesses, unless supported by arguments of a precise quantitative nature, are of little interest to physical science. For this reason no great importance is attributed to the vague atomistic speculations of Democritus or to the relativistic speculations of Mach, even though, in the light of subsequent developments, the latter may prove to be correct. In all cases of this sort, the credit goes (and rightly so) to the theoretical investigator who has succeeded in overcoming the major difficulty, that of working out the theory along rigid mathematical lines. To be sure, in many instances the thinker who makes the guess or advances the hypothesis also follows it up mathematically. Such was the case with Einstein when he formulated the postulate of equivalence for the purpose of interpreting the significance of the equality of the two masses. When a dual contribution of this type occurs, the credit is of course twofold.

would, in all probability, have been identified with Newton's absolute space. And it might have been claimed that absolute velocity, which had always escaped mechanical detection, had been revealed at last by electromagnetic and optical tests.

Yet, as we know, absolute velocity, even through the ether, obstinately refused to reveal itself. The situation was similar to the one we mentioned when discussing space. There also, absolute velocity through space appeared to elude us, in spite of the fact that, owing to the absoluteness of rotation, space could not help but be absolute. But, in the case of space, this duality entailed by the Newtonian principle of relativity was accounted for immediately by the mathematical form of the equations of mechanics. The fundamental law of mechanics, stating that a force is equal to a mass multiplied by an acceleration, makes no mention of velocity; hence, absolute velocity is obviously irrelevant to mechanical processes. In the case of the ether, the elusiveness of velocity was much more disquieting, because the equations of electromagnetics, contrary to those of mechanics, manifested a change when we passed from one Galilean frame to another.

And so, when account was taken of the supposedly semi-material nature the ether was thought to have, and when the lack of covariance of the equations of electromagnetics was considered, the course of least resistance obviously suggested that we assume velocity through the ether to be a reality, but that its effects were concealed by compensating influences. At any rate, we need not be surprised to find that chronologically this was the attitude first considered. It was, as we remember, the attitude championed by Lorentz. But when Lorentz had succeeded in accounting for the negative experiments under this view, his theory appeared so patently artificial that scientists recognised that something was wrong somewhere.

If, with Einstein, we adopt a very general attitude, neglecting to consider the why and wherefore of things, and if we restrict our attention solely to what experiment has revealed, we cannot fail to be struck by the following fact: Every time we have sought to detect absolute velocity, whether through space or through the ether, our attempts have failed. Even certain classical scientists, on the strength of mechanical experiments alone, had felt compelled to banish the thought that absolute velocity through space had any meaning. How much stronger, then, was the suspicion that some general principle was involved, when the same situation confronted us once more in the case of the ether! Under the circumstances, it can scarcely be said that Einstein followed a very revolutionary course when he postulated his *special principle of relativity,* claiming that *absolute velocity* through space or the ether was a meaningless concept. In so doing, he was merely stating in abstract form the result of experiment. Einstein's special principle can be expressed, as follows:

"No experiment, regardless of its nature, whether mechanical, optical or electromagnetic, can ever enable us to detect our absolute velocity through space or ether."

We see that the only difference between Einstein's principle and the Newtonian principle of relativity is that henceforth the relativity of velocity holds for all manner of experiments, or, again, that ether and space are identified. Apart from this difference nothing is changed. As before, space or ether *remains absolute,* in that though relative for velocity, it still manifests itself as absolute for accelerated or rotationary motions, just as in Newton's day.

Against Einstein's stand, the metaphysician may object to absolute velocity being cast aside as meaningless, merely because no experiment seems capable of demonstrating its existence. But science, as we know, is not metaphysics; it is based on experience. An *a priori* rejection of absolute velocity would sin against the scientific method; but an absolute velocity which, though supposedly present, no experiment can reveal, and for which, in addition, no useful function can be found, plays no part in the workings of nature. Were future experiment to detect this absolute velocity, it would of course have to be reintroduced; but to retain it on general principles against the verdict of experiment would be very poor science.

We must be careful not to confuse the present situation with that entailed by atomism, for example. Even to-day we can scarcely say that atoms have been observed in the same sense that stones and tables have been observed; nevertheless, the majority of scientists accepted the existence of atoms years ago, because by postulating their presence a number of phenomena could be co-ordinated with simplicity. Thus, though eluding direct detection, atoms were demanded by indirect mathematical reasoning. Nothing similar is observed in the case of absolute velocity. We insist on this point because it is sometimes thought that the theory of relativity is essentially phenomenological. In a wide sense this is true, but it is most certainly not true if by phenomenalism we understand the word in its narrower sense. An out-and-out phenomenologist, such as Mach, would go so far as to deny the existence of atoms merely because they had never been observed by human eyes, regardless of whether it was useful to conceive of them for the purpose of co-ordinating empirical facts. This is not the attitude of science. But if by phenomenalism we mean the desire to free our understanding of things from *unnecessary* metaphysical notions which are in no wise demanded by experiment, then we are undoubtedly justified in claiming that not only the theory of relativity, but modern science itself, is essentially phenomenological. If these points are clearly understood, we see why it was that Einstein considered it necessary to rid physics of absolute velocity. In short, the exclusion of absolute velocity from science appears to be imposed by experiment; and the only course to adopt is to pursue this train of enquiry in a logical way and see where it will lead us.

We remember that the relativity of velocity, taken in conjunction with the equations of electromagnetics, indicated the invariance of the velocity of light *in vacuo,* as measured in any Galilean frame. From this, the Lorentz-Einstein transformations followed with mathematical logic. It is these transformations that entail, as we know, the relativity of duration, length and simultaneity. We cannot, therefore, regard the new notions as the result of some pipe dream or some divine inspiration; they were forced on Einstein by his transformation equations, and these, in turn, were derived from the initial principles, hence from ultra-refined experiment. Needless to say, it would have been highly arbitrary, and in fact absurd, to lay down the special principle of relativity before such time as the negative experiments had driven us to it. To proceed on the strength of some divine inspiration prior to the disclosures of experiment would have been to start out on a wild-goose chase. Here it cannot be emphasised too strongly that a belief in the relativity of velocity *through the ether* does not impose itself *a priori;* quite the reverse. To put it differently, there is no *a priori* reason why the equations of electromagnetics should preserve the same form regardless of the velocity of our frame of reference. We may feel more sympathetic towards one doctrine or another, but in the last analysis it is experiment, and experiment alone, which can guide science in such matters.

It is important to understand that at this stage Einstein had discovered only the Lorentz-Einstein transformations together with their inevitable relativistic consequences. There is not the slightest hint, in his writings, of a world of space-time. This great discovery is due to the mathematician Minkowski, who, in the year 1908, proved that the Lorentz-Einstein transformations connoted the existence of a four-dimensional space-time continuum of events. From then on, those concepts of classical science, a space of points and a time of instants, considered by themselves, faded into shadows. Here again, Minkowski's discovery was purely mathematical. It issued from a simple application of the theory of groups, and no trace of philosophical prejudice can be found in his work.

In short, it was the mathematical equations, based on the ultra-precise experiments, that had rendered inevitable the new outlook of the world. As Minkowski tells us himself in his inaugural address: "The views of space and time which I wish to lay before you have sprung from the soil of experimental physics, and therein lies their strength." As for space-time, that amalgamated continuum, it had never even been suggested in the speculations either of scientists or of philosophers. To Einstein himself, the discovery of the four-dimensional space-time world was probably quite as much of a surprise as it was to the world at large; but there was no gainsaying the correctness of Minkowski's arguments, which any one with an elementary knowledge of mathematics could verify. Thus, although space-time constitutes one of the most astounding philosophic revolutions ever witnessed, yet the procedure which led to its discovery reduces to the customary mathematical treatment of the empirical facts yielded by the experimenters.

Let us examine what bearing these new ideas will have on Newton's absolute space and motion. Of course, to begin with, space-time taking the place of separate space and time, our outlook of the world is considerably modified. It is modified in the sense that distance, duration and simultaneity are no longer absolute. If, however, we view the problem of space from the standpoint of motion, that cornerstone of the Newtonian synthesis, we find very little change.

The same dynamical facts that had compelled Newton to recognise rotationary motion as absolute drove Einstein once more to the same conclusion. Hence, just as Newton, arguing from the standpoint of space, was compelled to accept absolute space, so now Einstein, arguing from the standpoint of space-time, was driven to absolute space-time. In short, whether we argue in terms of space or of space-time, the dynamical facts of motion appear to impose an absolute background in either case.

Yet, even when confining ourselves to the problem of motion, absolute space-time presents a great advantage over Newtonian absolute space. With absolute space, the difficulty was to account for the elusiveness, relativity or mythical nature of velocity through empty space (embodied in the Newtonian principle of relativity). But with absolute space-time this difficulty is overcome. For absolute space-time, while necessitating the absolute nature of rotation and acceleration, necessitates with equal inevitableness the relativity of velocity. And so the introduction of space-time has served to eliminate that displeasing dualistic feature which was characteristic of Newton's space. From all this, we see that the fundamental characteristic of the Newtonian synthesis, namely, the presence of a suprasensible absolute frame with respect to which rotation and acceleration would be measured (the inertial frame), remained unaffected. A vindication of Mach's ideas seemed to be as remote as ever.

Einstein, in the following words, summarises the situation as it stood before the general theory was elaborated:

"The principle of inertia, in particular, seems to compel us to ascribe physically objective properties to the space-time continuum. Just as it was necessary from the Newtonian standpoint to make both the statements, *tempus est absolutum, spatium est absolutum,* so from the standpoint of the special theory of relativity we must say, *continuum spatii et temporis est absolutum.* In this latter statement, *absolutum* means not only 'physically real,' but also 'independent in its physical properties,' having a physical effect, but not itself influenced by physical conditions.

"As long as the principle of inertia is regarded as the keystone of physics, this standpoint is certainly the only one which is justified." *

Now, when it was realised that the real world was one of flat four-dimensional absolute space-time and no longer one of separate space and time, it became necessary to adjust the laws of nature to the new mould. The laws of electrodynamics found a ready place, and this was only natural, since space-time had been moulded on those very laws. Einstein then succeeded in modifying the laws of mechanics, rendering them com-

* "The Meaning of Relativity."

patible with space-time. There still remained, however, that most important law, the law of gravitation. This law, as given by Newton, was incompatible with space-time, so that the next step was to modify Newton's law in a suitable way. Poincaré attacked the problem and obtained a solution in 1906. Further attempts were initiated by Abraham, Mie and Nordström.*

But all such attempts were soon to be overshadowed by Einstein's own brilliant generalisation. He succeeded in establishing the long-sought fusion between mechanics and gravitation, obtaining thereby the most beautiful theory known to science. However, before following Einstein in his solution, we may state that Poincaré, as far back as 1906, had succeeded in establishing a most important point.† Prior to his investigations it had been held that Newton's law of gravitation constituted a powerful argument against the space-time theory. For, on Laplace's authority, it was believed that the observed planetary motions required that gravitation should be propagated with a speed many times greater than that of light; a fact difficult to reconcile with the maximum velocity c required by relativity. Poincaré, however, proved that the force of gravitation could perfectly well be propagated with the speed of light and yet yield laws of planetary motions practically identical with those of Kepler. He proceeded to consider the various possible laws of gravitation compatible with the flat space-time theory though reducing to Newton's law for slow velocities. One of these laws, in particular, was found to account for the precessional advance of Mercury's perihelion.

The significance of these discoveries was to prove that the relativity theory had nothing to fear from the phenomenon of gravitation. But a still more important point had been established. Had Laplace's contention been correct, were it a fact that gravitation spreads instantaneously throughout space, we should be faced with gravitational *action' at a distance,* a most displeasing conception. Newton was averse to action at a distance, but it cannot be denied that his law of gravitation was a direct appeal to it. If, on the other hand, gravitation were propagated with a finite speed, we could conceive of it as a continuous action through a medium, similar to that of electromagnetic forces. Gravitation could then be connected with the methods of field physics inaugurated by Faraday and Maxwell. To-day, these investigations of Poincaré have none but a historical interest, for they have been superseded by Einstein's brilliant solution of the problem of gravitation, rendered possible by the introduction of a variable space-time curvature. Nevertheless, it is well to note that should Einstein's general theory be abandoned as a result of some crucial experiment, it would still be possible to preserve the special flat space-time theory by taking the law of gravitation given by Poincaré or Nordström. However, as we shall see, the

---

* Quite recently Dr. Whitehead has endeavoured to work out the same problem afresh.

† A very lucid exposition of Poincaré's attack is given in Cunningham's "Principle of Relativity" (1914), pp. 173 *ff.*

special theory drives us to the general theory in such a variety of ways that there is little fear of science having to suffer the severe setback which a reversion to flat space-time would entail in our understanding of the unity of nature.

Let us now pursue the trail of subsequent discoveries and see how Einstein was led to abandon the idea of a rigidly flat absolute space-time acting as a container for matter, but in no wise modified by the presence of matter. According to his own acknowledgment, this important discovery was reached by at least three different lines of reasoning. We shall examine these reasonings separately, for they throw an interesting light on the methodology of the theory.

In the first place, Einstein remarked that spatial measurements performed on the surface of a disk (rotating in flat space-time) would necessarily yield non-Euclidean results, since, owing to the FitzGerald contraction, the same rod placed radially or transversely would vary in length. But in a rotating frame forces of inertia are active; hence it was suggested that forces of inertia were related to a non-Euclideanism of the space (not space-time) of the frame in which the forces were active. We shall see how this discovery will be of use later on. Incidentally, let us note that this first result follows as a mathematical necessity from the special theory; no new assumptions have been introduced. Expressed geometrically, it means that the observer on the rotating disk splits up flat space-time into a curved space and a curved time.

Now we pass to the second step. The most delicate physical experiments had established the identity of the two types of masses, the inertial and the gravitational. Classical science had regarded this identity as accidental, or at least unexplained. Einstein assumed that we were in the presence of a fact of very great significance in nature, a fact that would have to be taken into consideration in any theoretical co-ordination of knowledge. If we analyse the reasons for Einstein's attitude we find that they reduce to a disinclination to believe that an identity of this sort could be a mere matter of chance. In the same way, if two men were to publish the same book, identical word for word, probability would suggest that one of them had copied the work of the other. If we accept the identity of the two masses as something more than a mere chance occurrence, we must assume that the great similarity between forces of inertia and of gravitation is due to the fact that these two types of forces are essentially the same. But we have seen that in regions of empty space where forces of inertia were active, space (though not space-time) was non-Euclidean. Hence it must follow that in a gravitational field near matter, space must also be non-Euclidean. But there is a marked difference between the distribution of forces of inertia and those of gravitation. Forces of inertia occurring far from matter can be cancelled by the observer's changing his motion. Thus, centrifugal force on the disk can be made to vanish; all we have to do is to arrest the disk's, hence the observer's, rotation. On the other hand, we cannot get rid of the force of gravitation. It is true that in a falling elevator the gravitational pull would vanish as a result of the elevator's motion; and we could no longer feel it, falling, as we should, together with the elevator (owing to the identity of the two masses). But we know that were the elevator extended enough the pull would reappear in distant places because of the radial distribution of the gravitational field round the earth. And so we must conclude that in the case of a gravitational field produced by matter, the non-Euclideanism of space can no longer be got rid of by merely varying our motion, hence by merely varying our method of splitting up space-time into our private space and time. It follows that the non-Euclideanism of space, present in a gravitational field, must come from a deeper source. In particular, it must arise from an intrinsic non-Euclideanism in the *space-time* which surrounds matter, since it is only a curved non-Euclidean space-time that can never be split up into a flat space and a flat time. And so we see that, around masses, space-time can no longer be flat, as it was in regions far removed from matter. The intimate connection between the presence of matter and a non-Euclideanism of space-time becomes apparent. Matter and, more generally, energy cause space-time to become curved.

And here we must note an important difference between the two arguments we have presented thus far. In the first case, that of the rotating frame or disk, the argument was, so to speak, irrefutable: it was imposed as a direct mathematical consequence of the Lorentz-Einstein transformations. But the second argument, that referring to the identity of the two types of masses, was based on an experimental fact. When we consider the wonderful discoveries that have issued at Einstein's hands from this identity of the two masses, we may find it strange that Newton or some other scientist should have failed to attach any theoretical importance to it. Of course, the reason for Newton's neglect to consider the matter is easily understood when we remember that he ignored space-time. Only with space-time, coupled with a knowledge of non-Euclidean geometry, could the full significance of this identity be understood. This same excuse, however, cannot hold for those scientists who in 1908 were just as well acquainted with the special theory as was Einstein himself. Why did three long years have to elapse between the discovery of space-time, in 1908, and Einstein's identification of the two types of masses, in 1911? This brings us to a different subject of discussion.

The point we have been attempting to explain is that the co-ordinations of theoretical physicists are based primarily on experimental facts, not on pipe dreams. But when this point is granted, the genius of the individual scientist consists in singling out those particular facts which he suspects will yield the most interesting consequences. That Einstein should have picked out this identity of the two masses, whereas no one else had thought of it, is a tribute to the genius of Einstein; and that is all that can be said. It is claimed that the idea came to him suddenly when, walking along the street, he saw a man fall from a house-top. This incident is said to have directed his attention to the conditions of observation that would hold for an observer falling freely. The story is reminiscent of Newton's apple. At any rate, regardless of how Einstein came to think of the identity of the two masses, the important point is that this identity constitutes an experimental fact—a fact which had till then appeared in the light of a miraculous coincidence. For this reason, if for no other, it might have been suspected that this marvellous coincidence concealed something important in nature, which we had not yet grasped.

Reverting to the motives which had driven Einstein to suspect that matter must modify the space-time structure, we have seen that they were of a totally different nature. The rotating-disk argument was exclusively mathematical; the equivalence of the two masses was based on a fact of experience. Neither of these two arguments was suggested by philosophy, theosophy or theology. To assert, then, that Einstein was influenced by the philosophy of Descartes or Leibnitz or Mme. Blavatsky or any one else would indicate a very poor knowledge of scientific method. Although the new views entail a marked change in our understanding of the relationships between extension and matter, yet there is no irresponsible guesswork in Einstein's procedure. In particular, there is no desire to impose some special matter-moulding philosophy. If the theory is drifting towards certain philosophical conclusions, it is drifting with the current; and the current is represented by a rational co-ordination of the facts of experience.

We now come to the third argument advanced by Einstein in favour of his matter-modifying space-time theory. We have left it to the last, as it is more philosophical than the two preceding ones. We ·may condense it into the statement that where there is action there must also be reaction. Now, in rigidly flat space-time, the courses of free bodies must be conceived of as directed by the geometry or structure of space-time. Hence there exists an action of the space-time structure on the bodies. But then there should also exist a reciprocal action of the bodies on the structure. In much the same way, the electron affects the electromagnetic field and, inversely, the field affects the electron. Einstein expresses, in the following words, his dislike for a conception which would conflict with the reciprocity of action:

"In the first place, it is contrary to the mode of thinking in science to conceive of a thing (the space-time continuum) which acts itself, but which cannot be acted upon. This is the reason why E. Mach was led to make the attempt to eliminate space as an active cause in the system of mechanics. According to him, a material particle does not move in unaccelerated motion relatively to space, but relatively to the centre of all the other masses in the universe; in this way the series of causes of mechanical phenomena was closed, in contrast to the mechanics of Newton and Galileo. In order to develop this idea within the limits of the modern theory of action through a medium, the properties of the space-time continuum which determine inertia must be regarded as field properties of space analogous to the electromagnetic field. The concepts of classical mechanics afford no way of expressing this. For this reason Mach's attempt at a solution failed for the time being."

If we review the position of the theory as it now stands, we may summarise it as follows: Space-time is primarily an absolute four-dimensional continuum of events. When devoid of matter and energy, its structure is flat. When matter is present, the flat structure yields gently both around matter and in its interior. The yielding, however, is exceedingly slight. Though sufficient to account for gravitation, it is far too insignificant to be detected through direct measurements with rods (this holds at least in the case of the space around the sun). Disturbances of structure, caused by the sudden arrival or passage of matter, are propagated from place to place with the velocity of light, so that gravitation manifests itself as a continuous action through a medium, just like the action of the electromagnetc field. Action at a distance is thus avoided.

And now, what is the bearing of these discoveries on the problem of absolute rotation and absolute space-time? We see that space-time, by yielding slightly in the presence of matter, has lost that absolute rigidity which characterised Newton's space. Owing to these varying distortions of the space-time background, brought about by variations in the matter distribution, the rotational velocity of a body suffers from a certain measure of indeterminateness. To this extent rotation is no longer absolute. But when we review the sequence of deductions that has led us to this partial rejection of Newton's belief in absolute rotation, we see that our discoveries have become possible only through the medium of space-time. Had space-time never been discovered, had we remained content with a separate space and time, it would have been quite impossible to establish this indeterminateness of rotation; *for it would have been impossible to account for gravitation in terms of the variations in the structure of space alone.* As a result, the spatial background would have remained rigid, and Newton's position would have stood secure.

Even now, though the fluctuating space-time background has been discovered, the absolute nature of rotation has not been fully disproved. For in an empty universe, space-time would still preserve a well-defined rigid structure; hence, if we conceive of one single body introduced into this otherwise empty world, an absolute rotation of this body could always be detected. It would be betrayed by centrifugal forces; and these forces would arise because the particles of the body would be following world-lines which departed from the geodesics, or lines of least resistance, of the space-time structure. For this reason, rotation would still be absolute, since it would preserve a real meaning in a world devoid of matter. In short, it would have nothing to do with a rotation relative to the star-masses. Obviously we are still a long way from having vindicated Mach's mechanics.

But when it is considered that Einstein's theory is based on the relativity of velocity, then that it establishes, in a partial way at least, the relativity of acceleration, it cannot be denied that a logical extension of its triumphs would be for it to succeed in establishing the complete relativity of all motion and of inertia. Although Einstein never conceals his hope that this will be the ultimate outcome, there is no desire on his part to force the issue in defiance of facts.

It may, however, be of interest to ascertain what modifications in the theory would be required for the relativity of all motion and of inertia to be established. Inasmuch as it is the inherent structure of space-time persisting even in the absence of matter which is responsible for centrifugal force in an empty world, hence for absolute rotation and inertia, a vindication of Mach's ideas would entail the following condition: It would be necessary to assume that in the absence of all matter, space-time would lose all trace of a structure and become amorphous. In other words, matter would have to create space-time and its structure (*i.e.*, the g_{ik}-distribution, or the metrical field), and *not merely modify locally a pre-existing structure*. It is interesting to note that the relativity of rotation is thus leading us to a concept of space which could not have been anticipated *a priori*. Mach himself, when defending the relativity of motion, does not appear to have realised that the abandonment of Newton's absolute space would entail a matter-created universe.

Now what Einstein has done has been to prove that it is mathematically possible to conceive of a universe in which the entire space-time structure (hence the entire g_{ik}-distribution, or metrical field) would be conditioned by matter. This, as we know, is the cylindrical universe. But, here again, we must remember that he has never asked us to accept this cylindrical universe merely because it entails the relativity of inertia and of rotation. He states expressly that the theory of relativity does not require it as an inevitable necessity; he merely suggests it as a possibility. If we assume that our stellar universe is concentrated into a nucleus surrounded by infinite space, the quasi-Euclidean infinite universe must take the place of the cylindrical one. With this alternative universe, matter still modifies space-time locally, but can never create it *in toto*. As Einstein tells us: "If the universe were quasi-Euclidean, then Mach was wholly in the wrong in his thought that inertia, as well as gravitation, depends upon a kind of mutual action between bodies."

Today the cylindrical universe is a discarded theory, having been superseded by Lemaitre's expanding universe. We have reverted to it in this chapter because our aim has merely been to show how ideas were arrived at.

These discussions on the form of the universe bring to light another important aspect of the methodology of science. We may notice that the accord which was practically unanimous in the earlier part of the theory now gives way to dissenting opinions. The reason for this is of course obvious. In the earlier part, experimental verifications were soon forthcoming and permitted us to decide matters one way or another. But when we are dealing with the form of the universe, verifications are extremely difficult; they depend on tedious astronomical observations which may require many years to be completed. Hence, the only method of advance is to explore all mathematical possibilities in the hope that somehow or other we may be led to an *experimentum crucis*.

But in all cases the driving force behind these attempts has been a desire to co-ordinate mathematically a maximum number of experimental facts in the simplest way possible. These facts have been collected from the various realms of physical science, and, indeed, it is owing to their wide variety that the conclusions obtained are compelling. Furthermore, it should be noted that the facts appealed to are the results of highly refined experiments. This again is important, since we have seen that a difference one way or another of a tiny fraction of an angle or a length might be sufficient to overthrow an entire theory.

In short, we see that this problem of the relativity of rotation, of the relationships between matter and extension, of the finiteness of the universe, is not one for philosophy and *a priori* arguments to solve. A mathematical co-ordination of the facts developed by ultra-precise experiment can alone yield us an answer. As Weyl tells us:

"The historical development of the problem of space teaches how difficult it is for us human beings entangled in external reality to reach a definite conclusion. A prolonged phase of mathematical development, the great expansion of geometry dating from Euclid to Riemann, the discovery of the physical facts of nature and their underlying laws from the time of Galileo together with the incessant impulses imparted by new empirical data, finally, the genius of individual great minds, Newton, Gauss, Riemann, Einstein, all these factors were necessary to set us free from the external, accidental, non-essential characteristics which would otherwise have held us captive."

To be sure, as our knowledge accumulates and our mathematical ability increases, we may be led to revise former conclusions. This is why the conclusions of Einstein differ from those of Newton, just as those of our descendants may differ from ours of to-day. But the main point to grasp is that these variations in our philosophical outlook are brought about by our increase of knowledge both mathematical and physical; and this increase requires centuries: it cannot be obtained overnight. Indeed, had Einstein lived in Newton's time and ignored the mathematical and physical discoveries of the present day, he could never have improved upon Newton's solution. Inversely, had Newton lived in our time and been acquainted with the facts known to modern science, it is very possible that he would have created the theory of relativity.

In our analysis of the methodology of scientific theories we have attempted to show that the deductions of the great scientists were always based on experiment and not on wild guesses. But this does not mean that other factors have not also come into play. For instance, we mentioned that a proper choice of facts was of supreme importance. Whereas a certain choice may lead to nothing of interest, some other choice may issue in a great discovery. A case in point was afforded by Einstein's attitude towards the well-known identity of the two masses.

But even this is not all. In mathematics, as in architecture, certain coordinations are beautiful, others top-heavy or unsymmetrical. When, therefore, physical phenomena are translated into mathematical formulæ, the theoretical physicist will always endeavour to obtain beautiful equations rather than awkward ones. Of course, since he must restrict himself to a slavish translation of physical results, his initiative in this respect may be extremely limited. Nevertheless, in certain cases it may be possible to add an unobtrusive term to the equations, though this term may not actually be demanded by experiment. If this additional term renders the equations more symmetric, there will be every incentive to retain it. It was an urge of this sort that guided Maxwell in his discovery of the equations of electromagnetics. The experimental data known in his day could not establish whether a certain additional term was necessary or not. Nevertheless, Maxwell introduced it because it beautified his equations, and was thus led to anticipate the existence of electromagnetic waves. As we know, subsequent experiment has justified Maxwell's deep vision. It is much the same with the theory of relativity. One of the reasons for its great appeal to mathematicians is the extreme harmony, beauty and simplicity which it permits us to bestow on many of the equations of physics.

However, if we leave these æsthetic urges aside, we may say that the methodology that has yielded us the theory of relativity is the same methodology that has yielded all the great scientific discoveries. It is the methodology of Galileo, Newton and Maxwell: First, ascertain experimental facts; then, as needs be, frame tentative hypotheses or scaffoldings for the sole purpose of co-ordinating these facts into a consistent whole with a maximum of simplicity. Only when the theory has succeeded in accumulating a sufficient number of facts can its philosophical implications be studied. It follows that any attempt to reverse the normal order and posit the philosophy before the science will result in hampering future discovery by subordinating accuracy of treatment to loose guesswork. The history of human thought is full of discarded philosophic prejudices swept away by the onward march of science. Every new scientific discovery reveals aspects of nature, or even of the mind, as in pure mathematics, which we never suspected before. To-day, both the discoveries of relativity and the quantum hypothesis are case in point. It is precisely because the philosophy of a theory of mathematical physics can only be attained *a posteriori,* coming as a crowning achievement, that the philosophy which is beginning to disentangle itself from Einstein's discoveries is still in an embryonic stage. It is not that a scientific philosopher of sufficient scope could not be found; it is because there are many scientific aspects of the subject which are still in doubt. The following passage from Einstein will explain what we mean:

"From the present state of theory it looks as if the electromagnetic field as opposed to the gravitational field rests upon an entirely new formal *motif*, as though nature might just as well have endowed the gravitational ether [metrical field *] with fields of quite another type, for example with fields of a scalar potential instead of fields of the electromagnetic type.

"Since according to our present conceptions the elementary particles of matter are also in their essence nothing else than condensations of the electromagnetic field, our present view of the universe presents two realities which are completely separated from each other conceptually, although connected causally, namely, gravitational ether [metrical field] and electromagnetic field, or—as they might also be called—space and matter.

"Of course it would be a great advance if we could succeed in comprehending the gravitational field and the electromagnetic field together as one unified conformation. Then, for the first time, the epoch of theoretical physics founded by Faraday and Maxwell would reach a satisfactory conclusion. The contrast between ether [metrical field] and matter would fade away and through the general theory of relativity the whole of physics would become a complete system of thought like geometry and the theory of gravitation."

We have quoted this passage at length, because it appears to us that by reading between the lines it is possible to realise the deeper philosophical problems pertaining to our knowledge of nature, hence to knowledge in general, which still obscure the significance of relativity.

* Inserted in order to conform to previous terminology.

CHAPTER XXXVIII

THE GENERAL SIGNIFICANCE OF THE THEORY OF RELATIVITY

THE theory of relativity, as we have seen, is one of mathematical physics; and its sole aim has been to co-ordinate the greatest number of natural phenomena and experimental results in as simple and as direct a manner as possible. In common with all other great developments of theoretical physics, it has adhered strictly to scientific methods. But now that the mathematico-physical co-ordination has been completed, it becomes permissible to investigate the changes which the theory may necessitate in our philosophical understanding of nature.

The most interesting aspect of the entire theory for philosophers would appear to be the discovery of space-time, with the paradoxes of feeling to which it leads. Thus, the duration of our life, the distance we cover, have no absolute significance. Two twins might both live seventy years in their own estimation, and yet if they met again before their death one might be younger than the other. Two men starting from the same point and travelling in the same direction might both cover what they would measure as the same distance, and yet might find when they came to a stop that they were many miles apart. The fact is that a distance and a duration are not absolute; they merely express a relation between something that is absolute, and the space-time mesh-system or motion of the observer. There is no more cause to be surprised at the curious results we have mentioned than at the relative character of the length of the shadow cast by a pole, or of the visual angle under which we see the pole.

As an immediate consequence of this relativity of distance and duration, we must accept the relativity of simultaneity. Classical science recognised that no absolute significance could be attached to the concept of the same point of space considered at different times; for since points of space could be determined only with respect to a system of reference, two events occurring in succession at the same point of our system would obviously occur at two different points of some other system, in motion with respect to the first. On the other hand, classical science refused to extend similar relativistic considerations to the concept of the same instant of time at two different points. It was assumed that a change of system of reference could have no effect on the rate of time-flow, which remained absolute, ever the same for all observers. It is this belief in the absolute nature of simultaneity that is shattered by Einstein's discoveries. Henceforth no greater significance can be attached to the absolute sameness of time in different places than to the absolute sameness of location at different times.

We may recall that all these so-called paradoxes of relativity are by no means antinomies of reason or of logic, which the theory could not possibly survive. They are merely paradoxes of feeling, which arise when from force of habit we continue to be haunted by the classical concepts of absolute duration and distance, and to credit these erroneously with some *a priori* inevitable virtue of logical necessity.

When we consider that there is scarcely a single field in our understanding of the physical universe upon which relativity has failed to throw some new light, we can realise that as yet the full significance of its discoveries is far from being established. Nevertheless, sufficient positive information has been gathered with respect to the concepts of space and time for certain philosophical conclusions to appear legitimate. What the theory has proved conclusively is that real space and real time can be approached only by empirical methods, and in particular by ultra-refined experiment.

We mentioned, when discussing space, that Kant's philosophy of the *a priori* nature of three-dimensional Euclidean space as representing the form of pure sensibility was exploded years ago by the mathematicians from Gauss onwards. So far as scientists were concerned, the matter was definitely settled. The theory of relativity merely confirms these views by justifying them on physical as well as on mathematical grounds. Henceforth, four-dimensional space-time, which, though flat, is non-Euclidean (in the extended sense), owing to the imaginary dimensionality of time, constitutes the fundamental continuum of the universe. Had it not been for the general theory, it might have been argued that space-time possessed no deep significance, and that it reduced to a mere convenient mode of mathematical representation, like the graph of a barometer needle. Though we may state (even confining ourselves to the special theory) that this opinion would have been untenable, the use which Einstein has made of space-time in the general theory excludes this view completely. With space-time we are in the presence of a new continuum which has exactly the same measure of reality as was formerly attributed to three-dimensional space.

Now we have seen that with the rejection of the *a priori* Kantian attitude, mathematicians long before the advent of Einstein's theory were thrown back on the empirical philosophy of space as arising from the simplest co-ordination of our sensory impressions. In particular, the apparent inevitability of Euclidean space was assumed to result from the co-ordination of sensory experience, whether that of the individual or that of the species. In other words, it was the behaviour of material bodies and of light rays which was held to be the fundamental source of our natural belief in the Euclideanism of space. But here an objection might be urged, for it might be contended with perfect truth that according to Einstein the behaviour of material bodies and light rays is such as to yield space-time, and not separate space and time. How, then, could we account for space-time having failed to impose itself ages ago; and why had it been necessary to await the arrival of Minkowski and Einstein?

The reason is, of course, obvious. Theoretically, there is a great difference in structure between space-time and separate space and time, and this difference arises, as we know, from the finiteness of the invariant velocity in space-time as contrasted with its infinite magnitude in separate space and time. The higher the relative speeds of bodies, and, in particular, the nearer they approximate to the critical speed of light, the more marked will become the physical discrepancies between the two worlds. And as the velocities of those bodies which enter into our daily experience are exceedingly small in contrast to that of light, crude perception cannot enable us to detect the minute differences which are theoretically present. To all intents and purposes, therefore, the world of common experience is one of separate space and time, and not one of space-time. In much the same way, the mere survey of a landscape would be incapable of revealing the rotundity of the earth. And so we understand why it was that the discovery of space-time had to await the performance of ultra-refined physical experiments.

We might, of course, conceive of beings whose senses were so far developed as to detect the minute differences that escape us. If such beings existed, there is every reason to suppose that their sensibility would accept space-time as a matter of course, just as ours accepts three-dimensional Euclidean space and time. In the words of Eddington: "It is merely the accident that we are not furnished with a pair of eyes in rapid relative motion, which has allowed our brains to neglect to develop a faculty for visualising this four-dimensional world as directly as we visualise its three-dimensional section." Being so poorly equipped as to sensory organs, being unable to discern the minute discrepancies that arise when C is very large though finite and when C is infinite, we have got into a rut and have believed from ages immemorial in a world of separate space and time. There is no reason to marvel at the paradoxes of feeling which we suffer when told of the strange disclosures that would ensue were relative velocities to become very great or our senses to become more acute. In much the same way, quite apart from the disclosures of relativity, were our eyes to be suddenly gifted with microscopic vision, our familiar world would become so terrifying as to baffle description.

Now the essential characteristic of space-time is that it exhibits no absolute separation between space and time which would hold for all observers, regardless of their motion. Each and every observer will split it up in a different way, so that the totality of space at an instant for one becomes a successive spread of points throughout space at a succession of times for another. Of course, so long as we are considering only our immediate perceptions, we shall always split up space-time into a separate space and time; and this fact has been construed by some as implying that the space-time amalgamation is obviously artificial, since it can be so easily separated into space and time. Needless to say, the argument is faulty. For the space and time that we obtain as a result of a splitting up of the space-time continuum correspond to a private view of the world, and it is never the private views considered in isolation that give us the objective world. It is solely a certain construct or synthesis of the private views which can be deemed to represent objective reality. To revert to an illustration from classical science, it is not the shape of the rock as seen from *here* or *there* that yields us a knowledge of the rock's shape, but solely a synthesis of the appearances of the rock as seen from here, there and anywhere. In relativity we must also consider the perceptions of observers moving with various relative motions. Then it is found that objective shape, size and duration can no longer be constructed by a synthesis of private views, in contrast to what classical science had erroneously supposed. For this reason, distance and time separation, or, what comes to the same thing, space and time, though remaining real from the standpoint of the private view, can no longer be claimed to constitute the scaffolding of the common objective world. From an objective point of view, such concepts are ambiguous, and we must substitute for them a space-time world and space-time configurations. In short, it must be realised that space-time is an entirely new concept, new both in science and in philosophy, and it must not be confused with the mere artificial representation of time and space directions in graphs such as classical science had often made use of.

Possibly further illustrations of space-time may be helpful. Thus, if we consider the three-dimensional extension occupied by the atmosphere and note the variations in temperature that occur from place to place, we are justified in speaking of a four-dimensional continuum space-temperature. But it is obvious that we are at liberty to divorce these two entities and study them separately, obtaining thereby two separate continua, one of space and one of temperature. In classical science, it was much the same with space and time. We could consider them jointly as constituting a four-dimensional continuum of events but then again, owing to the absolute nature of time and distance, we could divorce them. When this was done, both three-dimensional space and uni-dimensional time were separated into distinct self-supporting continua.

With the space-time of relativity, a separation of this sort is impossible. We cannot withdraw the time dimension, for it is no longer unique. If we cancel it so far as some definite observer is concerned, leaving him with an instantaneous space, we shall have mutilated the instantaneous space, without having cancelled time, in the estimation of another observer in relative motion. We see, then, that whereas in classical science the adjunction of time to space was artificial, in that time and space could always be separated like oil and vinegar, such is no longer the case in relativity. Space and time are now blended together and can be claimed to constitute one solid whole. In a very crude way, we may say that space-time is more akin to an opal emitting various colours in various directions. Just as with space-time it was impossible to annihilate time or space for all observers, so here, with the opal, it would be impossible to cancel all the points that reflected green light and only green light, since if we pursued this course from the standpoint of one observer we should have failed to achieve our object from the standpoint of another.

Let us now consider the characteristics of metrical continua. From a mathematical standpoint, a metrical continuum is one with which it is possible to associate a geometry, hence to obtain an invariant expression for the distance separating two points of the continuum, the value of this expression holding for all observers. In the case of space-time, this would imply that the distance between any two point-events should exhibit the same space-time separation holding for all observers regardless of their motion. Now, in the space and time of classical science, no such invariant space and time distance was possible. A simple example may make this point clear. Suppose that in classical science we were standing on a railroad embankment, and viewing the production of two instantaneous events—say, two light flashes emitted by two spatially separated signals on the line. By taking into consideration the speed with which the light signals had progressed towards us and by measuring our distance from the respective signals, we could determine the spatial distance between the two signals and the respective times at which the flashes had occurred. We should thus obtain the temporal separation and the spatial separation of the two events as referred to the embankment. But suppose now that we had viewed the same two events from a train moving uniformly along the track. In classical science, time being absolute, the temporal separation between the two events would, of course, suffer no modification. The spatial separation as referred to the train would, however, generally be different. We can understand this by supposing that the train passed before the two successive signals as their respective flashes were emitted; in this case, *as referred to the train,* the two events would have occurred at the same point of the train, and so their spatial separation would be zero. We see, then, that the time separation would be the same in either case, whereas the space separation would have varied. As a result, the space and time separation could not remain invariant. For the separation between the two point-events to manifest itself as invariant, we may surmise that the time separation should also vary in such a way as to offset the variation of the space separation. The absoluteness of time in classical science rendered this impossible, and this is what is meant by stating that space and time were not amalgamated into one continuum in any deep sense.

Now, what Einstein tells us is that had we performed our measurements with extreme accuracy, using our customary chronometers and rigid rods, we should have found that the time separation between the two events *had* varied slightly; and that as a result this slight variation would have cancelled the corresponding change of the space separation, leaving the space-time distance unmodified.

Calling x and t and x' and t' the space and time separations, as measured from the embankment and the train, respectively, we should have found that

$$x^2 - c^2 t^2 = x'^2 - c^2 t'^2.$$

Here, then, was the expression of the space-time distance between the two events, or rather of its square. It is the fact that our measurements would have verified the invariance of this distance, which permits us to consider space and time as amalgamated. It cannot be too strongly emphasised that our clocks and rods would have been the ordinary clocks and rods of classical science. They would not have been adjusted so as to force relativistic results, as is often erroneously supposed; for had this been the case the theory would have been artificial and worthless. All that the theory suggests, therefore, is that our ordinary rods and clocks do not behave quite as we used to believe they did.*

We see, then, that the juxtaposition of space and time in classical science was just as artificial as a juxtaposition of space and temperature would have been. In short, the reality of space-time arises from Minkowski's discovery that it was possible to define an invariant distance between two points in space-time, holding for all observers; and that it was impossible to define any such invariant distance in space alone or in time alone, showing that space and time by themselves were phantoms. These last two concepts must henceforth be considered jointly, and no longer as separate entities.

Now it is apparent that if the model of space-time is to spare us from a conflict with certain fundamental facts of consciousness it is imperative that some kind of mathematical difference should distinguish time directions from space directions. However, we need not anticipate any difficulties from the space-time model on this score; for we know that, owing to the heterogeneity which exists between time directions and space directions, space-time is anisotropic. Bearing this peculiarity in mind, we see that space-time around any point-event can be split up into a bundle of time directions and into bundles of space directions; these, however, can never overlap with the time-bundle. Then, from this bundle of possible time directions, every observer will select that particular line which constitutes his world-line through space-time. As for the flow of

---

* The difference is, however, exceedingly minute, as we can judge by reverting to the expression for the invariant space-time distance $s^2 = x^2 - c^2t^2$. If x is expressed in metres and t in seconds, c will be 300,000,000. Suppose, then, that the two signals, as measured from the embankment, are ten metres apart, and that the two light flashes are separated by an interval of one second. If these two events are viewed from a train travelling at the rate of ten metres a second, that is, at approximately 22 miles an hour, it will be possible for the observer in the train to pass before each of the two flashes just as they are produced, with the result that as referred to his train the two events will occur at the same spot. The time separation t' of the two events as referred to the train will then be given by $s^2 = c^2t^2 - x^2 = c^2t'^2$ (since $x' = 0$). Substituting our numerical values, we find $t' = 1 - \dfrac{100}{(300,000,000)^2}$, which is very nearly one second, as it would rigorously have been in classical science. We see, furthermore, that it is owing to the enormous value of c that the difference between the two sciences is so hard to detect in practice. It is for this reason that the fundamental continuum, though one of space-time, reduces for all practical purposes to the separate space and time of our forefathers, unless very high velocities are considered.

psychological time, or the duration between two psychological events measured by a clock at our side, it will be given by the space-time length of our world-line limited by those two events. Thus, two world-lines of equal length will always correspond to the same duration of time lived by the observers who follow these world-lines. For instance, of two identical sodium atoms, one near the sun and one far away, the solar atom will beat a slower time; but for both atoms the segments of the world-lines limited by successive vibrations will be the same in length. The existence of the Einstein shift-effect has verified this fact, and it is equivalent to stating that the atoms behave like perfect clocks.

So far as an observer's three space directions are concerned, they will be perpendicular to his world-line or time direction. If his time direction is curved owing to his acceleration or to the intrinsic curvature of space-time due to matter, the space and time directions will be properly determined as perpendicular to one another only in the immediate vicinity of the observer. Elsewhere a definition of orthogonality, or perpendicularity. will be impossible; as a result the very concepts of time and space become blurred in the general theory, and we have to content ourselves with co-ordinate systems which constitute measurements neither in space nor in time. We cannot dwell on this type of difficulty because it necessitates explanations of too technical a nature; suffice it to say that it is this complication that causes all problems connected with rotating disks and gravitational fields to suffer from a certain measure of obscurity when we attempt to interpret them in terms of space and time. We may also add that when large masses are moving about, the concept of simultaneity throughout space even for a given observer loses all precise significance.

The great difference between our present views and those of classical science is that whereas in classical science our lines of time-reckoning and of space-reckoning were unique, they now offer an indefinite number of possible alternatives, because of the plurality of time and space directions present in space-time. It is as though the flow of time were represented no longer by a single stream, but by a number of branches radiating from every point in different directions. In spite of this ambiguity in the possible space and time directions, we have seen that to each and every observer a definite space and time separation will correspond; so that all ambiguity disappears * as soon as we have defined the observer.

The strange results and concepts which the theory of relativity has disclosed have been appealed to by a number of writers in order to justify an ultra-idealistic philosophy of nature. These men have concluded that the real objective universe about which scientists were never weary of talking had been dashed to the ground and that realistic science had finally surrendered to the idealistic school of philosophy. This expression of opinion results from a confusion between the point of view of science and that of the metaphysical realist.

* Subject to the limitations mentioned in the previous paragraphs.

The vast majority of modern scientists are agnostics in that they reject the claim of the metaphysical realist who presumes to have discovered substance and true being in the outside world. They will claim that substance and the thing in itself are unknowable, or at least that these elude rational investigation, and that the objective world of science is nothing but a mental construct imagined for the purpose of co-ordinating our sense impressions. But, once this point is admitted, they will recognise that this mentally constructed objective universe must to all intents and purposes be treated as a reality pre-existing to the observer who discovers it bit by bit. This last expression of opinion is not the result of some philosophical system. It is imposed upon scientists as an inevitable conclusion; for had it been proved impossible to imagine a common objective universe, the same for all men, science could never have existed, since it would have reduced to individual points of view which could never have been co-ordinated. In other words, knowledge would have lacked generality; and without generality there could have been no such thing as science.

Had the theory of relativity, for instance, failed to furnish us with a common objective world, it could never have gone very far. All we can correctly state is that the objective world of absolute duration and distance has faded away, that space and time as the fundamental constituents of the objective world have sunk to mere shadows. Expressed in other words, the so-called primary qualities of shape, size, duration and mass erstwhile considered absolutes must now be regarded as relatives having no definite value until the observer is specified. But this work of destruction has been followed by one of reconstruction, as a result of which a new objective world, that of space-time with its invariant intervals, has taken the place of the discarded world of separate space and time.

Idealistic metaphysicians have a habit of passing rapidly over space-time as if it were of no particular importance. They lose sight thereby of the entire significance of Einstein's theory. As Einstein tells us himself: "Without space-time the general theory of relativity would perhaps have got no farther than its long clothes." Even though we were to limit our investigations to the bare facts of the theory, without studying it more deeply, the idealistic interpretation would be open to serious objections.

Consider, for example, the phenomenon of gravitation. It is obviously impossible to assert that the apparent bending of a ray of starlight as it passes near the sun is due entirely to the idiosyncrasies of the observer; for we know that were the sun to be removed suddenly during an eclipse the ray of starlight would automatically cease to bend, in spite of the fact that the purely subjective space and time idiosyncrasies of the astronomer at his telescope would have had no occasion to vary. Again, in the special theory, the space and time computations vary from one observer to another only when relative motion exists between the observers. The norms of Peter do not differ from those of Paul because Peter is Peter and Paul is Paul, but only because Peter and Paul are in relative motion. Were their relative motion to cease, their norms would cease to differ.

But this is not all. According to the idealistic views, the norms of the various observers are so adjusted, owing to some miraculous pre-established harmony, that the velocity of light appears the same to all. As a matter of fact, this statement is incorrect, for the velocity of light is invariant only for Galilean observers; it is variable from place to place when accelerated observers are considered, and it is also variable in the neighbourhood of matter. Again, if we revert to Michelson's experiment, we see that its essential feature is to have demonstrated that no matter what the velocity of a sphere as a whole may be, rays of light which leave the centre at the same instant of time all return to the centre at the same instant, after having suffered a reflection against the sphere's inner surface. In this experiment, we are therefore dealing with a *coincidence,* with waves of light which pass the same point at the same time. The space and time elongations produced by the idiosyncrasies of the observer could never create or disrupt a coincidence of this sort. They might cause a duration or a distance to appear longer or shorter; but they could never cause two rays of light to intersect on the surface of some object if they did not intersect there in the opinion of all other observers. Thus an intersection or a coincidence is an absolute, for the simple reason that no amount of magnification will ever cause a point to appear as two separate points. A point has no dimensions which can be torn apart by magnification. It follows that in an experiment such as Michelson's it is impossible to attribute the observed coincidence to the idiosyncrasies of the observer; there is obviously a mysterious something in nature which will have to be conceived of as existing independently of the observer and as situated in the objective world of science. This something is found to be represented by the intersections of the world-lines in absolute space-time, *i.e.,* in the objective world of science.

As a matter of fact, in the general theory the concepts of space and time become completely indeterminate, and it is only in the very simplest type of problems that it is feasible to reason in terms of space and time. As Bertrand Russell points out, the idealistic interpretation appears to have arisen from the continual reference to "the observer" in Einstein's theory. It should be realised, however, that "the observer" is a very loose term. A photographic camera and clock, or any other mechanical registering device, would be just as appropriate for purposes of observation as would a living human being. In short, the principal weakness of the exponents of the idealistic interpretation seems to be their desire to cling to space and time in spite of everything. They fail to see that the objective world of relativity cannot be built up of space and of time in the same simple way as that of classical science. Summing up, we may say that whether space-time be considered a metaphysical reality, or whether it be regarded as a mental construct devised for the purpose of co-ordinating our sensory experience, in either case it plays the part of an objective reality (in the scientific sense). We must conceive of it, therefore, as an entity pre-existing to the observer who explores it and locates events in its substance.

Now the criticism has often been raised that there is nothing in Einstein's space-time model which enables us to explain the reason for that most fundamental fact of consciousness, our awareness of the flowing of time. It is difficult to see why a criticism of this sort should be directed with any greater justification against space-time than against the classical theory. The fact that in classical science space and time were separate did not allow us to explain the phenomenon of the flowing of time any better than does the space-time theory. To state with Newton that time flows teaches us nothing about the cause of this flow. Hence, we must conclude that whatever merit there may be in the critic's objections, they would apply with equal force to classical science. But when we analyse the reason for these objections we find that the critic is starting from the assumption that the passage of time must be conceived of as pre-existing in the objective universe, and is thus denying its purely subjective nature. Speculations of this kind can never lead us very far in the present state of our knowledge; but, if need be, there is nothing to prevent us from assuming that space-time presents dynamic properties which urge natural processes on from past to future along the bundle of time directions. Those, however, who suggest views of this sort have not accomplished much thereby; they have merely removed the mystery of the passage of time from human consciousness and placed it elsewhere in the objective universe. It seems safer to concede that our awareness of the passage of time is in all probability of a subjective nature, connected with our consciousness of being alive. In this case we should have to assume that it is our consciousness which rises along a world-line, discovering, as it proceeds on its course, events situated in the cone of its absolute past. This has always been the view defended by the most competent authorities, in particular by Minkowski, Einstein, Weyl and Eddington. As Weyl tells us:

"However deep the chasm may be that separates the intuitive nature of space from that of time in our experience, nothing of this qualitative difference enters into the objective world which physics endeavours to crystallise out of direct experience. It is a four-dimensional continuum, which is neither 'time' nor 'space.' Only the consciousness that passes on in one portion of this world experiences the detached piece which comes to meet it and passes behind it, as *history*, that is, as a process that is going forward in time and takes place in space. . . . It is remarkable that the three-dimensional geometry of the statical world that was put into a complete axiomatic system by Euclid has such a translucent character, whereas we have been able to assume command over the four-dimensional geometry only after a prolonged struggle and by referring to an extensive set of physical phenomena and empirical data. Only now the theory of relativity has succeeded in enabling our knowledge of physical nature to get a full grasp of the fact of motion, of change in the world."

As can be seen from the preceding quotation, our awareness of the passage of time is ascribed to the passage of our consciousness through the space-time continuum. According to the views of Minkowski, a material body is then represented by a world-line, or continuous chain of events, the entire length of which is subjected to a modification of structure which we come to interpret as connoting the presence of matter. So long as we reason in terms of space-time, this is all we can say. But when in our habitual perception of things, *we,* as sentient observers, split the world up into an appropriate space and time, what we perceive of the world-line is its momentary intersection with our cone of the passive past. If our world-line and that of the body which is being observed are parallel, the body will be said to be at rest. But if the two world-lines are not parallel, then, when interpreting things in terms of space and time, the body will be said to be in relative motion.

We may illustrate these same views by considering the following example: We shall assume for reasons of convenience that space-time is reduced to three dimensions: two of space and one of time. Suppose,

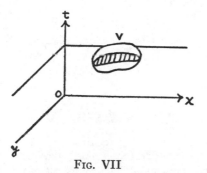

Fig. VII

then, that in this block of space-time we consider a volume of redness. If we assume that our own world-line is given by *ot,* our instantaneous present will at all times be given by an indefinite plane parallel to *xoy.* As our existence at every instant is represented by the successive points along *ot,* our instantaneous present will gradually rise, intersecting the red volume *V* along successive plane areas (Fig. VII). We may say, therefore, that from the standpoint of space and time, the red space-time volume appears as a splash of red colour squirming and varying in shape; in spite of the fact that from the space-time standpoint the volume is fixed and motionless. But the varying shape of the red patch as described above corresponds to the patch as it exists in our instantaneous present. It is not this that we perceive with our eyes, for whatever we perceive is already situated in our past. Hence, what we perceive at a given instant will be given, *not* by the intersection of the red volume with the plane of our instantaneous present, but by its intersection with the surface of our instantaneous cone of the passive past (Fig. VIII).

This is the light-cone which passes through those space-time events which are in our present range of vision. All other events are from a visual standpoint either in our visual past or future. (More detailed explanations will be given in the Appendix).

Looking back on what has been accomplished, we see that the motion and change with which we are familiar in our daily experience are now interpreted as arising from a passage of our consciousness through space-time. The cause of this passage remains utterly mysterious; but, as we said before, it would not solve our difficulties to attribute it to dynamic properties of the continuum itself—dynamic properties which would urge our consciousness on. Even were this solution to be adopted, the cause of the dynamic properties and their mode of action on our consciousness would remain as mysterious as ever.

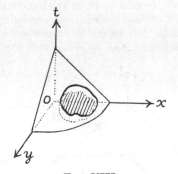

FIG. VIII

Now, in this interpretation of space-time, there is no room for free will, since everything is already predetermined and pre-exists in the future. To be sure, this is a most unsatisfactory aspect of the presentation. But how are we to avoid it? In classical science we could always reserve a part for free will by assuming that the present and the past alone were fully determined while the future might be modified at will. But in relativity this neat separation is no longer possible. For if we consider two observers passing each other, the present of the one comprises both the past and the future of the other. Hence, we cannot banish the future without annihilating a portion of the past together with the present, or at least that portion of the present which stands elsewhere than in our immediate vicinity. Neither may we assume boldly that the seat of free will resides in some suprasensible world transcending that of space-time, for the position of an object on the table is, after all, an effect of the will that placed it there, and the body as a material object occupies a position in space-time.

At any rate, there appears to be very little use in speculating on puzzles of this sort. The fact is that science is necessarily deterministic, not through an act of faith, not because it has convinced itself that free will is non-existent, but because determinism is for science a dire necessity if phenomena are to be co-ordinated and linked to one another. Regard-less of what the future may hold in store, the physicist is therefore compelled to act as though a rigidly deterministic scheme were existent in nature, and to consider space-time and its content as given, even though he may doubt whether such is really the case.

Interesting sidelights on causality are offered by the space-time theory. Since no effects can be propagated with a speed greater than that of light, the passage of time is an essential if causal connections are to operate. This was not necessarily true when instantaneous effects (or action at a distance), were contemplated. The theory of relativity thus imposes restrictions on causal connections; it allows us to assert that between certain space-time events no causal connections can ever exist. It is well to note, however, that the category of causality has no place in space-time as such. Only when we consider the perceptions of the observer can causality be considered, since only then is the passage of time introduced. When discussing space-time from an impersonal point of view as a four-dimensional continuum which is static, it would therefore seem preferable to refer to functional relationships.

The next most important aspect of the theory of relativity which we have to consider relates to our conception of real space. We remember that the geometrical space of mathematics was amorphous. It possessed no intrinsic metrics, no special geometry, no size, no shape; and the geometry which the mathematician credited to space was purely a matter of choice. On the other hand, the real space of physics appeared to possess a definite metrics. Furthermore, the amorphous nature of space was belied by the existence of centrifugal forces and the like, proving that not all motions through space were equivalent.

When it came to determining the precise geometry of real space, various methods could be considered. We might proceed by appealing to rigid bodies, that is, to bodies which visually and tactually appeared to be the same wherever we might carry them. More generally, rigid bodies were assumed to be exemplified by material bodies maintained as far as possible under permanent conditions of temperature and pressure. We could also investigate space by means of light rays, assuming that their courses *in vacuo* would define geodesics. Over a century ago Gauss attempted experiments of this sort between the summits of two mountains. A still more general method consisted in appealing to the laws of physics in general, and determining the type of geometry which would have to be credited to space in order to permit us to co-ordinate natural phenomena with the maximum of simplicity. This procedure was suggested by Riemann in 1854 in his inaugural address at Göttingen.

It will be noted that all these various methods of determining the geometry of space are essentially physical. Hence, the accusation could always be made, as it was indeed made by Poincaré, that all we could determine in this way would reduce to the geometrical properties of material bodies or of light rays, and that space itself was amorphous and would therefore escape us completely. On the other hand, it was an undeniable fact of experience that measurements in space, whether conducted with material rods or with light rays, always yielded the same Euclidean results. Furthermore, the same Euclidean geometry seemed to be imposed upon space when we sought to interpret the laws of physics in the simplest way possible. It was scarcely conceivable that this general concordance of results obtained by these various methods of approach should be attributed to chance. Hence, the problem was reversed and the belief arose that space itself possessed a Euclidean structure, and that it was this structure which controlled the behaviour of material solids, the paths of light rays, and the course of free bodies moving far from matter according to the law of inertia. When this attitude was adopted the significance of measurements with material rods was altered. The geometry of space was not merely the Euclidean geometry of material bodies; it was rather the reverse. The geometry of material bodies was Euclidean by reason of this pre-existing Euclidean structure of space, which moulded them into shape and thus gave them their geometrical properties. Under the circumstances, measurements in space with rigid bodies and light rays would resolve themselves into an exploration of the pre-existing geometry of space, and the rôle of the rigid body and the light ray became similar to that of the thermometer in detecting pre-existing variations of temperature.

Now this attitude of regarding the behaviour of rigid rods and light rays as symptoms, and not as causes, must not obscure the fact that rods, light rays and physical phenomena generally are an essential requirement for the discovery of the pre-existing structure of space. There is no sense, therefore, in attempting to determine this structure *a priori* on the plea that physical measurements, however perfect they may be, can never transcend a certain approximation, and are therefore of no value in permitting us to discover the precise structure of space. We must never forget that we are dealing here with a problem of physics, not with one of pure mathematics. For instance, in Euclidean geometry the mathematician does not have to appeal to measurements with rigid rods in order to determine the value of π, since in virtue of our basic Euclidean postulates and the rules of logical reasoning, π can be determined as accurately as we wish by purely mathematical means. But in the case of the geometry of physical space we are completely at sea. *A priori,* this geometry might be one of a variety of different types. All we might do would be to assert that the principle of sufficient reason compelled us to regard physical space as homogeneous and isotropic, hence as reducing to a constant-curvature geometry, whether Euclidean or not. Even this belief would be open to question until we knew more about the origin of this structure. At any rate, on *a priori* grounds the precise geometry of space can never be ascertained. Whether we like it or not, we are compelled to resort to rods, light rays and physical methods of exploration; and whatever geometry we obtain in this way will *ipso facto* become that of real physical space, to the order of precision of our measurements.

Up to this point we have limited ourselves to discussing the first step in our study of physical space. The structure is viewed as pre-existing; and our rods and light rays, by adjusting themselves to this structure, permit us to objectivise it, as it were. In much the same way a coloured liquid poured into an ideally transparent glass (so transparent as to be invisible) permits us to ascertain the inner shape of the glass. But we now come to the second step. What is the cause of this structure? Is it posited by the Creator, as an intrinsic property of space, or is it the mere manifestation of something more· fundamental?

Riemann refused to believe that a structure of the void could exist of itself. He recognised, as before, that our rods would adjust themselves and would thus reveal the pre-existing structure; but this structure in turn would have to be created and conditioned by something else foreign to the void. For this "something else" Riemann appealed to the forces generated by the enormous mass of the totality of the star distribution. It would follow that were there no stars and no matter in the universe aside from our exploration rods, no geometry would be revealed. The rods would not know how to act; light rays would not know where to go; physical space would be unthinkable. With this bold idea of Riemann's, the homogeneity and isotropy of space (whether Euclidean or non-Euclidean), formerly thought to be imposed by the principle of sufficient reason, was no longer inevitable; for star-matter, the generating cause of this structure, was not distributed equally throughout the universe. Riemann, therefore, considered that a varying non-Euclideanism might be present in space. Rods and light rays would, of course, adjust themselves to this heterogeneous structure, yielding different numerical results from place to place throughout space.

All these theories antedate Einstein's discoveries by many years, but with the advent of the theory of relativity the entire problem presented itself afresh, with this difference, however—*that the fundamental continuum to be explored was four-dimensional space-time and no longer three-dimensional space.* The introduction of space-time necessitated an appeal to chronometers or vibrating atoms as well as to material rods, although the progress of light rays through space could also serve to measure duration (owing to the postulate of the invariant velocity of light according to which a ray of light would describe equal spaces in equal times). At all events, the time determinations would have to yield the same results whether computed by light rays or by the vibrations of atoms, since otherwise we should have to assume that the structure of space-time itself could not be the governing and moulding influence regulating the behaviour of material bodies and of all processes of change. In this eventuality a belief in the structure of space or of space-time would be difficult to accept. This structure of space-time has received various names. Weyl calls it the *affine-relationship,* the *guiding field* or the *metrical field.* Einstein refers to it as the *ether of space-time* or the *gravitational ether.* In order to avoid confusion with the stagnant ether of classical science, from which it differs in so many respects, we shall refer to it as the "metrical field." Mathematically, as we know, it is represented by the g_{ik}-distribution.

It is to be noted that as space-time can be split up into a space and a time by any given observer (at least in his immediate vicinity), the space-time-structure will automatically be split up into a space-structure and a time-structure, or a spatial and a temporal metrical field. The behaviour of rods and bodies at rest with respect to the observer will be controlled by the space-structure, while the beatings of perfect clocks such as vibrating atoms will be controlled by the time-structure. When we consider space and time together, we may say that the difference between space-time and separate space and time can be expressed as follows: Whereas the three-dimensional Euclidean space of classical science could be considered as cleaved by parallel time-stratifications, defining the spatio-temporal situations of simultaneous events and separating neatly space from time, no such stratification exists in space-time. If we wish to adhere to the older conception, we may still speak of time-stratifications through space-time, but we must be careful to add that these stratifications manifest no fixed direction. This is due to the fact that those stratifications which would be time-stratifications defining simultaneous events for one observer would cease to be time-stratifications defining simultaneous events for another observer in relative motion. The relativity of simultaneity is but another way of expressing the same facts. But, of course, with this admission, the conception of time-stratifications loses all deep significance. It follows that, when discussing the space-time structure, it would be preferable to omit mentioning such relative concepts as those of space and time. Accordingly, we will say that the structure of space-time is defined by the light-cone, that is, by the locus of the world-lines of light rays issuing from any given point at any given time. It is this light-cone which will split up space-time into its permissible space and time directions.*

Let us revert to the metrical field, as defining the space-time structure. Although Riemann had attributed the existence of the structure, or metrical field, of space to the binding forces of matter, there is not the slightest indication in Einstein's special theory that any such view is going to be developed later on; in fact, it does not appear that Einstein was influenced in the slightest degree by Riemann's ideas. At any rate, in the special theory, the problem of determining whence the structure, or field, arises, what it is, what causes it, is not even discussed in a tentative manner. Space-time, with its flat structure, is assumed to be given or posited by the Creator.

But in the general theory the entire situation changes when Einstein accounts for gravitation, hence for a varying lay of the metrical field, in terms of a varying non-Euclidean structure of space-time around matter. We are then compelled to recognise not only that the metrical field regulates the behaviour of material bodies and clocks, as was also the case in the special theory, but, furthermore, that a reciprocal

* See Appendix.

action takes place and that matter and energy in turn must affect the lay of the metrical field. But we are still a long way from Riemann's view that the field is not alone affected but brought into existence by matter; and it is only when we consider the cosmological part of Einstein's theory that this idea of Riemann's may possibly be vindicated.

And here we come to a parting of the ways with de Sitter and Eddington on one side, Einstein and Thirring on the other, and Weyl somewhere between the two extremes. The differences of opinion arise from the views to be entertained on the subject of the origin of the metrical field. It is impossible to accept Einstein's general theory without admitting that the field regulates the behaviour of matter and all processes of change, and is in turn affected by matter yielding under its influence; but the point to be determined is whether this field or space-time structure is entirely created by matter, or whether some type of structure would subsist in the absence of matter. Inasmuch as these speculative questions are related to the form of the universe, which has already been dealth with in Chapter XXXIII, we shall not revert to them here.

In the theory of relativity the concept of matter also undergoes important modifications. In classical science, matter was distinguished by its attribute of mass and by its ability to create a gravitational field. It is true that the electromagnetic field was also known to possess momentum and hence inertia (as exemplified by the pressure of light), so that even in classical science the distinction between matter and ether was losing its definiteness. Owing, however, to the fundamental difference that was thought to exist between inertial mass, as representative of inertia, and gravitational mass, as representative of attraction, the electromagnetic field was never suspected of developing a gravitational field. Hence the difference between matter and energy remained fairly clear. Einstein's theory, by identifying mass and energy and by showing that all forms of energy are capable of producing a gravitational field, tends to render the distinction less legitimate. A further consequence of the theory has been to show that conservation of matter cannot hold. The result is that matter, as a substance, loses its meaning.

Passing to the laws of motion, we remember that there is but one law: All free bodies (when reduced to point-masses) follow time-geodesics in space-time regardless of whether space-time be flat, as it is (at least approximately) in interstellar space, or whether it be curved by the presence of matter. If space-time is flat, the geodesics are straight and the bodies describe straight courses with constant speeds as referred to a Galilean frame. Thus Newton's law of inertia is seen to express the flatness of space-time. When space-time, and hence its geodesics, are curved by the presence of matter, the courses of free bodies appear to be curved, or else their motion to be accelerated. But whereas, under those conditions, the law of inertia was at fault in classical science, and an additional gravitational influence had to be introduced, in Einstein's theory the general law of geodesic motion still holds good. Inasmuch as the structure of space-time determines the laws of our geometry, the beatings of natural clocks (atoms) and the motion of free bodies, we see that the theory has brought about a fusion between geometry and physics.

We may also remember that the existence of space-time has solved the old enigma of the dual nature of the structure of space in mechanics, seemingly relative for velocity and absolute for acceleration (see Chapter XVIII).

Force is another important concept on which the theory of relativity has shed new light; though to avoid any misunderstanding we may state that by force we refer not to muscular pulls, but to the forces of gravitation, of inertia and of the electromagnetic field. Let us recall how the new ideas were obtained. We saw that the fundamental continuum of the world was space-time, and that observers, according to their relative motions, carved it up into space and time in various ways. Now we know that even in flat space-time, far from matter, forces of inertia spring into existence when we place ourselves in a rotating frame or, more generally, in any accelerated frame; translated into terms of space-time, this means when we split up space-time into a curved space and a curved time.

It follows that in an empty region of flat space-time there is no sense in wondering whether forces are present or not. Forces can come into existence only after the splitting up of space-time into separate space and time has been accomplished. Then, according to the method selected, forces will appear to be distributed this way or that, or even may not arise at all; the last case would correspond to the choice of a Cartesian mesh-system. But this splitting-up process which may give rise to forces is obviously artificial and corresponds to nothing intrinsic in nature, any more than the splitting up of three-dimensional space into arbitrary directions, which we name above, below, left and right, reveals any intrinsic property of space. So we see that forces are introduced by the observer and that, in the same region of space-time, they may be present for one man and absent for another. We cannot, therefore, attribute to them any absolute existence, one transcending the observer.

Now the forces we have been discussing were those which might arise in flat space-time far from matter. They were the forces of inertia of classical science, and were often referred to as fictitious precisely because their existence depended on our conditions of observation. A distinction was then drawn between these fictitious forces and the force of gravitation, which was assumed to exist around matter regardless of the observer's frame of reference. But with the relativity theory we see that a discrimination of this sort is no longer permissible. Around matter, space-time manifests curvature, but, just as in the case of flat space-time, it is possible for an observer to carve it up into space and time in such a way that no force will be manifest, at least in his vicinity. This method of partitioning corresponds, as we know, to that of the observer falling freely, hence following a geodesic in space-time. If, then, forces of inertia must be regarded as fictitious, so must forces of gravitation. And it is much the same with electromagnetic forces. Here also, the intensity and nature of the field (whether electric, magnetic, or electromagnetic) depends in large part on the circumstances of observation.

The purport of all these discussions is that if we wish to form an impersonal picture of the universe, there is no place for forces of gravitation or of inertia, since these forces are introduced by the observer. Likewise, in the description of a monument there is no place for shadow, since shadow is introduced only by the sun, and, so far as the monument itself is concerned, the sun and light in general do not constitute essential conditions of its existence. The entire philosophy of the relativity theory is to separate entities which have no significance except as expressing relationships to the frame of the observer, from entities which can be conceived of as enduring regardless of the observer's presence. We recognise these absolute entities as those which remain unaltered for all observers. Mathematically, they are expressed by invariants and tensors in space-time. Action and the Einsteinian interval are representative of invariants, while the majority of types of space-time curvature, the electromagnetic-field tensors, and the matter tensor are representative of tensors. It is because gravitational force and potential energy are not represented by tensors that we can attribute no absolute significance to them.*

The considerations which we have just developed may be summarized in this statement: Only those magnitudes which can be represented by invariants, vectors, or tensors in four-dimensional space-time have an objective existence. A consequence of the foregoing is that all physical laws must reduce to relations between invariants, vectors, and tensors in space-time. Einstein refers to this restriction which all natural laws must satisfy, as the principle of the covariance of natural laws, or as the general principle of relativity. In more physical language the restriction implies that natural laws must not change in form according to the conditions of observation. From a certain point of view the principle of covariance is not as novel a departure as might appear, for classical physics also accepted the principle of covariance, but in classical physics the fundamental continuum was separate space and time, and so the invariants, vectors, and tensors involved pertained to three-dimensional space and not to space-time, which was unknown. The novelty of the relativity theory is thus merely to substitute space-time for space and time.

We can also approach the problem of the laws of nature in a deductive way by starting from space-time. As soon as we recognise the existence of an absolute universe underlying our spatial and temporal world of immediate perception, and as soon as we realise that variations in our motion correspond to mere differences in our methods of splitting up into space and time the space-time continuum, it becomes evident that all particularities of structure of the absolute universe will subsist regardless of our choice of mesh-system, and hence regardless of our motion. These absolute space-time relations correspond to the laws of nature,

* The fact that potential energy manifests itself as relative does not preclude the conservation of energy.

which transcend the motion of the observer. From this it follows that the existence of the absolute universe, in which our physical entities are represented by vectors and tensors, and the relativity of all motion so far as the expression of the laws of nature is concerned, are but alternative aspects of one same condition; so that, as we have had occasion to state, the theory of relativity might also be termed the "quest of the absolute."

The heuristic value of the space-time principle of covariance is considerable, for it narrows down our choice of possible laws in a remarkable degree. Maxwell's laws of electromagnetics satisfy the principle and so were left untouched. On the other hand, some of the best established laws of classical physics, notably the laws of mechanics and Newton's law of gravitation, fail to pass the acid test and had to be discarded in consequence. We must remember, however, that the space-time principle of covariance does not tell us what the laws of nature will be; it merely informs us on what they cannot be. Fortunately, there are other clues to guide us. Thus we know that for low velocities the classical laws of mechanics and Newton's law of gravitation are very approximately correct, and, therefore, the true laws must merge into the classical ones under the limiting conditions of low velocities. This clue, combined with the space-time principle of covariance, enabled Einstein to revise the laws of mechanics and later Newton's law of gravitation. On the basis of the principle of covariance and of the additional clue, it is impossible, however, to decide whether the cosmological term, λ, should be included in the new law of gravitation. That is a matter for astronomical explorations to decide.

We may also note that the general principle of relativity does not signify that all observers are on the same footing. The irrelevance of the observer holds only in so far as the mathematical expression of the laws of nature is concerned; and it would not be true to suppose that a rotating observer could ever beguile himself into believing that he was at rest in a Galilean frame, or that a man tossed about in a ship's cabin would assume that the sea was perfectly calm.

In short, we see that the general principle of relativity possesses essentially a mathematical and formal significance; the absoluteness of rotation remains, as before, an empirical fact which we cannot conjure away. It is only when we accept the hypothesis of the cylindrical universe, permitted but not obligated by the general principle, that the relativity of all motion in the Machian sense may be possible.

From the standpoint of method there is an interesting difference between the special and the general theory. The special theory consists in the coordination of certain known experimental results, chiefly electromagnetic. The general theory, on the other hand, is a work of rationalization which was in no wise imposed by the facts of observation that were known at the time. Its creation is due to the genius of Einstein. Only subsequently was the general theory confirmed by the discovery of new facts.

In this book we have attempted to enumerate the most salient features of Einstein's work. There are, of course, a number of aspects which it has been impossible to mention, owing to the highly technical arguments that they would involve.   Nevertheless, from what little has been said, certain general conclusions appear to be legitimate.

The theory of relativity has taught us that, strangely enough, there is nothing in the world of physics to justify our natural belief in the separateness of space and time.   It is this discovery, entailing that of the four-dimensional world of space-time, that constitutes the most important contribution of the theory to our knowledge of the universe.

In addition to the existence of space-time and its invariant intervals, the really novel views which the theory forces upon us consist in the relativity of such primary concepts as length, duration, simultaneity, mass, force, etc., to which temperature should also be added.   We also obtain new views on causal connections, and a better understanding of the principle of action and of the principles of conservation.   Matter becomes fused with energy and disappears as a substance.   Furthermore, matter possibly creates space-time, and all motion may be relative.   As for mass, it may be of an entirely relational nature, just like weight.

If we confine ourselves to the better-established philosophical aspects of the theory, we may say that it proves that nature is ruled by mathematical laws, at least to a first approximation; but this fact was already known to scientists many years ago, following Newton's discoveries in celestial mechanics.   So far as the unity of nature is concerned (a belief which was indissolubly connected with the very existence of science), the theory merely confirms the scientific attitude by revealing a degree of unity of sublime simplicity and beauty.   When it comes to deciding whether nature must be viewed in an idealistic or a realistic sense, relativity affords little new information; and, as before, those who wish to adopt a realistic attitude towards the objective world of science are at liberty to do so, though, as we have mentioned, there are a number of reasons which render this philosophy most unsatisfactory.   At any rate, the sole difference between relativity and classical science, in their bearing on realism and idealism, consists in the fact that the objective world of science is now one of space-time, no longer one of separate space and time.

The theory is undoubtedly opposed to the view that we can speculate to any advantage on things which are beyond the control of experiment and observation; so the theory is distinctly anti-metaphysical in this respect.   But this again brings no additional information to scientists; for the anti-metaphysical attitude was already that of science prior to Einstein's discoveries, and was expressed in Planck's statement: "What can be measured is what exists."

For similar reasons the theory brings home to us with increased force a truth already recognised by science, namely, that space, time and simultaneity are concepts which can be studied only by empirical methods, by the use of rods, clocks and physical processes, and, more generally, by investigating the simplest co-ordinations of physical laws. In particular, the problems of determining whether a simultaneity of external events has any meaning, and, if so, whether it is a relative or an absolute concept, or, again, whether the universe is finite or infinite, are questions which the theoretical physicist alone is in a position to discuss—and then only after he has been informed of the results of ultra-precise observation. *A priori* speculations are worthless.

Relativity has also brought about the fusion of two realms of knowledge which had hitherto been developed independently of each other; we refer to geometry and physics. This fusion is illustrated by the fundamental rôle that is played by the geometrical g_{ik}-quantities in the laws of mechanics, in that of gravitation, and in many other physical phenomena. As Weyl tells us, this synthesis may be indicated by the scheme

$$\underbrace{\text{Pythagoras} \qquad \text{Newton}}_{\text{Einstein}}$$

Although not much more can be said at the present time, one may realise that even this fragmentary accretion to our knowledge of nature is of an importance which it would be difficult to overestimate.

APPENDIX

APPENDIX

THE theory of relativity appeals to what is known as the space-time graphical representation of Minkowski, but aside from certain peculiarities which the relativity theory entails, the general method of graphical representation in space and in time was known to classical science. Indeed, the graph traced by a thermometer needle is an illustration of this method. In it we have a graphic description of the variations in height of the mercury as time passes by.

The essence of these space and time graphs is to select a frame of reference, then to represent the successive positions of a body moving through this frame, in terms of its successive space and time co-ordinates. As a simple case let us consider a railroad embankment which will serve as our frame of reference. We shall restrict ourselves to considering the graphical space and time representation of events occurring on the surface of the embankment; not above it or beneath it. In other words, the space we shall be dealing with will for all practical purposes be reduced to one line, hence to one dimension. We then select a fixed point (any one at all) on the embankment and call it our *origin O.* Thanks to this choice of an origin, the *spatial position* of any event occurring on the embankment may be specified by a number. Thus the position of an event occurring two units of distance to the right or to the left of the origin will be given by the number $+2$ or -2, and an event occurring at the origin itself will have zero for its number.

In order to represent these results on a sheet of paper, we shall draw a straight line, say a horizontal, called a *space axis;* this will represent the embankment. On this space axis we mark a point O which will represent our origin on the embankment. Then, in order to represent on our paper the positions of events occurring at, say, one mile, two miles, three miles, etc., to the right of the origin on the embankment, we mark off points along our space axis at one, two and three units of distance from the point O. Obviously, we cannot manipulate a sheet of paper miles in length; hence we agree to represent a distance of one mile along the embankment by a length of one foot or one inch or one centimetre along the space axis. It matters not what unit we select so long as, once specified, it is maintained consistently throughout. As the reader can understand, the procedure' is exactly the same as that followed in the plan of a city.

Thus far, our graph reduces to a space graph of the points situated along the embankment. But we have now to introduce time. Two events may happen at the same point of the embankment, but at widely different times, and our graph in its present form offers us no means of differentiating graphically between the occurrence of the two events. Accordingly, we shall agree to represent such differences in time on our sheet of paper by placing our representative points of the events at varying heights above or below the space axis. If, then, we assume that all points on our space axis represent the space and time positions of events occurring on the embankment at a time zero or at noon, it will follow that all points above or below the space axis will represent events occurring on the embankment either after or before the time zero. This is equivalent to considering a vertical axis called a *time axis*, along which durations, hence instants, of time will be measured. Of course, just as in the case of distances, we must agree on some unit of length in our graph, in order to represent one second in time. We may choose this unit as we please; we may, for example, represent one second by one foot, or by one inch, along the vertical Ot. However, for reasons which will become apparent in the theory of relativity, it is advantageous, though by no means imperative, to select the same unit of length in our graph

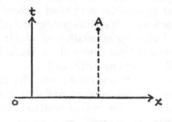

FIG. IX

in order to represent both one second in time and 186,000 miles in distance. Suppose, for instance, we agree to represent these magnitudes by a length of one inch on our graph; then a point such as A, one inch from Ox and one inch from Ot, will represent the space and time position of an event occurring on the embankment 186,000 miles to the right of the origin O and one second after the time zero (Fig. IX).

We see, then, that in our space and time graph, a point traced on our sheet of paper represents not merely a position in space along the embankment, but also an instant in time. For this reason such graph-points are known as *point-events*. Thus a point-event constitutes the graphical representation of an instantaneous event occurring anywhere and at any time along the embankment. The position of the point-event with respect to our space and time axes will then define without ambiguity the spatiotemporal position of the physical event with respect to the embankment and to the time zero, provided the units of measurement in the graph have been specified.

And now let us consider the representation of events that last and are not merely instantaneous. Here let us note that the existence of a body, say a stone on the embankment, constitutes an event, since the position of the stone can be defined in space and in time. But the stone endures: its existence is not merely momentary; hence its permanency is given by a continuous succession of point-events forming a continuous line. This line giving the successive positions of the stone both in space and in time is called a *world-line*. For the stone to possess a world-line, it is not necessary that it should be in motion along the embankment; it may just as well remain at rest. The sole difference will be that if the stone is at rest, its world-line will be a vertical, whereas if it is in motion along the embankment, its world-line will be slanting, since in this case the spatial position of the stone will vary as time passes by. If the speed of the stone along the embankment is constant, its world-line will be straight, whereas if the motion is uneven or accelerated, the world-line will be more or less curved. Assuming the motion to be uniform, the greater the speed of the body, the more will its world-line slant away from the vertical and tend to become horizontal.

Of course, as can easily be understood, the slants of the world-lines in the graph (aside from exceptional ones such as that of a body at rest) will be influenced by our choice of units. With the particular units

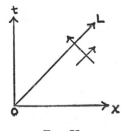

FIG. X

we have chosen, the world-lines of bodies moving along the embankment at a speed of 186,000 miles a second will possess a slant of 45° with respect to both space and time axes. In other words, the world-lines of such bodies, hence also of rays of light, will be inclined equally to our space and time axes. Thus any straight line inclined in this way, either to the right or to the left (Fig. X), will represent the world-line of a body moving with respect to the embankment with the speed of light, either to the right or to the left. The reason we selected our units of space and time as we did, was precisely in order to confer this symmetrical position of the world-lines of light, on account of the important rôle which the velocity of light plays in the theory of relativity. We may consider still another case, that of a body moving with infinite speed along the embankment (assuming, of course, that the existence of such a motion is physically possible). The body will obviously be everywhere along the embankment at the same instant of time; hence its world-line will be a horizontal. Thus the space axis is itself the world-line of a body moving along the embankment with infinite speed at the instant zero. Conversely, the time axis is the world-line of a body remaining motionless at the origin. Again, we may say that the space axis represents the totality of events occurring on the embankment at the instant zero, hence simultaneous with one another and with the instant zero. Likewise, any horizontal represents the totality of events occurring on the embankment at some given time, the height of this horizontal above or below the space axis defining the time.

Thus far, we have been considering happenings with reference to an observer at rest on the embankment, and everything we have said applies in an identical way to classical science and to relativity. It will only be when observers in relative motion are considered that differences in our graphical representation will arise.

Let us first consider the case of classical science. And here it is important to realise that our graph is nothing but a description: it merely

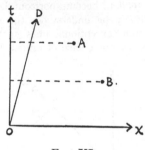

FIG. XI

describes graphically the relationships of duration, distance and motion which we wish to represent. Hence it is for us to discover through the medium of experiment what these relationships are going to be. Only after this preliminary work has been done can we represent these rela-tionships graphically. Now classical science, both as the result of crude experience and, later, of more refined measurements, held to the view that duration and distance were absolutes, in that their magnitudes would never be modified by our circumstances of motion. Accordingly, regard-less of whether we were at rest on the embankment or in motion, the dura-tion separating two events or the distance between two fixed points on the embankment was assumed to remain the same. Interpreted graphi-cally, this meant that the space and time axes would never have to be changed in our graph, regardless of the motion of the embankment observer whose measurements we were seeking to represent.

Consider two point-events, such as A and B (Fig. XI). As referred to the embankment, these two point-events represent two instantaneous events occurring on the embankment at a definite distance apart in space and in time. Suppose, now an observer leaves the origin at a time zero and moves along the embankment to the right. His world-line will be given by some such line as *OD*. Since, regardless of his motion, these two events *A* and *B* are to manifest exactly the same time separation as before, we must assume that the moving observer must continue to measure time computed along the same time-axis *Ot*.

Next consider two stones *M* and *N* lying on the embankment. Their world-lines will be *Mm* and *Nn*, respectively (Fig. XII), and their distance apart at any time *t* will be given by the length *GH* of the horizontal situated at the height *t* above *Ox* (the world-lines being vertical, this distance can of course never vary with time). Since the distance between the two stones must remain the same for the moving observer, he also must measure this distance along a horizontal, hence along the space axis *Ox*. In other words, the space axis and the time axis are unique, or absolute.

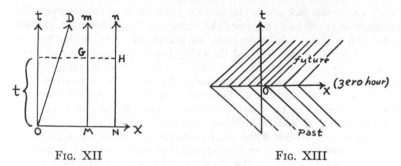

FIG. XII   FIG. XIII

It follows, of course, that since the space and time axes remain unchanged regardless of our motion along the embankment, all horizontals will represent events occurring simultaneously not only for an observer at rest on the embankment, but for all observers. *We thus get the absolute nature of simultaneity.*

Then again, since all observers measure time along the same direction, they will all recognise one same absolute distinction between past and future, regardless of their position and motion along the embankment. Thus, point-events below *Ox* (Fig. XIII), will represent events that occurred on the embankment *prior* to zero hour; those point-events lying above the line *Ox* will represent events that occurred *after* zero hour, while the point-events lying on *Ox* will give the events that occurred *at* zero hour.

As a final illustration (Fig. XIV), let us consider the case of an observer rushing after a light wave, both observer and light disturbance having left the origin *O* at the time zero. The world-line of the light ray will be represented by *OL*, and that of the observer by *OD*, less slanting than *OL*, at least if we consider the case of an observer moving along the embankment with a speed inferior to that of light. At a definite instant, say at one second after zero hour, the point-event defining the observer's position will be *A*, that is to say, a point on his world-line at a height of one inch above *Ox*. The point-event of the light wave at the same instant will

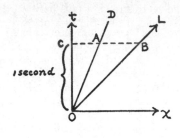

FIG. XIV

be B, also at one inch above Ox, and the point-event of the observer who remained at O will be C, the line CAB being, of course, horizontal. Hence the distance of the light wave from the moving observer, at that precise instant, will be AB, which is less than CB, its distance from the stationary observer. It follows that the light ray is moving with decreased speed with respect to the moving observer. We might have foreseen this directly since the world-line of light OL is no longer equally inclined to the world-line OD of the moving observer (hence to the successive positions of his body) and to his space line Ox.

We have now mentioned the chief characteristics of the space and time graph of classical science, and we should find that it reduced to a mere geometric representation of the classical Galilean space and time transformations of classical science, namely,

$$x' = x - vt; \quad t' = t.$$

Thus all the problems we have discussed could be solved either geometrically or analytically.

It is a matter of common knowledge that this space and time graph we have discussed was rarely mentioned in classical science; and such appellations as point-events and world-lines were unknown prior to Minkowski's discoveries. It may seem strange, therefore, that the theory of relativity should have made so extensive a use of a graphical method of representation. The reason is that in classical science the graph reduced to a mere geometrical superposition of space and time which was convenient in certain cases, but which had no profound significance. By reason of the separateness of space and time exemplified by the absoluteness of the space and time axes, there existed no amalgamation, no unity between space and time. They did not form one four-dimensional continuum of events *in any deep sense*. Let us endeavour to understand the meaning of these statements.

In classical science, space was regarded as a three-dimensional continuum of points, because it was possible to localise the position of a point in our frame by referring to its three co-ordinates. Suppose, then, that we wished to measure the distance between two points in space. All we should have to do would be to stretch a tape between the two points, and the length of the tape would define the distance between them.

And now let us consider the space *and* time of classical science. Here, again, we might claim that space and time constituted a four-dimensional continuum of events; for we could always localise the occurrence of an instantaneous event by measuring three spatial co-ordinates in our frame of reference, and by computing the instant at which the event occurred. But suppose we wished to measure the distance between two events occurring, say, one in New York on Monday, and the other in Washington on Tuesday. Obviously the problem would be meaningless. We might say that the distance in space between the two events was so many miles, and their distance in time so many hours; but we would be unable to measure the distance in space-cum-time directly, whereas, with the aid of the tape, in our previous example, we were able to measure immediately the spatial distance between the two points.

We see, then, that whereas both classical space and classical *space-cum-time* constituted a three-dimensional and a four-dimensional continuum, respectively, yet there was a vast difference between these two continua. The first one was a metrical continuum, whereas classical *space-cum-time* had no four-dimensional metrics.

We may present these arguments in a slightly different way, as follows: We say that a necessary condition for a continuum of points to form a metrical continuum possessing a definite geometry is that a definite unambiguous expression, invariant in magnitude, be found for the distance between any two points. Now this condition is certainly not realised in our classical space and time graph.

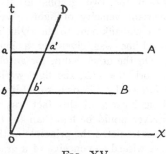

FIG. XV

Consider, as before, the embankment, an observer at rest at O (whose world-line is then Ot), an observer in motion (whose world-line is OD), and finally two point-events A and B (Fig. XV). We wish to find for the distance between the point-events A and B some mathematical expression which will remain the same for all observers regardless of their motion. Now it is obvious that whereas the time distance between

THE EVOLUTION OF SCIENTIFIC THOUGHT

A and *B,* given by the difference in height of these two point-events above *Ox,* remains the same for all observers, this is no longer true of their space distance. Thus, for the observer at rest, the space distance of *A* and *B* is *Aa – Bb,* whereas, for the moving observer, it is *Aa' – Bb',* *i.e.,* a different magnitude. Owing, then, to the fixedness of the time separation and the variableness of the spatial one, it becomes impossible to construct an invariant mathematical relation capable of expressing a distance. This is what is meant when we say that the space and time of classical science could not be regarded as a four-dimensional metrical continuum of events. Space and time, when considered jointly, reduced to the juxtaposition of a continuum of points in space and of a continuum of instants in time; for space alone and time alone constituted separate metrical continua.

To Minkowski belongs· the honour of having established the fusion between the two. Now and only now can we speak of the space-time distance, or Einsteinian interval, between the two events—say, one occurring in New York on Monday, and the other in Washington on Tuesday. Now and only now, thanks to ultra-precise experiment and to the genius of Einstein and Minkowski, is there any advantage in speaking of *space-cum-time* as a four-dimensional continuum of events which we call *space-time.* Prior to these achievements, the concept of space-time was as artificial as that of an *n*-dimensional continuum of space, time, pressure, temperature, colour, etc.

We shall now investigate in what measure the graphical representation of classical science will have to be modified in order to harmonise with the empirical facts revealed by ultra-refined experiment. There is no need to modify our understanding of point-events and world-lines; these will remain undisturbed.

The bifurcation between the two graphs arises when we consider the principle of the invariant velocity of light. We saw that if a ray of light was sent along the embankment to the right, leaving the origin *O* at a time zero, its world-line was given by a line *OL* bisecting the angle *tOx* (Fig. XIV). On the other hand, for an observer travelling to the right and having a world-line *OD,* the light-world-line *OL* would no longer bisect the angle formed by his own world-line *OD* and the space axis *Ox.* The physical significance of this fact was that the velocity of light for the moving observer would be less than for the embankment observer. Now this result contradicted the principle of the invariant velocity of light. If, therefore, we wished to conceive of a graph capable of yielding results in conformity with the principle, we should have to assume that for all observers moving along the embankment, the line *OL* would bisect the angle formed by their respective world-lines and space axes. It followed that the space axis of the moving observer could no longer be *Ox* but a new line *Ox',* such that *OL* would bisect the angle *DOx'.* In similar fashion, the observer's time would have to be measured along his own world-line *OD,* now called *Ot'* (Fig. XVI).

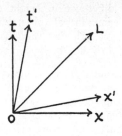

Fig. XVI

Hence we conclude that the space and time directions Ox and Ot are no longer absolute; every observer will have to measure time along his world-line and space along a line orthogonal to this world-line.* It follows that there exist an indefinite number of time directions given by the world-lines of the various observers, and a correspondingly indefinite number of space directions.

A first consequence of this novelty is that simultaneity can no longer be absolute. For whereas, in the classical graph, all events on the same horizontal or space direction were simultaneous for all observers, we now realise that with this variation in the space directions, or lines of simultaneous occurrence, the absoluteness of simultaneity must vanish. For instance, all the point-events lying on Ox' which are therefore simultaneous with the time zero for the moving observer, appear to be unfolding themselves in succession for the stationary observer.

Also it follows that as, according to relativity, no observer can travel faster than light, all the permissible world-lines of the observers (passing through O at a time zero) will be contained between the two light-world-lines OL and OL' (Fig. XVII). Any line passing through

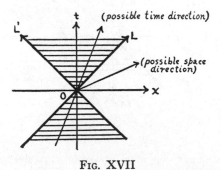

Fig. XVII

* Although they do not appear to be perpendicular to each other on the diagram, the time and space directions ot' and ox' are perpendicular when account is taken of the peculiar geometry of space-time with which we are dealing.

O and not contained between these two light-lines can never be a world-line, hence can never be a time direction; it will be a space direction of some possible observer. We may extend these results to two space dimensions. In this case our world-lines of light rays passing through O at the time zero form a cone called the *light-cone* with its vertex at O. Straight lines passing through O and contained within the light-cone are possible world-lines, or time directions; those passing through O but lying outside the cone constitute possible space directions for appropriate observers, or again possible loci of simultaneous events. Of course each point-event taken as point of departure gives rise to a light-cone, so that in a general way we may say that all lines parallel to lines contained within any given light-cone and passing through its vertex constitute possible world-lines or time directions. It is thus apparent that any light-cone defines limiting directions in space-time. All lines whose slant is less than that of the generators of the cone are possible time directions, while all those whose slant is greater are possible space directions. We thus realise the importance of the light-cone in defining the particularities of structure of space-time. In the case of four-dimensional space-time the cone becomes a three-dimensional surface, which is not easy to visualise, but this, of course, need not trouble us when we reason analytically.

Suppose our present existence be represented by the point O in the graph (Fig. XVIII). This point O is then the vertex of our light-cone at the instant considered. The only events of which we may be conscious, at the instant considered, will be represented by point-events situated inside, or on the surface of, the light-cone below us, *i.e.* in the direction of the past. Events which we perceive visually at an instant will always lie on the cone's surface. Thus, if a star suddenly becomes visible in the sky,

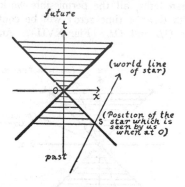

FIG. XVIII

the point-event of that star when it burst into prominence is situated on the surface of our instantaneous light-cone somewhere in the past (Fig. XVIII). In a similar way the only events which we can affect are represented by point-events that lie inside or on the upper part of our cone.

When the line joining two point-events is a possible time-direction, the order of succession of the two events will remain the same for all observers; hence, in this case, it will be possible to conceive of a causal relation as existing between them. But if the line joining the two point-events is a possible space direction, the events may be simultaneous, or subsequent and antecedent, or antecedent and subsequent, according to our motion. Events represented by such point-events can never be considered as manifesting any causal relationship one with the other. We see, then, that inasmuch as it is the light-cone which differentiates space from time directions, and events which may be causally connected from those which may not, the irreversibility of time and the problem of causality are linked with the existence of the light-cone.

In a general way, we may say that whereas the classical graph showed an absolute past, present and future for all observers, in the new graph this statement must be modified as follows: If we consider all the observers situated at a definite point at some definite time, that is to say, all observers whose spatio-temporal position is given by the same common point-event, then, regardless of their relative motions, the same instantaneous light-cone holds for all. The vertex of the cone is given by the common point-event, and all point-events situated within or on the surface of the cone stand in the instantaneous past or future for all the observers. There exist, therefore, an absolute past and an absolute future even in the theory of relativity.

We may also mention that, thanks to the indefiniteness of the space and time directions, our graph is now dealing with a veritable continuum, the *space-time continuum,* which cannot be separated in any unique way into space and time.

We should also find that the space-time distance between any two point-events, A and B, for which no absolute expression existed in the classical graph, would now assume an absolute value, the same for all observers. This distance is the *Einsteinian interval* discovered by Minkowski, which permits us to associate a definite geometry with the four-dimensional continuum of events. The geometry is not Euclidean, but semi-Euclidean. When account is taken of Minkowski's discoveries, it is seen that a space-time distance, when taken along a line which can be traced inside a light-cone, hence on a world-line, is an imaginary magnitude; whereas, when taken along a line lying outside any light-cone, the distance becomes real. However, no great importance need be attributed to time being imaginary and space being real; for we could just as well have conceived of space as imaginary and time as real. All that it is important to note is that there exists a mathematical difference between the various directions in space-time, those which lie inside the cone and are time-like, and those which lie without and are space-like. The former alone can be followed by physical disturbances and material bodies.

It may be instructive to look into this strange geometry of space-time a little more closely. We shall restrict ourselves to two dimensions, that is to say, to one space dimension and to the time ·dimension. In other words, our space-time graph will refer to the measurements of observers moving along the embankment.

Suppose, then, that a number of observers, moving with various uniform speeds to the right or to the left, pass the observer on the embankment at O, at the time zero. These observers carry ordinary clocks in

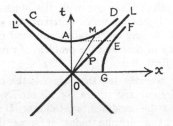

FIG. XIX

their hands, and these all mark zero hour as the observers pass O simultaneously. The question we wish to decide is as follows:

"Where will the point-events of the various observers be situated on the graph when their respective clocks mark one second past zero hour?"

We know that the point-event of the observer, who does not move from the point O on the embankment, will be situated on the vertical Ot, at a distance $OA =$ one second or one centimetre on our graph (Fig. XIX). Calculation then shows that the point-events of all the other observers will be situated on the equilateral hyperbola which passes through A and which has for asymptotes the two light-world-lines OL and OL'. For instance, the observer whose world-line is OM will find himself at the point M when his clock marks zero hour plus one second, that is to say, when he has travelled along the embankment away from O during a time of one second *as measured by his clock*. In classical science, of course, the various observers' point-events would have been situated at the intersection of their world-lines with the horizontal AE, no longer with the hyperbola CAD. It is not that the relativistic clocks differ from the classical ones; we are always considering our ordinary chronometers. The novelty is solely due to the fact that, thanks to ultra-precise experiment, our understanding of the behaviour of clocks and rods is more accurate than it used to be.

From all this, we may infer that the space-time distances from O to any of the points on the hyperbola represent congruent space-time distances in the space-time geometry. We may also note that, were space-time Euclidean instead of semi-Euclidean, the locus of points defined by the hyperbola would be given by a circle with O as its centre.

And now let us consider the world-line of a light ray leaving O at a time zero. We see that this world-line (OL or OL') does not intersect the hyperbola, or, if we prefer, intersects it at infinity. It follows that the space-time distances OA, OM, etc., are all equal to OL, where L is infinitely distant. Hence we must conclude that since the point L must be infinitely distant for OL to have a finite space-time value, any such space-time distance as OP, where P is any point on OL, must be nil. In other words, from the standpoint of space-time geometry, the point-events on the same world-line of light are all at a zero distance apart. Accordingly, a world-line of light, *i.e.*, any line inclined at 45°, is called a *null-line*. Lines of this sort were well known to geometricians long before Einstein's theory, so we need not suppose that a null-line is one of the queer conceptions we owe to relativity. All Minkowski has done has been to give us a physical interpretation of a null-line. It would be illustrated by the world-line of an observer moving with the speed of light. For this observer the events of his life would present no temporal separation; though, of course, another observer would realise that these various events were separated in time.

Next, let us examine the measurements of spatial lengths. Suppose that the various observers, when they pass O, carry poles which they hold as lances parallel to the embankment. If these poles are of equal length when placed side by side at rest, the point-events of their further extremities, at the instant the observers pass O, will lie on a hyperbola GF. In this case OG gives the length of the rod at rest on the embankment. Of course, the positions of the extremities of the rods at the instant the observers pass O must be computed according to the simultaneity determinations of the respective observers. If the rods are 186,000 miles in length, $OG = OA =$ one centimetre, and the two hyperbolas will be geometrically alike. We may also infer that the space-time distances from O to any of the points on the second hyperbola are all equal to one another. In fact, we may repeat for the second hyperbola, or space hyperbola, the same arguments we made when discussing the time one.

We are now in a position to understand how the FitzGerald contraction arises. Consider, for instance, a rod, OG, lying on the embankment. The world-lines of its two extremities will be Ot and Gg, respectively (Fig. XX). If, now, an observer passes O at time zero and moves to the right with velocity V, his world-line will be Ot'; but then his space direction will be Ox'. For him, therefore, the length of the rod lying on the embankment is no longer OG, as for the embankment observer, but OH. The graph shows us that this length OH is shorter than the length OK of the rod the observer is carrying with him. And as these two rods, when placed side by side at rest, are equal we see that as measured by the moving observer, the rod, on the embankment past which he is moving, will have suffered a contraction.

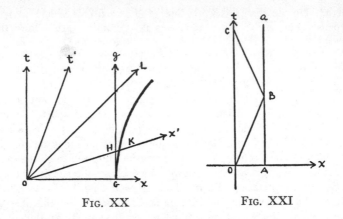

FIG. XX   FIG. XXI

We might, as a final example, consider the trip to the star (Fig. XXI). If *Aa* denotes the world-line of the star, and *OBC* the world-line of the travelling twin, we see that he will have lived a time *OB + BC* during his trip. His brother remaining on earth will have *OC* for world-line. Hence, he will have lived a time *OC*. As before, though *OB + BC* appears longer than *OC*, it is, in reality, shorter, as can be understood by referring to Fig. XIX. Incidentally, we see how the absolute nature of acceleration causes the traveller's world-line to bend; and it is this absolute bend in the world-line which differentiates the life-histories of the two twins and which is responsible for an *absolute, non-reciprocal* difference in their respective aging.

Let us also state that just as was the case with the classical graph which represented geometrically the classical Galilean transformations, so now the Minkowski graph merely translates the Lorentz-Einstein transformations.

And here a matter of some importance must be noted. We remember that before proceeding to draw our graph, we were compelled to settle on a co-ordination of units of measurement for space and time. We then chose the same length on our sheet of paper to represent a duration of one second and a distance of 186,000 miles. As a result of this choice the world-lines of light rays became diagonals. But suppose we had selected other units. Then, obviously, the slant of the world-lines, hence the angle of the light-cone, would have been modified and the entire appearance of the graph changed. Inasmuch as our choice of units is entirely arbitrary, we might be led to believe that the graph could not depict reality. But this opinion would be unfounded. While it is true that owing to the arbitrariness of our units, the graph cannot aspire to represent absolute shape, yet it does express certain definite relationships which a change of units could not disturb. In fact we might conceive the graph to be distorted by stretching, but still the relationships would endure; and relationships are all that science (or, we might even say, the human mind) can ever aspire to approach.

On the other hand, this question of units allows us to give a graphical solution of a point which is of ·great philosophical interest. Here we are living in a world which, theoretically at least, is vastly different from the world of separate space and time, and yet it is only thanks to ultra-refined experiment and to the genius of Einstein and Minkowski that we have finally realised it to be a four-dimensional continuum of events. How is it that ordinary perception is so blind to facts?

In order to understand this point, we must mention that though our choice of units for co-ordinating space and time measurements is arbitrary, since there is no rational connection between the magnitude of a distance in space and that of a duration in time, yet our daily activities suggest a common standard of comparison. The fact is that the distances which we ourselves and other material bodies cover in one second over the earth's surface are always comprised within certain narrow limits. This leads us to couple one second and one yard, rather than one second and 186,000 miles.

If, now, with these more homely units we were to set up Minkowski's graph, we should find that it was virtually identical with the classical one. The world-lines of the light rays would appear to coincide with the space axis Ox, and it would need a graph thousands of miles in length to detect their deviations from this line. A light-cone would cover the entire graph; hence the permissible space directions lying outside the cone would appear to be limited to the Ox axis. To all intents and purposes, there would be but one permissible space direction entailing the absoluteness of simultaneity and of time. We see, then, that it is by our immediate needs rather than by cosmic conditions, or, again, because slow velocities predominate around us, contrasted with which the velocity of light appears infinite, that we have been misled into believing in a world of separate space and time.

A CATALOGUE OF
SELECTED DOVER BOOKS
IN ALL FIELDS OF INTEREST

A CATALOGUE OF SELECTED DOVER BOOKS IN ALL FIELDS OF INTEREST

CELESTIAL OBJECTS FOR COMMON TELESCOPES, T. W. Webb. The most used book in amateur astronomy: inestimable aid for locating and identifying nearly 4,000 celestial objects. Edited, updated by Margaret W. Mayall. 77 illustrations. Total of 645pp. 5⅜ x 8½.
20917-2, 20918-0 Pa., Two-vol. set $9.00

HISTORICAL STUDIES IN THE LANGUAGE OF CHEMISTRY, M. P. Crosland. The important part language has played in the development of chemistry from the symbolism of alchemy to the adoption of systematic nomenclature in 1892. ". . . wholeheartedly recommended,"—Science. 15 illustrations. 416pp. of text. 5⅝ x 8¼. 63702-6 Pa. $6.00

BURNHAM'S CELESTIAL HANDBOOK, Robert Burnham, Jr. Thorough, readable guide to the stars beyond our solar system. Exhaustive treatment, fully illustrated. Breakdown is alphabetical by constellation: Andromeda to Cetus in Vol. 1; Chamaeleon to Orion in Vol. 2; and Pavo to Vulpecula in Vol. 3. Hundreds of illustrations. Total of about 2000pp. 6⅛ x 9¼.
23567-X, 23568-8, 23673-0 Pa., Three-vol. set $26.85

THEORY OF WING SECTIONS: INCLUDING A SUMMARY OF AIRFOIL DATA, Ira H. Abbott and A. E. von Doenhoff. Concise compilation of subatomic aerodynamic characteristics of modern NASA wing sections, plus description of theory. 350pp. of tables. 693pp. 5⅜ x 8½.
60586-8 Pa. $7.00

DE RE METALLICA, Georgius Agricola. Translated by Herbert C. Hoover and Lou H. Hoover. The famous Hoover translation of greatest treatise on technological chemistry, engineering, geology, mining of early modern times (1556). All 289 original woodcuts. 638pp. 6¾ x 11.
60006-8 Clothbd. $17.50

THE ORIGIN OF CONTINENTS AND OCEANS, Alfred Wegener. One of the most influential, most controversial books in science, the classic statement for continental drift. Full 1966 translation of Wegener's final (1929) version. 64 illustrations. 246pp. 5⅜ x 8½. 61708-4 Pa. $3.00

THE PRINCIPLES OF PSYCHOLOGY, William James. Famous long course complete, unabridged. Stream of thought, time perception, memory, experimental methods; great work decades ahead of its time. Still valid, useful; read in many classes. 94 figures. Total of 1391pp. 5⅜ x 8½.
20381-6, 20382-4 Pa., Two-vol. set $13.00

YUCATAN BEFORE AND AFTER THE CONQUEST, Diego de Landa. First English translation of basic book in Maya studies, the only significant account of Yucatan written in the early post-Conquest era. Translated by distinguished Maya scholar William Gates. Appendices, introduction, 4 maps and over 120 illustrations added by translator. 162pp. 5⅜ x 8½.
23622-6 Pa. $3.00

THE MALAY ARCHIPELAGO, Alfred R. Wallace. Spirited travel account by one of founders of modern biology. Touches on zoology, botany, ethnography, geography, and geology. 62 illustrations, maps. 515pp. 5⅜ x 8½.
20187-2 Pa. $6.95

THE DISCOVERY OF THE TOMB OF TUTANKHAMEN, Howard Carter, A. C. Mace. Accompany Carter in the thrill of discovery, as ruined passage suddenly reveals unique, untouched, fabulously rich tomb. Fascinating account, with 106 illustrations. New introduction by J. M. White. Total of 382pp. 5⅜ x 8½. (Available in U.S. only)   23500-9 Pa. $4.00

THE WORLD'S GREATEST SPEECHES, edited by Lewis Copeland and Lawrence W. Lamm. Vast collection of 278 speeches from Greeks up to present. Powerful and effective models; unique look at history. Revised to 1970. Indices. 842pp. 5⅜ x 8½.   20468-5 Pa. $8.95

THE 100 GREATEST ADVERTISEMENTS, Julian Watkins. The priceless ingredient; His master's voice; 99 44/100% pure; over 100 others. How they were written, their impact, etc. Remarkable record. 130 illustrations. 233pp. 7⅞ x 10 3/5.   20540-1 Pa. $5.00

CRUICKSHANK PRINTS FOR HAND COLORING, George Cruickshank. 18 illustrations, one side of a page, on fine-quality paper suitable for watercolors. Caricatures of people in society (c. 1820) full of trenchant wit. Very large format. 32pp. 11 x 16.   23684-6 Pa. $5.00

THIRTY-TWO COLOR POSTCARDS OF TWENTIETH-CENTURY AMERICAN ART, Whitney Museum of American Art. Reproduced in full color in postcard form are 31 art works and one shot of the museum. Calder, Hopper, Rauschenberg, others. Detachable. 16pp. 8¼ x 11.
23629-3 Pa. $2.50

MUSIC OF THE SPHERES: THE MATERIAL UNIVERSE FROM ATOM TO QUASAR SIMPLY EXPLAINED, Guy Murchie. Planets, stars, geology, atoms, radiation, relativity, quantum theory, light, antimatter, similar topics. 319 figures. 664pp. 5⅜ x 8½.
21809-0, 21810-4 Pa., Two-vol. set $10.00

EINSTEIN'S THEORY OF RELATIVITY, Max Born. Finest semi-technical account; covers Einstein, Lorentz, Minkowski, and others, with much detail, much explanation of ideas and math not readily available elsewhere on this level. For student, non-specialist. 376pp. 5⅜ x 8½.
60769-0 Pa. $4.00

THE COMPLETE BOOK OF DOLL MAKING AND COLLECTING, Catherine Christopher. Instructions, patterns for dozens of dolls, from rag doll on up to elaborate, historically accurate figures. Mould faces, sew clothing, make doll houses, etc. Also collecting information. Many illustrations. 288pp. 6 x 9. 22066-4 Pa. $4.00

THE DAGUERREOTYPE IN AMERICA, Beaumont Newhall. Wonderful portraits, 1850's townscapes, landscapes; full text plus 104 photographs. The basic book. Enlarged 1976 edition. 272pp. 8¼ x 11¼. 23322-7 Pa. $6.00

CRAFTSMAN HOMES, Gustav Stickley. 296 architectural drawings, floor plans, and photographs illustrate 40 different kinds of "Mission-style" homes from The Craftsman (1901-16), voice of American style of simplicity and organic harmony. Thorough coverage of Craftsman idea in text and picture, now collector's item. 224pp. 8⅛ x 11. 23791-5 Pa. $6.00

PEWTER-WORKING: INSTRUCTIONS AND PROJECTS, Burl N. Osborn. & Gordon O. Wilber. Introduction to pewter-working for amateur craftsman. History and characteristics of pewter; tools, materials, step-by-step instructions. Photos, line drawings, diagrams. Total of 160pp. 7⅞ x 10¾. 23786-9 Pa. $3.50

THE GREAT CHICAGO FIRE, edited by David Lowe. 10 dramatic, eyewitness accounts of the 1871 disaster, including one of the aftermath and rebuilding, plus 70 contemporary photographs and illustrations of the ruins—courthouse, Palmer House, Great Central Depot, etc. Introduction by David Lowe. 87pp. 8¼ x 11. 23771-0 Pa. $4.00

SILHOUETTES: A PICTORIAL ARCHIVE OF VARIED ILLUSTRATIONS, edited by Carol Belanger Grafton. Over 600 silhouettes from the 18th to 20th centuries include profiles and full figures of men and women, children, birds and animals, groups and scenes, nature, ships, an alphabet. Dozens of uses for commercial artists and craftspeople. 144pp. 8⅜ x 11¼. 23781-8 Pa. $4.00

ANIMALS: 1,419 COPYRIGHT-FREE ILLUSTRATIONS OF MAMMALS, BIRDS, FISH, INSECTS, ETC., edited by Jim Harter. Clear wood engravings present, in extremely lifelike poses, over 1,000 species of animals. One of the most extensive copyright-free pictorial sourcebooks of its kind. Captions. Index. 284pp. 9 x 12. 23766-4 Pa. $7.50

INDIAN DESIGNS FROM ANCIENT ECUADOR, Frederick W. Shaffer. 282 original designs by pre-Columbian Indians of Ecuador (500-1500 A.D.). Designs include people, mammals, birds, reptiles, fish, plants, heads, geometric designs. Use as is or alter for advertising, textiles, leathercraft, etc. Introduction. 95pp. 8¾ x 11¼. 23764-8 Pa. $3.50

SZIGETI ON THE VIOLIN, Joseph Szigeti. Genial, loosely structured tour by premier violinist, featuring a pleasant mixture of reminiscenes, insights into great music and musicians, innumerable tips for practicing violinists. 385 musical passages. 256pp. 5⅝ x 8¼. 23763-X Pa. $3.50

TONE POEMS, SERIES II: TILL EULENSPIEGELS LUSTIGE STREICHE, ALSO SPRACH ZARATHUSTRA, AND EIN HELDEN- LEBEN, Richard Strauss. Three important orchestral works, including very popular *Till Eulenspiegel's Marry Pranks*, reproduced in full score from original editions. Study score. 315pp. 9⅜ x 12¼. (Available in U.S. only)
23755-9 Pa. $7.50

TONE POEMS, SERIES I: DON JUAN, TOD UND VERKLARUNG AND DON QUIXOTE, Richard Strauss. Three of the most often per- formed and recorded works in entire orchestral repertoire, reproduced in full score from original editions. Study score. 286pp. 9⅜ x 12¼. (Avail- able in U.S. only)
23754-0 Pa. $7.50

11 LATE STRING QUARTETS, Franz Joseph Haydn. The form which Haydn defined and "brought to perfection." (*Grove's*). 11 string quartets in complete score, his last and his best. The first in a projected series of the complete Haydn string quartets. Reliable modern Eulenberg edition, otherwise difficult to obtain. 320pp. 8⅜ x 11¼. (Available in U.S. only)
23753-2 Pa. $6.95

FOURTH, FIFTH AND SIXTH SYMPHONIES IN FULL SCORE, Peter Ilyitch Tchaikovsky. Complete orchestral scores of Symphony No. 4 in F Minor, Op. 36; Symphony No. 5 in E Minor, Op. 64; Symphony No. 6 in B Minor, "Pathetique," Op. 74. Bretikopf & Hartel eds. Study score. 480pp. 9⅜ x 12¼.
23861-X Pa. $10.95

THE MARRIAGE OF FIGARO: COMPLETE SCORE, Wolfgang A. Mozart. Finest comic opera ever written. Full score, not to be confused with piano renderings. Peters edition. Study score. 448pp. 9⅜ x 12¼. (Available in U.S. only)
23751-6 Pa. $11.95

"IMAGE" ON THE ART AND EVOLUTION OF THE FILM, edited by Marshall Deutelbaum. Pioneering book brings together for first time 38 groundbreaking articles on early silent films from *Image* and 263 illustra- tions newly shot from rare prints in the collection of the International Museum of Photography. A landmark work. Index. 256pp. 8¼ x 11.
23777-X Pa. $8.95

AROUND-THE-WORLD COOKY BOOK, Lois Lintner Sumption and Marguerite Lintner Ashbrook. 373 cooky and frosting recipes from 28 countries (America, Austria, China, Russia, Italy, etc.) include Viennese kisses, rice wafers, London strips, lady fingers, hony, sugar spice, maple cookies, etc. Clear instructions. All tested. 38 drawings. 182pp. 5⅜ x 8.
23802-4 Pa. $2.50

THE ART NOUVEAU STYLE, edited by Roberta Waddell. 579 rare photographs, not available elsewhere, of works in jewelry, metalwork, glass, ceramics, textiles, architecture and furniture by 175 artists—Mucha, Seguy, Lalique, Tiffany, Gaudin, Hohlwein, Saarinen, and many others. 288pp. 8⅜ x 11¼.
23515-7 Pa. $6.95

CATALOGUE OF DOVER BOOKS

THE AMERICAN SENATOR, Anthony Trollope. Little known, long un-available Trollope novel on a grand scale. Here are humorous comment on American vs. English culture, and stunning portrayal of a heroine/villainess. Superb evocation of Victorian village life. 561pp. 5⅜ x 8½.
                                                          23801-6 Pa. $6.00

WAS IT MURDER? James Hilton. The author of *Lost Horizon* and *Good-bye, Mr. Chips* wrote one detective novel (under a pen-name) which was quickly forgotten and virtually lost, even at the height of Hilton's fame. This edition brings it back—a finely crafted public school puzzle resplendent with Hilton's stylish atmosphere. A thoroughly English thriller by the creator of Shangri-la. 252pp. 5⅜ x 8. (Available in U.S. only)
                                                          23774-5 Pa. $3.00

CENTRAL PARK: A PHOTOGRAPHIC GUIDE, Victor Laredo and Henry Hope Reed. 121 superb photographs show dramatic views of Central Park: Bethesda Fountain, Cleopatra's Needle, Sheep Meadow, the Blockhouse, plus people engaged in many park activities: ice skating, bike riding, etc. Captions by former Curator of Central Park, Henry Hope Reed, provide historical view, changes, etc. Also photos of N.Y. landmarks on park's periphery. 96pp. 8½ x 11.   23750-8 Pa. $4.50

NANTUCKET IN THE NINETEENTH CENTURY, Clay Lancaster. 180 rare photographs, stereographs, maps, drawings and floor plans recreate unique American island society. Authentic scenes of shipwreck, light-houses, streets, homes are arranged in geographic sequence to provide walking-tour guide to old Nantucket existing today. Introduction, captions. 160pp. 8⅞ x 11¾.   23747-8 Pa. $6.95

STONE AND MAN: A PHOTOGRAPHIC EXPLORATION, Andreas Feininger. 106 photographs by *Life* photographer Feininger portray man's deep passion for stone through the ages. Stonehenge-like megaliths, forti-fied towns, sculpted marble and crumbling tenements show textures, beauties, fascination. 128pp. 9¼ x 10¾.   23756-7 Pa. $5.95

CIRCLES, A MATHEMATICAL VIEW, D. Pedoe. Fundamental aspects of college geometry, non-Euclidean geometry, and other branches of mathematics: representing circle by point. Poincare model, isoperimetric property, etc. Stimulating recreational reading. 66 figures. 96pp. 5⅝ x 8¼.
                                                          63698-4 Pa. $2.75

THE DISCOVERY OF NEPTUNE, Morton Grosser. Dramatic scientific history of the investigations leading up to the actual discovery of the eighth planet of our solar system. Lucid, well-researched book by well-known historian of science. 172pp. 5⅜ x 8½.   23726-5 Pa. $3.00

THE DEVIL'S DICTIONARY. Ambrose Bierce. Barbed, bitter, brilliant witticisms in the form of a dictionary. Best, most ferocious satire America has produced. 145pp. 5⅜ x 8½.   20487-1 Pa. $1.75

HISTORY OF BACTERIOLOGY, William Bulloch. The only comprehensive history of bacteriology from the beginnings through the 19th century. Special emphasis is given to biography-Leeuwenhoek, etc. Brief accounts of 350 bacteriologists form a separate section. No clearer, fuller study, suitable to scientists and general readers, has yet been written. 52 illustrations. 448pp. 5⅝ x 8¼.   23761-3 Pa. $6.50

THE COMPLETE NONSENSE OF EDWARD LEAR, Edward Lear. All nonsense limericks, zany alphabets, Owl and Pussycat, songs, nonsense botany, etc., illustrated by Lear. Total of 321pp. 5⅜ x 8½. (Available in U.S. only)   20167-8 Pa. $3.00

INGENIOUS MATHEMATICAL PROBLEMS AND METHODS, Louis A. Graham. Sophisticated material from Graham *Dial*, applied and pure; stresses solution methods. Logic, number theory, networks, inversions, etc. 237pp. 5⅜ x 8½.   20545-2 Pa. $3.50

BEST MATHEMATICAL PUZZLES OF SAM LOYD, edited by Martin Gardner. Bizarre, original, whimsical puzzles by America's greatest puzzler. From fabulously rare *Cyclopedia*, including famous 14-15 puzzles, the Horse of a Different Color, 115 more. Elementary math. 150 illustrations. 167pp. 5⅜ x 8½.   20498-7 Pa. $2.50

THE BASIS OF COMBINATION IN CHESS, J. du Mont. Easy-to-follow, instructive book on elements of combination play, with chapters on each piece and every powerful combination team—two knights, bishop and knight, rook and bishop, etc. 250 diagrams. 218pp. 5⅜ x 8½. (Available in U.S. only)   23644-7 Pa. $3.50

MODERN CHESS STRATEGY, Ludek Pachman. The use of the queen, the active king, exchanges, pawn play, the center, weak squares, etc. Section on rook alone worth price of the book. Stress on the moderns. Often considered the most important book on strategy. 314pp. 5⅜ x 8½.   20290-9 Pa. $3.50

LASKER'S MANUAL OF CHESS, Dr. Emanuel Lasker. Great world champion offers very thorough coverage of all aspects of chess. Combinations, position play, openings, end game, aesthetics of chess, philosophy of struggle, much more. Filled with analyzed games. 390pp. 5⅜ x 8½.   20640-8 Pa. $4.00

500 MASTER GAMES OF CHESS, S. Tartakower, J. du Mont. Vast collection of great chess games from 1798-1938, with much material nowhere else readily available. Fully annotated, arranged by opening for easier study. 664pp. 5⅜ x 8½.   23208-5 Pa. $6.00

A GUIDE TO CHESS ENDINGS, Dr. Max Euwe, David Hooper. One of the finest modern works on chess endings. Thorough analysis of the most frequently encountered endings by former world champion. 331 examples, each with diagram. 248pp. 5⅜ x 8½.   23332-4 Pa. $3.50

SECOND PIATIGORSKY CUP, edited by Isaac Kashdan. One of the greatest tournament books ever produced in the English language. All 90 games of the 1966 tournament, annotated by players, most annotated by both players. Features Petrosian, Spassky, Fischer, Larsen, six others. 228pp. 5⅜ x 8½. 23572-6 Pa. $3.50

ENCYCLOPEDIA OF CARD TRICKS, revised and edited by Jean Hugard. How to perform over 600 card tricks, devised by the world's greatest magicians: impromptus, spelling tricks, key cards, using special packs, much, much more. Additional chapter on card technique. 66 illustrations. 402pp. 5⅜ x 8½. (Available in U.S. only) 21252-1 Pa. $3.95

MAGIC: STAGE ILLUSIONS, SPECIAL EFFECTS AND TRICK PHO-TOGRAPHY, Albert A. Hopkins, Henry R. Evans. One of the great classics; fullest, most authorative explanation of vanishing lady, levitations, scores of other great stage effects. Also small magic, automata, stunts. 446 illustrations. 556pp. 5⅜ x 8½. 23344-8 Pa. $5.00

THE SECRETS OF HOUDINI, J. C. Cannell. Classic study of Houdini's incredible magic, exposing closely-kept professional secrets and revealing, in general terms, the whole art of stage magic. 67 illustrations. 279pp. 5⅜ x 8½. 22913-0 Pa. $3.00

HOFFMANN'S MODERN MAGIC, Professor Hoffmann. One of the best, and best-known, magicians' manuals of the past century. Hundreds of tricks from card tricks and simple sleight of hand to elaborate illusions involving construction of complicated machinery. 332 illustrations. 563pp. 5⅜ x 8½. 23623-4 Pa. $6.00

MADAME PRUNIER'S FISH COOKERY BOOK, Mme. S. B. Prunier. More than 1000 recipes from world famous Prunier's of Paris and London, specially adapted here for American kitchen. Grilled tournedos with anchovy butter, Lobster a la Bordelaise, Prunier's prized desserts, more. Glossary. 340pp. 5⅜ x 8½. (Available in U.S. only) 22679-4 Pa. $3.00

FRENCH COUNTRY COOKING FOR AMERICANS, Louis Diat. 500 easy-to-make, authentic provincial recipes compiled by former head chef at New York's Fitz-Carlton Hotel: onion soup, lamb stew, potato pie, more. 309pp. 5⅜ x 8½. 23665-X Pa. $3.95

SAUCES, FRENCH AND FAMOUS, Louis Diat. Complete book gives over 200 specific recipes: bechamel, Bordelaise, hollandaise, Cumberland, apricot, etc. Author was one of this century's finest chefs, originator of vichyssoise and many other dishes. Index. 156pp. 5⅜ x 8.
23663-3 Pa. $2.50

TOLL HOUSE TRIED AND TRUE RECIPES, Ruth Graves Wakefield. Authentic recipes from the famous Mass. restaurant: popovers, veal and ham loaf, Toll House baked beans, chocolate cake crumb pudding, much more. Many helpful hints. Nearly 700 recipes. Index. 376pp. 5⅜ x 8½.
23560-2 Pa. $4.00

"OSCAR" OF THE WALDORF'S COOKBOOK, Oscar Tschirky. Famous American chef reveals 3455 recipes that made Waldorf great; cream of French, German, American cooking, in all categories. Full instructions, easy home use. 1896 edition. 907pp. 6⅝ x 9⅜.   20790-0 Clothbd. $15.00

COOKING WITH BEER, Carole Fahy. Beer has as superb an effect on food as wine, and at fraction of cost. Over 250 recipes for appetizers, soups, main dishes, desserts, breads, etc. Index. 144pp. 5⅜ x 8½. (Available in U.S. only)   23661-7 Pa. $2.50

STEWS AND RAGOUTS, Kay Shaw Nelson. This international cookbook offers wide range of 108 recipes perfect for everyday, special occasions, meals-in-themselves, main dishes. Economical, nutritious, easy-to-prepare: goulash, Irish stew, boeuf bourguignon, etc. Index. 134pp. 5⅜ x 8½.
23662-5 Pa. $2.50

DELICIOUS MAIN COURSE DISHES, Marian Tracy. Main courses are the most important part of any meal. These 200 nutritious, economical recipes from around the world make every meal a delight. "I . . . have found it so useful in my own household,"—N.Y. Times. Index. 219pp. 5⅜ x 8½.   23664-1 Pa. $3.00

FIVE ACRES AND INDEPENDENCE, Maurice G. Kains. Great back-to-the-land classic explains basics of self-sufficient farming: economics, plants, crops, animals, orchards, soils, land selection, host of other necessary things. Do not confuse with skimpy faddist literature; Kains was one of America's greatest agriculturalists. 95 illustrations. 397pp. 5⅜ x 8½.
20974-1 Pa. $3.95

A PRACTICAL GUIDE FOR THE BEGINNING FARMER, Herbert Jacobs. Basic, extremely useful first book for anyone thinking about moving to the country and starting a farm. Simpler than Kains, with greater emphasis on country living in general. 246pp. 5⅜ x 8½.
23675-7 Pa. $3.50

A GARDEN OF PLEASANT FLOWERS (PARADISI IN SOLE: PARADISUS TERRESTRIS), John Parkinson. Complete, unabridged reprint of first (1629) edition of earliest great English book on gardens and gardening. More than 1000 plants & flowers of Elizabethan, Jacobean garden fully described, most with woodcut illustrations. Botanically very reliable, a "speaking garden" of exceeding charm. 812 illustrations. 628pp. 8½ x 12¼.   23392-8 Clothbd. $25.00

ACKERMANN'S COSTUME PLATES, Rudolph Ackermann. Selection of 96 plates from the Repository of Arts, best published source of costume for English fashion during the early 19th century. 12 plates also in color. Captions, glossary and introduction by editor Stella Blum. Total of 120pp. 8⅜ x 11¼.   23690-0 Pa. $4.50

MUSHROOMS, EDIBLE AND OTHERWISE, Miron E. Hard. Profusely illustrated, very useful guide to over 500 species of mushrooms growing in the Midwest and East. Nomenclature updated to 1976. 505 illustrations. 628pp. 6½ x 9¼.   23309-X Pa. $7.95

AN ILLUSTRATED FLORA OF THE NORTHERN UNITED STATES AND CANADA, Nathaniel L. Britton, Addison Brown. Encyclopedic work covers 4666 species, ferns on up. Everything. Full botanical information, illustration for each. This earlier edition is preferred by many to more recent revisions. 1913 edition. Over 4000 illustrations, total of 2087pp. 6⅛ x 9¼.   22642-5, 22643-3, 22644-1 Pa., Three-vol. set $24.00

MANUAL OF THE GRASSES OF THE UNITED STATES, A. S. Hitchcock, U.S. Dept. of Agriculture. The basic study of American grasses, both indigenous and escapes, cultivated and wild. Over 1400 species. Full descriptions, information. Over 1100 maps, illustrations. Total of 1051pp. 5⅜ x 8½.   22717-0, 22718-9 Pa., Two-vol. set $12.00

THE CACTACEAE,, Nathaniel L. Britton, John N. Rose. Exhaustive, definitive. Every cactus in the world. Full botanical descriptions. Thorough statement of nomenclatures, habitat, detailed finding keys. The one book needed by every cactus enthusiast. Over 1275 illustrations. Total of 1080pp. 8 x 10¼.   21191-6, 21192-4 Clothbd., Two-vol. set $35.00

AMERICAN MEDICINAL PLANTS, Charles F. Millspaugh. Full descriptions, 180 plants covered: history; physical description; methods of preparation with all chemical constituents extracted; all claimed curative or adverse effects. 180 full-page plates. Classification table. 804pp. 6½ x 9¼.
23034-1 Pa. $10.00

A MODERN HERBAL, Margaret Grieve. Much the fullest, most exact, most useful compilation of herbal material. Gigantic alphabetical encyclopedia, from aconite to zedoary, gives botanical information, medical properties, folklore, economic uses, and much else. Indispensable to serious reader. 161 illustrations. 888pp. 6½ x 9¼. (Available in U.S. only)
22798-7, 22799-5 Pa., Two-vol. set $11.00

THE HERBAL or GENERAL HISTORY OF PLANTS, John Gerard. The 1633 edition revised and enlarged by Thomas Johnson. Containing almost 2850 plant descriptions and 2705 superb illustrations, Gerard's *Herbal* is a monumental work, the book all modern English herbals are derived from, the one herbal every serious enthusiast should have in its entirety. Original editions are worth perhaps $750. 1678pp. 8½ x 12¼.
23147-X Clothbd. $50.00

MANUAL OF THE TREES OF NORTH AMERICA, Charles S. Sargent. The basic survey of every native tree and tree-like shrub, 717 species in all. Extremely full descriptions, information on habitat, growth, locales, economics, etc. Necessary to every serious tree lover. Over 100 finding keys. 783 illustrations. Total of 986pp. 5⅜ x 8½.
20277-1, 20278-X Pa., Two-vol. set $10.00

AMERICAN BIRD ENGRAVINGS, Alexander Wilson et al. All 76 plates. from Wilson's *American Ornithology* (1808-14), most important ornithological work before Audubon, plus 27 plates from the supplement (1825-33) by Charles Bonaparte. Over 250 birds portrayed. 8 plates also reproduced in full color. 111pp. 9⅜ x 12½. 23195-X Pa. $6.00

CRUICKSHANK'S PHOTOGRAPHS OF BIRDS OF AMERICA, Allan D. Cruickshank. Great ornithologist, photographer presents 177 closeups, groupings, panoramas, flightings, etc., of about 150 different birds. Expanded *Wings in the Wilderness.* Introduction by Helen G. Cruickshank. 191pp. 8¼ x 11. 23497-5 Pa. $6.00

AMERICAN WILDLIFE AND PLANTS, A. C. Martin, et al. Describes food habits of more than 1000 species of mammals, birds, fish. Special treatment of important food plants. Over 300 illustrations. 500pp. 5⅜ x 8½. 20793-5 Pa. $4.95

THE PEOPLE CALLED SHAKERS, Edward D. Andrews. Lifetime of research, definitive study of Shakers: origins, beliefs, practices, dances, social organization, furniture and crafts, impact on 19th-century USA, present heritage. Indispensable to student of American history, collector. 33 illustrations. 351pp. 5⅜ x 8½. 21081-2 Pa. $4.00

OLD NEW YORK IN EARLY PHOTOGRAPHS, Mary Black. New York City as it was in 1853-1901, through 196 wonderful photographs from N.-Y. Historical Society. Great Blizzard, Lincoln's funeral procession, great buildings. 228pp. 9 x 12. 22907-6 Pa. $7.95

MR. LINCOLN'S CAMERA MAN: MATHEW BRADY, Roy Meredith. Over 300 Brady photos reproduced directly from original negatives, photos. Jackson, Webster, Grant, Lee, Carnegie, Barnum; Lincoln; Battle Smoke, Death of Rebel Sniper, Atlanta Just After Capture. Lively commentary. 368pp. 8⅜ x 11¼. 23021-X Pa. $8.95

TRAVELS OF WILLIAM BARTRAM, William Bartram. From 1773-8, Bartram explored Northern Florida, Georgia, Carolinas, and reported on wild life, plants, Indians, early settlers. Basic account for period, entertaining reading. Edited by Mark Van Doren. 13 illustrations. 141pp. 5⅜ x 8½. 20013-2 Pa. $4.50

THE GENTLEMAN AND CABINET MAKER'S DIRECTOR, Thomas Chippendale. Full reprint, 1762 style book, most influential of all time; chairs, tables, sofas, mirrors, cabinets, etc. 200 plates, plus 24 photographs of surviving pieces. 249pp. 9⅞ x 12¾. 21601-2 Pa. $6.50

AMERICAN CARRIAGES, SLEIGHS, SULKIES AND CARTS, edited by Don H. Berkebile. 168 Victorian illustrations from catalogues, trade journals, fully captioned. Useful for artists. Author is Assoc. Curator, Div. of Transportation of Smithsonian Institution. 168pp. 8½ x 9½. 23328-6 Pa. $5.00

THE SENSE OF BEAUTY, George Santayana. Masterfully written discussion of nature of beauty, materials of beauty, form, expression; art, literature, social sciences all involved. 168pp. 5⅜ x 8½.   20238-0 Pa. $2.50

ON THE IMPROVEMENT OF THE UNDERSTANDING, Benedict Spinoza. Also contains *Ethics, Correspondence*, all in excellent R. Elwes translation. Basic works on entry to philosophy, pantheism, exchange of ideas with great contemporaries. 402pp. 5⅜ x 8½.   20250-X Pa. $4.50

THE TRAGIC SENSE OF LIFE, Miguel de Unamuno. Acknowledged masterpiece of existential literature, one of most important books of 20th century. Introduction by Madariaga. 367pp. 5⅜ x 8½.
20257-7 Pa. $3.50

THE GUIDE FOR THE PERPLEXED, Moses Maimonides. Great classic of medieval Judaism attempts to reconcile revealed religion (Pentateuch, commentaries) with Aristotelian philosophy. Important historically, still relevant in problems. Unabridged Friedlander translation. Total of 473pp. 5⅜ x 8½.   20351-4 Pa. $5.00

THE I CHING (THE BOOK OF CHANGES), translated by James Legge. Complete translation of basic text plus appendices by Confucius, and Chinese commentary of most penetrating divination manual ever prepared. Indispensable to study of early Oriental civilizations, to modern inquiring reader. 448pp. 5⅜ x 8½.   21062-6 Pa. $4.00

THE EGYPTIAN BOOK OF THE DEAD, E. A. Wallis Budge. Complete reproduction of Ani's papyrus, finest ever found. Full hieroglyphic text, interlinear transliteration, word for word translation, smooth translation. Basic work, for Egyptology, for modern study of psychic matters. Total of 533pp. 6½ x 9¼. (Available in U.S. only)   21866-X Pa. $4.95

THE GODS OF THE EGYPTIANS, E. A. Wallis Budge. Never excelled for richness, fullness: all gods, goddesses, demons, mythical figures of Ancient Egypt; their legends, rites, incarnations, variations, powers, etc. Many hieroglyphic texts cited. Over 225 illustrations, plus 6 color plates. Total of 988pp. 6⅛ x 9¼. (Available in U.S. only)
22055-9, 22056-7 Pa., Two-vol. set $12.00

THE ENGLISH AND SCOTTISH POPULAR BALLADS, Francis J. Child. Monumental, still unsuperseded; all known variants of Child ballads, commentary on origins, literary references, Continental parallels, other features. Added: papers by G. L. Kittredge, W. M. Hart. Total of 2761pp. 6½ x 9¼.
21409-5, 21410-9, 21411-7, 21412-5, 21413-3 Pa., Five-vol. set $37.50

CORAL GARDENS AND THEIR MAGIC, Bronsilaw Malinowski. Classic study of the methods of tilling the soil and of agricultural rites in the Trobriand Islands of Melanesia. Author is one of the most important figures in the field of modern social anthropology. 143 illustrations. Indexes. Total of 911pp. of text. 5⅝ x 8¼. (Available in U.S. only)
23597-1 Pa. $12.95

THE PHILOSOPHY OF HISTORY, Georg W. Hegel. Great classic of Western thought develops concept that history is not chance but a rational process, the evolution of freedom. 457pp. 5⅜ x 8½. 20112-0 Pa. $4.50

LANGUAGE, TRUTH AND LOGIC, Alfred J. Ayer. Famous, clear introduction to Vienna, Cambridge schools of Logical Positivism. Role of philosophy, elimination of metaphysics, nature of analysis, etc. 160pp. 5⅜ x 8½. (Available in U.S. only) 20010-8 Pa. $1.75

A PREFACE TO LOGIC, Morris R. Cohen. Great City College teacher in renowned, easily followed exposition of formal logic, probability, values, logic and world order and similar topics; no previous background needed. 209pp. 5⅜ x 8½. 23517-3 Pa. $3.50

REASON AND NATURE, Morris R. Cohen. Brilliant analysis of reason and its multitudinous ramifications by charismatic teacher. Interdisciplinary, synthesizing work widely praised when it first appeared in 1931. Second (1953) edition. Indexes. 496pp. 5⅜ x 8½. 23633-1 Pa. $6.00

AN ESSAY CONCERNING HUMAN UNDERSTANDING, John Locke. The only complete edition of enormously important classic, with authoritative editorial material by A. C. Fraser. Total of 1176pp. 5⅜ x 8½.
20530-4, 20531-2 Pa., Two-vol. set $14.00

HANDBOOK OF MATHEMATICAL FUNCTIONS WITH FORMULAS, GRAPHS, AND MATHEMATICAL TABLES, edited by Milton Abramowitz and Irene A. Stegun. Vast compendium: 29 sets of tables, some to as high as 20 places. 1,046pp. 8 x 10½. 61272-4 Pa. $14.95

MATHEMATICS FOR THE PHYSICAL SCIENCES, Herbert S. Wilf. Highly acclaimed work offers clear presentations of vector spaces and matrices, orthogonal functions, roots of polynomial equations, conformal mapping, calculus of variations, etc. Knowledge of theory of functions of real and complex variables is assumed. Exercises and solutions. Index. 284pp. 5⅝ x 8¼. 63635-6 Pa. $4.50

THE PRINCIPLE OF RELATIVITY, Albert Einstein et al. Eleven most important original papers on special and general theories. Seven by Einstein, two by Lorentz, one each by Minkowski and Weyl. All translated, unabridged. 216pp. 5⅜ x 8½. 60081-5 Pa. $3.00

THERMODYNAMICS, Enrico Fermi. A classic of modern science. Clear, organized treatment of systems, first and second laws, entropy, thermodynamic potentials, gaseous reactions, dilute solutions, entropy constant. No math beyond calculus required. Problems. 160pp. 5⅜ x 8½.
60361-X Pa. $2.75

ELEMENTARY MECHANICS OF FLUIDS, Hunter Rouse. Classic undergraduate text widely considered to be far better than many later books. Ranges from fluid velocity and acceleration to role of compressibility in fluid motion. Numerous examples, questions, problems. 224 illustrations. 376pp. 5⅝ x 8¼. 63699-2 Pa. $5.00

AN AUTOBIOGRAPHY, Margaret Sanger. Exciting personal account of hard-fought battle for woman's right to birth control, against prejudice, church, law. Foremost feminist document. 504pp. 5⅜ x 8½.
20470-7 Pa. $5.50

MY BONDAGE AND MY FREEDOM, Frederick Douglass. Born as a slave, Douglass became outspoken force in antislavery movement. The best of Douglass's autobiographies. Graphic description of slave life. Introduction by P. Foner. 464pp. 5⅜ x 8½.   22457-0 Pa. $5.00

LIVING MY LIFE, Emma Goldman. Candid, no holds barred account by foremost American anarchist: her own life, anarchist movement, famous contemporaries, ideas and their impact. Struggles and confrontations in America, plus deportation to U.S.S.R. Shocking inside account of persecution of anarchists under Lenin. 13 plates. Total of 944pp. 5⅜ x 8½.
22543-7, 22544-5 Pa., Two-vol. set $9.00

LETTERS AND NOTES ON THE MANNERS, CUSTOMS AND CONDITIONS OF THE NORTH AMERICAN INDIANS, George Catlin. Classic account of life among Plains Indians: ceremonies, hunt, warfare, etc. Dover edition reproduces for first time all original paintings. 312 plates. 572pp. of text. 6⅛ x 9¼.   22118-0, 22119-9 Pa.. Two-vol. set $10.00

THE MAYA AND THEIR NEIGHBORS, edited by Clarence L. Hay, others. Synoptic view of Maya civilization in broadest sense, together with Northern, Southern neighbors. Integrates much background, valuable detail not elsewhere. Prepared by greatest scholars: Kroeber, Morley, Thompson, Spinden, Vaillant, many others. Sometimes called Tozzer Memorial Volume. 60 illustrations, linguistic map. 634pp. 5⅜ x 8½.
23510-6 Pa. $7.50

HANDBOOK OF THE INDIANS OF CALIFORNIA, A. L. Kroeber. Foremost American anthropologist offers complete ethnographic study of each group. Monumental classic. 459 illustrations, maps. 995pp. 5⅜ x 8½.
23368-5 Pa. $10.00

SHAKTI AND SHAKTA, Arthur Avalon. First book to give clear, cohesive analysis of Shakta doctrine, Shakta ritual and Kundalini Shakti (yoga). Important work by one of world's foremost students of Shaktic and Tantric thought. 732pp. 5⅜ x 8½. (Available in U.S. only)
23645-5 Pa. $7.95

AN INTRODUCTION TO THE STUDY OF THE MAYA HIEROGLYPHS, Syvanus Griswold Morley. Classic study by one of the truly great figures in hieroglyph research. Still the best introduction for the student for reading Maya hieroglyphs. New introduction by J. Eric S. Thompson. 117 illustrations. 284pp. 5⅜ x 8½.   23108-9 Pa. $4.00

A STUDY OF MAYA ART, Herbert J. Spinden. Landmark classic interprets Maya symbolism, estimates styles, covers ceramics, architecture, murals, stone carvings as artforms. Still a basic book in area. New introduction by J. Eric Thompson. Over 750 illustrations. 341pp. 8⅜ x 11¼.
21235-1 Pa. $6.95

THE STANDARD BOOK OF QUILT MAKING AND COLLECTING, Marguerite Ickis. Full information, full-sized patterns for making 46 traditional quilts, also 150 other patterns. Quilted cloths, lame, satin quilts, etc. 483 illustrations. 273pp. 6⅞ x 9⅝. 20582-7 Pa. $4.50

ENCYCLOPEDIA OF VICTORIAN NEEDLEWORK, S. Caulfield, Blanche Saward. Simply inexhaustible gigantic alphabetical coverage of every traditional needlecraft—stitches, materials, methods, tools, types of work; definitions, many projects to be made. 1200 illustrations; double-columned text. 697pp. 8⅛ x 11. 22800-2, 22801-0 Pa., Two-vol. set $12.00

MECHANICK EXERCISES ON THE WHOLE ART OF PRINTING, Joseph Moxon. First complete book (1683-4) ever written about typography, a compendium of everything known about printing at the latter part of 17th century. Reprint of 2nd (1962) Oxford Univ. Press edition. 74 illustrations. Total of 550pp. 6⅛ x 9¼. 23617-X Pa. $7.95

PAPERMAKING, Dard Hunter. Definitive book on the subject by the foremost authority in the field. Chapters dealing with every aspect of history of craft in every part of the world. Over 320 illustrations. 2nd, revised and enlarged (1947) edition. 672pp. 5⅜ x 8½. 23619-6 Pa. $7.95

THE ART DECO STYLE, edited by Theodore Menten. Furniture, jewelry, metalwork, ceramics, fabrics, lighting fixtures, interior decors, exteriors, graphics from pure French sources. Best sampling around. Over 400 photographs. 183pp. 8⅜ x 11¼. 22824-X Pa. $5.00